Neural Networks and Brain Function

Neural Networks and Brain Function

Edmund T. Rolls

University of Oxford,
Department of Experimental Psychology,
Oxford OX1 3UD, England

and

Alessandro Treves

International School of Advanced Studies
Programme in Neuroscience
34013 Trieste, Italy

OXFORD NEW YORK TOKYO
OXFORD UNIVERSITY PRESS
1998

Oxford University Press, Great Clarendon Street, Oxford OX2 6DP

Oxford New York
Athens Auckland Bangkok Bogota Bombay
Buenos Aires Calcutta Cape Town Dar es Salaam
Delhi Florence Hong Kong Istanbul Karachi
Kuala Lumpur Madras Madrid Melbourne
Mexico City Nairobi Paris Singapore
Taipei Tokyo Toronto Warsaw

and associated companies in
Berlin Ibadan

Oxford is a trade mark of Oxford University Press

Published in the United States
by Oxford University Press, Inc., New York

A catalogue record for this book is available from the British Library

Library of Congress Cataloging in Publication Data
(Data available)

ISBN 0 19 852433 1 (Hbk)
ISBN 0 19 852432 3 (Pbk)

Typeset by Hewer Text Composition Services, Edinburgh
Printed in Great Britain by Bookcraft Ltd., Midsomer North, Avon

Preface

This text provides an introduction to biologically plausible neural networks. Most of the networks (with the exception of those in Chapter 5) are biologically plausible in that they operate subject to the following constraints. First, they use a local learning rule, in which synaptic change is based on activity available locally at the synapse. Second, they use a simple rule to compute their output activity, in which the output reflects the sum of the activations produced in the neuron by each input axon operating through its own synapse which has its own synaptic strength or weight.

In the first main section, Chapters 2–5 introduce and describe the operation of several fundamental types of neural network, including pattern associators, autoassociators, competitive networks, perceptrons, and multilayer backpropagation. In the second main section, Chapters 6–10 describe real neural networks in several brain systems, show how it is starting to become possible to make theories about *how* some parts of the brain work, and provide an indication of the different techniques in neuroscience and neurocomputation that will continue to need to be combined for further rapid progress in understanding how parts of the brain work. The neural systems described include systems involved in memory (Chapter 6), in building representations in sensory systems of objects in the world (Chapter 8), and in learning the significance of these sensory representations in terms of associated rewards and punishments though pattern association, which is fundamental to much emotional and motivational behaviour (Chapter 7). The third section, Appendices A1–A5, introduces more formal quantitative approaches to many of the networks described.

The intention is that the three sections can be read in different combinations acording to the interests and level of the reader. Chapters 2–5 are intended to provide an introduction to the main classes of network that are likely to be of most interest to neuroscientists. This section should be accessible to undergraduates, and to neuroscientists who have not yet studied neural networks. It could provide the basis for a taught course.

Appendices A1–A5 are intended to guide readers towards more formal and quantitative approaches to computation in neural networks, but to keep what is described particularly within the area which is most likely to be useful to those studying the brain. This section includes approaches derived from theoretical physics and from information theory, and is intended to guide the reader beyond the graduate level.

Chapters 6–10 are intended to show how it is becoming possible to define the types of network present in some brain regions, and by combining knowledge of these networks with other approaches in neuroscience, to show how it is becoming possible to start to understand how some parts of the brain could function. We are clearly at an exciting early stage in understanding brain computation, and we believe that the tools are now available for rapid progress to be made by combining the approaches described in this section. Many different brain systems are described, and because this is a rapidly developing subject, this section must be somewhat in the nature of a monograph, which illustrates the application of neural network theory to understanding the operation of many different brain systems, rather than giving an exhaustive coverage. We are conscious, therefore, that there is much excellent work which was not included, but ask our colleagues to keep us informed of their work, so that we may include more in a future edition.

An excellent description of many of the formal, quantitative approaches to simple neural networks is provided by J. Hertz, A. Krogh and R. G. Palmer in their book 'Introduction to the Theory of Neural Computation' (Addison-Wesley, 1991). We commend this to readers, and in a number of places in this book on neural computation by the brain we refer to that book on theoretical approaches to neural computation.

It is our conviction that it is now possible to start to understand *how* the brain functions by analysing the computations it performs. This has become possible in the last few years because of advances in understanding quantitatively the connections made by neurons in the brain using modern anatomical techniques; the ways in which neurons operate based on advances in biophysics; the ways in which learning may be implemented in the brain using modification of synaptic strengths; analysis based on brain lesions, imagining, and single neuron recording of what function may be performed by each brain area; analysis using single and multiple single neuron recording of what information is represented in each brain area and how it is represented; and analysis of how neuronal networks with biologically plausible properties could perform the appropriate computations in each brain area. The aim of this book is to describe the types of computation that can be performed by biologically plausible networks, and to show how these may be implemented in different systems in the brain. The aim is thus to provide a foundation for understanding what is one of the most fascinating problems, which can now be approached with the methods described in this book, namely *how* the brain works, and how our behaviour is produced.

The material in this text is copyright E. T. Rolls (who was mainly responsible for Chapters 1–10) and A. Treves (who was mainly responsible for Appendices A2–A5). Part of the material presented in the book reflects work done over many years in collaboration with many colleagues, whose tremendous contribution is warmly appreciated. The contribution of many will be evident from references cited in the text. We are particularly grateful to our collaborators on the most recent research work, including F. Battaglia, M. Booth, H. Critchley. P. Georges-François, R. Lauro-Grotto, T. Milward, S. Panzeri and R. Robertson, who have agreed to include material in this book, at a time when part of it has not yet been published in research journals. In addition, we have benefited enormously from the discussions we have had with a large number of colleagues and friends, many of whom we hope will see areas of the text which they have been able to illuminate. Much of the work described would not have been possible without financial support from a number of sources, particularly the Medical Research Council of the UK, the McDonnell-Pew Foundation, The European Communities, and the Human Frontier Science Program.

Contents

Notation

We use the following notation throughout this book except where otherwise specified:

r	a postsynaptic firing rate or level of activity
r'	a presynaptic firing rate or level of activity
i	a postsynaptic neuron
j	a presynaptic neuron
r_i	the firing rate of postsynaptic neuron i
r'_j	the firing rate of presynaptic neuron j
w_{ij}	the synaptic strength onto postsynaptic neuron i from presynaptic neuron j
h	the activation of a neuron, generally calculated as the dot product between the set of presynaptic firings and the corresponding weights. h corresponds to the *field* in theoretical physics; and corresponds to *netinput* in some Connectionist models.
$f(h)$	activation function: this relates the firing r of a neuron to its activation h.
$\mathbf{r}$	the firing rate vector of a set of output neurons
$\mathbf{r'}$	the firing rate vector of a set of input neurons
$\mathbf{w}_i$	the vector of synaptic weights on the ith neuron

1 Introduction

1.1 Introduction

It is now becoming possible to develop an understanding not only of *what* functions are performed by each region of the brain, but of *how* they are performed. To understand what function is performed by a given brain region requires evidence from the deficits that follow damage to that region; evidence of which brain regions are selectively activated during the performance of particular tasks, as shown by brain imaging, using techniques such as Positron Emission Tomography (PET), functional magnetic resonance imaging, and optical imaging; and evidence of what inputs reach that region, as shown by recording the activity of single neurons in that region. Understanding what function is performed by a given brain region also involves knowing what is being performed in the regions that provide inputs to a given region, and that receive outputs from the region being considered. Knowledge of the connectional anatomy of the brain helps in this. This is the systems-level analysis of brain function. To understand *how* a brain region performs these functions involves quantitative neuroanatomy, to show how many inputs are received from each source by neurons in a given brain region, and where they terminate on a cell; studies of the rules and mechanisms (including their pharmacology) that determine synaptic connectivity and modifiability; analysis of the manner in which information is represented by neuronal responses in different brain regions; and computational approaches to *how* the system could operate. The latter, the computational approach, is essential to understanding how each part of the brain functions. Even if we had good systems-level evidence on the operation of a given brain region, and good evidence on what is represented in it, we would still not know how that brain region operated. For example, we would not know how many memories it could store, or how it might solve computationally difficult problems such as how objects can be recognized, despite presentation in different positions on the retina, in different sizes, and from different views. To understand these problems, some notion of how each area of the brain computes is needed. It is the aim of this book to provide an introduction to how different areas of the brain may perform their computations. In order to understand this, we need to take into account the nature of the computation performed by single neurons, how the connections between neurons alter in order to store information about the problem to be solved, how neurons in a region interact, and how these factors result in useful computations being performed.

To approach this aim, we describe a number of fundamental operations that can be performed by networks in the brain. We take the approach of describing how networks that are biologically plausible operate. In Chapters 2–4 we describe networks that can learn to associate two inputs, that can store patterns (e.g. memories), and can recall the patterns from a fragment; and that can perform categorization of the inputs received, and thus perform feature analysis. In later chapters, we show how these principles can be applied to develop an understanding of how some regions of the brain operate.

Without an understanding of how the brain performs its computations, we will never understand *how* the brain works. Some of the first steps to this understanding are described in this book.

In the rest of this chapter, we introduce some of the background for understanding brain computation, such as how single neurons operate, how some of the essential features of this can be captured by simple formalisms, and some of the biological background to what it can be taken happens in the nervous system, such as synaptic modification based on information available locally at each synapse.

1.2 Neuronal network approaches versus connectionism

The approach taken in this book is to introduce how real neuronal networks in the brain may compute, and thus to achieve a fundamental and realistic basis for understanding brain function. This may be contrasted with connectionism, which aims to understand cognitive function by analysing processing in neuron-like computing systems. Connectionist systems are neuron-like in that they analyse computation in systems with large numbers of computing elements in which the information which governs how the network computes is stored in the connection strengths between the nodes (or 'neurons') in the network. However, in many connectionist models the individual units or nodes are not intended to model individual neurons, and the variables that are used in the simulations are not intended to correspond to quantities that can be measured in the real brain. Moreover, connectionist approaches use learning rules in which the synaptic modification (the strength of the connections between the nodes) is determined by algorithms which require information which is not local to the synapse, that is, evident in the pre- and postsynaptic firing rates (see further Chapter 5). Instead, in many connectionist systems, information about how to modify synaptic strengths is propagated backwards from the output of the network to affect neurons hidden deep within the network. Because it is not clear that this is biologically plausible, we have instead in this text concentrated on introducing neuronal network architectures which are more biologically plausible, and which use a local learning rule. Connectionist approaches (see e.g. McClelland and Rumelhart, 1986; McLeod, Plunkett and Rolls, 1998) are very valuable, for they show what can be achieved computationally with networks in which the connection strength determines the computation that the network achieves with quite simple computing elements. However, as models of brain function, many connectionist networks achieve almost too much, by solving problems with a carefully limited number of 'neurons' or nodes, which contributes to the ability of such networks to generalize successfully over the problem space. Connectionist schemes thus make an important start on understanding how complex computations (such as language) could be implemented in brain-like systems. In doing this, connectionist models often use simplified representations of the inputs and outputs, which are often crucial to the way in which the problem is solved. In addition, they may use learning algorithms that are really too powerful for the brain to perform, and therefore they can be taken only as a guide to how cognitive functions might be implemented by neuronal networks in the brain. In this book, we focus on more biologically plausible neuronal networks.

1.3 Neurons in the brain, and their representation in neuronal networks

Neurons in the vertebrate brain typically have large dendrites extending from the cell body, which receive inputs from other neurons through connections called synapses. The synapses operate by chemical transmission. When a synaptic terminal receives an all-or-nothing action potential from the neuron of which it is a terminal, it releases a transmitter which crosses the synaptic cleft and produces either depolarization or hyperpolarization in the postsynaptic neuron, by opening particular ionic channels. (A textbook such as Kandel *et al.*, 1991 gives further information on this process.) Summation of a number of such depolarization or excitatory inputs within the time constant of the receiving neuron, which is typically 20–30 ms, produces sufficient depolarization that the neuron fires an action potential. There are often 5000–20 000 inputs per neuron. An example of a neuron found in the brain is shown in Fig. 1.1, and there are further examples in Chapter 10. Once firing is initiated in the cell body (or axon initial segment of the cell body), the action potential is conducted in an all-or-nothing way to reach the synaptic terminals of the neuron, whence it may affect other neurons. Any inputs the neuron receives which cause it to become hyperpolarized make it less likely to fire (because the membrane potential is moved away from the critical threshold at which an action potential is initiated), and are described as inhibitory. The neuron can thus be thought of in a simple way as a computational element which sums its inputs within its time constant and, whenever this sum, minus any inhibitory effects, exceeds a threshold, produces an action potential which propagates to all of its outputs. This simple idea is incorporated in many neuronal network models using a formalism of a type described in the next section.

1.4 A formalism for approaching the operation of single neurons in a network

Let us consider a neuron i as shown in Fig. 1.2 which receives inputs from axons which we label j through synapses of strength w_{ij}. The first subscript (i) refers to the receiving neuron, and the second subscript (j) to the particular input[1]. j counts from 1 to C, where C is the number of synapses or connections received. The firing rate of the ith neuron is denoted r_i, and of the jth input to the neuron r'_j. (The prime is used to denote the input or presynaptic term. The letter r is used to indicate that the inputs and outputs of real neurons are firing rates.) To express the idea that the neuron makes a simple linear summation of the inputs it receives, we can write the activation of neuron i, denoted h_i, as

$$h_i = \Sigma_j r'_j w_{ij} \qquad (1.1)$$

where Σ_j indicates that the sum is over the C input axons indexed by j. The multiplicative form here indicates that activation should be produced by an axon only if it is firing, and depending

---

[1] This convention, that i refers to the receiving neuron, and j refers to a particular input to that neuron via a synapse of weight w_{ij}, is used throughout this book, except where otherwise stated.

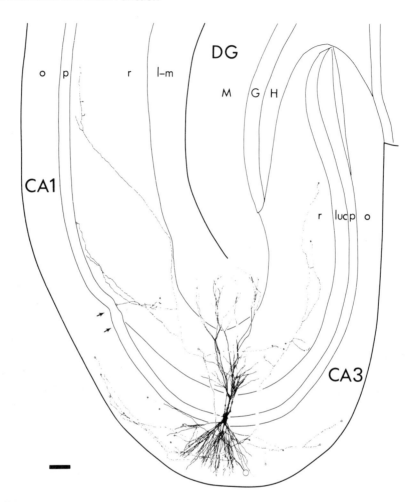

Fig. 1.1 An example of a real neuron found in the brain. The neuron is a CA3 neuron from the hippocampus. The thick extensions from the cell body or soma are the dendrites, which form an extensive dendritic tree receiving in this case approximately 12 000 synapses. The axon is the thin connection leaving the cell. It divides into a number of collateral branches. Two axonal branches can be seen in the plane of the section to travel to each end of the population of CA3 cells. One branch (on the left) continues to connect to the next group of cells, the CA1 cells. The junction between the CA3 and CA1 cells is shown by the two arrows. The diagram shows a camera lucida drawing of a single CA3 pyramidal cell intracellularly labelled with horseradish peroxidase. DG, dentate gyrus. The small letters refer to the different strata of the hippocampus. Scale bar = 100μm. (Reprinted with permission from Ishizuka, Weber and Amaral, 1990.)

on the strength of the synapse w_{ij} from input axon j onto the dendrite of the receiving neuron i. Equation 1.1 indicates that the strength of the activation reflects how fast the axon j is firing (that is r'_j), and how strong the synapse w_{ij} is. The sum of all such activations expresses the idea that summation (of synaptic currents in real neurons) occurs along the length of the dendrite, to produce activation at the cell body, where the activation h_i is converted into firing r_i. This conversion can be expressed as

$$r_i = f(h_i) \tag{1.2}$$

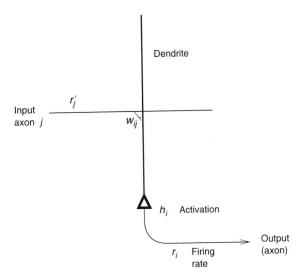

Fig. 1.2 Notation used to describe an individual neuron in a network model. By convention, we generally represent the dendrite as thick, and vertically oriented (as this is the normal way that neuroscientists view cortical pyramidal cells under the microscope), and the axon as thin. The cell body or soma is indicated between them. The firing rate we also call the (firing rate) activity of the neuron.

which indicates that the firing rate is a function of the postsynaptic activation. The function is called the activation function in this case. The function at its simplest could be linear, so that the firing rate would be proportional to the activation (see Fig. 1.3a). Real neurons have thresholds, with firing occurring only if the activation is above the threshold. A threshold linear activation function is shown in Fig. 1.3b. This has been useful in formal analysis of the properties of neural networks. Neurons also have firing rates which become saturated at a maximum rate, and we could express this as the sigmoid activation function shown in Fig. 1.3c. Another simple activation function, used in some models of neural networks, is the binary threshold function (Fig. 1.3d), which indicates that if the activation is below threshold, there is no firing, and that if the activation is above threshold, the neuron fires maximally. Some non-linearity in the activation function is an advantage, for it enables many useful computations to be performed in neuronal networks, including removing interfering effects of similar memories, and enabling neurons to perform logical operations, such as firing only if several inputs are present simultaneously.

A property implied by Eq. 1.1 is that the postsynaptic membrane is electrically short, and so summates its inputs irrespective of where on the dendrite the input is received. In real neurons, the transduction of current into firing frequency (the analogue of the transfer function of Eq. 1.2) is generally studied not with synaptic inputs but by applying a steady current through an electrode into the soma. An example of the resulting curves, which illustrate the additional phenomenon of firing rate adaptation, is reproduced in Fig. 1.3e.

1.5 Synaptic modification

For a neuronal network to perform useful computation, that is to produce a given output when it receives a particular input, the synaptic weights must be set up appropriately. This is often performed by synaptic modification occurring during learning.

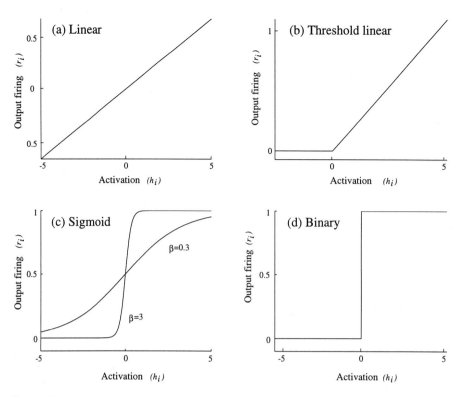

Fig. 1.3 Different types of activation function. The activation function relates the output activity (or firing rate), r_i, of the neuron (i) to its activation, h_i. (a) Linear. (b) Threshold linear. (c) Sigmoid. (One mathematical exemplar of this class of activation function is $r_i = 1/(1 + \exp(-2\beta h_i))$). The output of this function, also sometimes known as the logistic function, is 0 for an input of $-\infty$, 0.5 for 0, and 1 for $+\infty$. The function incorporates a threshold at the lower end, followed by a linear portion, and then an asymptotic approach to the maximum value at the top end of the function. The parameter β controls the steepness of the almost linear part of the function round $h_i = 0$. If β is small, the output goes smoothly and slowly from 0 to 1 as h_i goes from $-\infty$ to $+\infty$. If β is large, the curve is very steep, and approximates a binary threshold activation function.) (d) Binary threshold.

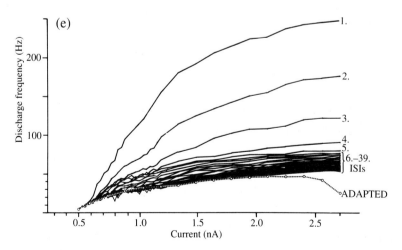

Fig. 1.3(e) Frequency–current plot (the closest experimental analogue of the activation function) for a CA1 pyramidal cell. The firing frequency (in Hz) in response to the injection of 1.5 s long, rectangular depolarizing current pulses has been plotted against the strength of the current pulses (in nA) (abscissa). The first 39 interspike intervals (ISIs) are plotted as instantaneous frequency (1/ISI, where ISI is the inter-stimulus interval), together with the average frequency of the adapted firing during the last part of the current injection (circles and broken line). The plot indicates a current threshold at approximately 0.5 nA, a linear range with a tendency to saturate, for the initial instantaneous rate, above approximately 200 Hz, and the phenomenon of adaptation, which is not reproduced in simple non-dynamical models (see further Appendix 5). (Reprinted with permission from Lanthorn, Storm and Andersen, 1984.)

A simple learning rule that was originally presaged by Donald Hebb (1949) proposes that synapses increase in strength when there is conjunctive presynaptic and postsynaptic activity. The Hebb rule can be expressed more formally as follows:

$$\delta w_{ij} = k\, r_i r'_j \qquad (1.3)$$

where δw_{ij} is the change of the synaptic weight w_{ij} which results from the simultaneous (or conjunctive) presence of presynaptic firing r'_j and postsynaptic firing r_i (or strong depolarization), and k is a learning rate constant which specifies how much the synapses alter on any one pairing. The presynaptic and postsynaptic activity must be present approximately simultaneously (to within perhaps 100–500 ms in the real brain).

The Hebb rule is expressed in this multiplicative form to reflect the idea that *both* presynaptic and postsynaptic activity must be present for the synapses to increase in strength. The multiplicative form also reflects the idea that strong pre- and postsynaptic firing will produce a larger change of synaptic weight than smaller firing rates. The Hebb rule thus captures what is typically found in studies of associative long-term potentiation (LTP) in the brain, described in Section 1.6.

One useful property of large neurons in the brain, such as cortical pyramidal cells, is that with their short electrical length, the postsynaptic term, r_i, is available on much of the dendrite of a cell. The implication of this is that once sufficient postsynaptic activation has been produced, any active presynaptic terminal on the neuron will show synaptic strengthening. This enables associations between coactive inputs, or correlated activity in input axons, to be learned by neurons using this simple associative learning rule.

If, in contrast, a group of coactive axons made synapses close together on a small dendrite, then the local depolarization might be intense, and these synapses only would modify onto the dendrite. (A single distant active synapse might not modify in this type of neuron, because of the long electronic length of the dendrite.) The computation in this case is described as Sigma–Pi ($\Sigma\Pi$), to indicate that there is a local product computed during learning, this allows a particular set of locally active synapses to modify together, and then the output of the neuron can reflect the sum of such local multiplications (see Rumelhart and McClelland, 1986). This idea is not used in most neuronal networks that have been studied.

1.6 Long-term potentiation and long-term depression as biological models of synaptic modifications that occur in the brain

Long-term potentiation (LTP) and long-term depression (LTD) provide useful models of some of the synaptic modifications that occur in the brain. The synaptic changes found appear to be synapse-specific, and to depend on information available locally at the synapse. LTP and LTD may thus provide a good model of biological synaptic modification involved in real neuronal network operations in the brain. We next therefore describe some of the properties of LTP and LTD, and evidence which implicates them in learning in at least some brain systems. Even if they turn out not to be the basis for the synaptic modifications that occur during learning, they have many of the properties that would be needed by some of the synaptic modification systems used by the brain.

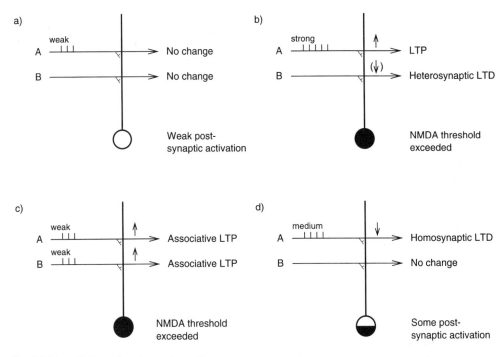

Fig. 1.4 Schematic illustration of synaptic modification rules as revealed by long-term potentiation (LTP) and long-term depression (LTD). The activation of the postsynaptic neuron is indicated by the extent to which its soma is black. There are two sets of inputs to the neuron: A and B. (a) A weak input (indicated by 3 spikes) on the set A of input axons produces little postsynaptic activation, and there is no change in synaptic strength. (b) A strong input (indicated by 5 spikes) on the set A of input axons produces strong postsynaptic activation, and the active synapses increase in strength. This is LTP. It is homosynaptic in that the synapses that increase in strength are the same as those through which the neuron is activated. LTP is synapse-specific, in that the inactive axons, B, do not show LTP. They either do not change in strength, or they may weaken. The weakening is called heterosynaptic LTD, because the synapses that weaken are other than those through which the neuron is activated (*hetero-* is Greek for other). (c) Two weak inputs present simultaneously on A and B summate to produce strong postsynaptic activation, and both sets of active synapses show LTP. (d) Intermediate strength firing on A produces some activation, but not strong activation, of the postsynaptic neuron. The active synapses become weaker. This is homosynaptic LTD, in that the synapses that weaken are the same as those through which the neuron is activated (*homo-* is Greek for same).

Long-term potentiation (LTP) is a use-dependent and sustained increase in synaptic strength that can be induced by brief periods of synaptic stimulation. It is usually measured as a sustained increase in the amplitude of electrically evoked responses in specific neural pathways following brief trains of high frequency stimulation (see Fig. 1.4b). For example, high frequency stimulation of the Schaffer collateral inputs to the hippocampal CA1 cells results in a larger response recorded from the CA1 cells to single test pulse stimulation of the pathway. LTP is *long-lasting*, in that its effect can be measured for hours in hippocampal slices, and in chronic *in vivo* experiments in some cases may last for months. LTP becomes evident rapidly, typically in less than 1 minute. LTP is in some brain systems *associative*. This is illustrated in Fig. 1.4c, in which a weak input to a group of cells (e.g. the commissural input to CA1) does not show LTP unless it is given at the same time as (i.e. associatively with) another input (which could be weak or strong) to the cells. The associativity arises because it is only when sufficient activation of the postsynaptic neuron to exceed the threshold of NMDA receptors (see below) is produced that any learning can occur. The two weak inputs summate to produce sufficient depolarization to exceed the threshold. This associative property is shown very clearly in experiments in which LTP of an input to a single cell only occurs if the cell membrane is depolarized by passing current through it at the same time as the input arrives at the cell. The depolarization alone or the input alone is not sufficient to produce the LTP, and the LTP is thus associative. Moreover, in that the presynaptic input and the postsynaptic depolarization must occur at about the same time (within approximately 500 ms), the LTP requires **temporal contiguity**. LTP is also **synapse-specific**, in that for example an inactive input to a cell does not show LTP even if the cell is strongly activated by other inputs (Fig. 1.4b, input B).

These spatiotemporal properties of LTP can be understood in terms of actions of the inputs on the postsynaptic cell, which in the hippocampus has two classes of receptor, NMDA (N-methyl-D-aspartate) and K–Q (kainate–quisqualate), activated by the glutamate released by the presynaptic terminals. Now the NMDA receptor channels are normally blocked by Mg^{2+}, but when the cell is strongly depolarized by strong tetanic stimulation of the type necessary to induce LTP, the Mg^{2+} block is removed, and Ca^{2+} entering via the NMDA receptor channels triggers events that lead to the potentiated synaptic transmission (see Fig. 1.5). Part of the evidence for this is that NMDA antagonists such as AP5 (D-2-amino-5-phosphonopentanoate) block LTP. Further, if the postsynaptic membrane is voltage clamped to prevent depolarization by a strong input, then LTP does not occur. The voltage-dependence of the NMDA receptor channels introduces a threshold and thus a non-linearity which contributes to a number of the phenomena of some types of LTP, such as cooperativity (many small inputs together produce sufficient depolarization to allow the NMDA receptors to operate), associativity (a weak input alone will not produce sufficient depolarization of the postsynaptic cell to enable the NMDA receptors to be activated, but the depolarization will be sufficient if there is also a strong input), and temporal contiguity between the different inputs that show LTP (in that if inputs occur non-conjunctively, the depolarization shows insufficient summation to reach the required level, or some of the inputs may arrive when the depolarization has decayed). Once the LTP has become established (which can be within one minute of the strong input to the cell), the LTP is expressed through the K–Q receptors, in that AP5 blocks only the establishment of LTP, and not its subsequent expression (see further Bliss and Collingridge, 1993; Nicoll and Malenka, 1995; Fazeli and Collingridge, 1996).

There are a number of possibilities about what change is triggered by the entry of Ca^{2+} to the postsynaptic cell to mediate LTP. One possibility is that somehow a messenger reaches the presynaptic terminals from the postsynaptic membrane and, if the terminals are active, causes them to release more transmitter in future whenever they are activated by an action potential. Consistent with this possibility is the observation that after LTP has been induced, more transmitter appears to be released from the presynaptic endings. Another possibility is that the postsynaptic membrane changes just where Ca^{2+} has entered, so that K–Q receptors become more responsive to glutamate released in future. Consistent with this possibility is the observation that after LTP, the postsynaptic cell may respond more to locally applied glutamate (using a microiontophoretic technique).

The rule which underlies associative LTP is thus that synapses connecting two neurons become stronger if there is conjunctive presynaptic and (strong) postsynaptic activity. This

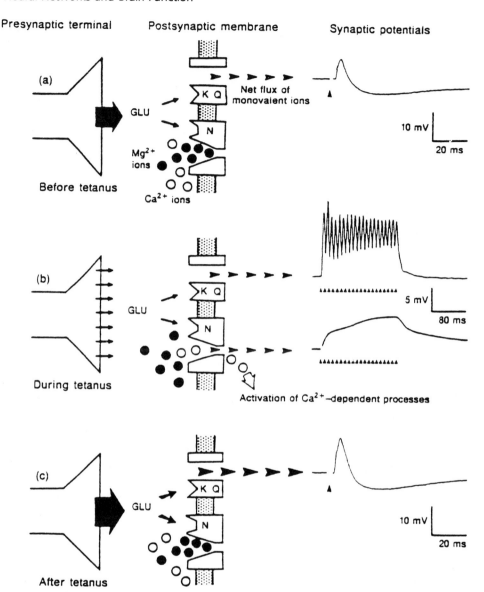

Fig. 1.5 The mechanism of induction of LTP in the CA1 region of the hippocampus. (a) Neurotransmitter (e.g. L-glutamate) is released and acts upon both K/Q (kainate/quisqualate) and NMDA (N) receptors. The NMDA receptors are blocked by magnesium and the excitatory synaptic response (EPSP) is therefore mediated primarily by ion flow through the channels associated with K/Q receptors. (b) During high-frequency activation, the magnesium block of the ion channels associated with NMDA receptors is released by depolarization. Activation of the NMDA receptor by transmitter now results in ions moving through the channel. In this way, calcium enters the postsynaptic region to trigger various intracellular mechanisms which eventually result in an alteration of synaptic efficacy. (c) Subsequent low-frequency stimulation results in a greater EPSP. See text for further details. (Reprinted with permission from Collingridge and Bliss, 1987.)

learning rule for synaptic modification is sometimes called the Hebb rule, after Donald Hebb of McGill University who drew attention to this possibility, and its potential importance in learning, in 1949.

In that LTP is long-lasting, develops rapidly, is synapse-specific, and is in some cases associative, it is of interest as a potential synaptic mechanism underlying some forms of memory. Evidence linking it directly to some forms of learning comes from experiments in which it has been shown that the drug AP5 infused so that it reaches the hippocampus to block NMDA receptors blocks spatial learning mediated by the hippocampus (see Morris, 1989). The task learned by the rats was to find the location relative to cues in a room of a platform submerged in an opaque liquid (milk). Interestingly, if the rats had already learned where the platform was, then the NMDA infusion did not block performance of the task. This is a close parallel to LTP, in that the learning, but not the subsequent expression of what had been learned, was blocked by the NMDA antagonist AP5. Although there is still some uncertainty about the experimental evidence that links LTP to learning (see e.g. Morris, 1996), there is a need for a synapse-specific modifiability of synaptic strengths on neurons if neuronal networks are to learn (see Rolls, 1996a and examples throughout this book), and if LTP is not always an exact model of the synaptic modification that occurs during learning, then something with many of the properties of LTP is nevertheless needed, and is likely to be present in the brain given the functions known to be implemented in many brain regions (see Chapters 6–10).

In another model of the role of LTP in memory, Davis (1992) has studied the role of the amygdala in learning associations to fear-inducing stimuli. He has shown that blockade of NMDA synapses in the amygdala interferes with this type of learning, consistent with the idea that LTP provides a useful model of this type of learning too (see further Chapter 7).

Long-term depression (LTD) can also occur. It can in principle be associative or non-associative. In associative LTD, the alteration of synaptic strength depends on the pre- and postsynaptic activities. There are two types. Heterosynaptic LTD occurs when the post-synaptic neuron is strongly activated, and there is low presynaptic activity (see Fig. 1.4b input B, and Table 2.1). Heterosynaptic LTD is so-called because the synapse that weakens is other than (hetero-) the one through which the postsynaptic neuron is activated. Heterosynaptic LTD is important in associative neuronal networks (see Chapters 2 and 3), and in competitive neuronal networks (see Chapter 4). In competitive neural networks it would be helpful if the degree of heterosynaptic LTD depended on the existing strength of the synapse, and there is some evidence that this may be the case (see Chapter 4). Homosynaptic LTD occurs when the presynaptic neuron is strongly active, and the postsynaptic neuron has some, but low, activity (see Fig. 1.4d and Table 2.1). Homosynaptic LTD is so-called because the synapse that weakens is the same as (homo-) the one that is active. Heterosynaptic and homosynaptic LTD are found in the neocortex (Artola and Singer, 1993; Singer, 1995) and hippocampus (Christie, 1996 and other papers in *Hippocampus* (1996) 6(1)), and in many cases are dependent on activation of NMDA receptors (see also Fazeli and Collingridge, 1996). LTD in the cerebellum is evident as weakening of active parallel fibre to Purkinje cell synapses when the climbing fibre connecting to a Purkinje cell is active (see Ito, 1984, 1989, 1993a,b).

1.7 Distributed representations

When considering the operation of many neuronal networks in the brain, it is found that many useful properties arise if each input to the network (arriving on the axons, r') is encoded in the activity of an ensemble or population of the axons or input lines (distributed encoding), and is not signalled by the activity of a single input, which is called local encoding. We start off with some definitions, and then highlight some of the differences, and summarize some evidence which shows the type of encoding used in some brain regions. Then in Chapter 2 (e.g. Table 2.2) on, we show how many of the useful properties of the neuronal networks described depend on distributed encoding. In Chapter 10, we review evidence on the encoding actually found in different regions of the cerebral cortex.

1.7.1 Definitions

A *local* representation is one in which all the information that a particular stimulus or event occurred is provided by the activity of one of the neurons. In a famous example, a single neuron might be active only if one's grandmother was being seen. An implication is that most neurons in the brain regions where objects or events are represented would fire only very rarely. A problem with this type of encoding is that a new neuron would be needed for every object or event that has to be represented. There are many other disadvantages of this type of encoding, many of which will become apparent in this book. Moreover, there is evidence that objects are represented in the brain by a different type of encoding.

A **fully distributed** representation is one in which all the information that a particular stimulus or event occurred is provided by the activity of the full set of neurons. If the neurons are binary (e.g. either active or not), the most distributed encoding is when half the neurons are active for any one stimulus or event.

A **sparse distributed** representation is a distributed representation in which a small proportion of the neurons is active at any one time. In a sparse representation with binary neurons, less than half of the neurons are active for any one stimulus or event. For binary neurons, we can use as a measure of the sparseness the proportion of neurons in the active state. For neurons with real, continuously variable, values of firing rates, the sparseness of the representation a can be measured, by extending the binary notion of the proportion of neurons that are firing, as

$$a = (\Sigma_{i=1,N} r_i/N)^2 / \Sigma_{i=1,N} (r_i^2/N) \tag{1.4}$$

where r_i is the firing rate of the ith neuron in the set of N neurons (Treves and Rolls, 1991).

Coarse coding utilizes overlaps of receptive fields, and can compute positions in the input space using differences between the firing levels of coactive cells (e.g. colour-tuned cones in the retina). The representation implied is distributed. **Fine coding** (in which for example a neuron may be 'tuned' to the exact orientation and position of a stimulus) implies more local coding.

1.7.2 Advantages of different types of coding

One advantage of distributed encoding is that the similarity between two representations can be reflected by the correlation between the two patterns of activity which represent the different stimuli. We have already introduced the idea that the input to a neuron is represented by the activity of its set of input axons r'_j, where j indexes the axons, numbered from $j = 1, C$ (see Fig. 1.2 and Eq. 1.1). Now the set of activities of the input axons is a vector (a vector is an ordered set of numbers; Appendix 1 provides a summary of some of the concepts involved). We can denote as $\mathbf{r'}^1$ the vector of axonal activity that represents stimulus 1, and $\mathbf{r'}^2$ the vector that represents stimulus 2. Then the similarity between the two vectors, and thus the two stimuli, is reflected by the correlation between the two vectors. The correlation will be high if the activity of each axon in the two representations is similar; and will become more and more different as the activity of more and more of the axons differs in the two representations. Thus the similarity of two inputs can be represented in a graded or continuous way if (this type of) distributed encoding is used. This enables generalization to similar stimuli, or to incomplete versions of a stimulus (if it is for example partly seen or partly remembered), to occur. With a local representation, either one stimulus or another is represented, and similarities between different stimuli are not encoded.

Another advantage of distributed encoding is that the number of different stimuli that can be represented by a set of C components (e.g. the activity of C axons) can be very large. A simple example is provided by the binary encoding of an 8-element vector. One component can code for which of two stimuli has been seen, 2 components (or bits in a computer byte) for 4 stimuli, 3 components for 8 stimuli, 8 components for 256 stimuli, etc. That is, the number of stimuli increases exponentially with the number of components (or in this case, axons) in the representation. (In this simple binary illustrative case, the number of stimuli that can be encoded is 2^C.) Put the other way round, even if a neuron has only a limited number of inputs (e.g. a few thousand), it can nevertheless receive a great deal of information about which stimulus was present. This ability of a neuron with a limited number of inputs to receive information about which of potentially very many input events is present is probably one factor that makes computation by the brain possible. With local encoding, the number of stimuli that can be encoded increases only linearly with the number C of axons or components (because a different component is needed to represent each new stimulus). (In our example, only 8 stimuli could be represented by 8 axons.)

In the real brain, there is now good evidence that in a number of brain systems, including the high-order visual and olfactory cortices, and the hippocampus, distributed encoding with the above two properties, of representing similarity, and of exponentially increasing encoding capacity as the number of neurons in the representation increases, is found (Rolls and Tovee, 1995a; Abbott, Rolls and Tovee, 1996; Rolls, Treves and Tovee, 1997; Rolls, Treves, Robertson, Georges-François and Panzeri, 1998). For example, in the high-order visual cortex in the temporal lobe of the primate brain, the number of faces that can be represented increases approximately exponentially with the number of neurons in the population (see Fig. 1.6a). If we plot instead the information about which stimulus is seen, we see that this rises approximately linearly with the number of neurons in the representation (Fig. 1.6b). This corresponds to an exponential rise in the number of stimuli encoded, because information is a

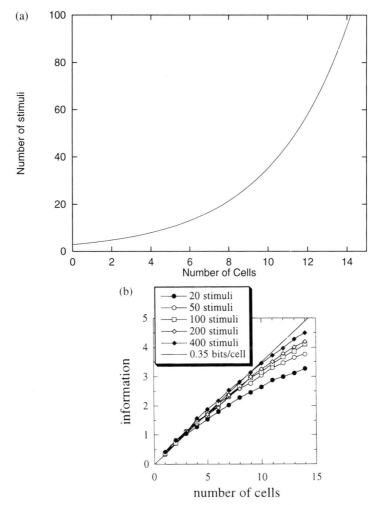

Fig. 1.6 (a) The number of stimuli (in this case from a set of 20 faces) that are encoded in the responses of different numbers of neurons in the temporal lobe visual cortex (after Rolls, Treves and Tovee, 1997; Abbott, Rolls and Tovee, 1996; see Section 10.4.3 and Appendix A2.) (b) The amount of information (in bits) available from the responses of different numbers of temporal cortex neurons about which of 20 faces has been seen (from Abbott, Rolls and Tovee, 1996). In (b), it is shown by simulation that for numbers of stimuli greater than 20, the information rises more and more linearly as the number of cells in the sample increases (see Appendix A2). In both (a) and (b), the analysis time was 500 ms.

log measure (see Appendix A2). A similar result has been found for the encoding of position in space by the primate hippocampus (Rolls, Treves, Robertson, Georges-François and Panzeri, 1998). It is particularly important that the information can be read from the ensemble of neurons using a simple measure of the similarity of vectors, the correlation (or dot product, see Appendix 1) between two vectors. The importance of this is that it is essentially vector similarity operations that characterize the operation of many neuronal networks (see e.g. Chapters 2–4, and Appendix 1). The neurophysiological results show that both the ability to reflect similarity by vector correlation, and the utilization of exponential coding capacity, are a property of real neuronal networks found in the brain.

To emphasize one of the points being made here, although the binary encoding used in the 8-bit vector described above has optimal capacity for binary encoding, it is not optimal for vector similarity operations. For example, the two very similar numbers 127 and 128 are represented by 01111111 and 10000000 with binary encoding, yet the correlation or bit overlap of these vectors is 0. The brain in contrast uses a code which has the attractive property of exponentially increasing capacity with the number of neurons in the representation, though it is different from the simple binary encoding of numbers used in computers; and at the same time codes stimuli in such a way that the code can be read off with simple dot product or correlation-related decoding, which is what is specified for the elementary neuronal network operation shown in Eq. 1.1.

1.8 Introduction to three simple neuronal network architectures

With neurons of the type outlined in Section 1.4, and an associative learning rule of the type described in Section 1.5, three neuronal network architectures arise which appear to be used in many different brain regions. The three architectures will be described in Chapters 2–4, and a brief introduction is provided here.

In the first architecture (see Fig. 1.7a,b), pattern associations can be learned. The output neurons are driven by an unconditioned stimulus. A conditioned stimulus reaches the output neurons by associatively modifiable synapses w_{ij}. If the conditioned stimulus is paired during learning with activation of the output neurons produced by the unconditioned stimulus, then later, after learning, due to the associative synaptic modification, the conditioned stimulus alone will produce the same output as the conditioned stimulus. Pattern associators are described in Chapter 2.

In the second architecture, the output neurons have recurrent associatively modifiable synaptic connections w_{ij} to other neurons in the network (see Fig. 1.7c). When an external input causes the output neurons to fire, then associative links are formed through the modifiable synapses that connect the set of neurons that is active. Later, if only a fraction of the original input pattern is presented, then the associative synaptic connections or weights allow the whole of the memory to be retrieved. This is called completion. Because the components of the pattern are associated with each other as a result of the associatively modifiable recurrent connections, this is called an autoassociative memory. It is believed to be used in the brain for many purposes, including episodic memory in which the parts of a memory of an episode are associated together, and helping to define the response properties of cortical neurons, which have collaterals between themselves within a limited region.

In the third architecture, the main input to the output neurons is received through associatively modifiable synapses w_{ij} (see Fig. 1.7d). Because of the initial values of the synaptic strengths, or because every axon does not contact every output neuron, different patterns tend to activate different output neurons. When one pattern is being presented, the most strongly activated neurons tend via lateral inhibition to inhibit the other neurons. For this reason the network is called competitive. During the presentation of that pattern, associative modification of the active axons onto the active postsynaptic neuron takes place. Later, that or similar patterns will have a greater chance of activating that neuron or set of neurons. Other neurons learn to respond to other input patterns. In this way, a network is built which can

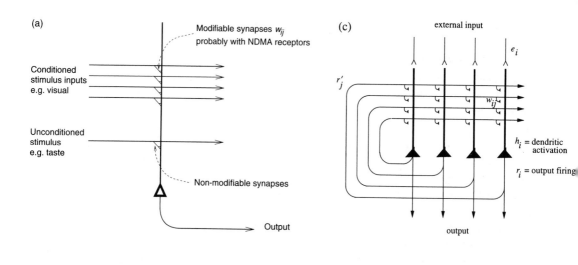

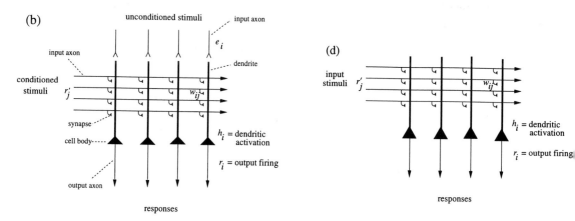

Fig. 1.7 Three network architectures that use local learning rules: (a) Pattern association introduced with a single output neuron; (b) Pattern association network; (c) Autoassociation network; (d) Competitive network.

categorize patterns, placing similar patterns into the same category. This is useful as a preprocessor for sensory information, and finds use in many other parts of the brain too.

These are three fundamental building blocks for neural architectures in the brain. They are often used in combination with each other. Because they are some of the building blocks of some of the architectures found in the brain, they are described in Chapters 2–4.

1.9 Systems-level analysis of brain function

To understand the neuronal network operations of any one brain region, it is useful to have an idea of the systems-level organization of the brain, in order to understand how the networks in each region provide a particular computational function as part of an overall computational scheme.

Some of the pathways followed by sensory inputs to reach memory systems and eventually motor outputs are shown in Fig. 1.8. Some of these regions are shown in the drawings of the primate brain in Figs 1.9–1.13. Each of these routes is described in turn. The description is based primarily on studies in non-human primates, for they have well-developed cortical areas which in many cases correspond to those found in humans, and it has been possible to analyse their connectivity and their functions by recording the activity of neurons in them. Further evidence on these systems is provided in Section 10.5.

Information on the ventral visual cortical processing stream projects after the primary visual cortex, area V1, to the secondary visual cortex (V2), and then via area V4 to the posterior and then to the anterior inferior temporal visual cortex (lower stream in Fig. 1.8;

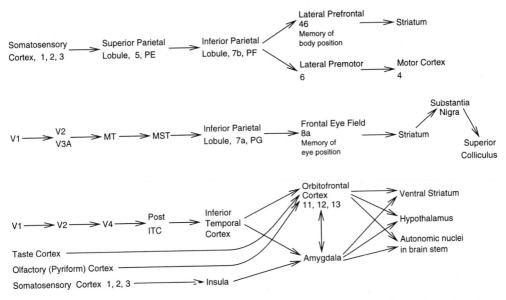

Fig. 1.8 Schematic diagram showing the major connection of three separate processing systems in the brain (see text). The top pathway, also shown in Fig. 1.13 on a lateral view of the macaque brain, shows the connections from the primary somatosensory cortex, areas 1, 2 and 3, via area 5 in the parietal cortex, to area 7b. The middle pathway, also shown in Fig 1.11, shows the connections in the 'dorsal visual system' from V1 to V2, MST, etc., with some connections reaching the frontal eye fields. The lower pathway, also shown in Fig. 1.9, shows the connections in the 'ventral visual system' from V1 to V2, V4, the inferior temporal visual cortex, etc., with some connections reaching the amygdala and orbitofrontal cortex.

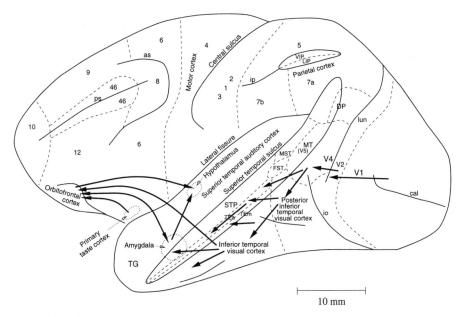

Fig. 1.9 Lateral view of the macaque brain showing the connections in the 'ventral visual system' from V1, V2 and V4, the inferior temporal visual cortex, etc., with some connections reaching the amygdala and orbitofrontal cortex. as, arcuate sulcus; cal, calcarine sulcus; cs, central sulcus; lf, lateral (or Sylvian) fissure; lun, lunate sulcus; ps, principal sulcus; io, inferior occipital sulcus; ip, intraparietal sulcus (which has been opened to reveal some of the areas it contains); sts, superior temporal sulcus (which has been opened to reveal some of the areas it contains). AIT, anterior inferior temporal cortex; FST, visual motion processing area; LIP, lateral intraparietal area; MST, visual motion processing area; MT, visual motion processing area (also called V5); PIT, posterior inferior temporal cortex; STP, superior temporal plane; TA, architectonic area including auditory association cortex; TE, architectonic area including high order visual association cortex, and some of its subareas TEa and TEm; TG, architectonic area in the temporal pole; V1–V4, visual areas 1–4; VIP, ventral intraparietal area; TEO, architectonic area including posterior visual association cortex. The numerals refer to architectonic areas, and have the following approximate functional equivalence: 1, 2, 3, somatosensory cortex (posterior to the central sulcus); 4, motor cortex; 5, superior parietal lobule; 7a, inferior parietal lobule, visual part; 7b, inferior parietal lobule, somatosensory part; 6, lateral premotor cortex; 8, frontal eye field; 12, part of orbitofrontal cortex; 46, dorsolateral prefrontal cortex.

Fig. 1.9; Fig. 1.12). Information processing along this stream is primarily unimodal, as shown by the fact that inputs from other modalities (such as taste or smell) do not anatomically have significant inputs to these regions, and by the fact that neurons in these areas respond primarily to visual stimuli, and not to taste or olfactory stimuli, etc. (see Rolls, 1997d; Baylis, Rolls and Leonard, 1987; Ungerleider, 1995). The representation built along this pathway is mainly about what object is being viewed, independently of exactly where it is on the retina, of its size, and even of the angle with which it is viewed (Rolls, 1994a, 1995a, 1997d). The representation is also independent of whether the object is associated with reward or punishment, that is the representation is about objects *per se* (Rolls *et al.*, 1977). The computation which must be performed along this stream is thus primarily to build a representation of objects which shows invariance. After this processing, the visual representation is interfaced to other sensory systems in areas in which simple associations must be learned between stimuli in different modalities (see Fig. 1.10). The representation must thus be in a form in which the simple generalization properties of associative networks can be useful. Given that the association is about which object is present (and not where it is on the retina), the representation computed in sensory systems must be in a form which allows the simple correlations computed by associative networks to reflect similarities between objects, and not between their positions on the retina. The way in which such invariant sensory representations could be built in the brain is the subject of Chapter 8. A similar mainly unimodal analysis of taste inputs is carried out by the primary taste cortex, to represent what taste is present, independently of the pleasantness or aversiveness of the taste (Rolls, 1995b, 1997c).

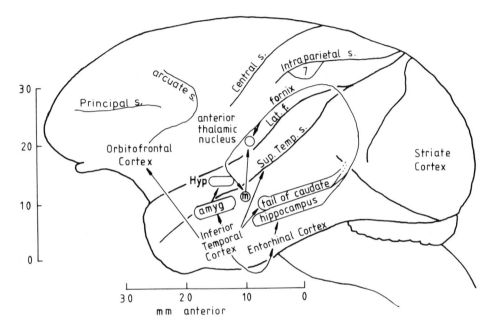

Fig. 1.10 Lateral view of the primate (rhesus monkey) brain showing some of the brain regions described. This diagram shows some of the outputs of the inferior temporal visual cortex. Abbreviations: Amyg, amygdala; central s, central sulcus; Hyp, hypothalamus/substantia innominata/basal forebrain; Lat f, lateral (or Sylvian) fissure; m, mammillary body; Sup Temp s, superior temporal sulcus; 7, posterior parietal cortex, area 7.

After mainly unimodal processing stages, these information processing streams converge together into a number of areas, particularly the amygdala and orbitofrontal cortex (see Figs 1.8, 1.9 and 1.10). These areas appear to be necessary for learning to associate sensory stimuli with other reinforcing (rewarding or punishing) stimuli. For example, the amygdala is involved in learning associations between the sight of food and its taste. (The taste is a primary or innate reinforcer.) The orbitofrontal cortex is especially involved in rapidly relearning these associations, when environmental contingencies change (see Rolls, 1990a, 1995b, 1996b; Chapter 7). They thus are brain regions in which the computation at least includes simple pattern association (e.g. between the sight of an object and its taste). In the orbitofrontal cortex, this association learning is also used to produce a representation of flavour, in that neurons are found in the orbitofrontal cortex which are activated by both olfactory and taste stimuli (Rolls and Baylis, 1994), and in that the neuronal responses in this region reflect in some cases olfactory to taste association learning (Rolls, Critchley, Mason and Wakeman, 1996; Critchley and Rolls, 1996b). In these regions too, the representation is concerned not only with what sensory stimulus is present, but for some neurons, with its hedonic or reward-related properties, which are often computed by association with stimuli in other modalities. For example, many of the visual neurons in the orbitofrontal cortex respond to the sight of food only when hunger is present. This probably occurs because the visual inputs here have been associated with a taste input, which itself in this region only occurs to a food if hunger is present, that is when the taste is rewarding (see Chapter 7 and Rolls, 1993, 1994b, 1995b, 1996b). The outputs from these associative memory systems, the amygdala and orbitofrontal cortex, project onwards to structures such as the hypothalamus, through which they control autonomic and endocrine responses such as salivation and insulin release to the sight of food; and to the striatum, including the ventral striatum, through which behaviour to learned reinforcing stimuli is produced. The striatal output pathways for the cerebral cortex are described in Chapter 9, on motor systems. Somatosensory (touch) inputs also reach the amygdala and orbitofrontal cortex, via projections from the somatosensory cortical areas (S1, 2 and 3) to the insula (see Fig. 1.8).

Another processing stream shown in Figs 1.8, 1.11 and 1.12 is that from V1 to MT, MST and thus to the parietal cortex (see Ungerleider, 1995; Ungerleider and Haxby, 1994; Section 10.5). This 'where' pathway for primate vision is involved in representing where stimuli are relative to the animal, and the motion of these stimuli. Neurons here respond for example to stimuli in visual space around the animal, including the distance from the observer, and also respond to optic flow or to moving stimuli. The outputs of this system control eye movements to visual stimuli (both slow pursuit and saccadic eye movements). One output of these regions is to the frontal eye fields which are important as a short term memory for where fixation should occur next, as shown by the effects of lesions to the frontal eye fields on saccades to

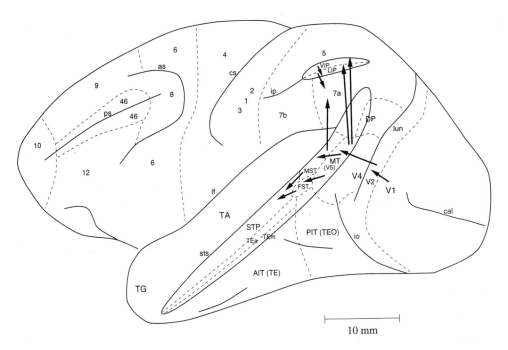

Fig. 1.11 Lateral view of the macaque brain showing the connections in the 'dorsal visual system' from V1 to V2, MST, etc. with some connections reaching the frontal eye fields. Abbreviations as in Fig. 1.9.

PARIETAL - "WHERE"

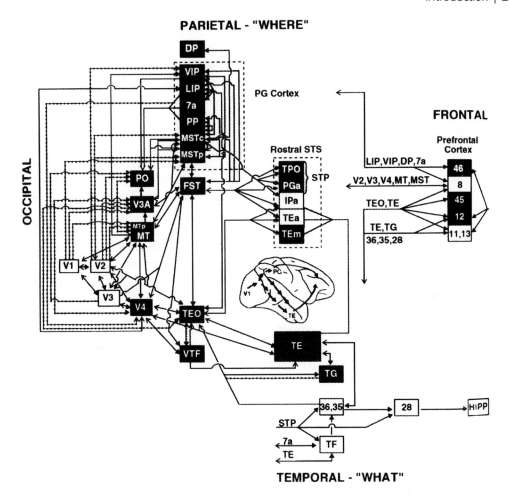

Fig. 1.12 Visual processing pathways in monkeys. Solid lines indicate connections arising from both central and peripheral visual field representations; dotted lines indicate connections restricted to peripheral visual field representations. Shaded boxes in the ventral (lower) stream indicate visual areas related primarily to object vision; shaded boxes in the dorsal stream indicate areas related primarily to spatial vision; and white boxes indicate areas not clearly allied with only one stream. The shaded region on the lateral view of the brain represents the extent of the cortex included in the diagram. Abbreviations: DP, dorsal prelunate area; FST, fundus of the superior temporal area; HIPP, hippocampus; LIP, lateral intraparietal area; MSTc, medial superior temporal area, central visual field representation; MSTp, medial superior temporal area, peripheral visual field representation; MT, middle temporal area; MTp, middle temporal area, peripheral visual field representation; PO, parieto-occipital area; PP, posterior parietal sulcal zone; STP, superior temporal polysensory area; V1, primary visual cortex; V2, visual area 2; V3, visual area 3; V3A, visual area 3, part A; V4, visual area 4; and VIP, ventral intraparietal area. Inferior parietal area 7a; prefrontal areas 8, 11 to 13, 45 and 46 are from Brodmann (1925). Inferior temporal areas TE and TEO, parahippocampal area TF, temporal pole area TG, and inferior parietal area PG are from Von Bonin and Bailey (1947). Rostral superior temporal sulcal (STS) areas are from Seltzer and Pandya (1978) and VTF is the visually responsive portion of area TF (Boussaoud, Desimone and Ungerleider, 1991). (Reprinted with permission from Ungerleider, 1995.)

remembered targets, and by neuronal activity in this region (see Section 10.5.2). Outputs from the frontal eye fields again reach the striatum, and then progress through the basal ganglia (that is, via the substantia nigra) to reach the superior colliculus.

A related processing stream is that from the somatosensory cortical areas to parietal cortex

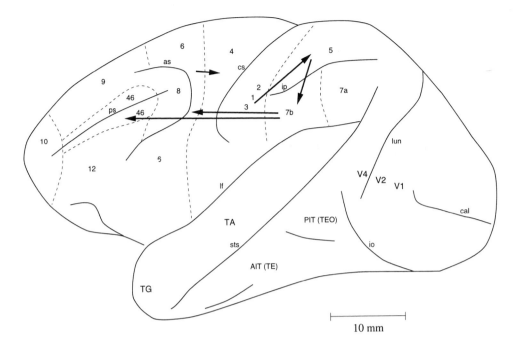

Fig. 1.13 Lateral view of the macaque brain showing the connections from the primary somatosensory cortex, areas 1, 2 and 3, via area 5 in the parietal cortex, to area 7b. Abbreviations as in Fig. 1.9.

area 5, and thus to parietal cortex area 7b (see Figs 1.8 and 1.13, and Section 10.5). Along this pathway more complex representations are formed of objects touched (Iwamura, 1993), and of where the limbs are in relation to the body (by combining information from different joint proprioceptors). In area 7, some neurons respond to visual and to related somatosensory stimuli. Outputs from the parietal cortex project to the premotor areas and to the basal ganglia, which both provide routes to behavioural output. In addition, the parietal cortex projects to the dorsolateral prefrontal cortex, which provides a short term or working memory for limb responses, as shown by the effects of lesions to the dorsolateral prefrontal cortex, and by recordings of neuronal activity in it during delayed response tasks (Goldman-Rakic, 1996). The dorsolateral prefrontal cortex can influence behaviour through basal ganglia outputs, and through outputs to premotor areas.

The hippocampus receives inputs from both the 'what' and the 'where' systems (see Chapter 6 and Fig. 1.12). By rapidly learning associations between conjunctive inputs in these systems, it is able to form memories of particular events occurring in particular places at particular times. To do this, it needs to store whatever is being represented in each of many cortical areas at a given time, and to later recall the whole memory from a part of it. The types of network it contains which are involved in this simple memory function are described in Chapter 6.

With this overview of some of the main processing streams in the cerebral cortex, it is now time to consider in Chapters 2–5 the operation of some fundamental types of biologically plausible network. Then in Chapter 6–10 we will consider how these networks may contribute to the particular functions being performed by different brain regions.

2 Pattern association memory

A fundamental operation of most nervous systems is to learn to associate a first stimulus with a second which occurs at about the same time, and to retrieve the second stimulus when the first is presented. The first stimulus might be the sight of food, and the second stimulus the taste of food. After the association has been learned, the sight of food would enable its taste to be retrieved. In classical conditioning, the taste of food might elicit an unconditioned response of salivation, and if the sight of the food is paired with its taste, then the sight of that food would by learning come to produce salivation. More abstractly, if one idea is associated by learning with a second, then when the first idea occurs again, the second idea will tend to be associatively retrieved.

2.1 Architecture and operation

The essential elements necessary for pattern association, forming what could be called a prototypical pattern associator network, are shown in Fig. 2.1. What we have called the second or unconditioned stimulus pattern is applied through unmodifiable synapses generating an input to each unit which, being external with respect to the synaptic matrix we focus on, we can call the external input e_i for the ith neuron. (We can also treat this as a vector, $\mathbf{e}$, as indicated in the legend to Fig. 2.1. Vectors and simple operations performed with them are summarized in Appendix A1). This unconditioned stimulus is dominant in producing or forcing the firing of the output neurons (r_i for the ith neuron, or the vector $\mathbf{r}$). At the same time, the first or conditioned stimulus pattern r'_j for the jth axon (or equivalently the vector $\mathbf{r}'$) present on the horizontally running axons in Fig. 2.1 is applied through *modifiable* synapses w_{ij} to the dendrites of the output neurons. The synapses are modifiable in such a way that if there is presynaptic firing on an input axon r'_j paired during learning with postsynaptic activity on neuron i, then the strength or weight w_{ij} between that axon and the dendrite increases. This simple learning rule is often called the Hebb rule, after Donald Hebb who in 1949 formulated the hypothesis that if the firing of one neuron was regularly associated with another, then the strength of the synapse or synapses between the neurons should increase in strength. (In fact, the terms in which Hebb put the hypothesis were a little different from an association memory, in that he stated that if one neuron regularly comes to elicit firing in another, then the strength of the synapses should increase. He had in mind the building of what he called cell assemblies. In a pattern associator, the conditioned stimulus need not produce before learning any significant activation of the output neurons.

The connections must simply increase if there is associated pre- and postsynaptic firing when, in pattern association, most of the postsynaptic firing is being produced by a different input.) After learning, presenting the pattern **r'** on the input axons will activate the dendrite through the strengthened synapses. If the cue or conditioned stimulus pattern is the same as that learned, the postsynaptic neurons will be activated, even in the absence of the external or unconditioned input, as each of the firing axons produces through a strengthened synapse some activation of the postsynaptic element, the dendrite. The total activation h_i of each postsynaptic neuron i is then the sum of such individual activations. In this way, the 'correct' output neurons, that is those activated during learning, can end up being the ones most strongly activated, and the second or unconditioned stimulus can be effectively recalled. The recall is best when only strong activation of the postsynaptic neuron produces firing, that is if there is a threshold for firing, just like real neurons. The reasons for this arise when many associations are stored in the memory, as will soon be shown.

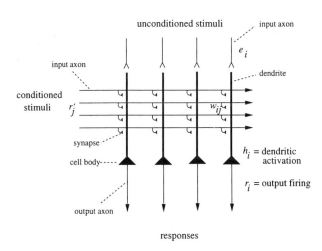

Fig. 2.1 A pattern association memory. An unconditioned stimulus has activity or firing rate e_i for the ith neuron, and produces firing r_i of the ith neuron. An unconditioned stimulus may be treated as a vector, across the set of neurons indexed by i, of activity **e**. The firing rate response can also be thought of as a vector of firing **r**. The conditioned stimuli have activity or firing rate r'_j for the jth axon, which can also be treated as a vector **r'**.

Next we introduce a more precise description of the above by writing down explicit mathematical rules for the operation of the simple network model of Fig. 2.1, which will help us to understand how pattern association memories in general operate. (In this description we introduce simple vector operations, and, for those who are not familiar with these, have provided in Appendix A1 a concise summary for later reference.) We have denoted above a conditioned stimulus input pattern as **r'**. Each of the axons has a firing rate, and if we count or index through the axons using the subscript j, the firing rate of the first axon is r'_1, of the second r'_2, of the jth r'_j, etc. The whole set of axons forms a vector, which is just an ordered (1, 2, 3, etc.) set of elements. The firing rate of each axon r'_j is one element of the firing rate vector **r'**. Similarly, using i as the index, we can denote the firing rate of any output neuron as r_i, and the firing rate output vector as **r**. With this terminology, we can then identify any synapse onto neuron i from neuron j as w_{ij} (see Fig. 2.1). In this book, the first index, i, always refers to the receiving neuron (and thus signifies a dendrite), while the second index, j, refers to the sending neuron (and thus signifies a conditioned stimulus input axon in Fig. 2.1). We can now specify the learning and retrieval operations as follows:

2.1.1 Learning

The firing rate of every output neuron is forced to a value determined by the unconditioned (or external or forcing stimulus) input. In our simple model this means that for any one neuron i,

$$r_i = f(e_i) \tag{2.1}$$

which indicates that the firing rate is a function of the dendritic activation, taken in this case to reduce essentially to that resulting from the external forcing input (see Fig. 2.1). The function f is called the activation function (see Fig. 1.3), and its precise form is irrelevant, at least during this learning phase. For example, the function at its simplest could be taken to be linear, so that the firing rate would be just proportional to the activation.

The Hebb rule can then be written as follows:

$$\delta w_{ij} = k \, r_i r'_j \tag{2.2}$$

where δw_{ij} is the change of the synaptic weight w_{ij} which results from the simultaneous (or conjunctive) presence of presynaptic firing r'_j and postsynaptic firing or activation r_i, and k is a learning rate constant which specifies how much the synapses alter on any one pairing.

The Hebb rule is expressed in this multiplicative form to reflect the idea that *both* presynaptic and postsynaptic activity must be present for the synapses to increase in strength. The multiplicative form also reflects the idea that strong pre- and postsynaptic firing will produce a larger change of synaptic weight than smaller firing rates. It is also assumed for now that before any learning takes place, the synaptic strengths are small in relation to the changes that can be produced during Hebbian learning. We will see that this assumption can be relaxed later when a modified Hebb rule is introduced that can lead to a reduction in synaptic strength under some conditions.

2.1.2 Recall

When the conditioned stimulus is present on the input axons, the total activation h_i of a neuron i is the sum of all the activations produced through each strengthened synapse w_{ij} by each active neuron r'_j. We can express this as

$$h_i = \Sigma_j r'_j w_{ij} \tag{2.3}$$

where Σ_j indicates that the sum is over the C input axons (or connections) indexed by j. The multiplicative form here indicates that activation should be produced by an axon only if it is firing, and only if it is connected to the dendrite by a strengthened synapse. It also indicates that the strength of the activation reflects how fast the axon r'_j is firing, and how strong the synapse w_{ij} is. The sum of all such activations expresses the idea that summation (of synaptic currents in real neurons) occurs along the length of the dendrite, to produce activation at the cell body, where the activation h_i is converted into firing r_i. This conversion can be expressed as

$$r_i = f(h_i) \tag{2.4}$$

where the function f is again the activation function. The form of the function now becomes more important. Real neurons have thresholds, with firing occurring only if the activation is above the threshold. A threshold linear activation function is shown in Fig. 1.3b. This has been useful in formal analysis of the properties of neural networks. Neurons also have firing rates which become saturated at a maximum rate, and we could express this as the sigmoid activation function shown in Fig. 1.3c. Yet another simple activation function, used in some models of neural networks, is the binary threshold function (Fig. 1.3d), which indicates that if the activation is below threshold, there is no firing, and that if the activation is above threshold, the neuron fires maximally. Whatever the exact shape of the activation function, some non-linearity is an advantage, for it enables small activations produced by interfering memories to be minimized, and it can enable neurons to perform logical operations, such as to fire or respond only if two or more sets of inputs are present simultaneously.

2.2 A simple model

An example of these learning and recall operations is provided in a very simple form as follows. The neurons will have simple firing rates, which can be 0 to represent no activity, and 1 to indicate high firing. They are thus binary neurons, which can assume one of two firing rates. If we have a pattern associator with six input axons and four output neurons, we could represent the network before learning, with the same layout as Fig. 2.1, as

		UCS		
CS	1	1	0	0
	↓	↓	↓	↓
1→	0	0	0	0
0→	0	0	0	0
1→	0	0	0	0
0→	0	0	0	0
1→	0	0	0	0
0→	0	0	0	0

Fig. 2.2

where **r'** or the conditioned stimulus (CS) is 101010, and **r** or the firing produced by the unconditioned stimulus (UCS) is 1100. (The arrows indicate the flow of signals.) The synaptic weights are initially all 0. After pairing the CS with the UCS during one learning trial, some of the synaptic weights will be incremented according to Eq. 2.2, so that after learning this pair the synaptic weights will become

		UCS		
CS	1	1	0	0
	↓	↓	↓	↓
1→	1	1	0	0
0→	0	0	0	0
1→	1	1	0	0
0→	0	0	0	0
1→	1	1	0	0
0→	0	0	0	0

Fig. 2.3

We can represent what happens during recall, when for example we present the CS that has been learned, as follows:

		CS			
1→	1	1	0	0	
0→	0	0	0	0	
1→	1	1	0	0	
0→	0	0	0	0	
1→	1	1	0	0	
0→	0	0	0	0	
	↓	↓	↓	↓	
	3	3	0	0	Activation h_i
	1	1	0	0	Firing r_i

Fig. 2.4

The activation of the four output neurons is 3300, and if we set the threshold of each output neuron to 2, then the output firing is 1100 (where the binary firing rate is 0 if below threshold, and 1 if above). The pattern associator has thus achieved recall of the pattern 1100, which is correct.

We can now illustrate how a number of different associations can be stored in such a pattern associator, and retrieved correctly. Let us associate a new CS pattern 110001 with the UCS 0101 in the same pattern associator. The weights will become as shown next in Fig. 2.5 after learning:

	UCS			
CS	0	1	0	1
	↓	↓	↓	↓
1→	1	2	0	1
1→	0	1	0	1
0→	1	1	0	0
0→	0	0	0	0
0→	1	1	0	0
1→	0	1	0	1

Fig. 2.5

If we now present the second CS, the retrieval is as follows:

	CS				
1→	1	2	0	1	
1→	0	1	0	1	
0→	1	1	0	0	
0→	0	0	0	0	
0→	1	1	0	0	
1→	0	1	0	1	
	↓	↓	↓	↓	
	1	4	0	3	Activation h_i
	0	1	0	1	Firing r_i

Fig. 2.6

The binary output firings were again produced with the threshold set to 2. Recall is perfect. This illustration shows the value of some threshold non-linearity in the activation function of the neurons. In this case, the activations did reflect some small cross-talk or interference from the previous pattern association of CS1 with UCS1, but this was removed by the threshold operation, to clean up the recall firing. The example also shows that when further associations are learned by a pattern associator trained with the Hebb rule, Eq. 2.2, some synapses will reflect increments above a synaptic strength of 1. It is left as an exercise to the reader to verify that recall is still perfect to CS1, the vector 101010. (The output activation vector **h** is 3401, and the output firing vector **r** with the same threshold of 2 is 1100, which is perfect recall.)

2.3 The vector interpretation

The way in which recall is produced, Eq. 2.3, consists for each output neuron i of multiplying each input firing rate r'_j by the corresponding synaptic weight w_{ij} and summing the products to obtain the activation h_i. Now we can consider the firing rates r'_j where j varies from 1 to N', the number of axons, to be a vector. (A vector is simply an ordered set of numbers—see

Appendix A1.) Let us call this vector $\mathbf{r}'$. Similarly, on a neuron i, the synaptic weights can be treated as a vector, $\mathbf{w}_i$. (The subscript i here indicates that this is the weight vector on the ith neuron.) The operation we have just described to obtain the activation can now be seen to be a simple multiplication operation of two vectors to produce a single output value (called a scalar output). This is the inner product or dot product of two vectors, and can be written

$$h_i = \mathbf{r}' \cdot \mathbf{w}_i \tag{2.5}$$

The inner product of two vectors indicates how similar they are. If two vectors have corresponding elements the same, then the dot product will be maximal. If the two vectors are similar but not identical, then the dot product will be high. If the two vectors are completely different, the dot products will be 0, and the vectors are described as orthogonal. (The term orthogonal means at right angles, and arises from the geometric interpretation of vectors, which is summarized in Appendix A1.) Thus the dot product provides a direct measure of how similar two vectors are. It can now be seen that a fundamental operation many neurons perform is effectively to compute how similar an input pattern vector $\mathbf{r}'$ is to a stored weight vector. The similarity measure they compute, the dot product, is a very good measure of similarity, and indeed, the standard (Pearson product-moment) correlation coefficient used in statistics is the same as a normalized dot product with the mean subtracted from each vector, as shown in Appendix 1. (The normalization used in the correlation coefficient results in the coefficient varying always between $+1$ and -1, whereas the actual scalar value of a dot product clearly depends on the length of the vectors from which it is calculated.)

With these concepts, we can now see that during learning a pattern associator adds to its weight vector a vector $\delta\mathbf{w}_i$ that has the same pattern as the input pattern $\mathbf{r}'$, if the postsynaptic neuron i is strongly activated. Indeed, we can express Eq. 2.2 in vector form as

$$\delta\mathbf{w}_i = k\, r_i\, \mathbf{r}' \tag{2.6}$$

We can now see that what is recalled by the neuron depends on the similarity of the recall cue vector $\mathbf{r}'_r$ to the originally learned vector $\mathbf{r}'$. The fact that during recall the output of each neuron reflects the similarity (as measured by the dot product) of the input pattern $\mathbf{r}'_r$ to each of the patterns used originally as $\mathbf{r}'$ inputs (conditioned stimuli in Fig. 2.1) provides a simple way to appreciate many of the interesting and biologically useful properties of pattern associators, as described next.

2.4 Properties

2.4.1 Generalization

During recall, pattern associators generalize, and produce appropriate outputs if a recall cue vector $\mathbf{r}'_r$ is similar to a vector that has been learned already. This occurs because the recall operation involves computing the dot (inner) product of the input pattern vector $\mathbf{r}'_r$ with the synaptic weight vector $\mathbf{w}_i$, so that the firing produced, r_i, reflects the similarity of the current input to the previously learned input pattern **r'**. (Generalization will occur to input cue or conditioned stimulus patterns **r'**, which are incomplete versions of an original conditioned stimulus **r'**, although the term completion is usually applied to the autoassociation networks described in Chapter 3.)

This is an extremely important property of pattern associators, for input stimuli during recall will rarely be absolutely identical to what has been learned previously, and automatic generalization to similar stimuli is extremely useful, and has great adaptive value in biological systems.

Generalization can be illustrated with the simple binary pattern associator considered above. (Those who have appreciated the vector description just given may wish to skip this illustration.) Instead of the second CS, pattern vector 110001, we will use the similar recall cue 110100.

$$
\begin{array}{ccccc}
 & & \text{CS} & & \\
1\rightarrow & 1 & 2 & 0 & 1 \\
1\rightarrow & 0 & 1 & 0 & 1 \\
0\rightarrow & 1 & 1 & 0 & 0 \\
1\rightarrow & 0 & 0 & 0 & 0 \\
0\rightarrow & 1 & 1 & 0 & 0 \\
0\rightarrow & 0 & 1 & 0 & 1 \\
 & \downarrow & \downarrow & \downarrow & \downarrow \\
 & & & & \\
 & 1 & 3 & 0 & 2 \quad \text{Activation } h_i \\
 & 0 & 1 & 0 & 1 \quad \text{Firing } r_i \\
\end{array}
$$

Fig. 2.7

It is seen that the output firing rate vector, 0101, is exactly what should be recalled to CS2 (and not to CS1), so correct generalization has occurred. Although this is a small network trained with few examples, the same properties hold for large networks with large numbers of stored patterns, as described more quantitatively in the section on capacity below and in Appendix A3.

2.4.2 Graceful degradation or fault tolerance

If the synaptic weight vector w_i (or the weight matrix, which we can call **W**) has synapses missing (e.g. during development), or loses synapses, then the output activation h_i or **h** is still reasonable, because h_i is the dot product (correlation) of **r'** with w_i. The result, especially after passing through the output activation function, can frequently be perfect recall. The same property arises if for example one or some of the CS input axons are lost or damaged. This is a very important property of associative memories, and is not a property of conventional computer memories, which produce incorrect data if even only 1 storage location (for 1 bit or binary digit of data) of their memory is damaged or cannot be accessed. This property of graceful degradation is of great adaptive value for biological systems.

We can illustrate this with a simple example. If we damage two of the synapses in Fig. 2.2 to produce the synaptic matrix shown in Fig. 2.8 (where x indicates a damaged synapse which has no effect, but was previously 1), and now present the second CS, the retrieval is as follows:

$$
\begin{array}{cccccl}
 & & & \text{CS} & & \\
1\rightarrow & 1 & 2 & 0 & 1 & \\
1\rightarrow & 0 & 1 & 0 & x & \\
0\rightarrow & 1 & 1 & 0 & 0 & \\
0\rightarrow & 0 & 0 & 0 & 0 & \\
0\rightarrow & 1 & x & 0 & 0 & \\
1\rightarrow & 0 & 1 & 0 & 1 & \\
 & \downarrow & \downarrow & \downarrow & \downarrow & \\
 & 1 & 4 & 0 & 2 & \text{Activation } h_i \\
 & 0 & 1 & 0 & 1 & \text{Firing } r_i \\
\end{array}
$$

Fig. 2.8

The binary output firings were again produced with the threshold set to 2. The recalled vector, 0101, is perfect. This illustration again shows the value of some threshold non-linearity in the activation function of the neurons. It is left as an exercise to the reader to verify that recall is still perfect to CS1, the vector 101010. (The output activation vector **h** is 3301, and the output firing vector **r** with the same threshold of 2 is 1100, which is perfect recall.)

2.4.3 The importance of distributed representations for pattern associators

A distributed representation is one in which the firing or activity of all the elements in the vector is used to encode a particular stimulus. For example, in the vector CS1 which has the value 101010, we need to know the state of all the elements to know which stimulus is being represented. Another stimulus, CS2, is represented by the vector 110001. We can represent many different events or stimuli with such overlapping sets of elements, and because in general any one element cannot be used to identify the stimulus, but instead the information about which stimulus is present is distributed over the population of elements or neurons, this is called a distributed representation. If, for binary neurons, half the neurons are in one state (e.g. 0), and the other are in the other state (e.g. 1), then the representation is described as **fully distributed**. The CS representations above are thus fully distributed. If only a smaller proportion of the neurons is active to represent a stimulus, as in the vector 100001, then this is a **sparse representation**. For binary representations, we can quantify the sparseness by the proportion of neurons in the active (1) state.

In contrast, a **local representation** is one in which all the information that a particular stimulus or event has occurred is provided by the activity of one of the neurons, or elements in the vector. One stimulus might be represented by the vector 100000, another stimulus by the vector 010000, and a third stimulus by the vector 001000. The activity of neuron or element 1 would indicate that stimulus 1 was present, and of neuron 2, that stimulus 2 was present. The representation is local in that if a particular neuron is active, we know that the stimulus represented by that neuron is present. In neurophysiology, if such cells were present, they might be called 'grandmother cells' (cf. Barlow, 1972, 1995), in that one neuron might represent a stimulus in the environment as complex and specific as one's grandmother. Where the activity of a number of cells must be taken into account in order to represent a stimulus (such as an individual taste), then the representation is sometimes described as using ensemble encoding.

Now, the properties just described for associative memories, generalization and graceful degradation, are only implemented if the representation of the CS or r' vector is distributed. This occurs because the recall operation involves computing the dot (inner) product of the input pattern vector r'_r with the synaptic weight vector w_i. This allows the output activation h_i to reflect the similarity of the current input pattern to a previously learned input pattern r' only if several or many elements of the r' and r'_r vectors are in the active state to represent a pattern. If local encoding were used, e.g. 100000, then if the first element of the vector (which might be the firing of axon 1, i.e. r'_1, or the strength of synapse $i1$, w_{i1}) is lost, then the resulting vector is not similar to any other CS vector, and the output activation is 0. In the case of local encoding, the important properties of associative memories, generalization and graceful degradation, do not thus emerge. Graceful degradation and generalization are dependent on distributed representations, for then the dot product can reflect similarity even when some elements of the vectors involved are altered. If we think of the correlation between Y and X in a graph, then this correlation is affected only little if a few X,Y pairs of data are lost (see Appendix A1).

2.4.4 Prototype extraction, extraction of central tendency, and noise reduction

If a set of similar conditioned stimulus vectors r' are paired with the same unconditioned stimulus e_i, the weight vector w_i becomes (or points towards) the sum (or with scaling the average) of the set of similar vectors r'. This follows from the operation of the Hebb rule in Eq. 2.2. When tested at recall, the output of the memory is then best to the average input pattern vector denoted $<r'>$. If the average is thought of as a prototype, then even though the prototype vector $<r'>$ itself may never have been seen, the best output of the neuron or network is to the prototype. This produces 'extraction of the prototype' or 'central tendency'. The same phenomenon is a feature of human memory performance (see McClelland and Rumelhart, 1986 Ch. 17, and Section 3.4.4), and this simple process with distributed representations in a neural network accounts for the phenomenon.

If the different exemplars of the vector r' are thought of as noisy versions of the true input pattern vector $<r'>$ (with incorrect values for some of the elements), then the pattern associator has performed 'noise reduction', in that the output produced by any one of these vectors will represent the true, noiseless, average vector $<r'>$.

2.4.5 Speed

Recall is very fast in a real neuronal network, because the conditioned stimulus input firings r'_j ($j = 1, C$ axons) can be applied simultaneously to the synapses w_{ij}, and the activation h_i can be accumulated in one or two time constants of the dendrite (e.g. 10–20 ms). Whenever the threshold of the cell is exceeded, it fires. Thus, in effectively one step, which takes the brain no more than 10–20 ms, all the output neurons of the pattern associator can be firing with rates that reflect the input firing of every axon. This is very different from a conventional digital computer, in which computing h_i in Eq. 2.2 would involve C multiplication and addition operations occurring one after another, or $2C$ time steps. The brain performs parallel computation in at least two senses in even a pattern associator. One is that for a single neuron, the separate contributions of the firing r'_j rate of each axon j multiplied by the synaptic weight w_{ij} are computed in parallel and added in the same time step. The second is that this can be performed in parallel for all neurons $i = 1, N$ in the network, where there are N output neurons in the network. It is these types of parallel processing which enable these classes of neuronal network in the brain to operate so fast, in effectively so few steps.

Learning is also fast ('one-shot') in pattern associators, in that a single pairing of the conditioned stimulus **r'** and the unconditioned stimulus **e** which produces the unconditioned output firing **r** enables the association to be learned. There is no need to repeat the pairing in order to discover over many trials the appropriate mapping. This is extremely important for biological systems, in which a single co-occurrence of two events may lead to learning which could have life-saving consequences. (For example, the pairing of a visual stimulus with a potentially life-threatening aversive event may enable that event to be avoided in future.) Although repeated pairing with small variations of the vectors is used to obtain the useful properties of prototype extraction, extraction of central tendency, and noise reduction, the essential properties of generalization and graceful degradation are obtained with just one pairing. The actual time scales of the learning in the brain are indicated by studies of associative synaptic modification using long-term potentiation paradigms (LTP, see Chapter 1). Co-occurrence or near simultaneity of the CS and UCS is required for periods of as little as 100 ms, with expression of the synaptic modification being present within typically a few seconds.

2.4.6 Local learning rule

The simplest learning rule used in pattern association neural networks, a version of the Hebb rule, is as in Eq. 2.2 above

$$\delta w_{ij} = k\, r_i r'_j \, .$$

This is a local learning rule in that the information required to specify the change in synaptic weight is available locally at the synapse, as it is dependent only on the presynaptic firing rate r'_j available at the synaptic terminal, and the postsynaptic activation or firing r_i available on the dendrite of the neuron receiving the synapse (see Fig. 2.9b). This makes the learning rule biologically plausible, in that the information about how to change the synaptic weight does not have to be carried from a distant source, where it is computed, to every synapse. Such a non-local learning rule would not be biologically plausible, in that there are no appropriate connections known in most parts of the brain to bring in the synaptic training or teacher signal to every synapse.

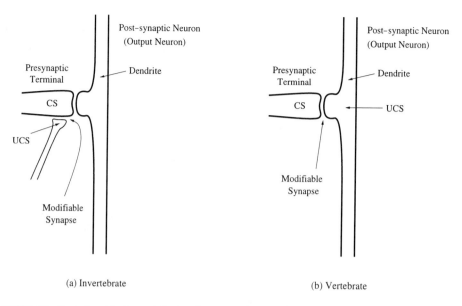

Fig. 2.9 (a) In at least some invertebrate association learning systems, the unconditioned stimulus (UCS) or teaching input makes a synapse onto the presynaptic terminal carrying the conditioned stimulus (CS). (b) In contrast, in vertebrate pattern association learning, the UCS may be made available at all the CS terminals onto the output neuron because the dendrite of the postsynaptic neuron is electrically short, so that the effect of the UCS spreads for long distances along the dendrite.

Evidence that a learning rule with the general form of Eq. 2.2 is implemented in at least some parts of the brain comes from studies of long-term potentiation, described in Chapter 1. Long-term potentiation (LTP) has the synaptic specificity defined by Eq. 2.2, in that only synapses from active afferents, not those from inactive afferents, become strengthened. Synaptic specificity is important for a pattern associator, and most other types of neuronal network, to operate correctly.

Another useful property of real neurons in relation to Eq. 2.2 is that the postsynaptic term, r_i, is available on much of the dendrite of a cell, because the electrotonic length of the dendrite is short. Thus if a neuron is strongly activated with a high value for r_i, then any active synapse onto the cell will be capable of being modified. This enables the cell to learn an association between the pattern of activity on all its axons and its postsynaptic activation, which is stored as an addition to its weight vector w_i. Then later on, at recall, the output can be produced as a vector dot product operation between the input pattern vector $\mathbf{r}'$ and the weight vector w_i, so that the output of the cell can reflect the correlation between the current input vector and what has previously been learned by the cell.

It is interesting that at least many invertebrate neuronal systems may operate differently from those described here. There is evidence (see e.g. Kandel, 1991) that the site of synaptic plasticity in, for example, Aplysia involves the UCS or teaching input having a connection onto the presynaptic terminal which carries the CS (Fig. 2.9a). Conjunctive activity between the UCS and the CS presynaptically alters the effect of the CS in releasing transmitter from the presynaptic terminal. Such a learning scheme is very economical in terms of neurons and synapses, for the conditioning occurs at a single synapse. The problem with such a scheme is that the UCS or teaching input is not made explicit in terms of postsynaptic activation and so is not available at other synapses on the cell. Thus the cell cannot learn an association between a pattern of presynaptic activity **r'** and a UCS; and dot product operations, with all their computationally desirable properties such as generalization, completion, and graceful degradation, will not occur in such invertebrate associators. If this arrangement is typical of what is found in associative learning systems in invertebrates, it is a real limitation on their processing, and indeed their computational style would be very different to that of vertebrates. If the associative modification in invertebrates typically involved conjunctions and learning between presynaptic elements, then to arrange a pattern association one might need the rather implausible connectivity shown in Fig. 2.10a, in which the UCS input neuron systematically makes presynaptic teaching connections onto every presynaptic terminal on a particular output neuron, to ensure that the teaching term is expressed globally across the set of inputs to a particular neuron. The implication of this arrangement is that computations and learning in invertebrate learning systems may operate on very different computational lines to those possible in vertebrates, in that the fundamental dot product operation so useful in large neuronal networks with distributed representations may not apply, at least in many invertebrate neuronal systems.

In some other invertebrate neuronal systems, synaptic plasticity may occur between presynaptic and postsynaptic neurons, so a little further comment on differences between at least some invertebrate neuronal systems and the systems found in for example the mammalian cortex may be useful. In some invertebrate systems with relatively small numbers of neurons, the neurons are sufficiently close that each neuron may not only have multiple inputs, but may also have multiple outputs, due to there being many different output

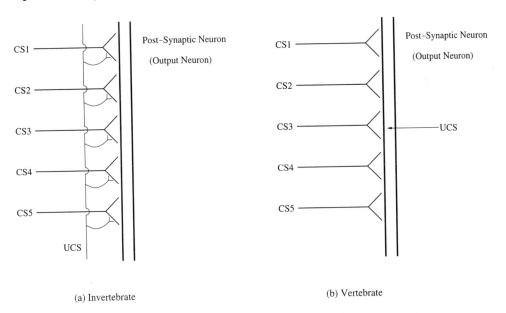

(a) Invertebrate   (b) Vertebrate

Fig. 2.10 (a) To make a pattern associator from an invertebrate learning system of the type shown in Fig. 2.9a, the UCS neurons would have to make presynaptic teaching terminals onto every CS input terminal on the cell. (b) The situation in vertebrates for a similar function is shown.

connections from different parts of the neuron to different neighbouring neurons. This makes it very difficult to analyse the computations in the networks, as each neuron not only performs local computation on its dendrites (which may occur in vertebrates too), but has many outputs, each difficult to measure. So each invertebrate neuron may operate effectively as a number of different computational units. In contrast, cortical pyramidal cells have one output, which is distributed by (measurable) all-or-none action potentials (spikes) to a large number (typically 10 000–20 000) of other neurons in the network. Thus the computational style of cortical neuronal systems is that there is a large number of neurons with the same type of general connectivity (e.g. neurons in the hippocampal CA3 network), and the computation is performed by a network composed of similar neurons. This enables the computation to be understood and to operate at the whole network level. It enables for example the statistical and analytic approaches to the network properties described in this book to be used. In contrast, in at least the invertebrate systems with relatively small numbers of specialized neurons each with multiple and different outputs, the computations will be much more specific, will not have the same distributed representations and the properties which arise from dot product operation with these, and will be difficult to analyse, especially dynamically. There are in the vertebrate retina (in which the topology makes local computation between neighbouring neurons useful), and olfactory bulb, neurons which do have multiple output connections with graded potentials to neighbouring neurons (see Shepherd, 1990), but these systems are not at all like those in most of the mammalian brain, in for example cortical systems.

2.4.7 Capacity

The question of the storage capacity of a pattern associator is considered in detail in Appendix A3. It is pointed out there that, for this type of associative network, the number of memories that it can hold simultaneously in storage has to be analysed together with the retrieval quality of each output representation, and then only for a given quality of the representation provided in the input. This is in contrast with autoassociative nets, in which a critical number of stored memories exists (as a function of various parameters of the network) beyond which attempting to store additional memories results in it becoming impossible to retrieve essentially anything. With a pattern associator, instead, one will always retrieve *something*, but this something will be very little (in information or correlation terms) if too many associations are simultaneously in storage and/or if too little is provided as input.

The conjoint quality–capacity–input analysis can be carried out, for any specific instance of a pattern associator, by using formal mathematical models and established analytical procedures (see e.g. Treves, 1995). This, however, has to be done case by case. It is anyway useful to develop some intuition for how a pattern associator operates, by considering what its capacity would be in certain well-defined simplified cases.

Linear associative neuronal networks

These networks are made up of units with a linear activation function, which appears to make them unsuitable to represent real neurons with their positive-only firing rates. However, even purely linear units have been considered as provisionally relevant models of real neurons, by assuming that the latter operate sometimes in the linear regime of their transfer function. (This implies a high level of spontaneous activity, and may be closer to conditions observed early on in sensory systems rather than in areas more specifically involved in memory.) As usual, the connections are trained by a Hebb (or similar) associative learning rule. The capacity of these networks can be defined as the total number of associations that can be learned independently of each other, given that the linear nature of these systems prevents anything more than a linear transform of the inputs. This implies that if input pattern C can be written as the weighted sum of input patterns A and B, the output to C will be just the same weighted sum of the outputs to A and B. If there are N' input axons, only at most N' input patterns are all mutually independent (i.e. none can be written as a weighted sum of the others), and therefore the capacity of linear networks, defined above, is just N', or equal to the number of input lines. In general, a random set of less than N' vectors (the CS input pattern vectors) will tend to be mutually independent but not mutually orthogonal (at 90° to each other). If they are not orthogonal (the normal situation), then the dot product of them is not 0°, and the output pattern activated by one of the input vectors will be partially activated by other input pattern vectors, in accordance with how similar they are (see Eqs 2.5 and 2.6). This amounts to interference, which is therefore the more serious the less orthogonal, on the whole, is the set of input vectors.

Since input patterns are made of elements with positive values, if a simple Hebbian learning rule like the one of Eq. 2.2 is used (in which the input pattern enters directly with no subtraction term), the output resulting from the application of a stored input vector will be the sum of contributions from all other input vectors that have a non-zero dot product with it (see Appendix A1), and interference will be disastrous. The only situation in which this would not occur is when different input patterns activate completely different input lines, but this is clearly an uninteresting circumstance for networks operating with distributed representations. A solution to this issue is to use a modified learning rule of the following form

$$\delta w_{ij} = k\, r_i (r'_j - x) \qquad (2.7)$$

where x is a constant, approximately equal to the average value of r'_j. This learning rule includes (in proportion to r_i) increasing the synaptic weight if $(r'_j - x) > 0$ (long-term potentiation), and decreasing the synaptic weight if $(r'_j - x) < 0$ (heterosynaptic long-term depression). It is useful for x to be roughly the average activity of an input axon r'_j across patterns, because then the dot product between the various patterns stored on the weights and the input vector will tend to cancel out with the subtractive term, except for the pattern equal to (or correlated with) the input vector itself. Then up to N' input vectors can still be learned by the network, with only minor interference (provided of course that they are mutually independent, as they will in general tend to be).

This modified learning rule can also be described in terms of a contingency table (Table 2.1) showing the synaptic strength modifications produced by different types of learning rule, where LTP indicates an increase in synaptic strength (called 'long-term potentiation' in neurophysiology), and LTD indicates a decrease in synaptic strength (called 'long-term depression' in neurophysiology). Heterosynaptic long-term depression is so-called because it is the decrease

Table 2.1 Effects of pre- and postsynaptic activity on synaptic modification

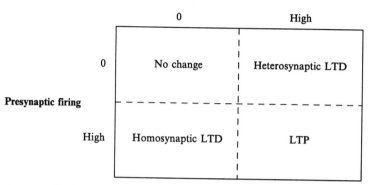

in synaptic strength that occurs to a synapse which is other than that through which the postsynaptic cell is being activated. This heterosynaptic long-term depression is the type of change of synaptic strength that is required (in addition to LTP) for effective subtraction of the average presynaptic firing rate, in order, as it were, to make the CS vectors appear more orthogonal to the pattern associator. The rule is sometimes called the Singer–Stent rule, after work by Singer (1987) and Stent (1973), and was discovered in the brain by Levy (Levy, 1985; Levy and Desmond, 1985; see Brown *et al.*, 1990). Homosynaptic long-term depression is so-called because it is the decrease in synaptic strength that occurs to a synapse which is (the same as that which is) active. For it to occur, the postsynaptic neuron must simultaneously be inactive, or have only low activity. (This rule is sometimes called the BCM rule after the paper of Bienenstock, Cooper and Munro, 1982, see Chapter 4, on competitive networks.)

Associative neuronal networks with non-linear neurons

With non-linear neurons, that is with at least a threshold in the activation function so that the output firing r_i is 0 when the activation h_i is below the threshold, the capacity can be measured in terms of the number of different clusters of output pattern vectors that the network produces. This is because the non-linearities now present (one per output unit) result in some clustering of all possible (conditioned stimulus) input patterns r'. Input patterns that are similar to a stored input vector can result due to the non-linearities in output patterns even closer to the stored output; and vice versa sufficiently dissimilar inputs can be assigned to different output clusters thereby increasing their mutual dissimilarity. As with the linear counterpart, in order to remove the correlation that would otherwise occur between the patterns because the elements can take only positive values, it is useful to use a modified Hebb rule

$$\delta w_{ij} = k\, r_i\, (r'_j - x).$$

With fully distributed output patterns the number p of associations that leads to different clusters is of order C, the number of input lines (axons) per output unit (that is, of order N' for a fully connected network), as shown in Appendix A3. If sparse patterns are used in the output, or alternatively if the learning rule includes a non-linear postsynaptic factor that is effectively equivalent to using sparse output patterns, the coefficient of proportionality between p and C can be much higher than one, that is many more patterns can be stored than inputs per unit (see Appendix A3). Indeed, the number of different patterns or prototypes p that can be stored can be derived for example in the case of binary units (Gardner, 1988) to be

$$p \approx C / [a_o \log(1/a_o)] \tag{2.8}$$

where a_o is the sparseness of the *output* firing pattern $\mathbf{r}$ produced by the unconditioned stimulus. p can in this situation be much larger than C (see Rolls and Treves, 1990, and Appendix A3). This is an important result for encoding in pattern associators, for it means that provided that the activation functions are non-linear (which is the case with real neurons), there is a very great advantage to using sparse encoding, for then many more than C pattern associations can be stored. Sparse representations may well be present in brain regions involved in associative memory (see Chapters 6 and 7) for this reason.

2.4.8 Interference

Interference occurs in linear pattern associators if two vectors are not orthogonal, and is simply dependent on the angle between the originally learned vector and the recall cue or CS vector, for the activation of the output neuron depends simply on the dot product of the recall vector and the synaptic weight vector (Eq. 2.5). Also in non-linear pattern associators (the interesting case for all practical purposes), interference may occur if two CS patterns are not orthogonal, though the effect can be controlled with sparse encoding of the UCS patterns, effectively by setting high thresholds for the firing of output units. In other words the CS vectors need not be strictly orthogonal, but if they are too similar, some interference will still be likely to occur.

The fact that interference is a property of neural network pattern associator memories is of interest, for interference is a major property of human memory. Indeed, the fact that interference is a property of human memory and of network association memories is entirely consistent with the hypothesis that human memory is stored in associative memories of the type described here, or at least that network associative memories of the type described represent a useful exemplar of the class of parallel distributed storage network used in human memory. It may also be suggested that one reason that interference is tolerated in biological memory is that it is associated with the ability to generalize between stimuli, which is an invaluable feature of biological network associative memories, in that it allows the memory to cope with stimuli which will almost never on different occasions be identical, and in that it allows useful analogies which have survival value to be made.

2.4.9 Expansion recoding

If patterns are too similar to be stored in associative memories, then one solution which the brain seems to use repeatedly is to expand the encoding to a form in which the different stimuli are less correlated, that is more orthogonal, before they are presented as CS stimuli to a pattern associator. The problem can be highlighted by a non-linearly separable mapping (which captures part of the eXclusive OR (XOR) problem), in which the mapping that is desired is as follows. The neuron has two inputs, A and B (see Fig. 2.11).

Input A	1	0	1
Input B	0	1	1
Required Output	1	1	0

Fig. 2.11

This is a mapping of patterns that is impossible for a one-layer network, because the patterns are not linearly separable (see Appendix A1. There is no set of synaptic weights in a one-layer net that could solve the problem shown in Fig. 2.12. Two classes of patterns are not linearly separable if no hyperplane can be positioned in their N-dimensional space so as to separate them, see Appendix A1. The XOR problem has the additional constraint that A = 0,

Competitive Network   Pattern Associator

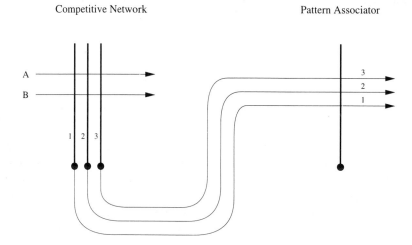

Fig. 2.12 Expansion recoding. A competitive network followed by a pattern associator that can enable patterns that are not linearly separable to be learned correctly.

B = 0 must be mapped to Output = 0.) A solution is to remap the two input lines A and B to three input lines 1–3, that is to use expansion recoding, as shown in Fig. 2.12. This can be performed by a competitive network. The synaptic weights on the dendrite of the output neuron could then learn the following values using a simple Hebb rule, Eq. 2.2, and the problem could be solved as in Fig. 2.13.

Synaptic weight

Input 1 (A = 1, B = 0)	1
Input 2 (A = 0, B = 1)	1
Input 3 (A = 1, B = 1)	0

Fig. 2.13

The whole network would look like that shown in Fig. 2.12.

Expansion encoding which maps vectors with N' inputs to a set of neurons that is larger than N' appears to be present in several parts of the brain, and to precede networks which perform pattern association. Marr (1969) suggested that one such expansion recoding was performed in the cerebellum by the mapping from the mossy fibres to the granule cells, which provide the associatively modifiable inputs to the Purkinje cells in the cerebellum (see Chapter 9). The expansion in this case is in the order of 1000 times. Marr (1969) suggested that the expansion recoding was performed by each cerebellar granule cell responding to a low-order combination of mossy fibre activity, implemented by each granule cell receiving 5–7 mossy fibre inputs. Another example is in the hippocampus, in which there is an expansion from the perforant path fibres originating in the entorhinal cortex to the dentate granule cells, which are thought to decorrelate the patterns which are stored by the CA3 cells, which form an autoassociative network (see Chapter 6). The suggestion is that this expansion, performed by the dentate granule cells, helps to separate patterns so that overlapping patterns in the entorhinal cortex are made separate in CA3, to allow separate episodic memories with overlapping information to be stored and recalled separately (see Chapter 6). A similar principle was probably being used as a preprocessor in Rosenblatt's original perceptron (see Chapter 5).

2.4.10 Implications of different types of coding for storage in pattern associators

Throughout this chapter, we have made statements about how the properties of pattern associators, such as the number of patterns that can be stored, and whether generalization and graceful degradation occur, depend on the type of encoding of the patterns to be associated. (The types of encoding considered, local, sparse distributed, and fully distributed, are described in Chapter 1.) We draw together these points in Table 2.2. The amount of information that can be stored in each pattern in a pattern associator is considered in Appendix A3, and some of the relevant information theory itself is described in Appendix A2.

Table 2.2 Coding in associative memories*

	Local	Sparse distributed	Fully distributed
Generalization, completion, graceful degradation	No	Yes	Yes
Number of patterns that can be stored	N (large)	of order $C/[a_o \log(1/a_o)]$ (can be larger)	of order C (usually smaller than N)
Amount of information in each pattern	Minimal ($\log(N)$ bits)	Intermediate ($N a_o \log(1/a_o)$ bits)	Large (N bits)

*N refers here to the number of output units, and C to the average number of inputs to each output unit. a_o is the sparseness of output patterns, or roughly the proportion of output units activated by a UCS pattern. Note: logs are to the base 2.

3 Autoassociation memory

Autoassociative memories, or attractor neural networks, store memories, each one of which is represented by a pattern of neural activity. They can then recall the appropriate memory from the network when provided with a fragment of one of the memories. This is called completion. Many different memories can be stored in the network and retrieved correctly. The network can learn each memory in one trial. Because of its 'one-shot' rapid learning, and ability to complete, this type of network is well suited for episodic memory storage, in which each past episode must be stored and recalled later from a fragment, and kept separate from other episodic memories. An autoassociation memory can also be used as a short term memory, in which iterative processing round the recurrent collateral connection loop keeps a representation active until another input cue is received. In this short term memory role, it appears to be used in the temporal visual cortical areas with their connections to the ventrolateral prefrontal cortex for the short term memory of visual stimuli (in delayed match to sample tasks, see Chapters 6 and 8); and in the dorsolateral prefrontal cortex for short term memory of spatial responses (see Chapter 10). A feature of this type of memory is that it is content addressable; that is, the information in the memory can be accessed if just the contents of the memory (or a part of the contents of the memory) are used. This is in contrast to a conventional computer, in which the *address* of what is to be accessed must be supplied, and used to access the contents of the memory. Content addressability is an important simplifying feature of this type of memory, which makes it suitable for use in biological systems. The issue of content addressability will be amplified below.

3.1 Architecture and operation

The prototypical architecture of an autoassociation memory is shown in Fig. 3.1. The external input e_i is applied to each neuron i by unmodifiable synapses. This produces firing r_i of each neuron, or a vector of firing on the output neurons $\mathbf{r}$. Each output neuron i is connected by a recurrent collateral connection to the other neurons in the network, via modifiable connection weights w_{ij}. This architecture effectively enables the output firing vector $\mathbf{r}$ to be associated during learning with itself. Later on, during recall, presentation of part of the external input will force some of the output neurons to fire, but through the recurrent collateral axons and the modified synapses, other neurons in $\mathbf{r}$ can be brought into activity. This process can be repeated a number of times, and recall of a complete pattern may be perfect. Effectively, a pattern can be recalled or recognized because of associations formed between its parts. This of course requires distributed representations.

Next we introduce a more precise and detailed description of the above, and describe the properties of these networks. Ways to analyse formally the operation of these networks are introduced in Appendix A4.

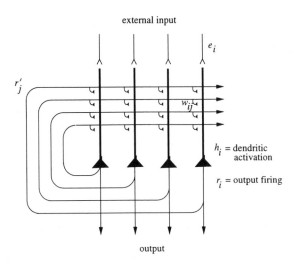

Fig. 3.1 The architecture of an autoassociative neural network.

3.1.1 Learning

The firing of every output neuron i is forced to a value r_i determined by the external input e_i. Then a Hebb-like associative local learning rule is applied to the recurrent synapses in the network:

$$\delta w_{ij} = k \, r_i r_j \qquad \text{(Hebb rule)} \qquad (3.1)$$

It is notable that in a fully connected network, this will result in a symmetric matrix of synaptic weights, that is the strength of the connection from neuron 1 to neuron 2 will be the same as the strength of the connection from neuron 2 to neuron 1 (both implemented via recurrent collateral synapses).

It is a factor which is sometimes overlooked that there must be a mechanism for ensuring that during learning r_i does approximate e_i, and must not be influenced much by activity in the recurrent collateral connections, otherwise the new external pattern **e** will not be stored in the network, but instead something will be which is influenced by the previously stored memories. It is thought that in some parts of the brain, such as the hippocampus, there are processes which help the external connections to dominate the firing during learning (see Chapter 6 and Rolls, 1989b–e; Treves and Rolls, 1992).

3.1.2 Recall

During recall the external input e_i is applied, and produces output firing, operating through the non-linear activation function described below. The firing is fed back by the recurrent collateral axons shown in Fig. 3.1 to produce activation of each output neuron through the modified synapses on each output neuron. The internal activation h_i produced by the recurrent collateral effect on the ith neuron is in the standard way the sum of the activations produced in proportion to the firing rate of each axon r_j operating through each modified synapse w_{ij}, that is

$$h_i = \Sigma_j r_j w_{ij} \qquad (3.2)$$

where Σ_j indicates that the sum is over the C input axons to each neuron, indexed by j.

The output firing r_i is a function of the activation produced by the recurrent collateral effect (internal recall) and by the external input (e_i):

$$r_i = \mathrm{f}(h_i + e_i) \qquad (3.3)$$

The activation function should be non-linear, and may be for example binary threshold, linear threshold, sigmoid, etc. (see Fig. 1.3). A non-linear activation function can minimize interference between the pattern being recalled and other patterns stored in the network, and can also be used to ensure that what is a positive feedback system remains stable. The network can be allowed to repeat this recurrent collateral loop a number of times. Each time the loop operates, the output firing becomes more like the originally stored pattern, and this progressive recall is usually complete within 5–15 iterations.

3.2 Introduction to the analysis of the operation of autoassociation networks

With complete connectivity in the synaptic matrix, and the use of a Hebb rule, the matrix of synaptic weights formed during learning is symmetric. The learning algorithm is fast, 'one-shot', in that a single presentation of an input pattern is all that is needed to store that pattern.

During recall, a *part* of one of the originally learned stimuli can be presented as an external input. The resulting firing is allowed to iterate repeatedly round the recurrent collateral system, gradually on each iteration recalling more and more of the originally learned pattern. *Completion* thus occurs. If a pattern is presented during recall that is *similar* but not identical to any of the previously learned patterns, then the network settles into a stable recall state in which the firing corresponds to that of the previously learned pattern. The network can thus *generalize* in its recall to the most similar previously learned pattern. The activation function of the neurons should be non-linear, since a purely linear system would not produce any categorization of the input patterns it receives, and therefore would not be able to effect anything more than a trivial (i.e. linear) form of completion and generalization.

Recall can be thought of in the following way, relating it to what occurs in pattern associators. The external input **e** is applied, produces firing **r**, which is applied as a recall cue on the recurrent collaterals as **r'**. The activity on the recurrent collaterals is then multiplied with the synaptic weight vector stored during learning on each neuron to produce the new activation h_i which reflects the similarity between $\mathbf{r}'$ and one of the stored patterns. Partial recall has thus occurred as a result of the recurrent collateral effect. The activations h_i after thresholding (which helps to remove interference from other memories stored in the network, or noise in the recall cue) result in firing r_i, or a vector of all neurons $\mathbf{r}$, which is already more like one of the stored patterns than, at the first iteration, the firing resulting from the recall cue alone, $\mathbf{r} = f(\mathbf{e})$. This process is repeated a number of times to produce progressive recall of one of the stored patterns.

Autoassociation networks operate by effectively storing associations between the elements of a pattern. Each element of the pattern vector to be stored is simply the firing of a neuron. What is stored in an autoassociation memory is a set of pattern vectors. The network operates to recall one of the patterns from a fragment of it. Thus, although this network implements recall or recognition of a pattern, it does so by an association learning mechanism, in which associations between the different parts of each pattern are learned. These memories have sometimes been called autocorrelation memories, because they learn correlations between the units of the network, in the sense that each pattern learned is defined by a set of simultaneously active neurons; this is provided for by the Hebb-like learning rule.

The system formally resembles spin glass systems of magnets analysed quantitatively in statistical mechanics. This has led to the analysis of (recurrent) autoassociative networks as dynamical systems made up of many interacting elements, in which the interactions are such as to produce a large variety of basins of attraction of the dynamics. Each basin of attraction corresponds to one of the originally learned patterns, and once the network is within a basin it keeps iterating until a recall state is reached which is the learned pattern itself or a pattern closely similar to it (interference effects may prevent an exact identity between the recall state and a learned pattern). This type of system is contrasted with other, simpler, systems of magnets (e.g. ferromagnets), in which the interactions are such as to produce only a limited number of related basins, since the magnets tend to be, for example, all aligned with each other. The states reached within each basin of attraction are called attractor states, and the analogy between autoassociator neural networks and physical systems with multiple attractors was drawn by Hopfield (1982) in a very influential paper. He was able to show that the recall state can be thought of as the local minimum in an energy landscape, where the energy would be defined, in our notation, as

$$E = -1/2 \Sigma_{ij} w_{ij} (r_i - <r>) (r'_j - <r'>) \tag{3.4}$$

This equation can be understood in the following way. If two neurons are both firing above their mean rate (denoted by $<r>$), and are connected by a weight with a positive value, then the firing of these two neurons is consistent with each other, and they mutually support each other, so that they contribute to the system's tendency to remain stable. If across the whole network such mutual support is generally provided, then no further change will take place, and the system will indeed remain stable. If, on the other hand, either of our pair of neurons were not firing, or if the connecting weight had a negative value, the neurons would not support each other, and indeed the tendency would be for the neurons to try and alter ('flip' in the case of binary units) the state of the other. This would be repeated across the whole network until a situation in which most mutual support, and least 'frustration', was reached. What makes it possible to define an energy function and to make these simple considerations is that the matrix is symmetric (see Hopfield, 1982; Hertz, Krogh and Palmer, 1991; Amit, 1989).

Physicists have generally analysed a system in which the input pattern is presented and then immediately removed, so that the network then 'falls' without further assistance (in what is referred to as the *unclamped* condition) towards the minimum of its basin of attraction. A more biologically realistic system is one in which the external input is left on contributing to the recall during the fall into the recall state. In this *clamped* condition, recall is usually faster, and more reliable, so that more memories may be usefully recalled from the network. The approach using methods developed in theoretical physics has led to rapid advances in the understanding of autoassociative networks, and its basic elements are described in Appendix A4.

3.3 Properties

The internal recall in autoassociation networks involves multiplication of the firing vector of neuronal activity by the vector of synaptic weights on each neuron. This inner product vector multiplication allows the similarity of the firing vector to previously stored firing vectors to be provided by the output (as effectively a correlation), if the patterns learned are distributed. As a result of this type of correlation computation performed if the patterns are distributed, many important properties of these networks arise, including pattern completion (because part of a pattern is correlated with the whole pattern), and graceful degradation (because a damaged synaptic weight vector is still correlated with the original synaptic weight vector). Some of these properties are described next.

3.3.1 Completion

Perhaps the most important and useful property of these memories is that they complete an incomplete input vector, allowing recall of a whole memory from a small fraction of it. The memory recalled in response to a fragment is that stored in the memory that is closest in pattern similarity (as measured by the dot product, or correlation). Because the recall is iterative and progressive, the recall can be perfect.

This property and the associative property of pattern associator neural networks are very similar to the properties of human memory. This property may be used when we recall a part of a recent memory of a past episode from a part of that episode. The way in which this could be implemented in the hippocampus is described in Chapter 6.

3.3.2 Generalization

The network generalizes in that an input vector similar to one of the stored vectors will lead to recall of the originally stored vector, provided that distributed encoding is used. The principle by which this occurs is similar to that described for a pattern associator.

3.3.3 Graceful degradation or fault tolerance

If the synaptic weight vector $\mathbf{w}_i$ on each neuron (or the weight matrix) has synapses missing (e.g. during development), or loses synapses (e.g. with brain damage or ageing), then the output activation h_i (or vector of output activations $\mathbf{h}$) is still reasonable, because h_i is the dot product (correlation) of $\mathbf{r}'$ with $\mathbf{w}_i$. The same argument applies if whole input axons are lost. If an output neuron is lost, then the network cannot itself compensate for this, but the next network in the brain is likely to be able to generalize or complete if its input vector has some elements missing, as would be the case if some output neurons of the autoassociation network were damaged.

3.3.4 Prototype extraction, extraction of central tendency, and noise reduction

These arise when a set of similar input pattern vectors $\{\mathbf{r}\}$ are learned by the network. The weight vectors $\mathbf{w}_i$ become (or point towards) the average of that set of similar vectors. This produces 'extraction of the prototype' or 'extraction of the central tendency', and 'noise reduction'. This process can result in better recognition or recall of the prototype than of any of the exemplars, even though the prototype may never itself have been presented. The general principle by which the effect occurs is similar to that by which it occurs in pattern associators. It of course only occurs if each pattern uses a distributed representation.

There has been intense debate about whether when human memories are stored, a prototype of what is to be remembered is stored, or whether all the instances or the exemplars are each stored separately so that they can be individually recalled (McClelland and Rumelhart, 1986, Ch. 17, p. 172). Evidence favouring the prototype view is that if a number of different examples of an object are shown, then humans may report that they have seen the prototype more confidently than they report having seen other exemplars, even though the prototype has never been shown (Posner and Keele, 1968; Rosch, 1975). Evidence favouring the view that exemplars are stored is that in categorization and perceptual identification tasks the responses made are often sensitive to the congruity between particular training stimuli and particular test stimuli (Brooks, 1978; Medin and Schaffer, 1978; Jacoby, 1983a,b; Whittlesea, 1983). It is of great interest that both types of phenomena can arise naturally out of distributed information storage in a neuronal network such as an autoassociator. This can be illustrated by the storage in an autoassociation memory of sets of stimuli which are all somewhat different examples of the same pattern. These can be generated, for example, by randomly altering each of the input vectors from the input stimulus. After many such randomly altered exemplars have been learned by the network, recall can be tested, and it is found that the network responds best to the original input vector, with which it has never been presented. The reason for this is that the autocorrelation components which build up in the synaptic matrix with repeated presentations of the exemplars represent the average correlation between the different elements of the vector, and this is highest for the prototype. This effect also gives the storage some noise immunity, in that variations in the input which are random noise average out, while the signal which is constant builds up with repetition.

3.3.5 Speed

The *recall* operation is fast on each neuron on a single iteration, because the pattern $\mathbf{r}'$ on the axons can be applied simultaneously to the synapses w_i, and the activation h_i can be accumulated in one or two time constants of the dendrite (e.g. 10–20 ms). If a simple implementation of an autoassociation net such as that described by Hopfield (1982) is simulated on a computer, then 5–15 iterations are typically necessary for completion of an incomplete input cue $\mathbf{e}$. This might be taken to correspond to 50–200 ms in the brain, rather too slow for any one local network in the brain to function. However, recent work (Treves, 1993; Appendix A5) has shown that if the neurons are treated not as McCulloch–Pitts neurons which are simply 'updated' at each iteration, or cycle of time steps (and assume the active state if the threshold is exceeded), but instead are analysed and modelled as 'integrate-and-fire' neurons in real continuous time, then the network can effectively 'relax' into its recall state very rapidly, in one or two time constants of the synapses. This corresponds to perhaps 20 ms in the brain. One factor in this rapid dynamics of autoassociative networks with brain-like 'integrate-and-fire' membrane and synaptic properties is that with some spontaneous activity, some of the neurons in the network are close to threshold already before the recall cue is applied, and hence some of the neurons are very quickly pushed by the recall cue into firing, so that information starts to be exchanged very rapidly (within 1–2 ms of brain time) through the modified synapses by the neurons in the network. The progressive exchange of information starting early on within what would otherwise be thought of as an iteration period (of perhaps 20 ms, corresponding to a neuronal firing rate of 50 spikes/s), is the mechanism accounting for rapid recall in an autoassociative neuronal network made biologically realistic in this way. Further analysis of the fast dynamics of these networks if they are implemented in a biologically plausible way with 'integrate-and-fire' neurons, is provided in Appendix A5. The general approach applies to other networks with recurrent connections, not just autoassociators, and the fact that such networks can operate much faster than it would seem from simple models, that follow instead discrete time dynamics, is probably a major factor in enabling these networks to provide some of the building blocks of brain function.

Learning is fast, 'one-shot', in that a single presentation of an input pattern $\mathbf{e}$ or $\mathbf{r}$ enables the association between the activation of the dendrites (the post-synaptic term h_i) and the firing of the recurrent collateral axons $\mathbf{r}'$, to be learned. Repeated presentation with small variations of a pattern vector is used to obtain the properties of prototype extraction, extraction of central tendency, and noise reduction, because these arise from the averaging process produced by storing very similar patterns in the network.

3.3.6 Local learning rule

The simplest learning used in autoassociation neural networks, a version of the Hebb rule, is as in Eq. 3.1

$$\delta w_{ij} = k \, r_i r_j$$

The rule is a local learning rule in that the information required to specify the change in synaptic weight is available locally at the synapse, as it is dependent only on the presynaptic firing rate r_j available at the synaptic terminal, and the postsynaptic activation or firing r_i available on the dendrite of the neuron receiving the synapse. This makes the learning rule biologically plausible, in that the information about how to change the synaptic weight does not have to be carried to every synapse from a distant source where it is computed. As with pattern associators, since firing rates are positive quantities, a potentially interfering correlation is induced between different pattern vectors. This can be removed by subtracting the mean of the presynaptic activity from each presynaptic term, using a type of long-term depression. This can be specified as

$$\delta w_{ij} = k\, r_i (r_j - x) \tag{3.5}$$

where k is a learning rate constant. This learning rule includes (in proportion to r_i) increasing the synaptic weight if $(r_j - x) > 0$ (long-term potentiation), and decreasing the synaptic weight if $(r_j - x) < 0$ (heterosynaptic long-term depression). This procedure works optimally if x is the average activity $<r_j>$ of an axon across patterns.

Evidence that a learning rule with the general form of Eq. 3.1 is implemented in at least some parts of the brain comes from studies of long-term potentiation, described in Chapter 1. One of the important potential functions of heterosynaptic long-term depression is its ability to allow in effect the average of the presynaptic activity to be subtracted from the presynaptic firing rate (see Chapter 2, Appendix A3 and Rolls and Treves, 1990).

Autoassociation networks can be trained with the error-correction or delta learning rule described in Chapter 5. Although a delta rule is less biologically plausible than a Hebb-like rule, a delta rule can help to store separately patterns that are very similar (see McClelland and Rumelhart, 1988; Hertz, Krogh and Palmer, 1991).

3.3.7 Capacity

One measure of storage capacity is to consider how many orthogonal patterns could be stored, as with pattern associators. If the patterns are orthogonal, there will be no interference between them, and the maximum number p of patterns that can be stored will be the same as the number N of output neurons (in a fully connected network). Although in practice the patterns that have to be stored will hardly be orthogonal, this is not a purely academic speculation, since it was shown how one can construct a synaptic matrix that effectively orthogonalizes any set of (linearly independent) patterns (Kohonen, 1984; see also Personnaz et al., 1985; Kanter and Sompolinsky, 1987). However, this matrix cannot be learned with a local, one-shot learning rule, and therefore its interest for autoassociators in the brain is limited. The more general case of random non-orthogonal patterns, and of Hebbian learning rules, is considered next.

With non-linear neurons used in the network, the capacity can be measured in terms of the number of input patterns $\mathbf{r}$ (produced by the external input $\mathbf{e}$, see Fig. 3.1) that can be stored in the network and recalled later whenever the net falls within their basin of attraction. The first quantitative analysis of storage capacity (Amit, Gutfreund and Sompolinsky, 1987) considered a fully connected Hopfield (1982) autoassociator model, in which units are binary elements with an equal probability of being 'on' or 'off' in each pattern, and the number C of inputs per unit is the same as the number N of output units (actually, it is equal to $N-1$, since a unit is taken not to connect to itself). Learning is taken to occur by clamping the desired patterns on the network and using a modified Hebb rule, in which the mean of the presynaptic and postsynaptic firings is subtracted from the firing on any one learning trial (this amounts to a covariance learning rule, and is described more fully in Appendix A4). With such fully distributed random patterns, the number of patterns that can be learned is (for C large) $p \approx 0.14C = 0.14N$, hence well below what could be achieved with orthogonal patterns or with an 'orthogonalizing' synaptic matrix. Many variations of this 'standard' autoassociator model have been analysed subsequently.

Treves and Rolls (1991) have extended this analysis to autoassociation networks which are much more biologically relevant in the following ways. First, some or many connections between the recurrent collaterals and the dendrites are missing (this is referred to as diluted connectivity, and results in a non-symmetric synaptic connection matrix in which w_{ij} does not equal w_{ji}, one of the original assumptions made in order to introduce the energy formalism in the Hopfield model). Second, the neurons need not be restricted to binary threshold neurons, but can have a threshold–linear activation function (see Fig. 1.3). This enables the neurons to assume real continuously variable firing rates, which are what is found in the brain (Rolls and Tovee, 1994). Third, the representation need not be fully distributed (with half the neurons 'on', and half 'off'), but instead can have a small proportion of the neurons firing above the spontaneous rate, which is what is found in parts of the brain such as the hippocampus that are involved in memory (see Treves and Rolls, 1994, and Chapter 6). Such a representation is defined as being sparse, and the sparseness of the representation a can be measured, by extending the binary notion of the proportion of neurons that are firing, as

$$a = (\Sigma_{i=1,N} r_i/N)^2 / \Sigma_{i=1,N} (r_i^2/N) \tag{3.6}$$

where r_i is the firing rate of the ith neuron in the set of N neurons. Treves and Rolls (1991) have shown that such a network does operate efficiently as an autoassociative network, and can store (and recall correctly) a number of different patterns p as follows

$$p \approx \frac{C^{\mathrm{RC}}}{a \ln(1/a)} k \tag{3.7}$$

where C^{RC} is the number of synapses on the dendrites of each neuron devoted to the recurrent collaterals from other neurons in the network, and k is a factor that depends weakly on the detailed structure of the rate distribution, on the connectivity pattern, etc., but is roughly in the order of 0.2–0.3.

The main factors that determine the maximum number of memories that can be stored in an autoassociative network are thus the number of connections on each neuron devoted to the recurrent collaterals, and the sparseness of the representation. For example, for $C^{\mathrm{RC}} = 12\,000$ and $a = 0.02$, p is calculated to be approximately $36\,000$. This storage capacity can be realized, with little interference between patterns, if the learning rule includes some form of heterosynaptic long-term depression that counterbalances the effects of associative long-term potentiation (Treves and Rolls, 1991; see Chapter 2 and Appendix A4). It should be noted that the number of neurons N (which is greater than C^{RC}, the number of recurrent collateral inputs received by any neuron in the network from the other neurons in the network) is not a parameter which influences the number of different memories that can be stored in the network. The implication of this is that increasing the number of neurons (without increasing the number of connections per neuron) does not increase the number of different patterns that can be stored (see Appendix A4), although it may enable simpler encoding of the firing patterns, for example more orthogonal encoding, to be used.

3.3.8 Context

The environmental context in which learning occurs can be a very important factor which affects retrieval in humans and other animals. Placing the subject back into the same context in which the original learning occurred can greatly facilitate retrieval.

Context effects arise naturally in association networks if some of the activity in the network reflects the context in which the learning occurs. Retrieval is then better when that context is present, for the activity contributed by the context becomes part of the retrieval cue for the memory, increasing the correlation of the current state with what was stored. (A strategy for retrieval arises simply from this property. The strategy is to keep trying to recall as many fragments of the original memory situation, including the context, as possible, as this will provide a better cue for complete retrieval of the memory than just a single fragment.)

The very well-known effects of context in the human memory literature could arise in the simple way just described. An implication of the explanation is that context effects will be especially important at late stages of memory or information processing systems in the brain, for there information from a wide range of modalities will be mixed, and some of that information could reflect the context in which the learning takes place. One part of the brain where such effects may be strong is the hippocampus, which is implicated in the memory of recent episodes, and which receives inputs derived from most of the cortical information processing streams, including those involved in space (see Chapter 6).

3.3.9 Mixture states

If an autoassociation memory is trained on pattern vectors A, B, and A + B (i.e. A and B are both included in the joint vector A + B; that is if the vectors are not linearly independent), then the autoassociation memory will have difficulty in learning and recalling these three memories as separate, because completion from either A or B to A + B tends to occur during recall. (This is referred to as configurational learning in the animal learning literature, see e.g. Sutherland and Rudy, 1991.) This problem can be minimized by re-representing A, B, and A + B in such a way that they are different vectors before they are presented to the autoassociation memory. This can be performed by recoding the input vectors to minimize overlap using, for example, a competitive network, and possibly involving expansion recoding, as described for pattern associators (see Chapter 2, Fig. 2.13). It is suggested that this is a function of the dentate granule cells in the hippocampus, which precede the CA3 recurrent collateral network (Treves and Rolls, 1992, 1994; and see Chapter 6).

3.3.10 Memory for sequences

One of the first extensions of the standard autoassociator paradigm that has been explored in the literature is the capability to store and retrieve not just individual patterns, but whole *sequences* of patterns. Hopfield, in the same 1982 paper, suggested that this could be achieved by adding to the standard connection weights, which associate a pattern with itself, a new, asymmetric component, that associates a pattern with the next one in the sequence. In practice this scheme does not work very well, unless the new component is made to operate on a slower time scale that the purely autoassociative component (Kleinfeld, 1986; Sompolinsky and Kanter, 1986). With two different time scales, the autoassociative component can stabilize a pattern for a while, before the heteroassociative component moves the network, as it were, into the next pattern. The heteroassociative retrieval cue for the next pattern in the sequence is just the previous pattern in the sequence. A particular type of 'slower' operation occurs if the asymmetric component acts after a delay τ. In this case the network sweeps through the sequence, staying for a time of order τ in each pattern.

One can see how the necessary ingredient for the storage of sequences is only a minor departure from purely Hebbian learning: in fact, the (symmetric) autoassociative component of the weights can be taken to reflect the Hebbian learning of strictly simultaneous conjunctions of pre- and postsynaptic activity, whereas the (asymmetric) heteroassociative component can be implemented by Hebbian learning of each conjunction of postsynaptic activity with presynaptic activity *shifted* a time τ in the past. Both components can then be seen as resulting from a generalized Hebbian rule, which increase the weight whenever postsynaptic activity is paired with presynaptic activity occurring within a given time range, that may extend from a few hundred milliseconds in the past up to include strictly simultaneous activity. This is similar to a trace rule (see Chapter 8), which itself matches very well the observed conditions for induction of long-term potentiation, and appears entirely plausible; the learning rule necessary for learning sequences, though, is more complex than a simple trace rule in that the time-shifted conjunctions of activity that are encoded in the weights must in retrieval produce activations that are time-shifted as well (otherwise one falls back in the Hopfield (1982) proposal, which does not quite work). The synaptic weights should therefore keep separate 'traces' of what was simultaneous and what was time-shifted during the original experience, and this is not very plausible.

A series of recent investigations of the storage of temporal sequences in autoassociators is that by Levy and colleagues (Levy, Wu and Baxter, 1995; Wu, Baxter and Levy, 1996).

Another way in which a delay could be inserted in a recurrent collateral path in the brain is by inserting another cortical area in the recurrent path. This could fit in with the cortico-cortical backprojection connections described below and in Chapters 6 and 10, which would introduce some conduction delay.

3.4 Use of autoassociation networks in the brain

Because of its 'one-shot' rapid learning, and ability to complete, this type of network is well suited for episodic memory storage, in which each episode must be stored and recalled later from a fragment, and kept separate from other episodic memories. It does not take a long time (the 'many epochs' of backpropagation networks) to train this network, because it does not have to 'discover the structure' of a problem. Instead, it stores information in the form in which it is presented to the memory, without altering the representation. An autoassociation network may be used for this function in the CA3 region of the hippocampus (see Chapter 6).

An autoassociation memory can also be used as a short term memory, in which iterative processing round the recurrent collateral loop keeps a representation active until another input cue is received. This may be used to implement many types of short term memory in the brain. For example, it may be used in the perirhinal cortex and adjacent temporal lobe cortex to implement short term visual object memory (Miyashita and Chang, 1988; Amit, 1995); in the dorsolateral prefrontal cortex to implement a short term memory for spatial responses (Goldman-Rakic, 1996); and in the frontal eye fields to implement a short term memory for where eye movements should be made in space. Such an autoassociation memory in the temporal lobe visual cortical areas may be used to implement the firing which continues for often 300 ms after a very brief (16 ms) presentation of a visual stimulus (Rolls and Tovee, 1994), and may be one way in which a short memory trace is implemented to facilitate invariant learning about visual stimuli (see Chapter 8). In all these cases, the short term memory may be implemented by the recurrent excitatory collaterals which connect nearby pyramidal cells in the cerebral cortex. The connectivity in this system, that is the probability that a neuron synapses on a nearby neuron, may be in the region of 10% (Braitenberg and Schuz, 1991; Abeles, 1991).

The recurrent connections between nearby neocortical pyramidal cells may also be important in defining the response properties of cortical cells, which may be triggered by external inputs (from for example the thalamus or a preceding cortical area), but may be considerably dependent on the synaptic connections received from nearby cortical pyramidal cells.

The cortico-cortical backprojection connectivity described in Chapters 6, 7, and 10 can be interpreted as a system which allows the forward-projecting neurons in one cortical area to be linked autoassociatively with the backprojecting neurons in the next cortical area (see Figs 6.1, 7.12, 7.13, and 10.7). This particular architecture may be especially important in constraint satisfaction (as well as recall), that is it may allow the networks in the two cortical areas to settle into a mutually consistent state. This would effectively enable information in higher cortical areas, which would include information from more divergent sources, to influence the response properties of neurons in earlier processing stages.

4 Competitive networks, including self-organizing maps

4.1 Function

Competitive neural networks learn to categorize input pattern vectors. Each category of inputs activates a different output neuron (or set of output neurons—see below). The categories formed are based on similarities between the input vectors. Similar, that is correlated, input vectors activate the same output neuron. In that the learning is based on similarities in the input space, and there is no external teacher which forces classification, this is an unsupervised network. The term categorization is used to refer to the process of placing vectors into categories based on their similarity. The term classification is used to refer to the process of placing outputs in particular classes as instructed or taught by a teacher. Examples of classifiers are pattern associators, one-layer delta-rule perceptrons, and multilayer perceptrons taught by error backpropagation (see Chapters 2, 5, and 6). In supervised networks there is usually a teacher for each output neuron.

The categorization produced by competitive nets is of great potential importance in perceptual systems. Each category formed reflects a set or cluster of active inputs r'_j which occur together. This cluster of coactive inputs can be thought of as a feature, and the competitive network can be described as building *feature analysers*, where a feature can now be defined as a correlated set of inputs. During learning, a competitive network gradually discovers these features in the input space, and the process of finding these features without a teacher is referred to as self-organization. Another important use of competitive networks is to *remove redundancy* from the input space, by allocating output neurons to reflect a set of inputs which co-occur. Another important aspect of competitive networks is that they separate patterns which are somewhat correlated in the input space, to produce outputs for the different patterns which are less correlated with each other, and may indeed easily be made orthogonal to each other. This has been referred to as *orthogonalization*. Another important function of competitive networks is that partly by removing redundancy from the input information space, they can produce sparse output vectors, without losing information. We may refer to this as *sparsification*.

4.2 Architecture and algorithm

4.2.1 Architecture

The basic architecture of a competitive network is shown in Fig. 4.1. It is a one-layer network with a set of inputs which make modifiable excitatory synapses w_{ij} with the output neurons. The output cells compete with each other (for example by mutual inhibition) in such a way that the most strongly activated neuron or neurons win the competition, and are left firing strongly. The synaptic weights, w_{ij}, are initialized to random values before learning starts. If some of the synapses are missing, that is if there is randomly diluted connectivity, that is not a problem for such networks, and can even help them (see below).

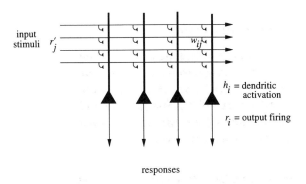

input
stimuli r'_j

w_{ij}

h_i = dendritic
activation

r_i = output firing

responses

Fig. 4.1 The architecture of a competitive network.

In the brain, the inputs arrive through axons, which make synapses with the dendrites of the output or principal cells of the network. The principal cells are typically pyramidal cells in the cerebral cortex. In the brain the principal cells are typically excitatory, and mutual inhibition between them is implemented by inhibitory interneurons, which receive excitatory inputs from the principal cells. The inhibitory interneurons then send their axons to make synapses with the pyramidal cells, typically using GABA (gamma-aminobutyric acid) as the inhibitory transmitter.

4.2.2 Algorithm

1. Apply an input vector $\mathbf{r}'$ and calculate the output activation h_i of each neuron

$$h_i = \Sigma_j r'_j \, w_{ij} \qquad (4.1)$$

where the sum is over the C input axons, indexed by j. (It is useful to normalize the length of each input vector $\mathbf{r}'$. In the brain, a scaling effect is likely to be achieved both by feedforward inhibition, and by feedback inhibition among the set of input cells that give rise to the axons conveying $\mathbf{r}'$.)

The output firing $\check{r}_i$ is a function of the activation of the neuron

$$\check{r}_i = f\,(h_i) \tag{4.2}$$

This function can be linear, sigmoid, monotonically increasing, etc. (see Fig. 1.3).

2. Allow competitive interaction between the output neurons by a mechanism like lateral or mutual inhibition (possibly with self-excitation), to produce a contrast-enhanced version of the firing rate vector $\check{\mathbf{r}}$

$$\mathbf{r} = g\,(\check{\mathbf{r}}) \tag{4.3}$$

Function g is typically a non-linear operation, and in its most extreme form may be a winner-take-all function, in which after the competition one neuron may be 'on', and the others 'off'. Algorithms which produce softer competition without a single winner are described in Section 4.8.5 below.

3. Apply an associative Hebb-like learning rule

$$\delta w_{ij} = k\,r_i\,r'_j \tag{4.4}$$

4. Normalize the length of the synaptic weight vector on each dendrite to prevent the same few neurons always winning the competition.

$$\Sigma_j (w_{ij})^2 = 1 \tag{4.5}$$

(A less efficient alternative is to scale the sum of the weights to a constant, e.g. 1.)

5. Repeat steps 1–4 for each different input stimulus $\mathbf{r}'$ in random sequence a number of times.

4.3 Properties

4.3.1 Feature discovery by self-organization

Each neuron in a competitive network becomes activated by a set of consistently coactive, that is correlated, input axons, and gradually learns to respond to that cluster of coactive inputs. We can think of competitive networks as discovering features in the input space, where features can now be defined by a set of consistently coactive inputs. Competitive networks thus show how feature analysers can be built, with no external teacher. The feature analysers respond to correlations in the input space, and the learning occurs by self-organization in the competitive network. Competitive networks are therefore well suited to the analysis of sensory inputs. Ways in which they may form fundamental building blocks of sensory systems are described in Chapter 8.

The operation of competitive networks can be visualized with the help of Fig. 4.2. The input patterns are represented as dots on the surface of a sphere. (The patterns are on the surface of a sphere because they are normalized to the same length.) The directions of the weight vectors of the three neurons are represented by crosses. The effect of learning is to move the weight vector of each of the neurons to point towards the centre of one of the clusters of inputs. If the neurons are winner-take-all, the result of the learning is that although there are correlations between the input stimuli, the outputs of the three neurons are orthogonal. In this sense, orthogonalization is performed. At the same time, given that each of the patterns within a cluster produces the same output, the correlations between the patterns within a cluster become higher. In a winner-take-all network, the within-pattern correlation becomes 1, and the patterns within a cluster have been placed within the same category.

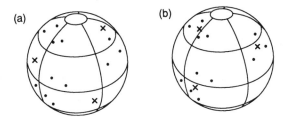

Fig. 4.2 Competitive learning. The dots represent the directions of the input vectors, and the crosses the weights for each of three neurons. (a) Before learning. (b) After learning. (After Rumelhart and Zipser, 1986.)

4.3.2 Removal of redundancy

In that competitive networks recode sets of correlated inputs to one or a few output neurons, then redundancy in the input representation is removed. Identifying and removing redundancy in sensory inputs is an important part of processing in sensory systems (cf. Barlow, 1989), in part because a compressed representation is more manageable as an output of sensory systems. The reason for this is that neurons in the receiving systems, for example pattern associators in the orbitofrontal cortex or autoassociation networks in the hippocampus, can then operate with the limited numbers of inputs that each neuron can receive. For example, although the information that a particular face is being viewed is present in the 10^6 fibres in the optic nerve, the information is unusable by associative networks in this form, and is compressed through the visual system until the information about which of many hundreds of faces is present can be represented by less than 100 neurons in the temporal cortical visual areas (Rolls, Treves and Tovee, 1997; Abbott, Rolls and Tovee, 1996). (Redundancy can be defined as the difference between the maximum information content of the input data stream (or channel capacity) and its actual content; see Appendix A2.)

The recoding of input pattern vectors into a more compressed representation that can be conveyed by a much reduced number of output neurons of a competitive network is referred to in engineering as vector quantization. With a winner-take-all competitive network, each output neuron points to or stands for one of or a cluster of the input vectors, and it is more efficient to transmit the states of the few output neurons than the states of all the input elements. (It is more efficient in the sense that the information transmission rate required, that is the capacity of the channel, can be much smaller.) Vector quantization is of course possible when the input representation contains redundancy.

4.3.3 Orthogonalization and categorization

Figure 4.2 shows visually how competitive networks reduce the correlation between different clusters of patterns, by allocating them to different output neurons. This is described as *orthogonalization*. It is a process which is very usefully applied to signals before they are used as inputs to associative networks (pattern associators and autoassociators) trained with Hebbian rules (see Chapters 2 and 3), because it reduces the interference between patterns stored in these memories. The opposite effect in competitive networks, of bringing closer together very similar input patterns, is referred to as *categorization*.

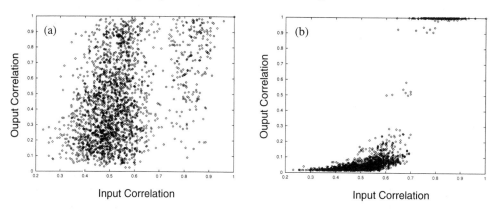

Fig. 4.3 Orthogonalization and categorization in a competitive network: (a) before learning; (b) after learning. The correlations between pairs of output vectors (abscissa) are plotted against the correlations of the corresponding pairs of input vectors that generated the output pair, for all possible pairs in the input set. The competitive net learned for 16 cycles. One cycle consisted of presenting the complete input set of stimuli in a renewing random sequence. The correlation measure shown is the cosine of the angle between two vectors (i.e. the normalized dot product). The network used had 64 input axons to each of 8 output neurons. The net was trained with 64 stimuli, made from 8 initial random binary vectors with each bit having a probability of 0.5 of being 1, from each of which 8 noisy exemplars were made by randomly altering 10% of the 64 elements. Soft competition was used between the output neurons. (The normalized exponential activation function was used to implement the soft competition.) The sparseness *a* of the input patterns thus averaged 0.5; and the sparseness *a* of the output firing vector after learning was close to 0.17 (i.e. after learning, primarily one neuron was active for each input pattern. Before learning, the average sparseness of the output patterns produced by each of the inputs was 0.39).

These two processes are also illustrated in Fig. 4.3, which shows that in a competitive network, very similar input patterns (with correlations higher in this case than approximately 0.8) produce more similar outputs (close to 1.0), whereas the correlation between pairs of patterns which are smaller than approximately 0.7 become less correlated in the output representation. (This simulation used soft competition between neurons with graded firing rates.)

4.3.4 Sparsification

Competitive networks can produce more sparse representations than those that they receive, depending on the degree of competition. With the greatest competition, winner-take-all, only one output neuron remains active, and the representation is at its most sparse. This effect can be understood further using Figs 4.2 and 4.3. This sparsification is useful to apply to representations before input patterns are applied to associative networks, for sparse representations allow many different pattern associations or memories to be stored in these networks (see Chapters 2 and 3).

4.3.5 Capacity

In a competitive net with N output neurons and a simple winner-take-all rule for the competition, it is possible to learn up to N output categories, in that each output neuron may be allocated a category. When the competition acts in a less rudimentary way, the number of categories that can be learned becomes a complex function of various factors, including the number of modifiable connections per cell and the degree of dilution, or incompleteness, of the connections. Such a function has not yet been described analytically in general, but an upper bound on it can be deduced for the particular case in which the learning is fast, and can be achieved effectively in one shot, or one presentation of each pattern. In that case, the number of categories that can be learned (by the self-organizing process) will at most be equal to the number of associations that can be formed by the corresponding pattern associators, a process which occurs with the additional help of the driving inputs, which effectively determine in the pattern associator the categorization.

Separate constraints on the capacity result if the output vectors are required to be strictly orthogonal. Then, if the output firing rates can assume only positive values, the maximum number p of categories arises, obviously, in the case when only one output neuron is firing for any stimulus, so that up to N categories are formed. If ensemble encoding of output neurons is used (soft competition), again under the orthogonality requirement, then the number of output categories that can be learned will be reduced according to the degree of ensemble encoding. The p categories in the ensemble-encoded case reflect the fact that the between-cluster correlations in the output space are lower than those in the input space. The advantages of ensemble encoding are that dendrites are more evenly allocated to patterns (see Section 4.8.5), and that correlations between different input stimuli can be reflected in correlations between the corresponding output vectors, so that later networks in the system can generalize usefully. This latter property is of crucial importance, and is utilized for example when an input pattern is presented which has not been learned by the network. The relative similarity of the input pattern to previously learned patterns is indicated by the relative activation of the members of an ensemble of output neurons. This makes the number of different representations that can be reflected in the output of competitive networks with ensemble encoding much higher than with winner-take-all representations, even though with soft competition all these representations cannot strictly be learned.

4.3.6 Separation of non-linearly separable patterns

A competitive network can not only separate (e.g. by activating different output neurons) pattern vectors that overlap in almost all elements, but can also lead to the separation of vectors that are not linearly separable. An example is that three patterns A, B, and A + B will lead to three different output neurons being activated (see Fig. 4.4). For this to occur, the length of the synaptic weight vectors must be normalized (to, for example, unit length), so that they lie on the surface of a sphere or hypersphere (see Fig. 4.2). (If the weight vectors of each neuron are scaled to the same sum, then the weight vectors do not lie on the surface of a hypersphere, and the ability of the network to separate patterns is reduced.)

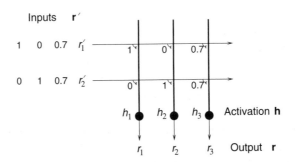

Fig. 4.4 Separation of linearly dependent patterns by a competitive network. The network was trained on patterns 10, 01 and 11, applied on the inputs r'_1 and r'_2. After learning, the network allocated output neuron 1 to pattern 10, neuron 2 to pattern 01, and neuron 3 to pattern 11. The weights in the network produced during the learning are shown. Each input pattern was normalized to unit length, and thus for pattern 11, $r'_1 = 0.7$ and $r'_2 = 0.7$, as shown. Because the weight vectors were also normalized to unit length, $w_{31} = 0.7$ and $w_{32} = 0.7$.

The property of pattern separation makes a competitive network placed before an autoassociator network very valuable, for it enables the autoassociator to store the three patterns separately, and to recall A + B separately from A and B. This is referred to as the configuration learning problem in animal learning theory (Sutherland and Rudy, 1991). Placing a competitive network before a pattern associator will enable a linearly inseparable problem to be solved. For example, three different output neurons of a two-input competitive network could respond to the patterns 01, 10, and 11, and a pattern associator can learn different outputs to neurons 1–3, which are orthogonal to each other (see Fig. 2.12). This is an example of expansion recoding (cf. Marr, 1969, who used a different algorithm to obtain the expansion). The sparsification that can be produced by the competitive network can also be advantageous in preparing patterns for presentation to a pattern associator or autoassociator, because the sparsification can increase the number of memories that can be associated or stored.

4.3.7 Stability

These networks are generally stable if the input statistics are stable. If the input statistics keep varying, then the competitive network will keep following the input statistics. If this is a problem, then a critical period in which the input statistics are learned, followed by stabilization, may be useful. This appears to be a solution used in developing sensory systems, which have critical periods beyond which further changes become more difficult. An alternative approach taken by Carpenter and Grossberg in their Adaptive Resonance Theory is to allow the network to learn only if it does not already have categorizers for a pattern (see Hertz *et al.*, 1991, p. 228).

Diluted connectivity can help stability, by making neurons tend to find inputs to categorize in only certain parts of the input space, and then making it difficult for the neuron to wander randomly throughout the space later.

4.3.8 Frequency of presentation

If some stimuli are presented more frequently than others, then there will be a tendency for the weight vectors to move more rapidly towards frequently presented stimuli, and more neurons may become allocated to the frequently presented stimuli. If winner-take-all competition is used, the result is that the neurons will tend to become allocated during the learning process to the more frequently presented patterns. If soft competition is used, the tendency of neurons to move from patterns that are infrequently or never presented can be reduced by making the competition fairly strong, so that only few neurons show any learning when each pattern is presented. Provided that the competition is moderately strong (see Section 4.8.4), the result is that more neurons are allocated to frequently presented patterns, but one or some neurons are allocated to infrequently presented patterns. These points can all be easily demonstrated in simulations.

4.3.9 Comparison to principal component analysis (PCA) and cluster analysis

Although competitive networks find clusters of features in the input space, they do not perform hierarchical cluster analysis as typically performed in statistics. In hierarchical cluster analysis, input vectors are joined starting with the most correlated pair, and the level of the joining of vectors is indicated. Competitive nets produce different outputs (i.e. activate different output neurons) for each cluster of vectors (i.e. perform vector quantization), but do not compute the level in the hierarchy, unless the network is redesigned (see Hertz *et al.*, 1991).

The feature discovery can also be compared to principal component analysis (PCA). (In PCA, the first principal component of a multidimensional space points in the direction of the vector which accounts for most of the variance, and subsequent principal components account for successively less of the variance, and are mutually orthogonal.) In competitive learning with a winner-take-all algorithm, the outputs are mutually orthogonal, but are not in an ordered series according to the amount of variance accounted for, unless the training algorithm is modified. The modification amounts to allowing each of the neurons in a winner-take-all network to learn one at a time, in sequence. The first neuron learns the first principal component. (Neurons trained with a modified Hebb rule learn to maximize the variance of their outputs—see Hertz *et al.*, 1991.) The second neuron is then allowed to learn, and because its output is orthogonal to the first, it learns the second principal component. This process is repeated. Details are given by Hertz *et al.*, 1991, but as this is not a biologically plausible process, it is not considered in detail here. We note that simple competitive learning is very helpful biologically, because it can separate patterns, but that a full ordered set of principal components as computed by PCA would probably not be very useful in biologically plausible networks. Our point here is that biological neuronal networks may operate well if the variance in the input representation is distributed across many input neurons, whereas principal component analysis would tend to result in most of the variance being allocated to a few neurons, and the variance being unevenly distributed across the neurons.

4.4 Utility of competitive networks in information processing by the brain

4.4.1 Orthogonalization

The orthogonalization performed by competitive networks is very useful for preparing signals for presentation to pattern associators and autoassociators, for this re-representation decreases interference between the patterns stored in such networks. Indeed, this can be essential if patterns are overlapping and non-linearly independent, e.g. 01, 10, and 11. If three such binary patterns were presented to an autoassociative network, it would not form separate representations of them, for either of the patterns 01 or 10 would result by completion in recall of the 11 pattern. A competitive network allows a separate neuron to be allocated to each of the three patterns, and this set of orthogonal representations can be learned by associative networks (see Fig. 4.4).

4.4.2 Sparsification

The sparsification performed by competitive networks is very useful for preparing signals for presentation to pattern associators and autoassociators, for this re-representation increases the number of patterns that can be associated or stored in such networks (see Chapters 2 and 3).

4.4.3 Brain systems in which competitive networks may be used for orthogonalization and sparsification

One system is the hippocampus, in which the dentate granule cells are believed to operate as a competitive network in order to prepare signals for presentation to the CA3 autoassociative network (see Chapter 6). In this case, the operation is enhanced by expansion recoding, in that (in the rat) there are approximately three times as many dentate granule cells as there are cells in the preceding stage, the entorhinal cortex (see Chapter 6). This expansion recoding will itself tend to reduce correlations between patterns (cf. Marr, 1970, 1969).

Also in the hippocampus, the CA1 neurons are thought to act as a competitive network that recodes the separate representations of each of the parts of an episode that must be separately represented in CA3, into a form more suitable for the recall using pattern association performed by the backprojections from the hippocampus to the cerebral cortex (see Chapter 6).

The granule cells of the cerebellum may perform a similar function, but in this case the principle may be that each of the very large number of granule cells receives a very small random subset of inputs, so that the outputs of the granule cells are decorrelated with respect to the inputs (Marr, 1969; see Chapter 9).

4.4.4 Removal of redundancy

The removal of redundancy by competition is thought to be a key aspect of how sensory systems operate. Competitive networks can also be thought of as performing dimension reduction, in that a set of correlated inputs may be responded to as one category or dimension by a competitive network. Although networks with anti-Hebbian synapses between the principal cells (in which the anti-Hebbian learning forces neurons with initially correlated activity to effectively inhibit each other) (Foldiak, 1990), and networks which perform independent component analysis (Bell and Sejnowski, 1995), could in principle remove redundancy more effectively, it is not clear that they are implemented biologically. In contrast, competitive networks are more biologically plausible, and illustrate redundancy reduction. The more general use of an unsupervised competitive preprocessor is discussed below (see Fig. 4.10).

4.4.5 Feature analysis and preprocessing

Neurons which respond to correlated combinations of their inputs can be described as feature analysers. Neurons which act as feature analysers perform useful preprocessing in many sensory systems (see e.g. Chapter 8). The power of competitive networks in multistage hierarchical processing to build combinations of what is found at earlier stages, and thus effectively to build higher-order representations, is also described in Chapter 8.

4.5 Guidance of competitive learning

Although competitive networks are primarily unsupervised networks, it is possible to influence the categories found by supplying a second input, as follows (Rolls, 1989e). Consider a competitive network as shown in Fig. 4.5 with the normal set of inputs A to be categorized, and with an additional set of inputs B from a different source. Both sets of inputs work in the normal way for a competitive network, with random initial weights, competition between the output neurons, and a Hebb-like synaptic modification rule which normalizes the lengths of the synaptic weight vectors onto each neuron. The idea then is to use the B inputs to influence the categories formed by the A input vectors. The influence of the B vectors works best if they are orthogonal to each other. Consider any two A vectors. If they occur together with the same B vector, then the categories produced by the A vectors will be more similar than they would be without the influence of the B vectors. The categories will be pulled closer together if soft competition is used, or will be more likely to activate the same neuron if winner-take-all competition is used. Conversely, if any two A vectors are paired with two different, preferably orthogonal, B vectors, then the categories formed by the A vectors will be drawn further apart than they would be without the B vectors. The differences in categorization remain present after the learning when just the A inputs are used.

This guiding function of one of the inputs is one way in which the consequences of sensory stimuli could be fed back to a sensory system to influence the categories formed when the A inputs are presented. This could be one function of backprojections in the cerebral cortex (Rolls, 1989b, 1989e). In this case, the A inputs of Fig. 4.5 would be the forward inputs from a preceding cortical area, and the B inputs backprojecting axons from the next cortical area, or from a structure such as the amygdala or hippocampus. If two A vectors were both associated with positive reinforcement which was fed back as the same B vector from another part of the brain, then the two A vectors would be brought closer together in the

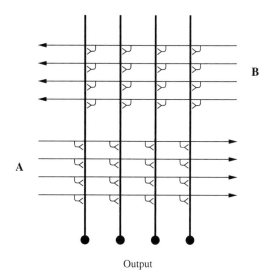

Output

Fig. 4.5 Competitive net receiving a normal set of inputs A, but also another set of inputs B which can be used to influence the categories formed in response to A inputs.

representational space provided by the output of the neurons. If one of the A vectors was associated with positive reinforcement, and the other with negative reinforcement, then the output representations of the two A vectors would be further apart. This is one way in which external signals could influence in a mild way the categories formed in sensory systems. Another is that if any B vector only occurred for important sensory A inputs (as shown by the immediate consequences of receiving those sensory inputs), then the A inputs would simply be more likely to have any representation formed than otherwise, due to strong activation of neurons only when combined A and B inputs are present.

A similar architecture could be used to provide mild guidance for one sensory system (e.g. olfaction) by another (e.g. taste), as shown in Fig. 4.6. The idea is that the taste inputs would be more orthogonal to each other than the olfactory inputs, and that the taste inputs would influence the categories formed in the olfactory input categorizer in layer 1, by feedback from a convergent net in layer 2. The difference from the previous architecture is that we now have a two-layer net, with unimodal or separate networks in layer 1, each feeding forward to a single competitive network in layer 2. The categories formed in layer 2 reflect the co-occurrence of particular taste with particular odours (which together form flavour in layer 2). Layer 2 then provides feedback connections to both the networks in layer 1. It can be shown in such a network that the categories formed in, for example, the olfactory net in layer 1 are influenced by the tastes with which the odours are paired. The feedback signal is built only in layer 2, after there has been convergence between the different modalities. This architecture captures some of the properties of sensory systems, in which there are unimodal processing cortical areas followed by multimodal cortical areas. The multimodal cortical areas can build representations which represent the unimodal inputs that tend to co-occur, and the higher level representations may in turn by the highly developed cortico-cortical backprojections be able to influence sensory categorization in earlier cortical processing areas (Rolls, 1989e).

Another such example might be the effect by which the phonemes heard are influenced by the visual inputs produced by seeing mouth movements (cf McGurk and MacDonald, 1976). This could be implemented by auditory inputs coming together in the cortex in the superior temporal sulcus onto neurons activated by the sight of the lips moving (recorded during experiments of Baylis, Rolls and Leonard, 1987, and Hasselmo, Rolls, Baylis and Nalwa, 1989), using Hebbian learning with coactive inputs. Backprojections from such multimodal areas to the early auditory cortical areas could then influence the responses of auditory cortex neurons to auditory inputs (cf. Calvert *et al.*, 1997).

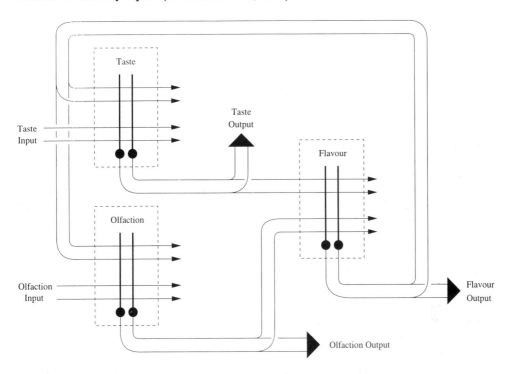

Fig. 4.6 A two-layer set of competitive nets in which feedback from layer 2 can influence the categories formed in layer 1. In the example, taste and olfactory inputs are received by separate competitive nets in layer 1, and converge into a single competitive net in layer 2. The categories formed in layer 2 (which may be described as representing 'flavour') may be dominated by the relatively orthogonal set of a few tastes that are received by the net. When these layer 2 categories are fed back to layer 1, they may produce in layer 1 categories in, for example, the olfactory network that reflect to some extent the flavour categories of layer 2, and are different from the categories that would otherwise be formed to a large set of rather correlated olfactory inputs.

The idea that the statistical correlation between the inputs received by neighbouring processing streams can be used to guide unsupervised learning within each stream has also been developed by Becker and Hinton (1992) and others (see Phillips *et al.*, 1995). The networks considered by these authors self-organize under the influence of collateral connections, such as may be implemented by cortico-cortical connections between parallel processing systems in the brain. They use learning rules that, although somewhat complex, are still local in nature, and tend to optimize specific objective functions. The locality of the learning rule, and the simulations performed so far, raise some hope that, once the operation of these types of networks is better understood, they might achieve similar computational capabilities to backpropagation networks (see Chapter 5) while retaining biological plausibility.

4.6 Topographic map formation

A simple modification to the competitive networks described so far enables them to develop topological maps. In such maps, the closeness in the map reflects the similarity (correlation) between the features in the inputs. The modification which allows such maps to self-organize is to add short-range excitation and long-range inhibition between the neurons. The function to be implemented has a spatial profile which is described as having a Mexican hat shape (see Fig. 4.7). The effect of this connectivity between neurons, which need not be modifiable, is to encourage neurons which are close together to respond to similar features in the input space, and to encourage neurons which are far apart to respond to different features in the input space. When these response tendencies are present during learning, the feature analysers which are built by modifying the synapses from the input onto the activated neurons tend to be similar if they are close together, and different if far apart. This is illustrated in Figs 4.8 and 4.9. Feature maps built in this way were described by von der Malsburg (1973) and Willshaw and von der Malsburg (1976). It should be noted that the learning rule needed is simply the modified Hebb rule described above for competitive networks, and is thus local and biologically plausible. For computational convenience, the algorithm that Kohonen (1982, 1989, 1995) has mainly used does not use Mexican hat connectivity between the neurons, but instead arranges that when the weights to a winning neuron are updated, so to a smaller extent are those of its neighbours (see further Hertz *et al.*, 1991).

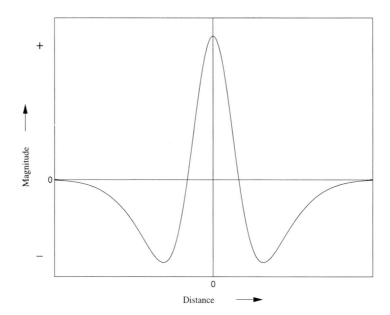

Fig. 4.7 Mexican hat lateral spatial interaction profile.

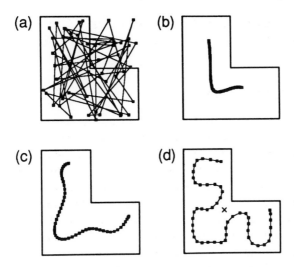

Fig. 4.8 Kohonen feature mapping from a two-dimensional L-shaped region to a linear array of 50 units. Each unit has 2 inputs. The input patterns are the X,Y coordinates of points within the L-shape shown. In the diagrams, each point shows the position of a weight vector. Lines connect adjacent units in the 1D (linear) array of 50 neurons. The weights were initialized to random values within the unit square (a). During feature mapping training, the weights evolved through stages (b) and (c) to (d). By stage (d) the weights have formed so that the positions in the original input space are mapped to a 1D vector in which adjacent points in the input space activate neighbouring units in the linear array of output units. (Reproduced with permission from Hertz *et al.*, 1991, Fig. 9.13.)

A very common characteristic of connectivity in the brain, found for example throughout the neocortex, consists of short-range excitatory connections between neurons, with inhibition mediated via inhibitory interneurons. The density of the excitatory connectivity even falls gradually as a function of distance from a neuron, extending typically a distance in the order of 1 mm from the neuron (Braitenberg and Schuz, 1991), contributing to a spatial function quite like that of a Mexican hat. (Longer-range inhibitory influences would form the negative part of the spatial response profile.) This supports the idea that topological maps, though in some cases probably seeded by chemoaffinity, could develop in the brain with the assistance of the processes just described. It is noted that some cortico-cortical connections even within an area may be longer, skipping past some intermediate neurons, and then making connections after some distance with a further group of neurons. Such longer-range connections are found for example between different columns with similar orientation selectivity in the primary visual cortex. The longer range connections may play a part in stabilizing maps, and again in the exchange of information between neurons performing related computations, in this case about features with the same orientations.

If a low-dimensional space, for example the orientation sensitivity of cortical neurons in the primary visual cortex (which is essentially one-dimensional, the dimension being angle), is mapped to a two-dimensional space such as the surface of the cortex, then the resulting map can have long spatial runs where the value along the dimension (in this case angle) alters gradually, and continuously. If a high-dimensional information space is mapped to the two-dimensional cortex, then there will be only short runs of groups of neurons with similar

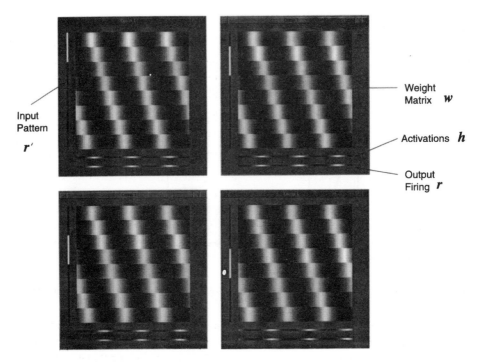

Fig. 4.9 Example of a one-dimensional topological map that self-organized from inputs in a low-dimensional space. The network has 64 neurons (vertical elements in the diagram) and 64 inputs per neuron (horizontal elements in the diagram). The four different diagrams represent the net tested with different input patterns. The input patterns **r'** are displayed at the left of each diagram, with white representing firing and black not firing for each of the 64 inputs. The central square of each diagram represents the synaptic weights of the neurons, with white representing a strong weight. The row vector below each weight matrix represents the activations of the 64 output neurons, and the bottom row vector the output firing **r**. The network was trained with a set of 8 binary input patterns, each of which overlapped in 8 of its 16 'on' elements with the next pattern. The diagram shows that as one moves through correlations in the input space (top left to top right to bottom left to bottom right), so the output neurons activated move steadily across the output array of neurons. Closely correlated inputs are represented close together in the output array of neurons. The way in which this occurs can be seen by inspection of the weight matrix. The network architecture was the same as for a competitive net, except that the output activations were converted linearly into output firings, and then each neuron excited its neighbours and inhibited neurons further away. This lateral inhibition was implemented for the simulation by a spatial filter operating on the output firings with the following filter weights (cf. Fig. 4.7):

5,–5,–5,–5,–5,–5,–5,–5,–5,10,10,10,10,10,10,10,10,10,–5,–5,–5,–5,–5,–5,–5,–5,–5

which operated on the 64-element firing rate vector.

feature responsiveness, and then the map must fracture, with a different type of feature mapped for a short distance after the discontinuity. This is exactly what Rolls suggests is the type of topology found in the inferior temporal visual cortex. Here, visual stimuli are not represented with reference to their position on the retina, for here the neurons are relatively translation invariant. Instead, when recording here small clumps of neurons with similar responses may be encountered close together, and then one moves into a group of neurons with quite different feature selectivity. This topology will arise naturally, given the anatomical connectivity of the cortex with its short-range excitatory connections, because there are very many different objects in the world and different types of feature which describe objects, with no special continuity between the different combinations of features possible. This last type of map implies that topological self-organization is an important way in which maps in the brain are formed, for it seems most unlikely that the locations in the map of different types of object seen in an environment could be specified genetically.

The biological utility of developing such topology-preserving feature maps may be that if the computation requires neurons with similar types of response to exchange information more than neurons involved in different computations (which is more than reasonable), then the total length of the connections between the neurons is minimized if the neurons which need to exchange information are close together (cf. Cowey, 1979; Durbin and Mitchison, 1990). Minimizing the total connection length between neurons in the brain is very important in order to keep the size of the brain relatively small. Placing neurons close to each other which need to exchange information, or which need to receive information from the same source, or which need to project towards the same destination, may also help to minimize the complexity of the rules required to specify cortical (and indeed brain) connectivity.

The rules of information exchange just described could also tend to produce more gross topography in cortical regions. For example, neurons which respond to animate objects may have certain visual feature requirements in common, and may need to exchange information about these features. Other neurons which respond to inanimate objects might have somewhat different visual feature requirements for their inputs, and might need to exchange information strongly. (For example, selection of whether an object is a chisel or a screwdriver may require competition by mutual (lateral) inhibition to produce the contrast enhancement necessary to result in unambiguous neuronal responses.) The rules just described would account for neurons with responsiveness to inanimate and animate objects tending to be grouped in separate parts of a cortical map or representation, and thus separately susceptible to brain damage (see e.g. Farah, 1990).

4.7 Radial basis function networks

As noted above, a competitive network can act as a useful preprocessor for other networks. In the neural examples above, competitive networks were useful preprocessors for associative networks. Competitive networks are also used as preprocessors in artificial neural networks, for example in hybrid two-layer networks such as that illustrated in Fig. 4.10. The competitive network is advantageous in this hybrid scheme, because as an unsupervised network, it can relatively quickly (in perhaps 100 trials) discover the main features in the input space, and code for them. This leaves the second layer of the network to act as a supervised network (taught for example by the delta rule, see Chapter 5), which learns to map the features found by the first layer into the output required. This learning scheme is very much faster than that of a (two-layer) backpropagation network, which learns very slowly because it takes it a long time to perform the credit assignment to build useful feature analyzers in layer one (the hidden layer) (see Chapter 5).

The general scheme shown in Fig. 4.10 is used in radial basis function (RBF) neural networks. The main difference from what has been described is that in an RBF network, the hidden neurons do not use a winner-take-all function (as in some competitive networks), but instead use a normalized Gaussian activation function:

$$r_i = \frac{\exp[-(\mathbf{r}' - \mathbf{w}_i)^2 / 2\sigma_i^2]}{\Sigma_k \exp[-(\mathbf{r}' - \mathbf{w}_k)^2 / 2\sigma_k^2]} \qquad (4.6)$$

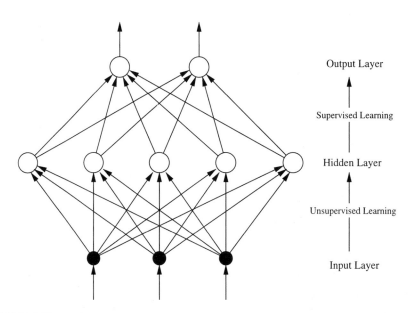

Output Layer

Supervised Learning

Hidden Layer

Unsupervised Learning

Input Layer

Fig. 4.10 A hybrid network, in which for example unsupervised learning rapidly builds relatively orthogonal representations based on input differences, and this is followed by a one-layer supervised network (taught for example by the delta rule) that learns to classify the inputs based on the categorizations formed in the hidden/intermediate layer.

The effect is that the response r_i of neuron i is a maximum if the input stimulus vector $\mathbf{r}'$ is centred at $\mathbf{w}_i$, the weight vector of neuron i (this is the upper term in Eq. 4.6). The magnitude is normalized by dividing by the sum of the activations of all the k neurons in the network. If the input vector $\mathbf{r}'$ is not at the centre of the receptive field of the neuron, then the response is decreased according to how far the input vector is from the weight vector $\mathbf{w}_i$ of the neuron, with the weighting decreasing as a Gaussian function with a standard deviation of σ. The idea is like that implemented with soft competition, in that the relative response of different neurons provides an indication of where the input pattern is in relation to the weight vectors of the different neurons. The rapidity with which the response falls off in a Gaussian radial basis function neuron is set by σ_i, which is adjusted so that for any given input pattern vector, a number of RBF neurons are activated. The position in which the RBF neurons are located (i.e. their weight vectors, $\mathbf{w}$) are determined usually by unsupervised learning, e.g. the vector quantization which is produced by the normal competitive learning algorithm. The first layer of a RBF network is not different in principle from a network with soft competition, and it is not clear how biologically a Gaussian activation function would be implemented, so the treatment is not developed further here (see Hertz *et al.*, 1991, for further details).

4.8 Appendix on the algorithms used in competitive networks

4.8.1 Normalization of the inputs

Normalization is useful because in step 1 of the training algorithm described in Section 4.2.2, the output activations, formed by the inner product of the pattern and the normalized weight vector on each neuron, are scaled in such a way that they have a maximum value of 1. This helps different input patterns to be equally effective in the learning process. A way in which this normalization could be achieved by a layer of input neurons is given by Grossberg (1976a). In the brain, a number of factors may contribute to normalization of the inputs. One factor is that a set of input axons to a neuron will come from another network in which the firing is controlled by inhibitory feedback, and if the numbers of axons involved is large (hundreds or thousands), then the inputs will be in a reasonable range. Second, there is increasing evidence that the different classes of input to a neuron may activate different types of inhibitory interneuron (e.g. Buhl et al., 1994), which terminate on separate parts of the dendrite, usually close to the site of termination of the corresponding excitatory afferents. This may allow separate feedforward inhibition for the different classes of input. In addition, the feedback inhibitory interneurons also have characteristic termination sites, often on or close to the cell body, where they may be particularly effective in controlling firing of the neuron, rather than scaling a class of input.

4.8.2 Normalization of the length of the synaptic weight vector on each dendrite

This is necessary to ensure that one or a few neurons do not always win the competition. (If the weights on one neuron were increased by simple Hebbian learning, and there was no normalization of the weights on the neuron, then it would tend to respond strongly in the future to patterns with some overlap with patterns to which that neuron has previously learned, and gradually that neuron would capture a large number of patterns.) A biologically plausible way to achieve this weight adjustment is to use a modified Hebb rule:

$$\delta w_{ij} = k\, r_i (r'_j - w_{ij}) \tag{4.7}$$

where k is a constant, and r'_j and w_{ij} are in appropriate units. In vector notation,

$$\delta \mathbf{w}_i = k\, r_i\, (\mathbf{r'} - \mathbf{w}_i)$$

where $\mathbf{w}_i$ is the synaptic weight vector on neuron i. This implements a Hebb rule which increases synaptic strength according to conjunctive pre- and postsynaptic activity, and also allows the strength of each synapse to decay in proportion to the firing rate of the postsynaptic neuron (as well as in proportion to the existing synaptic strength). This results in a decrease in synaptic strength for synapses from weakly active presynaptic neurons onto strongly active postsynaptic neurons. Such a modification in synaptic strength is termed heterosynaptic long-term depression in the neurophysiological literature, referring to the fact that the synapses that weaken are other than those that activate the neuron. This is an important computational use of heterosynaptic long-term depression. This rule can maintain the sums of the synaptic weights on each dendrite to be very similar without any need for explicit normalization of the synaptic strengths, and is useful in competitive nets. This rule was used by Willshaw and von der Malsburg (1976). As is made clear with the vector notation above, the modified Hebb rule moves the direction of the weight vector $\mathbf{w}_i$ towards the current input pattern vector $\mathbf{r}'$ in proportion to the difference between these two vectors and the firing rate r_i of neuron i.

If explicit weight normalization is needed, the appropriate form of the modified Hebb rule is:

$$\delta w_{ij} = k\, r_i (r'_j - r_i w_{ij}).\tag{4.8}$$

This rule, formulated by Oja (1982), makes weight decay proportional to r_i^2, normalizes the synaptic weight vector (see Hertz, Krogh and Palmer, 1991), is still a local learning rule, and is known as the Oja rule.

4.8.3 Non-linearity in the learning rule

Non-linearity in the learning rule can assist competition (Rolls, 1989c, 1996a). For example, in the brain, long-term potentiation typically occurs only when strong activation of a neuron has produced sufficient depolarization for the voltage-dependent NMDA receptors to become unblocked, allowing Ca^{2+} to enter the cell (see Chapter 1). This means that synaptic modification occurs only on neurons that are strongly activated, effectively assisting competition to select few winners. The learning rule can be written:

$$\delta \mathbf{w}_{ij} = k\, m_i (r'_j - w_{ij})\tag{4.9}$$

where m_i is a (e.g. threshold) non-linear function of the post-synaptic firing r_i which mimics the operation of the NMDA receptors in learning.

4.8.4 Competition

In a simulation of a competitive network, a single winner can be selected by searching for the neuron with the maximum activation. If graded competition is required, this can be achieved by an activation function which increases greater than linearly. In some of the networks we have simulated (Rolls, 1989c, 1989e; Wallis and Rolls, 1997), raising the activation to a fixed power, typically in the range 2–5, and then rescaling the outputs to a fixed maximum (e.g. 1) is simple to implement. In a real network, winner-take-all competition can be implemented using mutual (lateral) inhibition between the neurons with non-linear activation functions, and self-excitation of each neuron (see e.g. Grossberg, 1976a, 1988; Hertz et al., 1991).

Perhaps the most generally useful method to implement soft competition in simulations is to use the normalized exponential or softmax activation function for the neurons (Bridle, 1990; see Bishop, 1995):

$$r = \exp(h) / \Sigma_i \exp(h_i)\tag{4.10}$$

This function specifies that the firing rate of each neuron is an exponential function of the activation, scaled by the whole vector of activations h_i, $i = 1,N$. The exponential function (in increasing supralinearly) implements soft competition, in that after the competition the faster firing neurons are firing relatively much faster than the slower firing neurons. In fact, the strength of the competition can be adjusted by using a 'temperature' T greater than 0 as follows:

$$r = \exp(h/T) / \Sigma_i \exp(h_i/T) \tag{4.11}$$

Very low temperatures increase the competition, until with $T \to 0$, the competition becomes 'winner-take-all'. At high temperatures, the competition becomes very soft. (When using the function in simulations, it may be advisable to prescale the firing rates to for example the range 0–1, both to prevent machine overflow, and to set the temperature to operate on a constant range of firing rates, as increasing the range of the inputs has an effect similar to decreasing T.)

The softmax function has the property that activations in the range $-\infty$ to $+\infty$ are mapped into the range 0 to 1.0, and the sum of the firing rates is 1.0. This facilitates interpretation of the firing rates under certain conditions as probabilities, for example that the competitive network firing rate of each neuron reflects the probability that the input vector is within the category or cluster signified by that output neuron (see Bishop, 1995).

4.8.5 Soft competition

The use of graded (continuous valued) output neurons in a competitive network, and soft competition rather than winner-take-all competition, has the value that the competitive net generalizes more continuously to an input vector which lies between input vectors that it has learned. Also, with soft competition, neurons with only a small amount of activation by any of the patterns being used will nevertheless learn a little, and move gradually towards the patterns that are being presented. The result is that with soft competition, the output neurons all tend to become allocated to one of the input patterns or one of the clusters of input patterns.

4.8.6 Untrained neurons

In competitive networks, especially with winner-take-all or finely tuned neurons, it is possible that some neurons remain unallocated to patterns. This may be useful, in case patterns in the unused part of the space occur in future. Alternatively, unallocated neurons can be made to move towards the parts of the space where patterns are occurring by allowing such losers in the competition to learn a little. Another mechanism is to subtract a bias term μ_i from r_i, and to use a 'conscience' mechanism that raises μ_i if a neuron wins frequently, and lowers μ_i if it wins infrequently (Grossberg, 1976b; Bienenstock, Cooper and Munro, 1982; DeSieno, 1988).

4.8.7 Large competitive nets: further aspects

If a large neuronal network is considered, with the number of synapses on each neuron in the region of 10 000, as occurs on large pyramidal cells in some parts of the brain, then there is a potential disadvantage in using neurons with synaptic weights which can take on only positive values. This difficulty arises in the following way. Consider a set of positive normalized input firing rates and synaptic weight vectors (in which each element of the vector can take on any value between 0.0 and 1.0). Such vectors of random values will on average be more highly aligned with the direction of the central vector $(1,1,1,\ldots,1)$ than with any other vector. An example can be given for the particular case of vectors evenly distributed on the positive 'quadrant' of a high-dimensional hypersphere: the average overlap (i.e. dot product) between two random vectors (e.g. a random pattern vector and a random dendritic weight vector) will be approximately 0.637 while the average overlap between a random vector and the central vector will be approximately 0.798. A consequence of this will be that if a neuron begins to learn towards several input pattern vectors it will get drawn towards the average of these input patterns which will be closer to the $1,1,1,\ldots,1$ direction than to any one of the patterns. As a dendritic weight vector moves towards the central vector, it will become more closely aligned with more and more input patterns so that it is more rapidly drawn towards the central vector. The end result is that in large nets of this type, many of the dendritic weight vectors will point towards the central vector. This effect is not seen so much in small systems since the fluctuations in the magnitude of the overlaps are sufficiently large that in most cases a dendritic weight vector will have an input pattern very close to it and thus will not learn towards the centre. In large systems the fluctuations in the overlaps between random vectors become smaller by a factor of $1/\sqrt{N}$ so that the dendrites will not be particularly close to any of the input patterns.

One solution to this problem is to allow the elements of the synaptic weight vectors to take negative as well as positive values. This could be implemented in the brain by feedforward inhibition. A set of vectors taken with random values will then have a reduced mean correlation between any pair, and the competitive net will be able to categorize them effectively. A system with synaptic weights which can be negative as well as positive is not physiologically plausible, but we can instead imagine a system with weights lying on a hypersphere in the positive quadrant of space but with additional inhibition which results in the cumulative effects of some input lines being effectively negative. This can be achieved in a network by using positive input vectors, positive weight vectors, and thresholding the output neurons at their mean activation. A large competitive network of this general nature does categorize well, and has been described more fully elsewhere (Bennett, 1990). In a large network with inhibitory feedback neurons, and principal cells with thresholds, the network could achieve at least in part an approximation to this type of thresholding useful in large competitive networks.

A second way in which nets with positive-only values of the elements could operate is by making the input vectors sparse and initializing the weight vectors to be sparse, or to have a reduced contact probability. (A measure of sparseness is defined in Eq. 1.4.) For relatively small net sizes simulated ($N = 100$) with patterns with a sparseness a of, for example, 0.1 or 0.2, learning onto the average vector can be avoided. However, as the net size increases, the sparseness required does become very low. In large nets, a greatly reduced contact probability between neurons (many synapses kept identically zero) would prevent learning of the average vector, thus allowing categorization to occur. Reduced contact probability will, however, prevent complete alignment of synapses with patterns, so that the performance of the network will be affected.

5 Error-correcting networks: perceptrons, the delta rule, backpropagation of error in multilayer networks, and reinforcement learning algorithms

The networks described in this chapter are capable of mapping a set of inputs to a set of required outputs using correction when errors are made. Although some of the networks are very powerful in the types of mapping they can perform, the power is obtained at the cost of learning algorithms which do not use local learning rules. A local learning rule specifies that synaptic strengths should be altered on the basis of information available locally at the synapse, for example the activity of the presynaptic and the postsynaptic neurons. Because the networks described here do not use local learning rules, their biological plausibility remains at present uncertain, although it has been suggested that perceptron learning may be implemented by the special neuronal network architecture of the cerebellum (see Chapter 9). One of the aims of future research must be to determine whether comparably difficult problems to those solved by the networks described in this chapter can be solved by biologically plausible neuronal networks.

5.1 Perceptrons and one-layer error-correcting networks

Under this heading, we describe one-layer networks taught by an error-correction algorithm. The term perceptron refers strictly to networks with binary threshold activation functions. The outputs might take the values only 1 or 0 for example. The term perceptron arose from networks designed originally to solve perceptual problems (Rosenblatt, 1961; Minsky and Papert, 1969), and these networks are referred to briefly below. If the output neurons have continuous-valued firing rates, then a more general error-correcting rule called the delta rule is used, and is introduced in this chapter. For such networks, the activation function may be linear, or it may be non-linear but monotonically increasing, without a sharp threshold, as in the sigmoid activation function (see Fig. 1.3).

5.1.1 Architecture and general description

The one-layer error-correcting network has a set of inputs which it is desired to map or classify into a set of outputs (see Fig. 5.1). During learning, an input pattern is selected, and produces output firing by activating the output neurons through modifiable synapses, which then fire as a function of their typically non-linear activation function. The output of each neuron is then compared to a target output for that neuron given that input pattern, an error between the actual output and the desired output is determined, and the synaptic weights on that neuron are then adjusted to minimize the error. This process is then repeated for all patterns until the average error across patterns has reached a minimum. A one-layer error-correcting network can thus produce output firing for each pattern in a way that has similarities to a pattern associator. It can perform more powerful mappings than a pattern associator, but requires an error to be computed for each neuron, and for that error to affect the synaptic strength in a way that is not altogether local. A more detailed description follows.

These one-layer networks have a target for each output neuron (for each input pattern). They are thus an example of a supervised network. With the one-layer networks taught with the delta rule or perceptron learning rule described next, there is a separate teacher for each output neuron.

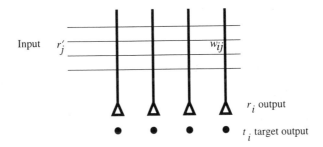

Fig. 5.1 One-layer perceptron.

5.1.2 Generic algorithm (for a one-layer network taught by error correction)

For each input pattern and desired target output:

1. Apply an input pattern to produce input firing **r'**, and obtain the output activation of each neuron in the standard way by computing the dot product of the input pattern and the synaptic weight vector. The synaptic weight vector can be initially zero, or (small) random values.

$$h_i = \Sigma_j r'_j w_{ij} \qquad (5.1)$$

where the sum is over the C input axons, r'_j.

2. Apply an activation function to produce the output firing r_i.

$$r_i = f(h_i)$$

This activation function f may be sigmoid, linear, binary threshold, linear threshold, etc.

If the activation function is non-linear, this helps to classify the inputs into distinct output patterns, but a linear activation function may be used if an optimal linear mapping is desired (see Adaline and Madaline, below).

3. Calculate the difference for each cell i between the target output t_i and the actual output r_i produced by the input, which is the error Δ_i

$$\Delta_i = t_i - r_i.$$

4. Apply the following learning rule, which corrects the (continuously variable) weights according to the error and the input firing r'_j

$$\delta w_{ij} = k(t_i - r_i) r'_j \tag{5.2}$$

where k is a constant which determines the learning rate. This is often called the delta rule, the Widrow–Hoff rule, or the LMS (least mean squares) rule (see below).

5. Repeat steps 1–4 for all input pattern/output target pairs until the root mean square error becomes zero or reaches a minimum.

5.1.3 Variations on the single-layer error-correcting network

5.1.3.1 Rosenblatt's (1961) perceptron

The perceptron was developed by Rosenblatt (1961), and its properties and constraints were extensively analysed by Minsky and Papert (1969) (see also Nilsson, 1965). The original perceptron was conceived as a model of the eye, and consists (Fig. 5.2) of sensory cells, which connect to association or feature detector cells, which in turn connect by modifiable weights to the response cells in the output layer. The part which corresponds to the delta-rule perceptron described above is the association cell-to-response cell mapping through the modifiable weights. The association cell layer can be thought of as a preprocessor.

Each sensory cell receives as input stimulus either 1 or 0. This excitation is passed on to the association cells (or feature detector cells as in Fig. 5.2) with either a 1 or –1 multiplying factor (equivalent to a synaptic weight). If the input to the association cell exceeds 0, the cell fires and outputs 1; if not, it outputs 0. The association cell layer output is passed on to the response cells in the output layer, through modifiable weights w_{ij}, which can take any value, positive or negative. (Association cell A_j connects to response cell R_i through weight w_{ij}.) Each response cell sums its total input and if it exceeds a threshold, the response cell outputs a 1; if not, it outputs 0. Sensory input patterns are in class 1 for response cell R_i if they cause the response cell to fire, in class 0 if they do not.

The algorithm given for training the original perceptron as a pattern classifier was as follows. (The delta rule given above is a general formulation of what the following training algorithm achieves.) The perceptron is given a set of input patterns, and for each input pattern the output of the perceptron is compared with what is desired, and the weights are adjusted according to a rule of the following type (formulated for response cell R_i):

1. If a pattern is incorrectly classified (by cell R_i) in class 0 when it should be in class 1, increase by a constant all the weights (to cell R_i) from the association cells that are active.
2. If a pattern is incorrectly classified in class 1 when it should be in class 0, decrease by a constant all the weights coming from association cells that are active.
3. If a pattern is correctly classified, do not change any weights.

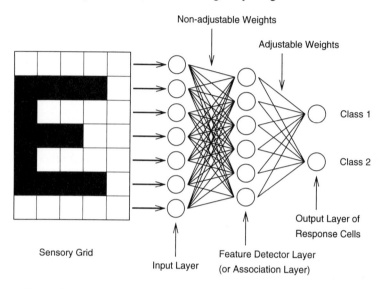

Fig. 5.2 A schematic diagram of Rosenblatt's conceptualization of a perceptron applied to vision. The sensory stimulus activates an input layer which is connected to a feature detector (or association) layer formed from combinations of the inputs. The feature detector layer is connected to the output layer (which in this case has two neurons, which indicate whether the input stimulus is in Class 1 or Class 2) by a layer of adjustable weights trained by the perceptron learning rule.

Operating under these rules, the perceptron will gradually make fewer and fewer wrong classifications and (under certain restricted conditions—see Minsky and Papert, 1969) will classify every pattern in the set correctly. In this scheme, each response cell must be told how to adjust its weights for each input pattern. The learning scheme works with single and with multiple output cells (see Fig. 5.2, in which there are two output cells), and can learn to produce different output response patterns to different input stimulus patterns. Each cell needs its own teacher to supervise the learning of that cell.

5.1.3.2 Adaline

The adaline (<u>ada</u>ptive <u>lin</u>ear <u>e</u>lement) of Widrow and Hoff (1960) used binary (1, −1) inputs and outputs, and a binary threshold output activation function. The learning rule was expressed in the more general form of Eq. 5.2, and this rule is thus often called the Widrow–Hoff rule.

5.1.3.3 Adaptive filter

The adaptive filter of Widrow *et al.* (1976) (see Widrow and Lehr, 1990) used continuous inputs and outputs, and a linear activation function. Its rule is equivalent to Eq. 5.2 above,

$$\delta w_{ij} = k\,(t_i - h_i)\,r'_j$$

This approach was developed in engineering, and enables an optimal linear filter to be set up by learning. With multiple output cells, the filter was known as a Madaline.

In general, networks taught by the delta rule may have linear, binary threshold, or non-linear but monotonically increasing (e.g. sigmoid) activation functions, and may be taught with binary or continuous input patterns. The properties of these variations are made clear next.

5.1.3.4 Capability and limitations of single-layer error-correcting networks

Perceptrons perform pattern classification. That is, each neuron classifies the input patterns it receives into classes determined by the teacher. This is thus an example of a supervised network, with a separate teacher for each output neuron. The classification is most clearly understood if the output neurons are binary, or are strongly non-linear, but the network will still try to obtain an optimal mapping with linear or near-linear output neurons.

When each neuron operates as a binary classifier, we can consider how many input patterns p can be classified by each neuron, and the classes of pattern that can be correctly classified. The result is that the maximum number of patterns that can be correctly classified by a neuron with C inputs is

$$p_{\max} = 2C$$

when the inputs have random continuous-valued inputs, but the patterns must be linearly separable (see Hertz *et al.*, 1991). The linear separability requirement can be made clear by considering a geometric interpretation of the logical AND problem, which is linearly separable, and the XOR problem, which is not linearly separable. The truth tables for the AND and XOR functions are shown first (there are two inputs, r'_1 and r'_2, and one output):

Inputs		Output	
r'_1	r'_2	AND	XOR
0	0	0	0
1	0	0	1
0	1	0	1
1	1	1	0

For the AND function, we can plot the mapping required in a 2D graph as shown in Fig. 5.3. A line can be drawn to separate the input coordinates for which 0 is required as the output from those for which 1 is required as the output. The problem is thus linearly separable. A neuron with two inputs can set its weights to values which draw this line through this space, and such a one-layer network can thus solve the AND function.

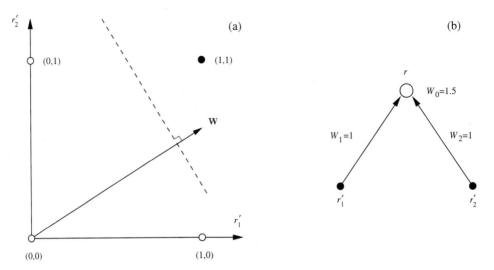

Fig. 5.3 (a) The AND function shown in a 2D space. Input values for the two neurons are shown along the two axes of the space. The outputs required are plotted at the coordinates where the inputs intersect, and the values of the output required are shown as an open circle for 0, and a filled circle for 1. The AND function is linearly separable, in that a line can be drawn in the space which separates the coordinates for which 0 output is required from those from which a 1 output is required. **w** shows the direction of the weight vector. (b) A one-layer neural network can set its two weights w_1 and w_2 to values which allow the output neuron to be activated only if both inputs are present. In this diagram, w_0 is used to set a threshold for the neuron, and is connected to an input with value 1. The neuron thus fires only if the threshold of 1.5 is exceeded, which happens only if both inputs to the neuron are 1.

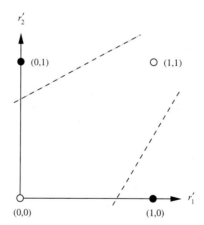

Fig. 5.4 The XOR function shown in a 2D space. Input values for the two neurons are shown along the two axes of the space. The outputs required are plotted at the coordinates where the inputs intersect, and the values of the output required are shown as an open circle for 0, and a filled circle for 1. The XOR function is not linearly separable, in that a line cannot be drawn in the space to separate the coordinates for those from which a 0 output is required from those from which a 1 output is required. A one-layer neural network cannot set its two weights to values which allow the output neuron to be activated appropriately for the XOR function.

For the XOR function, we can plot the mapping required in a 2D graph as shown in Fig. 5.4. No straight line can be drawn to separate the input coordinates for which 0 is required as the output from those for which 1 is required as the output. The problem is thus not linearly separable. For a one-layer network, no set of weights can be found that will perform the XOR, or any other non-linearly separable function.

Although the inability of one-layer networks with binary neurons to solve non-linearly separable problems is a limitation, it is not in practice a major limitation on the processing which can be performed in a neural network for a number of reasons. First, if the inputs can take continuous values, then if the patterns are drawn from a random distribution, the one-layer network can map up to $2C$ of them. Second, as described for pattern associators, the perceptron could be preceded by an expansion recoding network such as a competitive network with more output than input neurons. This effectively provides a two-layer network for solving the problem, and (non-linear) multilayer networks are in general capable of solving arbitrary mapping problems. Ways in which such multilayer networks might be trained are discussed later in this chapter.

We now return to the issue of the capacity of one-layer perceptrons, that is how many patterns p can be correctly mapped to correct binary outputs if the input patterns are linearly separable.

Output neurons with continuous values, random patterns

Before treating this case we note that if the inputs are orthogonal, then just as in the pattern associator, C patterns can be correctly classified, where there are C inputs, r'_j, $(j = 1,C)$, per neuron. The argument is the same as for a pattern associator.

We consider next the capacity of a one-layer error-correcting network which learns patterns drawn from a random distribution. For neurons with continuous output values, whether the activation function is linear or not, the capacity (for fully distributed inputs) is set by the criterion that the set of input patterns must be linearly independent (see Hertz et al., 1991). (Three patterns are linearly independent if any one cannot be formed by addition (with scaling allowed) of the other two patterns—see Appendix A1.) Given that there can be a maximum of C linearly independent patterns in a C-dimensional space (see Appendix A1), the capacity of the perceptron with such patterns is C patterns. If we choose p random patterns with continuous values, then they will be linearly independent for $p \leqslant C$ (except for cases with very low probability when the randomly chosen values may not produce linearly independent patterns). (With random continuous values for the input patterns, it is very unlikely that the addition of any two, with scaling allowed, will produce a third pattern in the set.) Thus with continuous valued input patterns,

$$p_{max} = C.$$

If the inputs are not linearly independent, networks trained with the delta rule produce a least mean squares (LMS) error (optimal) solution (see below).

Output neurons with binary threshold activation functions

Let us consider here strictly defined perceptrons, that is networks with (binary) threshold output neurons, and taught by the perceptron learning procedure described above.

Capacity with fully distributed output patterns

The condition here for correct classification is that described above for the AND and XOR functions, that the patterns must be linearly separable. If we consider random continuous-valued inputs, then the capacity is

$$p_{max} = 2C$$

(see Cover, 1965; Hertz *et al.*, 1991; this capacity is the case with *C* large, and the number of output neurons small). The interesting point to note here is that, even with fully distributed inputs, a perceptron is capable of learning more (fully distributed) patterns than there are inputs per neuron. This formula is in general valid for large *C*, but happens to hold also for the AND function illustrated above.

Sparse encoding of the patterns

If the output patterns **r** are sparse (but still distributed), then just as with the pattern associator, it is possible to map many more than *C* patterns to correct outputs. Indeed, the number of different patterns or prototypes *p* that can be stored is

$$p \approx C/a$$

where *a* is the sparseness of the target pattern **t**. *p* can in this situation be much larger than *C* (cf. Rolls and Treves, 1990, and Appendix A3).

Perceptron convergence theorem

It can be proved that such networks will learn the desired mapping in a finite number of steps (Block, 1962; Minsky and Papert, 1969; see Hertz *et al.*, 1991). (This of course depends on there being such a mapping, the condition for this being that the input patterns are linearly separable.) This is important, for it shows that single-layer networks can be proved to be capable of solving certain classes of problem.

As a matter of history, Minsky and Papert (1969) went on to emphasize the point that no one-layer network can correctly classify non-linearly separable patterns. Although it was clear that multilayer networks can solve such mapping problems, Minsky and Papert were pessimistic that an algorithm for training such a multilayer network would be found. Their emphasis that neural networks might not be able to solve general problems in computation, such as computing the XOR, which is a non-linearly separable mapping, resulted in a decline in research activity in neural networks. In retrospect, this was unfortunate, for humans are rather poor at solving parity problems such as the XOR (Thorpe *et al.*, 1989), yet can perform many other useful neural network operations very fast. Algorithms for training multilayer perceptrons were gradually discovered by a number of different investigators, and became widely known after the publication of the algorithm described by Rumelhart, Hinton and Williams (1986a,b). Even before this, interest in neural network pattern associators, autoassociators and competitive networks was developing (see Hinton and Anderson, 1981; Kohonen, 1977, 1988), but the acceptance of the algorithm for training multilayer perceptrons led to a great rise in interest in neural networks, partly for use in connectionist models of cognitive function (McClelland and Rumelhart, 1986; McLeod, Plunkett and Rolls, 1998), and partly for use in applications (see Bishop, 1995).

In that perceptrons can correctly classify patterns provided only that they are linearly separable, but pattern associators are more restricted (see Chapter 2), perceptrons are more powerful learning devices than Hebbian pattern associators.

Gradient descent for neurons with continuous-valued outputs

We now consider networks trained by the delta (error correction) rule (Eq. 5.2), and having continuous valued outputs. The output activation function may be linear or non-linear, but provided that it is differentiable (in practice, does not include a sharp threshold), the network can be thought of as gradually decreasing the error on every learning trial, that is as performing some type of gradient descent down a continuous error function. The concept of gradient descent arises from defining an error ε for a neuron as

$$\varepsilon = \Sigma_\mu (t^\mu - r^\mu)^2 \tag{5.3}$$

where μ indexes the patterns learned by the neuron. The error function for a neuron in the direction of a particular weight would have the form shown in Fig. 5.5. The delta rule can be conceptualized as performing gradient descent of this error function, in that for the jth synaptic weight on the neuron

$$\delta w_j = -k\, \partial \varepsilon / \partial w_j \tag{5.4}$$

where $\partial \varepsilon / \partial w_j$ is just the slope of the error curve in the direction of w_j in Fig. 5.5. This will decrease the weight if the slope is positive and increase the weight if the slope is negative. Given (5.3), and recalling that $h = \Sigma_j r'_j w_j$, Eq. 5.4 becomes

$$\delta w_j = -k\, \partial / \partial w_j \Sigma_\mu [(t^\mu - f(h^\mu))^2] = 2k \Sigma_\mu [(t^\mu - r^\mu)]\, f'(h)\, r'_j \tag{5.5}$$

where $f'(h)$ is the derivative of the activation function. Provided that the activation function is monotonically increasing, its derivative will be positive, and the sign of the weight change will only depend on the mean sign of the error. Equation 5.5 thus shows one way in which, from a gradient descent conceptualization, Eq. 5.2 can be derived.

With linear output neurons, this gradient descent is proved to reach the correct mapping (see Hertz et al., 1991). (As with all single-layer networks with continuous-valued output neurons, a perfect solution is only found if the input patterns are linearly independent. If they are not, an optimal mapping is achieved, in which the sum of the squares of the errors is a minimum.) With non-linear output neurons (for example with a sigmoid activation function), the error surface may have local minima, and is not guaranteed to reach the optimal solution, although typically a near-optimal solution is achieved. Part of the power of this gradient descent conceptualization is that it can be applied to multilayer networks with neurons with non-linear but differentiable activation functions, for example with sigmoid activation functions (see Hertz et al., 1991).

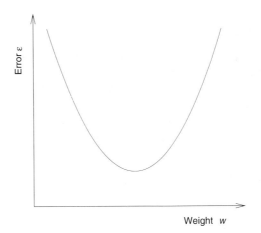

Fig. 5.5 The error (ε) function for a neuron in the direction of a particular weight w.

5.1.4 Properties

The properties of single-layer networks trained with a delta rule (and of perceptrons) are similar to those of pattern associators trained with a Hebbian rule in many respects (see Chapter 2). In particular, the properties of generalization and graceful degradation are similar, provided that (for both types of network) distributed representations are used. The main differences are in the types of pattern that can be separated correctly, the learning speed (in that delta-rule networks can take advantage of many training trials to learn to separate patterns that could not be learned by Hebbian pattern associators), and in that the delta-rule network needs an error term to be supplied for each neuron, whereas an error term does not have to be supplied for a pattern associator, just an unconditioned or forcing stimulus. Given these overall similarities and differences, the properties of one-layer delta-rule networks are considered here briefly.

5.1.4.1 Generalization

During recall, delta-rule one-layer networks with non-linear output neurons produce appropriate outputs if a recall cue vector r'_r is similar to a vector that has been learned already. This occurs because the recall operation involves computing the dot (inner) product of the input pattern vector r'_r with the synaptic weight vector w_i, so that the firing produced, r_i, reflects the similarity of the current input to the previously learned input pattern r'. Distributed representations are needed for this property. If two patterns that a delta-rule network has learned to separate are very similar, then the weights of the network will have been adjusted to force the different outputs to occur correctly. At the same time, this will mean that the way in which the network generalizes in the space between these two vectors will be very sharply defined. (Small changes in the input vector will force it to be classified one way or the other.)

5.1.4.2 Graceful degradation or fault tolerance

One-layer delta-rule networks show graceful degradation provided that the input patterns $\mathbf{r}'$ are distributed. Just as for generalization, graceful degradation may appear to be limited if the network has had to learn to separate two very similar patterns which differ in only a few inputs r'_j by a small amount. However, in this situation the input patterns can be thought of as being, in so far as their difference is concerned, rather non-distributed, so this is not a real exception to this general property of graceful degradation.

5.1.4.3 Prototype extraction, extraction of central tendency, and noise reduction

This occurs as for pattern autoassociators.

5.1.4.4 Speed

Recall is very fast in a one-layer pattern associator or perceptron, because it is a feedforward network (with no recurrent or lateral connections). Recall is also fast if the neuron has cell-like properties, because the stimulus input firings r'_j ($j = 1,C$ axons) can be applied simultaneously to the synapses w_{ij}, and the activation h_i can be accumulated in one or two time constants of the dendrite (e.g. 10–20 ms). Whenever the threshold of the cell is exceeded, it fires. Thus, in effectively one time step, which takes the brain no more than 10–20 ms, all the output neurons of the delta-rule network can be firing with rates that reflect the input firing of every axon.

Learning is as fast ('one-shot') in perceptrons as in pattern associators if the input patterns are orthogonal. If the patterns are not orthogonal, so that the error-correction rule has to work to separate patterns, then the network may take many trials to achieve the best solution (which will be perfect under the conditions described above).

5.1.4.5 Non-local learning rule

The learning rule is not truly local, as it is in pattern associators, autoassociators, and competitive networks, in that with one-layer delta-rule networks, the information required to change each synaptic weight is not available in the presynaptic terminal (reflecting the presynaptic rate) and the postsynaptic activation. Instead, an error for the neuron must be computed, possibly by another neuron, and then this error must be conveyed back to the postsynaptic neuron to provide the postsynaptic error term, which together with the presynaptic rate determines how much the synapse should change, as in Eq. 5.2:

$$\delta w_{ij} = k\,(t_i - r_i)\,r'_j$$

where $(t_i - r_i)$ is the error.

A rather special architecture would be required if the brain were to utilize delta-rule error-correcting learning. One such architecture might require each output neuron to be supplied with its own error signal by another neuron. The possibility that this is implemented in one part of the brain, the cerebellum, is introduced in Chapter 9. Another functional architecture would require each neuron to compute its own error by subtracting its current activation by its r' inputs from another set of afferents providing the target activation for that neuron. A neurophysiological architecture and mechanism for this is not currently known.

5.1.4.6 Interference

Interference is less of a property of single-layer delta rule networks than of pattern autoassociators and autoassociators, in that delta rule networks can learn to separate patterns even when they are highly correlated. However, if patterns are not linearly independent, then the delta rule will learn a least mean squares solution, and interference can be said to occur.

5.1.4.7 Expansion recoding

As with pattern associators and autoassociators, expansion recoding can separate input patterns into a form which makes them learnable, or which makes learning more rapid with only a few trials needed, by delta rule networks. It has been suggested that this is the role of the granule cells in the cerebellum, which provide for expansion recoding by 1000: 1 of the mossy fibre inputs before they are presented by the parallel fibres to the cerebellar Purkinje cells (Marr, 1969; Albus, 1971; see Chapter 9).

5.1.5 Utility of single-layer error-correcting networks in information processing by the brain

In the cerebellum, each output cell, a Purkinje cell, has its own climbing fibre, which distributes from its inferior olive cell its terminals throughout the dendritic tree of the Purkinje cell. It is this climbing fibre which controls whether learning of the r' inputs supplied by the parallel fibres onto the Purkinje cell occurs, and it has been suggested that the function of this architecture is for the climbing fibre to bring the error term to every part of the postsynaptic neuron (see Chapter 9). This rather special arrangement with each output cell apparently having its own teacher is probably unique in the brain, and shows the lengths to which the brain might need to go to implement a teacher for each output neuron. The requirement for error-correction learning is to have the neuron forced during a learning phase into a state which reflects its error while presynaptic afferents are still active, and rather special arrangements are needed for this.

5.2 Multilayer perceptrons: backpropagation of error networks

5.2.1 Introduction

So far, we have considered how error can be used to train a one-layer network using a delta rule. Minsky and Papert (1969) emphasized the fact that one-layer networks cannot solve certain classes of input–output mapping problems (as described above). It was clear then that these restrictions would not apply to the problems that can be solved by feedforward multilayer networks, if they could be trained. A multilayer feedforward network has two or more connected layers, in which connections allow activity to be projected forward from one layer to the next, and in which there are no lateral connections within a layer. Such a multilayer network has an output layer (which can be trained with a standard delta rule using an error provided for each output neuron), and one or more hidden layers, in which the neurons do not receive separate error signals from an external teacher. (Because they do not provide the outputs of the network directly, and do not directly receive their own teaching error signal, these layers are described as hidden.) To solve an arbitrary mapping problem (in which the inputs are not linearly separable), a multilayer network could have a set of hidden neurons which would remap the inputs in such a way that the output layer can be provided with a linearly separable problem to solve using training of its weights with the delta rule. The problem was, how could the synaptic weights into the hidden neurons be trained in such a way that they would provide an appropriate representation? Minsky and Papert (1969) were pessimistic that such a solution would be found, and partly because of this, interest in computations in neural networks declined for many years. Although some work in neural networks continued in the following years (e.g. Marr, 1969, 1970, 1971; Willshaw and Longuet-Higgins, 1969, see Willshaw, 1981; von der Malsburg, 1973; Grossberg, 1976a,b; Arbib, 1964, 1987; Amari, 1982; Amari *et al.*, 1977), widespread interest in neural networks was revived by the type of approach to associative memory and its relation to human memory taken by the work described in the volume edited by Hinton and Anderson (1981), and by Kohonen (1984). Soon after this, a solution to training a multilayer perceptron using backpropagation of error became widely known (Rumelhart, Hinton and Williams, 1986a,b) (although earlier solutions had been found), and very great interest in neural networks and also in neural network approaches to cognitive processing (connectionism) developed (Rumelhart and McClelland, 1986; McClelland and Rumelhart, 1986).

5.2.2 Architecture and algorithm

An introduction to the way in which a multilayer network can be trained by backpropagation of error is described next. Then we consider whether such a training algorithm is biologically plausible. A more formal account of the training algorithm for multilayer perceptrons (sometimes abbreviated MLP) is given by Rumelhart, Hinton and Williams, 1986a,b).

Consider the two-layer network shown in Fig. 5.6. Inputs to the hidden neurons in layer A feed forward activity to the output neurons in layer B. The neurons in the network have a sigmoid activation function. One reason for such an activation function is that it is non-linear, and non-linearity is needed to enable multilayer networks to solve difficult (non-linearly separable) problems. (If the neurons were linear, the multilayer network would be equivalent to a one-layer network, which cannot solve such problems.) Neurons B1 and B2 of the output layer, B, are each trained using a delta rule and an error computed for each output neuron from the target output for that neuron when a given input pattern is being applied to the network. Consider the error which needs to be used to train neuron A1 by a delta rule. This error clearly influences the error of neuron B1 in a way which depends on the magnitude of the synaptic weight from neuron A1 to B1; and on the error of neuron B2 in a way that depends on the magnitude of the synaptic weight from neuron A1 to B2. In other words, the error for neuron A1 depends on

the weight from A1 to B1 (w_{11}) · error of neuron B1
+ the weight from A1 to B2 (w_{21}) · error of neuron B2.

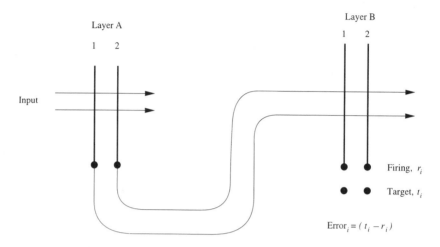

Fig. 5.6 A two-layer perceptron. Inputs are applied to layer A through modifiable synapses. The outputs from layer A are applied through modifiable synapses to layer B. Layer B can be trained using a delta rule to produce firing (r_i) which will approach the target t_i. It is more difficult to modify the weights in layer A, because appropriate error signals must be backpropagated from layer B.

In this way, the error calculation can be propagated backwards through the network to any neuron in any hidden layer, so that each neuron in the hidden layer can be trained, once its error is computed, by a delta rule. For this to work, the way in which each neuron is activated and sends a signal forward must be continuous (not binary), so that the extent to which there is an error in, for example, neuron B1 can be related back in a graded way to provide a continuously variable correction signal to previous stages. This is one of the requirements for enabling the network to descend a continuous error surface. The activation function must be non-linear (e.g. sigmoid) for the network to learn more than could be learned by a single-layer network. (Remember that a multilayer linear network can always be made equivalent to a single-layer linear network, and that there are some problems that cannot be solved by single-layer networks.) For the way in which the error of each output neuron should be taken into account to be specified in the error-correction rule, the position at which the output neuron is operating on its output activation function must also be taken into account. For this, the slope of the activation function is needed, and because the slope is needed, the activation function must be differentiable. Although we indicated use of a sigmoid activation function, other activation functions which are non-linear and monotonically increasing (and differentiable) can be used. (For further details, see Rumelhart, Hinton and Williams, 1986a,b and Hertz *et al.*, 1991).

5.2.3 Properties of multilayer networks trained by error backpropagation

5.2.3.1 Arbitrary mappings

Arbitrary mappings of non-linearly separable patterns can be achieved. For example, such networks can solve the XOR problem, and parity problems in general of which XOR is a special case. (The parity problem is to determine whether the sum of the (binary) bits in a vector is odd or even.) Multilayer feedforward backpropagation of error networks are not guaranteed to converge to the best solution, and may become stuck in local minima in the error surface. However, they generally perform very well.

5.2.3.2 Fast operation

The network operates as a feedforward network, without any recurrent or feedback processing. Thus (once it has learned) the network operates very fast, with a time proportional to the number of layers.

5.2.3.3 Learning speed

The learning speed can be very slow, taking many thousands of trials. The network learns to gradually approximate the correct input–output mapping required, but the learning is slow because of the credit assignment problem for neurons in the hidden layers. The credit assignment problem refers to the issue of how much to correct the weights of each neuron in the hidden layer. As the example above shows, the error for a hidden neuron could influence the errors of many neurons in the output layers, and the error of each output neuron reflects the error from many hidden neurons. It is thus difficult to assign credit (or blame) on any single trial to any particular hidden neuron, so an error must be estimated, and the net run until the weights of the crucial hidden neurons have become altered sufficiently to allow good performance of the network. Another factor which can slow learning is that if a neuron operates close to a horizontal part of its activation function, then the output of the neuron will depend rather little on its activation, and correspondingly the error computed to backpropagate will depend rather little on the activation of that neuron, so learning will be slow.

More general approaches to this issue suggest that the number of training trials for such a network will (with a suitable training set) be of the same order of magnitude as the number of synapses in the network (see Cortes *et al.*, 1996).

5.2.3.4 Number of hidden neurons and generalization

Backpropagation networks are generally intended to discover regular mappings between the input and output, that is mappings in which generalization will occur usefully. If there were one hidden neuron for every combination of inputs that had to be mapped to an output, then this would constitute a look-up table, and no generalization between similar inputs (or inputs not yet received) would occur. The best way to ensure that a backpropagation network learns the structure of the problem space is to set the number of neurons in the hidden layers close to the minimum that will allow the mapping to be implemented. This forces the network not to operate as a look-up table. A problem is that there is no general rule about how many hidden neurons are appropriate, given that this depends on the types of mappings required. In practice, these networks are sometimes trained with different numbers of hidden neurons, until the minimum number required to perform the required mapping has been approximated.

A problem with this approach in a biological context is that in order to achieve their competence, backpropagation networks use what is almost certainly a learning rule which is much more powerful than those which could be implemented biologically, and achieves its excellent performance by performing the mapping though a minimal number of hidden neurons. In contrast, real neuronal networks in the brain probably use much less powerful learning rules, in which errors are not propagated backwards, and at the same time have very large numbers of hidden neurons, without the bottleneck which helps to provide back-propagation networks with their good performance. A consequence of these differences between backpropagation and biologically plausible networks may be that the way in which biological networks solve difficult problems may be rather different from the way in which backpropagation networks find mappings. Thus the solutions found by connectionist systems may not always be excellent guides to how biologically plausible networks may perform on similar problems. Part of the challenge for future work is to discover how more biologically plausible networks than backpropagation networks can solve comparably hard problems, and then to examine the properties of these networks, as a perhaps more accurate guide to brain computation.

5.2.3.5 Non-local learning rule

Given that the error for a hidden neuron is calculated by propagating backwards information based on the errors of all the output neurons to which a hidden neuron is connected, and all the relevant synaptic weights, and the activations of the output neurons to define the part of the activation function on which they are operating, it is implausible to suppose that the correct information to provide the appropriate error for each hidden neuron is propagated backwards between real neurons. A hidden neuron would have to 'know', or receive information about, the errors of all the neurons to which it is connected, and its synaptic weights to them, and their current activations. If there were more than one hidden layer, this would be even more difficult. To expand on the difficulties: first, there would have to be a mechanism in the brain for providing an appropriate error signal to each output neuron in the network. With the possible exception of the cerebellum, an architecture where a separate error signal could be provided for each output neurons is difficult to identify in the brain. Second, any retrograde passing of messages across multiple-layer forward-transmitting pathways in the brain that could be used for backpropagation seems highly implausible, not only because of the difficulty of getting the correct signal to be backpropagated, but also because retrograde signals in a multilayer net would take days to arrive, long after the end of any feedback given in the environment indicating a particular error. Third, as noted in Section 10.2, the backprojection pathways that are present in the cortex seem suited to perform recall, and this would make it difficult for them also to have the correct strength to carry the correct error signal.

As stated above, it is a major challenge for brain research to discover whether there are algorithms that will solve comparably difficult problems to backpropagation, but with a local learning rule. Such algorithms may be expected to require many more hidden neurons than backpropagation networks, in that the brain does not appear to use information bottlenecks to help it solve difficult problems. The issue here is that much of the power of back-propagation algorithms arises because there is a minimal number of hidden neurons to perform the required mapping using a final one-layer delta-rule network. Useful generalization arises in such networks because with a minimal number of hidden neurons, the net sets the representation they provide to enable appropriate generalization. The danger with more hidden neurons is that the network becomes a look-up table, with one hidden neuron for every required output, and generalization when the inputs vary becomes poor. The challenge is to find a more biologically plausible type of network which operates with large numbers of neurons, and yet which still provides useful generalization.

5.3 Reinforcement learning

One issue with perceptrons and multilayer perceptrons that makes them generally biologically implausible for many brain regions is that a separate error signal must be supplied for each output neuron. When operating in an environment, usually a simple binary or scalar signal representing success or failure is received. Partly for this reason, there has been some interest in networks that can be taught with such a single reinforcement signal. In this section, we describe one such approach to such networks. We note that such networks are classified as supervised networks in which there is a single teacher, and that these networks attempt to perform an optimal mapping between an input vector and an output neuron or set of neurons. They thus solve the same class of problems as single and (if multilayer) multilayer perceptrons. They should be distinguished from pattern association networks in the brain, which might learn associations between previously neutral stimuli and primary reinforcers such as taste (signals which might be interpreted appropriately by a subsequent part of the brain), but do not attempt to produce arbitrary mappings between an input and an output, using a single reinforcement signal. A class of problems to which such networks might be applied are motor control problems. It was to such a problem that Barto and Sutton (Barto, 1985; Sutton and Barto, 1981) applied their reinforcement learning algorithm.

5.3.1 Associative reward–penalty algorithm of Barto and Sutton

The terminology of Barto and Sutton is followed here (see Barto, 1985).

5.3.1.1 Architecture

The architecture, shown in Fig. 5.7, uses a single reinforcement signal, r, $= +1$ for reward, and -1 for penalty. The inputs x_i take real (continuous) values. The output of a neuron, y, is binary, $+1$ or 1. The weights on the output neuron are designated w_i.

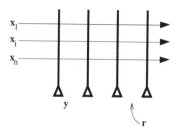

Fig. 5.7 A network trained by a single reinforcement input r. The inputs to each neuron are in the terminology used by Barto and Sutton x_i, i = 1,n; and y is the output of one of the output neurons.

5.3.1.2 Operation

1. An input vector is applied to the network, and produces output activation, s, in the normal way as follows:

$$s = \Sigma \, w_i \, x_i$$

where x_i is the firing of the input axon i.
2. The output is calculated from the activation with a noise term η included. The principle of the network is that if the added noise on a particular trial helps performance, then whatever change it leads to should be incorporated into the synaptic weights, in such a way that the next time that input occurs, the performance is improved.

$$\text{Output } y = +1 \text{ if } s + \eta > 0$$
$$= -1 \text{ else}$$

where η = the noise added on each trial.
3. Learning rule. The weights are changed as follows:

$$\delta w_i = \rho(y - E[y|s])x_i \quad \text{if } r = +1$$
$$\delta w_i = \rho\lambda(-y - E[y|s])x_i \text{ if } r = -1.$$

ρ and λ are learning rate constants. (They are set so that the learning rate is higher when positive reinforcement is received than when negative reinforcement is received.) E[y|s] is the expectation of y given s (usually a sigmoidal function of s with the range ±1). E[y|s] is a (continuously varying) indication of how the unit usually responds to the current input pattern, i.e. if the actual output y is larger than normally expected, by computing s = Σ $w_i x_i$, because of the noise term, and the reinforcement is +1, increase the weight from x_i; and vice versa. The expectation could be the prediction generated before the noise term is incorporated.

This network combines an associative capacity with its properties of generalization and graceful degradation, with a single 'critic' or error signal for the whole network (Barto, 1985). The network can solve difficult problems (such as balancing a pole by moving a trolley which supports the pole from side to side, as the pole starts to topple). Although described for single-layer networks, the algorithm can be applied to multilayer networks. The learning rate is very slow, for there is a single reinforcement signal on each trial for the whole network, not a separate error signal for each neuron in the network.

An important advance in the area of reinforcement learning was the introduction of algorithms that allow for learning to occur when the reinforcement is delayed or received over a number of time steps. A solution to this problem is the addition of an adaptive critic that learns through a time difference (TD) algorithm how to predict the future value of the reinforcer. The output of the critic is used as an effective reinforcer instead of the instantaneous reinforcement being received (which reflects previous performance) (see Sutton and Barto, 1990; Barto, 1995). This is a solution to the temporal credit assignment problem. The algorithm has been applied to modelling the time course of classical conditioning (Sutton and Barto, 1990).

The reinforcement learning algorithm is certainly a move towards biological relevance, in that learning with a single reinforcer can be achieved. That single reinforcer might be broadcast throughout the system by a general projection system, such as the dopamine pathways in the brain, which distribute to large parts of the striatum and the prefrontal cortex. It is not clear yet how a biological system might store the expected output E[y|s] for comparison with the actual output when noise has been added, and might take into account the sign and magnitude of the noise. Nevertheless, this is an interesting algorithm.

5.4 Contrastive Hebbian learning: the Boltzmann machine

In a move towards a learning rule which is more local than in backpropagation networks, yet which can solve similar mapping problems in a multilayer architecture, we describe briefly contrastive Hebbian learning. The multilayer architecture has forward connections through the network to the output layer, and a set of matching backprojections from the output layer through each of the hidden layers to the input layer. The forward connection strength between any pair of neurons has the same value as the backward connection strength between the same two neurons, resulting in a symmetric set of forward and backward connection strengths. An input pattern is applied to the multilayer network, and an output is computed using normal feedforward activation processing with neurons with a sigmoid (non-linear and monotonically increasing) activation function. The output firing then via the backprojections is used to create firing of the input neurons. This process is repeated until the firing rates settle down, in an iterative way (which is similar to the settling of the autoassociative nets described in Chapter 3). After settling, the correlations between any two neurons are remembered, for this type of unclamped operation, in which the output neurons fire at the rates that the process just described produces. The correlations reflect the normal presynaptic and postsynaptic terms used in the Hebb rule, e.g. $(r'_j r_i)^{uc}$, where uc refers to the unclamped condition, and as usual r'_j is the firing rate of the input neuron, and r is the activity of the receiving neuron. The output neurons are then clamped to their target values, and the iterative process just described is repeated, to produce for every pair of synapses in the network $(r'_j r_i)^c$, where the c refers now to the clamped condition. An error-correction term for each synapse is then computed from the difference between the remembered correlation of the unclamped and the clamped conditions, to produce a synaptic weight correction term as follows:

$$\delta w_{ij} = k[(r'_j r_i)^c - (r'_j r_i)^{uc}]$$ (5.6)

where k is a learning rate constant. This process is then repeated for each input pattern to output pattern to be learned. The whole process is then repeated many times with all patterns until the output neurons fire similarly in the clamped and unclamped conditions, that is until the errors have become small. Further details are provided by Hinton and Sejnowski (1986).

The version described above is the mean field (or deterministic) Boltzmann machine (Peterson and Anderson, 1987; Hinton, 1989). More traditionally, a Boltzmann machine updates one randomly chosen neuron at a time, and each neuron fires with a probability that depends on its activation (Ackley, Hinton and Sejnowski, 1985; Hinton and Sejnowski, 1986). The latter version makes fewer theoretical assumptions, while the former may operate an order of magnitude faster. An application of this type of network is described by Baddeley (1995).

In terms of biological plausibility, it certainly is the case that there are backprojections between adjacent cortical areas (see Rolls, 1989a–e). Indeed, there are as many backprojections between adjacent cortical areas as there are forward projections. The backward projections seem to be more diffuse than the forward projections, in that they connect to a wider region of the preceding cortical area than the region which sends the forward projections. If the backward and the forward synapses in such an architecture were Hebb-modifiable, then there is a possibility that the backward connections would be symmetric with the forward connections. Indeed, such a connection scheme would be useful to implement top-down recall, as described in Chapter 6. What seems less biologically plausible is that after an unclamped phase of operation, the correlations between all pairs of neurons would be remembered, there would then be a clamped phase of operation with *each* output neuron clamped to the required rate for that particular input pattern, and then the synapses would be corrected by an error-correction rule that would require a comparison of the correlations between the neuronal firing of every pair of neurons in the unclamped and clamped conditions.

5.5 Conclusion

The networks described in this chapter are capable of mapping a set of inputs to a set of required outputs using correction when errors are made. Although some of the networks are very powerful in the types of mapping they can perform, the power is obtained at the cost of learning algorithms which do not use local learning rules. Because the networks described in this chapter do not use local learning rules, their biological plausibility remains at present uncertain, although it has been suggested that perceptron learning may be implemented by the special neuronal network architecture of the cerebellum (see Chapter 9). One of the aims of future research must be to determine whether comparably difficult problems to those solved by the networks described in this chapter can be solved by biologically plausible neuronal networks (see e.g. Dayan and Hinton, 1996; O'Reilly, 1996).

6   The hippocampus and memory

In this chapter we show how it is becoming possible to link quantitative neuronal network approaches to other techniques in neuroscience to develop quantitative theories about how brain systems involved in memory operate. The particular brain system considered is the hippocampus and nearby structures in the temporal lobe of the brain, which are involved in learning about new events or facts. The primary aim of this chapter is to show how it is possible to link these approaches, and to produce a theory of the operation of a brain structure. Although of course the particular theory of hippocampal function described here is subject to revision with future empirical as well as theoretical research, the aim is nevertheless to show how evidence from many different disciplines can now be combined to produce a theory of *how* a part of the brain may work. The empirical evidence required comes from: lesion studies which show what function may be performed by a brain region; single and multiple single neuron recording in the hippocampus while it is performing its normal functions, to show what information reaches the hippocampus, and how the information is represented; the external connections of the hippocampus, to help define what information reaches the hippocampus, and which brain regions the hippocampus can influence; the internal connections of the hippocampus, to show what networks are present, and to give quantitative evidence which is important in defining the computational properties of the networks; biophysical studies to define the ways in which neurons operate; and from studies of synaptic modifiability in different parts of the hippocampus and its role in learning, to define the synaptic modification rules at different hippocampal synapses which should be incorporated into models of how the hippocampus works.

6.1 What functions are performed by the hippocampus? Evidence from the effects of damage to the brain

In humans with bilateral damage to the hippocampus and nearby parts of the temporal lobe, amnesia is produced. The amnesia is anterograde, that is for events which happen after the brain damage. Although there is usually some retrograde amnesia, that is for events before the brain damage, this may not be severe, and indeed its severity appears to depend on how much damage there is to the cortical areas adjacent and connected to the hippocampus such as the entorhinal, perirhinal, and parahippocampal cortex. These effects of damage to the hippocampus indicate that the very long-term storage of information is not in the hippocampus, at least in humans.[1] On the other hand, the hippocampus does appear to be necessary for the storage of certain types of information, that have been characterized by the description *declarative*, or *knowing that*, as contrasted with *procedural*, or *knowing how*, which is spared in amnesia. Declarative memory includes what can be declared or brought to mind as a proposition or an image. Declarative memory includes episodic memory, that is memory for particular episodes, and semantic memory, that is memory for facts (Squire, 1992; Squire and Knowlton, 1995; Squire, Shimamura and Amaral, 1989).

Because of the clear importance of the hippocampus and nearby connected structures in memory, there have been many studies to analyse its role in memory. Recent ideas on autoassociation within the hippocampus help to show how the computations being performed in the hippocampus of humans and other animals are closely related. In monkeys, damage to the hippocampus or to some of its connections such as the fornix (see Fig. 6.5) produces deficits in learning about where objects are and where responses must be made (see Rolls, 1990b,d, 1991a,b; 1996c,e). For example, macaques and humans with damage to the hippocampus or fornix are impaired in object–place memory tasks in which not only the objects seen, but where they were seen, must be remembered (Gaffan and Saunders, 1985; Parkinson, Murray and Mishkin, 1988; Smith and Milner, 1981). Such object–place tasks require a whole-scene or snapshot-like memory (Gaffan, 1994). Also, fornix lesions impair conditional left–right discrimination learning, in which the visual appearance of an object specifies whether a response is to be made to the left or the right (Rupniak and Gaffan, 1987). A comparable deficit is found in humans (Petrides, 1985). Fornix sectioned monkeys are also impaired in learning on the basis of a spatial cue which object to choose (e.g. if two objects are on the left, choose object A, but if the two objects are on the right, choose object B) (Gaffan and Harrison, 1989a). Further, monkeys with fornix damage are also impaired in using information about their place in an environment. For example, Gaffan and Harrison (1989b) found learning impairments when which of two or more objects the monkey had to choose depended on the position of the monkey in the room. Rats with hippocampal lesions are impaired in using environmental spatial cues to remember particular places (see O'Keefe and Nadel, 1978; Jarrard, 1993).

It appears that the deficits in 'recognition' memory (e.g. for visual stimuli seen recently) produced by damage to this brain region are related to damage to the perirhinal cortex, which receives from high order association cortex and has connections to the hippocampus (see Fig. 6.1) (Zola-Morgan, Squire, Amaral and Suzuki, 1989; Zola-Morgan *et al.*, 1994; Suzuki and Amaral, 1994a,b).

6.2 Relation between spatial and non-spatial aspects of hippocampal function, episodic memory, and long-term semantic memory

One way of relating the impairment of spatial processing to other aspects of hippocampal function (including the memory of recent events or episodes in humans) is to note that this spatial processing involves a snapshot type of memory, in which one whole scene with its often unique set of parts or elements must be remembered. This memory may then be a

---

[1] This general statement does not exclude, of course, the possibility that individual memories may never become permanently established outside the hippocampus, and thus may appear hippocampal-dependent throughout their retention time.

special case of episodic memory, which involves an arbitrary association of a set of spatial and/or non-spatial events which describe a past episode. Indeed, the non-spatial tasks impaired by damage to the hippocampal system may be impaired because they are tasks in which a memory of a particular episode or context rather than of a general rule is involved (Gaffan, Saunders *et al.*, 1984). Further, the deficit in paired associate learning in humans (see Squire, 1992) may be especially evident when this involves arbitrary associations between words, for example window–lake.

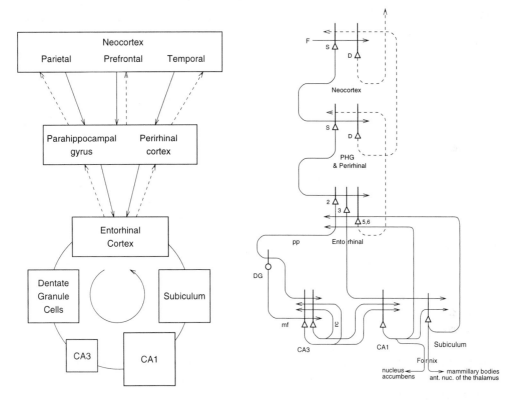

Fig. 6.1 Forward connections (solid lines) from areas of cerebral association neocortex via the parahippocampal gyrus and perirhinal cortex, and entorhinal cortex, to the hippocampus; and backprojections (dashed lines) via the hippocampal CA1 pyramidal cells, subiculum, and parahippocampal gyrus to the neocortex. There is great convergence in the forward connections down to the single network implemented in the CA3 pyramidal cells; and great divergence again in the backprojections. Left: block diagram. Right: more detailed representation of some of the principal excitatory neurons in the pathways. Abbreviations: D, Deep pyramidal cells; DG, dentate granule cells; F, forward inputs to areas of the association cortex from preceding cortical areas in the hierarchy. mf: mossy fibres; PHG, parahippocampal gyrus and perirhinal cortex; pp, perforant path; rc, recurrent collaterals of the CA3 hippocampal pyramidal cells; S, superficial pyramidal cells; 2, pyramidal cells in layer 2 of the entorhinal cortex; 3, pyramidal cells in layer 3 of the entorhinal cortex; 5, 6, pyramidal cells in the deep layers of the entorhinal cortex. The thick lines above the cell bodies represent the dendrites.

Rolls (1987, 1989a,b, 1990e, 1994c, 1996c–e) has suggested that the reason why the hippocampus is used for the spatial and non-spatial types of memory described above, and the reason that makes these two types of memory so analogous, is that the hippocampus contains one stage, the CA3 stage, which acts as an autoassociation memory. It is suggested that an autoassociation memory implemented by the CA3 neurons equally enables whole (spatial) scenes or episodic memories to be formed, with a snapshot quality which depends on the arbitrary associations which can be made and the short temporal window which characterizes the synaptic modifiability in this system (see below and Rolls, 1987, 1989a,b). The hypothesis is that the autoassociation memory enables arbitrary sets of concurrent activity, involving for example the spatial context where an episode occurred, the people present during the episode, and what was seen during the episode, to be associated together and stored as one event. Later recall of that episode from the hippocampus in response to a partial cue can then lead to reinstatement of the activity in the neocortex that was originally present during the episode. The readout of information from the hippocampus could be useful in a number of ways. One is in organizing action based on the stored information, as described later. Another is in providing information useful to build long-term memories, as outlined next.

The information about episodic events recalled from the hippocampus could be used to help form semantic memories (Rolls, 1989a,b, 1990b,e; Treves and Rolls, 1994). For example, remembering many particular journeys could help to build a geographic cognitive map in the neocortex. One important constraint here is to add information to an existing (neocortical, semantic) network without overwriting or disrupting its existing contents. This is probably best performed by making a succession of small adjustments to the existing semantic network (McClelland et al., 1995). In any case the process is very different from that involved in forming an episodic memory 'on the fly', in which the whole collection of subevents present in the episode must be stored together strongly so that what happened on a single occasion can later be recalled correctly, and must be kept separate from other episodes, even if the episodes are somewhat similar. Another way in which the hippocampus may be able to make a special contribution to learning new unique events or episodes is that it provides by its connectivity the opportunity to bring together in an association information originating from many different cortical areas (Marr, 1971; Rolls, 1989a–c, 1990a–c; Murre, 1996) (see Fig. 6.1). The hippocampus and neocortex would thus be complementary memory systems, with the hippocampus being used for rapid, 'on the fly', *unstructured* storage of information involving activity potentially arriving from many areas of the neocortex; while the neocortex would gradually build and adjust on the basis of much accumulating information, and within the constraints provided by its own limited connectivity, the semantic representation (Rolls, 1989; Treves and Rolls, 1994; McClelland et al., 1995; Murre, 1996; Treves, Panzeri et al., 1996). A suggested role of the hippocampus in this process is that it acts as a temporary buffer store, the contents of which can be recalled, and used to help long-term storage elsewhere, in the neocortex, if the information is useful. The idea that the hippocampus may act as a fast but temporary buffer store to complement long-term memory storage in the neocortex is an idea which is consistent with the fact that the hippocampus is necessary for the new learning but not (at least in humans) for the long-term storage of most memories, and has been developed by many authors (e.g. Marr, 1971; Rolls, 1989a–c, 1990a–c; McClelland et al., 1995).

This raises the issue of the possible gradient of retrograde amnesia following hippocampal damage. The issue of whether memories stored some time before hippocampal damage are less impaired than more recent memories, and whether the time course is minutes, hours, days, weeks, or years is still a debated issue (Squire, 1992; Gaffan, 1993). If there is a gradient of retrograde amnesia related to hippocampal damage, then this suggests that information may be retrieved from the hippocampus if it is needed, allowing the possibility of incorporating the retrieved information into neocortical memory stores. If, on the other hand, it does transpire that there is no gradient of retrograde amnesia related to hippocampal damage, but old as well as recent memories of the hippocampal type are stored in the hippocampus, and lost if it is damaged, then again this implies the necessity of a mechanism to retrieve information stored in the hippocampus, and to use this retrieved information to affect neural circuits elsewhere (for if this were not the case, information stored in the hippocampus could never be used for anything). The current perspective is thus that whichever view of the gradient of retrograde amnesia is correct, information stored in the hippocampus will need to be retrieved and affect other parts of the brain in order to be used. The present theory shows how information could be retrieved within the hippocampus, and how this retrieved information could enable the activity in neocortical areas that was present during the original storage of the episodic event to be reinstated, thus implementing recall. The backprojections from the hippocampus to the neocortex are one of the two major outputs of the hippocampus (see Fig. 6.1). The backprojections are most likely to be involved in what is described by humans as recall. As a result of such neocortical recall, information may be incorporated into long-term, neocortical stores, but also action may be initiated.

The other major set of outputs from the hippocampus projects via the fimbria/fornix system to the anterior nucleus of the thalamus (both directly and via the mammillary bodies), which in turn projects to the cingulate cortex. This may provide an output for more action-directed use of information stored in the hippocampus, for example in the initiation of conditional spatial responses in a visual conditional spatial response task (Rupniak and Gaffan, 1987; Miyashita et al., 1989). In such a task, a rapid mapping must be learned between a visual stimulus and a spatial response, and a new mapping must be learned each day. The hippocampus is involved in this rapid visual to spatial response mapping (Rupniak and Gaffan, 1987), and the way in which hippocampal circuitry may be appropriate for this is that the CA3 region enables signals originating from very different parts of the cerebral cortex to be associated rapidly together (see below).

Given that some topographic segregation is maintained in the afferents to the hippocampus through the perirhinal and parahippocampal cortices (Amaral and Witter, 1989; Suzuki and Amaral, 1994a,b), it may be that these areas are able to subserve memory within one of these topographically separated areas, whereas the final convergence afforded by the hippocampus into a single autoassociation network in CA3 (see Fig. 6.1 and below) is, especially for an episodic memory, typically involving arbitrary associations between any of the inputs to the hippocampus, e.g. spatial, visual object, and auditory (see below).

6.3 Neurophysiology of the hippocampus

In order to understand how a part of the brain performs its computation, it is essential to know what information reaches that part of the brain, and how it is represented. The issue of representation is extremely important, for what can be performed by neuronal networks is closely related to how the information is represented. In associative networks for example, the representation (local vs. sparse distributed vs. fully distributed, and the issues of linear separability, and overlap vs. orthogonality of different representations) influences very greatly what can be stored, and the generalization and capacity of the system (see Table 2.2). The only method available for providing evidence on how as well as what information is represented in a real neuronal network in the brain is analysis of the activity of single or multiple single neurons, and evidence from this is crucial to building a neuronal network theory of a part of the brain, that is a theory for how a part of the brain functions.

In the rat, many hippocampal pyramidal cells fire when the rat is in a particular place, as defined for example by the visual spatial cues in an environment such as a room (O'Keefe, 1990, 1991; Kubie and Muller, 1991). There is information from the responses of many such cells about the place where the rat is in the environment (Wilson and McNaughton, 1994). Cells with such place responses include many of the CA3 and CA1 pyramidal cells, and of the dentate granule cells. These cells have low spontaneous firing rates (typically less than 1 spike/s), low peak firing rates (10–20 spikes/s), and the representation is quite sparse, in that the neurons fire only when the rat is in one place, and in a large environment this can be quite infrequently. (Quantitative estimates are provided below.) Some neurons are found to be more task-related, responding for example to olfactory stimuli to which particular behavioural responses must be made (Eichenbaum *et al.*, 1992; some of these neurons may in different experiments show place-related responses). The interneurons in the hippocampus have generally faster firing rates, and fire particularly during locomotion, when their firing is phase-locked to the hippocampal theta (7–9 Hz) rhythm that is found in rats when they locomote or are in paradoxical sleep. The high peak firing rates, their rapid response dynamics, their use of the inhibitory transmitter GABA, and their synapses primarily on dendritic shafts as opposed to spines of pyramidal cells, are all consistent with the view that these are neurons for feedforward and feedback inhibition which keep the activity of the principal cells of the hippocampus, the pyramidal cells, low (Buhl *et al.*, 1994). This function may be especially important within the hippocampus, given the recurrent excitatory connectivity of particularly the CA3 pyramidal cells, and also the need to keep activity low and sparse in the hippocampus for it to function efficiently as a memory system (see below).

In the primate hippocampus, spatial cells have been found, but instead of having activity related to where the monkey is, the responses of many of the neurons have responses related to where in space the monkey is looking (see Rolls, 1996d,e). The cells are thus not place cells, but spatial view cells. The evidence for this is described next, and then its implications for understanding the hippocampus as a memory system are considered. It is important to know *what* function is being performed by a brain structure before a detailed model of *how* it performs that function is elaborated.

6.3.1 Primate hippocampal cells which respond to a combination of spatial ('where') and object ('what') information

Because the primate hippocampus is necessary for remembering the spatial positions where objects have been shown (see above), neuronal activity has been recorded in monkeys performing object–place memory tasks. The task required a memory for the position on a video monitor in which a given object had appeared previously (Rolls, Miyashita, Cahusac,

Kesner, Niki, Feigenbaum and Bach, 1989). It was found that 9% of neurons recorded in the hippocampus and parahippocampal gyrus had spatial fields in this and related tasks, in that they responded whenever there was a stimulus in some but not in other positions on the screen. 2.4% of the neurons responded to a combination of spatial information and information about the object seen, in that they responded more the first time a particular image was seen in any position. Six of these neurons were found which showed this combination even more clearly, in that they for example responded only to some positions, and only provided that it was the first time that a particular stimulus had appeared there. Part of the importance of this for understanding hippocampal function is that spatial information is represented in cerebral cortical areas such as the parietal cortex, and information about what objects are seen is represented in other cortical areas, such as the inferior temporal visual cortex (see Rolls, 1994a, 1995b). The hippocampal recordings show that this evidence converges together in the hippocampus, resulting in a small proportion of neurons responding to a combination of 'what' and 'where' information.

Further evidence for this convergence is that in another memory task for which the hippocampus is needed, learning where to make spatial responses when a picture is shown, a small proportion of hippocampal neurons responds to a combination of which picture is shown, and where the response must be made (Miyashita *et al.*, 1989; Cahusac *et al.*, 1993). The explanation for the small proportion of such combination-responding neurons may be that this increases the number of different memories that can be stored in the hippocampus, as described in the computational model below.

6.3.2 Spatial views are encoded by primate hippocampal neurons

These studies showed that some hippocampal neurons in primates have spatial fields, and that a small proportion combine information from the 'what' and 'where' systems. In order to investigate how space is represented in the hippocampus, Feigenbaum and Rolls (1991) investigated whether the spatial fields use egocentric or some form of allocentric coordinates. This was investigated by finding a neuron with a space field, and then moving the monitor screen and the monkey relative to each other, and to different positions in the laboratory. For 10% of the spatial neurons, the responses remained in the same position relative to the monkey's body axis when the screen was moved or the monkey was rotated or moved to a different position in the laboratory. These neurons thus represented space in egocentric coordinates. For 46% of the spatial neurons analysed, the responses remained in the same position on the screen or in the room when the monkey was rotated or moved to a different position in the laboratory. These neurons thus represented space in allocentric coordinates. Evidence for two types of allocentric encoding was found. In the first type, the field was defined by its position on the monitor screen independently of the position of the monitor relative to the monkey's body axis and independently of the position of the monkey and the screen in the laboratory. These neurons were called 'frame of reference' allocentric, in that their fields were defined by the local frame provided by the monitor screen. The majority of the allocentric neurons responded in this way. In the second type of allocentric encoding, the field was defined by its position in the room, and was relatively independent of position relative to the monkey's body axis or to position on the monitor screen face. These neurons were called 'absolute' allocentric, in that their fields were defined by position in the room.

Further evidence on the representation of information in the primate hippocampus was obtained in studies in which the monkey was moved to different places in a spatial environment. The aim was to test whether, as in the rat (see O'Keefe, 1983; McNaughton *et al.*, 1983), there are place cells in the hippocampus. The responses of hippocampal cells were recorded when macaques were moved in a small chair or robot on wheels in a cue-controlled testing environment (a 2 m × 2 m × 2 m chamber with matt black internal walls and floors) (Rolls and O'Mara, 1995). The most common type of cell responded to the part of space at which the monkeys were looking, independently of the place where the monkey was. These neurons were called 'view' neurons, and are similar to the space neurons described by Rolls, Miyashita *et al.* (1989), and Feigenbaum and Rolls (1991). (The main difference was that in the study of Rolls *et al.* (1989) and Feigenbaum and Rolls (1991), the allocentric representation was defined by where on a video monitor a stimulus was shown; whereas view cells respond when the monkey looks at a particular part of a spatial environment.) Some of these view neurons had responses which depended on the proximity of the monkey to what was being viewed. Thus in this study the neuronal representation of space found in the primate hippocampus was shown to be defined primarily by the view of the environment, and not by the place where the monkey was (Rolls and O'Mara, 1993, 1995). Ono *et al.* (1993) also performed studies on the representation of space in the primate hippocampus while the monkey was moved to different places in a room. They found that 13.4% of hippocampal formation neurons fired more when the monkey was at some than when at other places in the test area, but before being sure that these are place cells, it will be important to know whether their responses are independent of spatial view.

In rats, place cells fire best during active locomotion by the rat (Foster *et al.*, 1989). To investigate whether place cells might be present in monkeys if *active* locomotion was being performed, Rolls, Robertson and Georges-François (1995, 1997; Rolls, 1996d,e) recorded from single hippocampal neurons while monkeys moved themselves round the test environment by walking. Also, to bring out the responses of spatial cells in the primate hippocampus, the cue-controlled environment of Rolls and O'Mara (1995), which was matt black apart from four spatial cues, was changed to the much richer environment of the open laboratory, within which the monkey had a 2.5 m × 2.5 m area to walk. The position, head direction, and eye gaze angle of the monkey were tracked continuously. It was found that when the monkey was actively locomoting and looking at different parts of the environment from different places, the firing rate depended on where the monkey was looking, and not on where the monkey was (e.g. Fig. 6.2a, cell AV057). They were thus described as spatial view cells and not as place cells (see Rolls, 1996d,e).

In further experiments on these spatial view neurons, it was shown that their responses reflected quite an abstract representation of space, in that if the visual details of the view were completely obscured by black curtains, then the neurons could still respond when the monkey looked towards where the view had been (e.g. Fig. 6.2b, cell AV057). It was further possible to demonstrate a memory component to this representation, by placing the room in darkness (with in addition black curtains completely obscuring all views of the testing environment), and showing that at least some of these neurons still had responses (though of smaller selectivity) when the monkey was rotated to face towards the view to which the spatial view neuron being recorded normally responded. This shows that the spatial representation can be recalled in response to cues other than the sight of the spatial location. Such other cues might include vestibular and proprioceptive inputs (O'Mara *et al.*, 1994; McNaughton *et al.*, 1996). This recall of a representation in response to a subset of many different recall keys is consistent with the hypothesis that the hippocampus is a memory device which can perform recall from a fragment of the original information (see below).

These spatial view (or 'space' or 'spatial view') cells (Rolls, Robertson and Georges-François, 1995, 1997; Rolls, 1996d,e) are thus unlike place cells found in the rat (O'Keefe, 1979; Muller *et al.*, 1991). Primates, with their highly developed visual and eye movement control systems, can explore and remember information about what is present at places in the environment without having to visit those places. Such spatial view cells in primates would thus be useful as part of a memory system, in that they would provide a representation of a part of space which would not depend on exactly where the monkey was, and which could be associated with items that might be present in those spatial locations. An example of the utility of such a representation in monkeys might be in enabling a monkey to remember where it had seen ripe fruit, or a human to remember where a particular person had been seen. The representations of space provided by hippocampal spatial view-responsive neurons may thus be useful in forming memories of spatial environments. The representations of space provided by hippocampal spatial view-responsive neurons may also, together with other hippocampal neurons that respond during whole-body motion, be useful in remembering trajectories through environments, of use for example in short-range spatial navigation (O'Mara *et al.*, 1994).

The representation of space in the rat hippocampus, which is of the place where the rat is, may be related to the fact that with a much less developed visual system than the primate, the rat's representation of space may be defined more by the olfactory and tactile as well as distant visual cues present, and may thus tend to reflect the place where the rat is. Although the representation of space in rats therefore may be in some ways analogous to the representation of space in the primate hippocampus, the difference does have implications for theories, and modelling, of hippocampal function. In rats, the presence of place cells has led to theories that the rat hippocampus is a spatial cognitive map, and can perform spatial computations to implement navigation through spatial environments (O'Keefe and Nadel, 1978; O'Keefe, 1991; Burgess, Recce and O'Keefe, 1994). The details of such navigational theories could not apply in any direct way to what is found in the primate hippocampus. Instead, what is applicable to both the primate and rat hippocampal recordings is that hippocampal neurons contain a representation of space (for the rat, primarily where the rat is, and for the primate primarily of positions 'out there' in space) which is a suitable representation for an episodic memory system. In primates, this would enable one to remember, for example, where an object was seen. In rats, it might enable memories to be formed of where particular objects (for example those defined by olfactory, tactile, and taste inputs) were found. Thus at least in primates, and possibly also in rats, the neuronal representation of space in the hippocampus may be appropriate for forming memories of events (which usually in these animals have a spatial component). Such memories would be useful for spatial navigation, for which according to the present hypothesis the hippocampus would implement the memory component but not the spatial computation component.

(a)

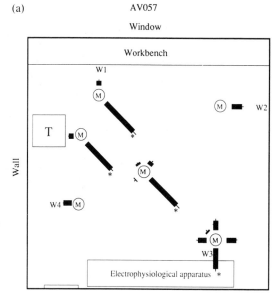

AV057

Window

Mean spontaneous firing rate = 1.5 + 0.8 spikes/sec.

(b) AV057: view obscured by curtain.

Window

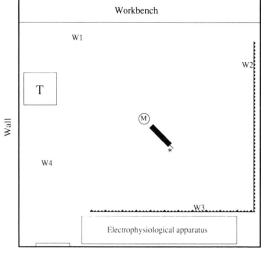

Mean spontaneous firing rate (active) = 1.4 + 0.4

* - increased firing rate, p < 0.001

(c)   trolley052.dat.filt

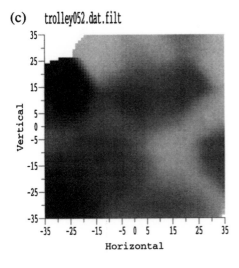

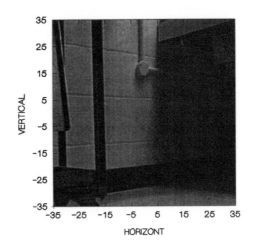

Grey Scale Firing Rate

Fig. 6.2 Examples of the responses of a hippocampal spatial view cell recorded in a macaque (cell AV057) in a 2.5 m 2.5 m open laboratory space in which he could walk. Abbreviations: T, trolley; W1–4, location of food wells; M, monkey; ——, firing rate in direction of view; *, increased firing rate, *P* < 0.001; ▲▲▲▲, curtain. (a) The firing rate of the same cell when the monkey was situated at different places in the laboratory, looking in different directions. It is clear that the firing rate of the cell increased whenever the monkey was looking towards Well 3. The firing rate did not depend on where the monkey was. The cell was thus classed as a spatial view and not as a place cell. Mean spontaneous rate = 1.5 spikes/s. (b) The same neuron still responded when the monkey faced towards the spatial position of well 3, even when the whole of the area near well 3 was totally covered by curtains from the ceiling to the floor. The cell did not fire when the monkey was rotated to face in different directions. Thus the representation of space that is sufficient to activate these cells can become quite abstract, and does not require (after learning) the details of the view to be present. Mean spontaneous rate (active) = 1.4 spikes/s. (c) The firing rate of another spatial view cell with the monkey stationary in the environment, but moving his eyes to look at different parts of the environment. The cell increased its firing rate when the monkey looked at some parts of the environment (e.g. towards the black trolley shown at the left of the picture), but not when he was looking at other parts of the environment (e.g. towards the sink and bench top shown at the top right of the picture). The firing rate of the cell in spikes/s is indicated by the shade, with the calibration bars shown below.

Another type of cell found in the primate hippocampus responds to whole body motion (O'Mara, Rolls, Berthoz and Kesner, 1994). For example, the cell shown in Fig. 6.3 responded when the monkey was rotated about the vertical axis, with a much larger response for clockwise than for anticlockwise rotation. By occluding the visual field, it was possible to show that in some cases the response of these cells required visual input. For other cells, visual input was not required, and it is likely that such cells responded on the basis of vestibular inputs. Some cells were found that responded to a combination of body motion and spatial view or place. In some cases these neurons respond to linear motion, in others to axial rotation. These whole-body motion cells may be useful in a memory system for remembering trajectories through environments, as noted above.

Another important type of information available from the hippocampal single neuron recordings is about the sparseness of the representation provided by hippocampal neurons. The sparseness, as defined below, indicates roughly the proportion of hippocampal neurons active at any one time in the hippocampus. The relevant time period for this measure is in

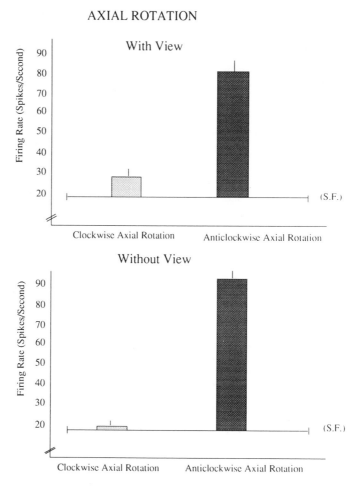

Fig. 6.3 The responses of a cell that responded to whole-body, axial rotation, with a strong preference for clockwise rotation (means ± sem of the firing rates are shown). S.F., spontaneous rate.

the order of 1 s. Two arguments support this. First, the time course of hippocampal synaptic modification is a few hundred milliseconds, in that the postsynaptic events involved in LTP operate over this time scale. Second, in normal environments in which stimuli might not change extremely rapidly, the stimuli forming an event would be expected to be co-occurrent for a period in the order of 1 s. In the monkey, in the different spatial and memory tasks in which recordings have been made, the evidence suggests that the sparseness is in the order of 1%. For example, in the object and place memory task, 9% of cells responded in the task in which four places were to be remembered, so that approximately 2% of the neurons responded to any one place (Rolls, Miyashita *et al.*, 1989). In a delayed spatial response task, 5.7% of the neurons had selective responses, and different neurons within this population responded to the different spatial responses required, in the sample, delay, and/or spatial response period, so that the proportion of neurons which responded in any one period in relation to any one spatial response was low (Cahusac *et al.*, 1989). In rats, data from McNaughton and Barnes' laboratory show that representations about place are very sparse in the dentate granule cells, are still sparse in CA3 neurons, and are fairly sparse in CA1 neurons (Leonard and McNaughton, 1990; Barnes *et al.*, 1990; Jung and McNaughton, 1993; Skaggs *et al.*, 1996). For example, values obtained for CA1 place cells in the rat were mainly in the region 0.1–0.4 (Skaggs *et al.*, 1996), although lower values would probably be obtained if all cells were sampled, not just those with place fields in one particular environment. Estimates of these parameters are very important for defining how many memories could be stored in associative networks in the brain, as shown below and elsewhere (Rolls and Treves, 1990; Treves and Rolls, 1991, 1994; Chapters 2 and 3, Appendices A3 and A4).

Although there are differences in the spatial representations in the primate and rat hippocampus, the quantitative theory that is described below and by Treves and Rolls (1994) would apply to both. The theory leads to predictions that can be tested in neurophysiological experiments. For example, Treves, Skaggs and Barnes (1996) show that the information content of rat place cell representations is very similar to that predicted by the theory. In particular, the information provided by each cell about where the rat was in a spatial environment was 0.1–0.4 bits, and the information from different cells recorded simultaneously added independently, within the constraint imposed by the maximum amount of information required to specify the place where the rat was in the environment. In monkeys, hippocampal cells provide on average 0.32 bits of information in 500 ms about the spatial view, and the information from an ensemble of these cells again adds almost independently (Rolls, Treves *et al.*, 1998).

6.4 Architecture of the hippocampus

In order to develop a neural network theory of how a part of the brain computes, it is necessary to know the quantitative details of its connectivity. This is described for the hippocampus next.

The hippocampus receives, via the adjacent parahippocampal gyrus and entorhinal cortex, inputs from virtually all association areas in the neocortex, including those in the parietal, temporal, and frontal lobes (Van Hoesen, 1982; Squire *et al.*, 1989; Suzuki and Amaral, 1994a,b). Therefore, the hippocampus has available highly elaborated multimodal information, which has already been processed extensively along different, and partially interconnected, sensory pathways. Although there is some mixing possible in entorhinal and perirhinal cortices, there is nonetheless some topographical segregation of the inputs, with for example olfactory inputs projecting primarily anteriorly in the entorhinal cortex, parietal inputs posteriorly, and temporal lobe visual inputs primarily to the perirhinal cortex (Insausti, Amaral, and Cowan, 1987; Suzuki and Amaral, 1994a,b) (see Fig. 6.1). Although there are recurrent collaterals within these systems, they do not provide for the extensive connection of neurons into a single network as appears to happen later in the system, in the CA3 region of the hippocampus proper. Indeed, it is argued below that the provision of a single network for association between any subset of neurons (the CA3 neurons) may be an important feature of what the hippocampal circuitry provides for memory. Additional inputs to the hippocampus come from the amygdala and, via a separate pathway, from the cholinergic and other regulatory systems.

To complement this massively convergent set of pathways into the hippocampus, there is a massively divergent set of backprojecting pathways from the hippocampus via the entorhinal cortex to the cerebral neocortical areas which provide inputs to the hippocampus (Van Hoesen, 1982; Amaral, 1993) (see Fig. 6.4).

Within the hippocampus, information is processed along a mainly unidirectional path, consisting of three major stages, as shown in Figs 6.5 and 6.1 (see the articles in Storm-Mathiesen *et al.*, 1990; Amaral and Witter, 1989; Amaral, 1993). Axonal projections mainly from layer 2 of entorhinal cortex reach the granule cells in the dentate gyrus via the perforant path (PP), and also proceed to make synapses on the apical dendrites of pyramidal cells in the next stage, CA3. A different set of fibres projects from entorhinal cortex (mainly layer 3) directly onto the third processing stage, CA1.

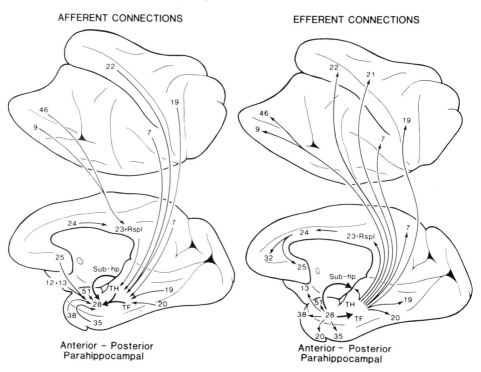

Fig. 6.4 Connections of the primate hippocampus with the neocortex (from Van Hoesen, 1982). A medial view of the macaque brain is shown below, and a lateral view is shown inverted above. The hippocampus receives its inputs via the parahippocampal gyrus, areas TF and TH, and the entorhinal cortex, area 28. The return projections to the neocortex (shown on the right) pass through the same areas. Cortical areas 19, 20, and 21 are visual association areas, 22 is auditory association cortex, 7 is parietal association cortex, and 9, 46, 12, and 13 are areas of frontal association cortex.

There are about 10^6 dentate granule cells in the rat, and more than 10 times as many in humans (more detailed anatomical studies are available for the rat) (Amaral *et al.*, 1990; West and Gundersen, 1990). They project to the CA3 cells via the mossy fibres (MF), which provide a small number of probably strong inputs to each CA3 cell: each mossy fibre makes, in the rat, about 15 synapses onto the proximal dendrites of CA3 pyramidal cells. As there are some 3×10^5 CA3 pyramidal cells in the rat (SD strain; 2.3×10^6 in man, Seress, 1988), each of

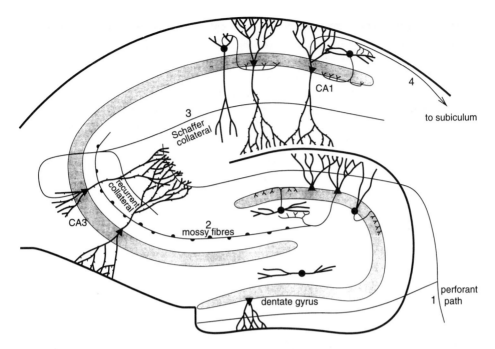

Fig. 6.5 Representation of connections within the hippocampus. Inputs reach the hippocampus through the perforant path (1) which makes synapses with the dendrites of the dentate granule cells and also with the apical dendrites of the CA3 pyramidal cells. The dentate granule cells project via the mossy fibres (2) to the CA3 pyramidal cells. The well-developed recurrent collateral system of the CA3 cells is indicated. The CA3 pyramidal cells project via the Schaffer collaterals (3) to the CA1 pyramidal cells, which in turn have connections (4) to the subiculum.

them receives no more than around 50 mossy synapses. (The sparseness of this connectivity is thus 0.005%.) By contrast, there are many more—but weaker—direct perforant path inputs onto each CA3 cell, in the rat of the order of 4×10^3. The largest number of synapses (about 1.2×10^4 in the rat) on the dendrites of CA3 pyramidal cells is, however, provided by the (recurrent) axon collaterals of CA3 cells themselves (RC) (see Fig. 6.6). A diagram showing a CA3 neuron, its extensive dendritic tree, and its axon which spreads from one end of the CA3 cells to the other, is given in Fig. 1.1. It is remarkable that the recurrent collaterals are distributed to other CA3 cells throughout the hippocampus (Ishizuka et al., 1990; Amaral and Witter, 1989), so that effectively the CA3 system provides a single network, with a connectivity of approximately 4% between the different CA3 neurons. The implication of this widespread recurrent collateral connectivity is that each CA3 cell can transmit information to every other CA3 cell within 2–3 synaptic steps. (This takes into account the fact that the CA3 recurrent collaterals are not quite uniform in density throughout the hippocampus, having some gradient of connectivity to favour relatively near rather than relatively far neurons; see Amaral, 1993). (In the rat, the CA3 fibres even connect across the midline, so that effectively in the rat there is a single CA3 network in the hippocampus, with a 2% connectivity on average between the 600 000 neurons; see Rolls, 1990b.) The CA3 system therefore is, far more than either DG or CA1, a system in which intrinsic, recurrent excitatory connections are, at least numerically, dominant with respect to excitatory afferents.

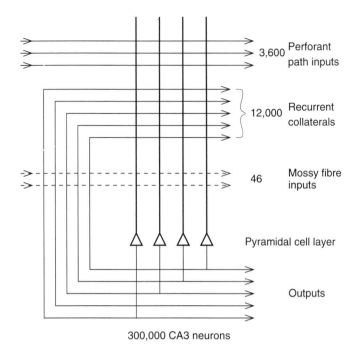

Fig. 6.6 The approximate numbers of connections onto each CA3 cell from three different sources in the rat (see text).

In addition, there are also intrinsic connections with a variety of numerically limited and mainly inhibitory populations of interneurons, as well as extrinsic connections with sublimbic structures; such connections are believed to subserve generic regulation of neural activity in CA3, as opposed to providing signals specific to the information being processed in the system.

The extrinsic axonal projections from CA3, the Schaffer collaterals, provide the major input to CA1 pyramidal cells, of which there are about 4×10^5 in the (SD) rat. The CA1 pyramidal cells are characteristically smaller than the CA3 ones and, across different species, come in larger numbers. In terms of cell numbers, therefore, information appears to be funnelled from DG through the CA3 bottleneck, and then spread out again into CA1. The output of CA1 returns directly, and via the subiculum, to the entorhinal cortex, from which it is redistributed to neocortical areas (see Fig. 6.1).

6.5 Plasticity

Neurophysiological evidence also indicates that many of the types of synapse within the hippocampus are modified as a result of experience, in a way explicitly related to the types of learning for which the hippocampus is necessary, as shown in studies in which the blocking of such modifiability with drugs results in specific learning impairments (Morris, 1989; see Chapter 1).

Studies on long-term potentiation (LTP) have shown that some synaptic systems (in DG and CA1, but possibly also PP and RC synapses in CA3) display a 'Hebbian', or associative, form of plasticity, whereby presynaptic activity concurrent with strong postsynaptic depolarization can result in a strengthening of the synaptic efficacy (Brown *et al..*, 1990; Miles, 1988; Larkman and Jack, 1995). Such strengthening appears to be associated with the activation of NMDA (*N*-methyl-D-aspartate) receptors (Collingridge and Singer, 1990), and it is possible that the same synapses display also (associative) long-term depression (Levy and Desmond, 1985; Levy *et al.*, 1990; Mulkey and Malenka, 1992; Christie, 1996; Fazeli and Collingridge, 1996). Also MF synapses are known to display long-term activity-dependent synaptic enhancement, and there is evidence both against and for the view that this is associative (Weisskopf, Zalutsky and Nicoll, 1993; Williams and Johnston, 1989; Nicoll and Malenka, 1995; Derrick and Martinez, 1996); that is, for the mossy fibre synapses it is not clear whether the potentiation is produced by strong presynaptic activity independently of whether there is postsynaptic activity.

6.6 Outline of a computational hypothesis of hippocampal operation

Marr (1971) developed a pioneering neural network theory of hippocampal function. In the theory he was not able to indicate different functions for individual networks within the hippocampus, but did analyse how modifiable connections between neurons in the hippocampus might enable memories to be stored, and later retrieved from a part. Gardner-Medwin (1976) elaborated how progressive recall in recurrent networks of interconnected simplified binary neurons might operate. Rolls (1987, 1989a–c, 1990a,b, 1991a,b) suggested that the reason why the hippocampus is used for the spatial and non-spatial types of memory described above, and the reason that makes these two types of memory so analogous, is that the hippocampus contains one stage, the CA3 stage, which acts as an autoassociation memory. An autoassociation memory implemented by the CA3 neurons would enable whole (spatial) scenes as well as episodic memories to be formed, with a snapshot quality which depends on the arbitrary associations which can be made and the short temporal window which characterizes the synaptic modifiability in this system. This hypothesis implies that any new event to be memorized is given a unitary representation as a firing pattern of CA3 pyramidal cells, that the pattern is stored into associatively modifiable synapses, and that subsequently the extensive RC connectivity allows for the retrieval of a whole representation to be initiated by the activation of some small part of the same representation (the cue). Cued retrieval of the CA3 representation might enable recall within the hippocampus, which in turn may allow recall of neuronal activity present during the original learning to be reinstated in the cortex, providing information to the neocortex which would be useful in setting up in the neocortex new long-term memories. Quantitative aspects of this hypothesis, leading to a detailed model of the functioning of some parts of the hippocampal circuitry, are described below.

It is suggested that this ability to link information originating from different brain regions in a single autoassociation matrix in the CA3 regions is a key feature of hippocampal architecture, which is unlikely to be implemented in input regions such as the entorhinal cortex and dentate granule cells, which not only do not have the required system of recurrent collaterals over the whole population of cells, but also appear to maintain some segregation of inputs originating from different parts of the cerebral cortex (Insausti, Amaral, and

Cowan, 1987; Suzuki and Amaral, 1994a,b). Indeed, the inputs which reach the hippocampus, via the parahippocampal gyrus and the entorhinal cortex (see Figs 6.1 and 6.4) originate from very different cortical areas, which are at the ends of the main cortical processing streams, and are not heavily interconnected. For example, these inputs come from the inferior temporal visual cortex (but not from early cortical visual areas), from the parietal cortex, from the superior temporal auditory cortex, and from the prefrontal cortex. In this anatomical sense, the hippocampus in its CA3 system provides a single net in which information from these different cortical areas could be combined to form a 'snapshot', event, or episodic, memory (Rolls, 1987, 1989a–c, 1990a,b, 1991a,b). Such an episodic memory might by combining inputs from these different cortical systems contain information about what one ate at lunch the previous day, where it was, and the faces and voices of the people one was with. Information about rewards, punishments, and emotional states which can be parts of episodic memories could be incorporated into the hippocampal representation of an episodic memory via the afferents to the entorhinal cortex from structures such as the amygdala, which are involved in processing information about rewards and emotions (Rolls, 1990c, 1992a). The relatively non-specific subcortical afferents to the hippocampus, for example the cholinergic inputs from the medial septum, may be involved in threshold setting in the hippocampus, and in particular may, by allowing arousal to affect the hippocampus, make it more likely that memories will be stored when in certain states, such as arousal (Rolls, 1989b). This would tend to economize in the use of the available storage space for episodic memories in the hippocampus, for new episodic memories would be less likely to be stored during low arousal when little may be happening in the environment, but would be stored strongly when significant and therefore arousing environmental events are occurring (cf. Wilson and Rolls, 1990a,b). The effects of cholinergic and other non-specific inputs to the hippocampus may be realized at least partly because of their role in making it more likely that postsynaptic activation will exceed the threshold for activation of the NMDA receptors involved in long-term potentiation.

6.7 The CA3 network as an autoassociative memory

6.7.1 Analysis of storage capacity

Our hypotheses concerning the operation of hippocampal circuits are based on the notion that the CA3 network operates as an autoassociative memory, that is it is able, when provided with a small cue, to selectively retrieve a specific distribution of firing activity from among several stored on the same synaptic efficacies (strengths). This property is in our view intimately related to the architecture of the CA3 network, in particular to the prominent recurrent collateral connections. Others, to begin with Marr (1971), have noted that cued retrieval of one of a set of stored firing patterns can be achieved also with an alternative architecture, consisting of a cascade of feedforward associative nets (Willshaw *et al.*, 1969), possibly assisted to some minor extent by the effect of recurrent collaterals (see Willshaw and Buckingham, 1990). We have shown, however, that the simple autoassociative architecture has at least two advantages over the multilayered one: first, it requires far fewer synapses (by a factor equal to the number of layers needed to complete retrieval in the feedforward system;

Treves and Rolls, 1991) and, second and more importantly, it is possible to describe the conditions appropriate for learning, that is the formation and storage of new patterns (Treves and Rolls, 1992), whereas no one has yet been able to describe how new firing patterns could be stored in a multi-layered associative net with a local learning rule. (The problem with the latter is that there in no simple way to set the firing of the cells appropriately in the intermediate layers of the 'autoassociator' or pattern associator; see Treves and Rolls, 1991.) In this section we briefly review our quantitative analyses of the storage and retrieval processes in the CA3 network (see Treves and Rolls, 1991, 1992, and Appendix A4).

To derive results applicable to CA3, we have analysed a formal model of an autoassociative network with graded response neurons, to represent more realistically the continuously variable rates at which neurons fire; with diluted recurrent collateral connectivity, as compared with the symmetric complete connectivity which is more easily analysed; and with sparse representations, in which only a small proportion of neurons is active at any one time, to analyse the quantitative implications of sparse representations in memory systems (see Appendices A3 and A4). In general, as discussed in the appendices, the maximum number p_{max} of firing patterns that can be retrieved is proportional to the number C^{RC} of associatively modifiable RC synapses per cell, by a factor that increases roughly with the inverse of the sparseness a of the neuronal representation. Approximately,

$$p_{max} \approx \frac{C^{RC}}{a \ln(1/a)} k \qquad (6.1)$$

where k is a factor that depends weakly on the detailed structure of the rate distribution, on the connectivity pattern, etc., but is roughly in the order of 0.2–0.3 (Treves and Rolls, 1991). The sparseness a is that measured from the firing rates during a period in which a memory is being formed (see Rolls, Treves, Robertson and Georges-François, 1997), that is over a period of the order of 1 s for the hippocampus.

The main factors that determine the maximum number of memories that can be stored in an autoassociative network are thus the number of connections on each neuron devoted to the recurrent collaterals, and the sparseness of the representation. For example, for $C^{RC} = 12\,000$ and $a = 0.02$, p_{max} is calculated to be approximately 36 000. This storage capacity can be realized, without interference between patterns, if the learning rule includes some form of heterosynaptic long-term depression that counterbalances the effects of associative long-term potentiation (Treves and Rolls, 1991).

We have also shown (see Appendix A4) how to estimate I, the total amount of information (in bits per synapse) that can be retrieved from the network. I is defined with respect to the information i_p (in bits per cell) contained in each stored firing pattern, by subtracting the amount i_l lost in retrieval and multiplying by p/C^{RC}:

$$I \equiv \frac{p}{C^{RC}} (i_p - i_l) \qquad (6.2)$$

The maximal value I_{max} of this quantity was found (Treves and Rolls, 1991) to be in several interesting cases around 0.2–0.3 bits per synapse, with only a mild dependence on parameters such as the sparseness of coding a, and whether the patterns are nearly binary or are continuously variable. An interesting implication of this is that if each pattern uses little information to represent it by using essentially binary (e.g. slow vs. fast) firing rates, then many patterns can be stored; but that if one attempts to store more information in each pattern by using continuously variable firing rates for the neurons, then either fewer different patterns can be stored with accurate retrieval, or much of the information in the pattern is lost in retrieval if many patterns are stored (Treves, 1990; Treves and Rolls, 1991; Rolls, Treves, Foster and Perez-Vicente, 1997).

We can next estimate how much information has to be stored in each pattern for the network to efficiently exploit its capacity I_{max} to retrieve information from the recurrent collateral associative memory (Treves and Rolls, 1992). The estimate is expressed as a requirement on i_p:

$$i_p > a \ln(1/a) \tag{6.3}$$

As the information content of each stored pattern i_p depends on the storage process, we see how the retrieval capacity analysis, coupled with the notion that the system is organized so as to be an efficient memory device in a quantitative sense, leads to a constraint on the storage process.

We note that although there is some spatial gradient in the CA3 recurrent connections, so that the connectivity is not fully uniform (Ishizuka et al., 1990), nevertheless the network will still have the properties of a single interconnected autoassociation network allowing associations between arbitrary neurons to be formed, given the presence of many long-range connections which overlap from different CA3 cells.

It is found that the spontaneous and peak firing rates of hippocampal pyramidal cells tend to be low, with values of the order of < 1 spike/s for the spontaneous rate, and 10 spikes/s for the peak rate, for both place cells in rats (Barnes et al., 1990) and spatial view cells in monkeys (see Rolls, 1996c, 1996d; Rolls, Treves, et al., 1998). One reason for the low rates may be that if neurons have a Gaussian distribution of input activations across stimulus conditions, a simple way to produce a sparse representation is to set a high threshold, for then just the upper tail of the Gaussian results in firing. If the threshold is set below the mean of the distribution, the resulting representations are very distributed. Another point is that in what is a system with massive positive feedback due to the recurrent collaterals, keeping the firing rates low may simplify the dynamics of the negative feedback produced through the inhibitory interneurons that is required to keep the cell population stable.

6.7.2 The requirement of the input systems for efficient storage of new information

By calculating the amount of information that would end up being carried by a CA3 firing pattern produced solely by the perforant path input and by the effect of the recurrent connections, we have been able to show (Treves and Rolls, 1992) that an input of the perforant path type, alone, is unable to direct efficient information storage. Such an input is too weak, it turns out, to drive the firing of the cells, as the 'dynamics' of the network are dominated by the effect of the recurrent collaterals, which reflects previously stored memories. This is the manifestation, in the CA3 network, of a general problem affecting storage (i.e. learning) in *all* autoassociative memories. The problem is that during storage a set of input axons must force a pattern of firing onto the principal neurons of the network, and their activity must not be dominated during learning by the recurrent connections. This is a problem which can be conveniently ignored in artificial networks by having the appropriate set of synaptic weights artificially stored in the synaptic matrix, but a solution is needed when such a network is to be used realistically or biologically.

We hypothesize that the mossy fibre inputs force efficient information storage by virtue of their strong and sparse influence on the CA3 cell firing rates (Treves and Rolls, 1992; Rolls, 1989a,b). The mossy fibre input appears to be particularly appropriate in several ways. First of all, the fact that mossy fibre synapses are large and located very close to the soma makes them relatively powerful in activating the postsynaptic cell. (This should not be taken to imply that a CA3 cell can be fired by a single mossy fibre EPSP.) Second, the firing activity of granule cells appears to be very sparse (Jung and McNaughton, 1993) and this, together with the small number of mossy fibre connections on each CA3 cell, produces a sparse signal, which can then be transformed into an even sparser firing activity in CA3 by a threshold effect.[2] Third, the plasticity of mossy fibres (whether associative or not) might have a useful effect in enhancing the signal-to-noise ratio, in that a consistently firing mossy fibre would produce non-linearly amplified currents in the postsynaptic cell, which would not happen with an occasionally firing fibre (Treves and Rolls, 1992). This effect might be helpful, but would not be essential for learning (Huang *et al.*, 1995). Rolls and Perez-Vicente (in preparation) have also shown how a few large inputs through the mossy fibres would enable the thresholds of the CA3 cells to be set to achieve a given sparseness of the representation without requiring great precision of the threshold-setting mechanism. In contrast, if the dominant input during learning were a very numerous set of weak inputs to each pyramidal cell, then great precision of the threshold-setting mechanism (presumably by inhibitory interneurons) would be needed to maintain the sparseness at the desired low level that enables good storage of many different memory patterns to occur.

These arguments suggest, then, that an input system with the characteristics of the mossy fibres is essential during learning, in that it may act as a sort of (unsupervised) teacher that effectively strongly influences which CA3 neurons fire based on the pattern of granule cell activity. This establishes an information-rich neuronal representation of the episode in the CA3 network (see further Treves and Rolls, 1992). The perforant path input would, the quantitative analysis shows, not produce a pattern of firing in CA3 that contains sufficient information for learning (Treves and Rolls, 1992). The hypothesis is then that the mossy fibre system produces firing in a different, arbitrary, small subset of CA3 neurons for every memory that is to be stored. This hypothesis is given support by the fact that the responses of

---

[2] For example, if only 1 granule cell in 100 were active in the dentate gyrus, and each CA3 cell received a connection from 50 randomly placed granule cells, then the number of active mossy fibre inputs received by CA3 cells would follow a Poisson distribution of average $50/100 = 1/2$, i.e. 60% of the cells would not receive any active input, 30% would receive only one, 7.5% two, little more than 1% would receive three, and so on. (It is easy to show from the properties of the Poisson distribution and our definition of sparseness, that the sparseness of the mossy fibre signal as seen by a CA3 cell would be $x/(1 + x)$, with $x = C^{MF}a_{DG}$, assuming equal strengths for all mossy fibre synapses.) If three mossy fibre inputs were required to fire a CA3 cell and these were the only inputs available, we see that the activity in CA3 would be roughly as sparse as in the dentate gyrus, in this example.

primate hippocampal pyramidal cells convey independent information, in that the information from a population of such cells about spatial view rises approximately linearly with the number of neurons in the population (Rolls, Treves, Robertson, Georges-François and Panzeri, 1998). This independence of primate hippocampal neuronal responses would help the CA3 system to store large amounts of information, and not to be restricted because of high correlations between the representations it stores.

6.7.3 The role of the perforant path input in initiating retrieval

An autoassociative memory network needs afferent inputs also in the other mode of operation, that is when it retrieves a previously stored pattern of activity. We have shown (Treves and Rolls, 1992) that the requirements on the organization of the afferents are in this case very different, implying the necessity of a second, separate, input system, which we have identified with the perforant path to CA3. In brief, the argument is based on the notion that the cue available to initiate retrieval might be rather small, that is the distribution of activity on the afferent axons might carry a small correlation, $q \ll 1$, with the activity distribution present during learning. In order not to lose this small correlation altogether, but rather transform it into an input current in the CA3 cells that carries a sizable signal—which can then initiate the retrieval of the full pattern by the recurrent collaterals—one needs an extensive number of associatively modifiable synapses. This is expressed by the formulas that give the specific signal S produced by sets of associatively modifiable synapses, or by non-associatively modifiable synapses: if C^{AFF} is the number of afferents per cell,

$$S_{\mathrm{ASS}} \approx \frac{\sqrt{C^{\mathrm{AFF}}}}{\sqrt{p}} q \tag{6.4a}$$

$$S_{\mathrm{NONASS}} \approx \frac{1}{\sqrt{C^{\mathrm{AFF}}}} q \tag{6.4b}$$

Associatively modifiable synapses are therefore needed, and are needed in a number C^{AFF} which is of the same order as the number of concurrently stored patterns p, so that small cues can be effective; whereas non-associatively modifiable synapses—or even more so, non-modifiable ones—produce very small signals, which decrease in size the larger the number of synapses. In contrast with the storage process, the average strength of these synapses does not now play a crucial role. This suggests that the perforant path system is the one involved in relaying the cues that initiate retrieval. Empirical evidence that the perforant path synapses onto CA3 neurons are indeed associatively modifiable, as predicted, has now been described (Berger, Yeckel and Thiels, 1996).

6.7.4 Predictions arising from the analysis of CA3

Given the hypothesis, supported by the formal analysis above, that the mossy fibre system is particularly important during storage of new information in the hippocampus, we have predicted that temporary inactivation of the mossy fibres should result, unless more generic effects complicate the picture, in amnesia specific to events occurring during the time window of inactivation, but not involving events learned before or after the inactivation. In contrast, inactivation of the perforant path to CA3, leaving intact, if possible, the initial part of the perforant path still connected to DG, should result in a deficit in retrieving information stored recently (while it is still in the hippocampus), and also in learning new events. (The deficits are predicted to occur in hippocampal-dependent tasks, particularly when the system is heavily loaded, for it is under conditions of storing and retrieving many different memories that hippocampal circuitry will be fully used.) The deficit in retrieval would be due to the perforant path to CA3 synapses not being available to trigger retrieval from the CA3 network. This suggests experimental testing procedures and, interestingly, is consistent with neurophysiological results already available, which showed that CA3 and CA1 neurons could respond in previously learned spatial environments even when the dentate was damaged (McNaughton *et al.*, 1989). The roles we propose for the input systems to CA3 are also consistent with the general hypothesis (Rolls, 1987, 1989a–c, 1990a,b) that one of the functions of the processing occurring in the dentate gyrus is to transform certain neuronal representations into convenient inputs for the autoassociative memory we hypothesize is implemented in the CA3 network.

The analytic approaches to the storage capacity of the CA3 network, the role of the mossy fibres and of the perforant path, the functions of CA1, and the operation of the back-projections in recall, have also been shown to be computationally plausible based on computer simulations of the circuitry shown in the lower part of Fig. 6.1. In the simulation, during recall partial keys are presented to the entorhinal cortex, completion is produced by the CA3 autoassociation network, and recall is produced in the entorhinal cortex of the original learned vector. The network, which has 1000 neurons at each stage, can recall large numbers which approach the calculated storage capacity of different sparse random vectors. The simulation is described in Section 6.10.4 below.

6.7.5 Dynamics and the speed of operation of autoassociation in the brain

The arguments above have been developed while abstracting for a while from the time domain, by considering each pattern of activity, be it a new one or a memory being retrieved, as a steady-state vector of firing rates. Reintroducing the temporal dimension poses a whole new set of issues. For example, how long does it take before a pattern of activity, originally evoked in CA3 by afferent inputs, becomes influenced by the activation of recurrent collaterals, and how long is it before recall from an incomplete cue has become complete? How does the time scale for recurrent activation compare, for example, with the period of the theta-rhythm in the rat? For how long are steady-state, or approximately steady-state, firing patterns sustained? Is this length of time compatible with the time requirements for associative synaptic plasticity as expressed via NMDA receptors?

A partial answer to the first questions can be inferred from recent theoretical developments based on the analysis of the collective dynamical properties of realistically modelled neuronal units (Treves, 1993; see Appendix A5). The analysis indicates that the evolution of distributions of activity in a recurrent network follows a plurality of different time scales; and that the most important of those time scales are only mildly dependent on single cell properties, such as prevailing firing rates or membrane time constants, but rather depend crucially on the time constants governing synaptic conductances. This result suggests that in the case of CA3 the activation of recurrent collaterals between pyramidal cells, through synapses whose time constant is rather short (a few milliseconds), will contribute to determine the overall firing pattern within a period of a very few tens of milliseconds (see Treves, 1993). (In animals in which there is a hippocampal theta rhythm, such as the rat, the relevant CA3 dynamics could take place within a theta period. In monkeys, no prominent theta rhythm has been found in the hippocampus.) The indication is thus that retrieval would be very rapid from the CA3 network, indeed fast enough for it to be biologically plausible. Simulations of the dynamics confirming fast retrieval are described in Appendix A5 and by Simmen, Rolls and Treves (1996), and Treves, Rolls and Simmen (1997).

6.8 The dentate granule cells

The theory is developed elsewhere that the dentate granule cell stage of hippocampal processing which precedes the CA3 stage acts in four ways to produce during learning the sparse yet efficient (i.e. non-redundant) representation in CA3 neurons which is required for the autoassociation to perform well (Rolls, 1989a–c, 1994c; see also Treves and Rolls, 1992).

The first way is that the perforant path–dentate granule cell system with its Hebb-like modifiability is suggested to act as a competitive learning network to remove redundancy from the inputs producing a more orthogonal, sparse, and categorized set of outputs (Rolls, 1987, 1989a–c, 1990a,b). The synaptic modification involved in this competitive learning could take place for each single event or episode being learned by the hippocampus during the single occurrence of that event or episode. The time-scale of this synaptic modification on the dentate granule cells would thus be as fast as that in the CA3 recurrent collateral system. The non-linearity in the NMDA receptors may help the operation of such a competitive net, for it ensures that only the most active neurons left after the competitive feedback inhibition have synapses that become modified and thus learn to respond to that input (Rolls, 1989c, 1996a). We note that if the synaptic modification produced in the dentate granule cells lasts for a period of more than the duration of learning the episodic memory, then it could reflect the formation of codes for regularly occurring combinations of active inputs that might need to participate in different episodic memories. Because of the non-linearity in the NMDA receptors, the non-linearity of the competitive interactions between the neurons (produced by feedback inhibition and non-linearity in the activation function of the neurons) need not be so great (Rolls, 1989c). Because of the feedback inhibition, the competitive process may result in a relatively constant number of strongly active dentate neurons relatively independently of the number of active perforant path inputs to the dentate cells. The operation of the dentate granule cell system as a competitive network may also be facilitated by a Hebb rule of the general form:

$$\delta w_{ij} = k \, r_i (r'_j - w_{ij}) \qquad (6.5)$$

where k is a constant, r_i is the activation of the dendrite (the postsynaptic term), r'_j is the presynaptic firing rate, w_{ij} is the synaptic weight, and r'_j and w_{ij} are in appropriate units (see

Rolls, 1989c). Incorporation of a rule such as this which implies heterosynaptic long-term depression (LTD) as well as long-term potentiation (LTP) (see Levy and Desmond, 1985; Levy et al., 1990) makes the sum of the synaptic weights on each neuron remain roughly constant during learning (cf. Oja, 1982; see Rolls, 1989c; Chapter 4). Because of the necessity for some scaling of synaptic weights in competitive networks, it is a prediction that a rule of this type should be present in the dentate (and in other putative competitive networks in the brain). In practice, what is predicted is that LTP should be easier to obtain if the synapses are currently weak, e.g. have not been potentiated yet; and that LTD should be easier to obtain if the synapses are currently strong, e.g. after LTP (see Rolls, 1996a).

The second way is also a result of the competitive learning hypothesized to be implemented by the dentate granule cells (Rolls, 1987, 1989a–c, 1990a,b, 1994c). It is proposed that this allows overlapping inputs to the hippocampus to be separated, in the following way (see also Rolls, 1994c). Consider three patterns B, W, and BW where BW is a linear combination of B and W. (To make the example very concrete, we could consider binary patterns where B = 10, W = 01, and BW = 11.) Then the memory system is required to associate B with reward, W with reward, but BW with punishment. This is one of the configural learning tasks of Sutherland and Rudy (1989), and for them is what characterizes the memory functions performed by the hippocampus. Without the hippocampus, rats might have more difficulty in solving such problems. (This is the more recent hypothesis of Rudy and Sutherland, 1995.) However, it is a property of competitive neuronal networks that they can separate such overlapping patterns perfectly, as has been shown elsewhere (Rolls, 1989c; see Chapter 4). (To be precise: to separate non-linearly separable patterns, the synaptic weight vectors on each neuron in competitive nets must be normalized. Even if competitive nets in the brain do not use explicit weight normalization, they may still be able to separate similar patterns; see Chapter 4.) It is thus an important part of hippocampal neuronal network architecture that there is a competitive network that precedes the CA3 autoassociation system. Without the dentate gyrus, if a conventional autoassociation network were presented with the mixture BW having learned B and W separately, then the autoassociation network would produce a mixed output state, and would therefore be incapable of storing separate memories for B, W, and BW. It is suggested, therefore, that competition in the dentate gyrus is one of the powerful computational features of the hippocampus, and could enable it to contribute to the solution of what have been called configural types of learning task (Sutherland and Rudy, 1989; Rudy and Sutherland, 1995). Our view is that computationally the separation of overlapping patterns before storage in memory is a function to which the hippocampus may be particularly able to contribute because of the effect just described. It is not inconsistent with this if configural learning can take place without the hippocampus; one might just expect it to be better with the hippocampus, particularly when not just one, but many different memories must be learned and kept separate. Given that the analysis we have provided indicates that the hippocampus could act as a store for many thousands of different memories, it might be expected that particularly when many different memories, not just one memory, must be learned quickly, the effects of hippocampal lesions would be most evident.

The third way arises because of the very low contact probability in the mossy fibre–CA3 connections, and has been explained above and by Treves and Rolls (1992).

A fourth way is that, as suggested and explained above, the dentate granule cell–mossy fibre input to the CA3 cells may be powerful and its use particularly during learning would be efficient in forcing a new pattern of firing onto the CA3 cells during learning.

6.9 The CA1 network

6.9.1 Recoding in CA1

Consider the CA3 autoassociation effect. In this, several arbitrary patterns of firing occur together on the CA3 neurons, and become associated together to form an episodic or 'whole scene' memory. It is essential for this operation that several different sparse representations are present conjunctively in order to form the association. Moreover, when completion operates in the CA3 autoassociation system, all the neurons firing in the original conjunction can be brought into activity by only a part of the original set of conjunctive events. For these reasons, a memory in the CA3 cells consists of several different simultaneously active ensembles of activity. To be explicit, the parts A, B, C, D, and E of a particular episode would each be represented, roughly speaking, by its own population of CA3 cells, and these five populations would be linked together by autoassociation. It is suggested that the CA1 cells, which receive these groups of simultaneously active ensembles, can detect the conjunctions of firing of the different ensembles which represent the episodic memory, and allocate by rapid competitive learning neurons to represent at least larger parts of each episodic memory (Rolls, 1987, 1989a–c, 1990a,b). In relation to the simple example above, some CA1 neurons might code for ABC, and others for BDE, rather than having to maintain independent representations in CA1 of A, B, C, D, and E. This implies a more efficient representation, in the sense that when eventually, after many further stages, neocortical neuronal activity is recalled (as discussed below), each neocortical cell need not be accessed by all the axons carrying each component A, B, C, D, and E, but instead by fewer axons carrying larger fragments, such as ABC and BDE. Concerning the details of operation of the CA1 system, we note that although competitive learning may capture part of how it is able to recode, the competition is probably not global, but instead would operate relatively locally within the domain of the connections of inhibitory neurons. This simple example is intended to show how the coding may become less componential and more conjunctive in CA1 than in CA3, but should not be taken to imply that the representation produced becomes more sparse.

6.9.2 Preserving the information content of CA3 patterns

The amount of information about each episode retrievable from CA3 has to be balanced against the number of episodes that can be held concurrently in storage. This is because the total amount of information that can be stored in an autoassociation memory is approximately constant, so that there must be a trade-off between the number of patterns stored and the amount of information in each pattern (see Appendix A4). More patterns can be stored if each pattern involves a sparse representation, because there is then less information in each pattern. Whatever the amount of information per episode in CA3, one may hypothesize that the organization of the structures that follow CA3 (i.e. CA1, the various subicular fields, and the return projections to neocortex) should be optimized so as to preserve and use this information content in its entirety. This would prevent further loss of information, after the massive but necessary reduction in information content that has taken place along the sensory pathways and before the autoassociation stage in CA3. We have proposed that the need to preserve the full information content present in the output of an autoassociative memory requires an intermediate recoding stage (CA1) with special characteristics (Treves and Rolls, 1994). In fact, a calculation (see Appendix A3) of the information present in the CA1 firing pattern, elicited by a pattern of activity retrieved from CA3, shows that a considerable fraction of the information is lost if the synapses are non-modifiable, and that this loss can be prevented only if the CA3 to CA1 synapses are associatively modifiable (Treves, 1995). Their modifiability should match the plasticity of the CA3 recurrent collaterals. In addition, if the total amount of information carried by CA3 cells is redistributed over a larger number of CA1 cells, less information can be loaded onto each CA1 cell, rendering the code more robust to information loss in the next stages. For example, if each CA3 cell had to code for 2 bits of information, for example by firing at one of four equiprobable activity levels, then each CA1 cell (if there were twice as many as there are CA3 cells) could code for just 1 bit, for example by firing at one of only two equiprobable levels. Thus the same information content could be maintained in the overall representation while reducing the sensitivity to noise in the firing level of each cell. Another aspect of CA1 is that the number of cells increases relative to the number of CA3 cells in monkeys as compared to rats, and even more in humans (see Treves and Rolls, 1992). This may be related to the much greater divergence in the backprojections to the neocortex required in monkeys and humans, because the neocortex expands so much in monkeys and humans. The hypothesis here is that while it is not beneficial to increase the number of cells in CA3 because it must remain as one network to allow arbitrary associations and its storage capacity is limited by the number of inputs that a single hippocampal CA3 cell can receive and integrate over, there is no such limit once the information has passed the CA3 bottleneck, and to prepare for the great divergence in the backprojections it may be helpful in primates to increase the number of CA1 cells.

6.9.3 The perforant path projection to CA1

One major feature of the CA1 network is its double set of afferents, with each of its cells receiving most synapses from the Schaffer collaterals coming from CA3, but also a definite proportion (about one-third in the rat; Amaral *et al.*, 1990) from direct perforant path projections from entorhinal cortex. Such projections appear to originate mainly in layer 3 of entorhinal cortex (Witter *et al.*, 1989), from a population of cells only partially overlapping with that (mainly in layer 2) giving rise to the perforant path projections to DG and CA3. The existence of the double set of inputs to CA1 appears to indicate that there is a need for a system of projections that makes available, after the output of CA3, some information closely related to that which was originally given as input to the hippocampal formation.

One approach based on a quantitative analysis suggests an explanation for this. During retrieval, the information content of the firing pattern produced by the hypothesized CA3 autoassociative net is invariably reduced both with respect to that originally present during learning, and, even more, with respect to that available, during learning, at the input to CA3. The perforant path projection may thus serve, during retrieval, to integrate the information-reduced description of the full event recalled from CA3, with the information-richer description of only those elements used as a cue provided by the entorhinal/perforant path signal. This integration may be useful both in the consolidation of longer-term storage, and in the immediate use of the retrieved memory.

Another approach taken by McClelland *et al.* (1995) to the possible function of the direct perforant path input to the apical dendrites of the CA1 cells is to suggest that this input forces the CA1 cells to fire in such a way that they produce, through relatively unmodifiable synapses, a long-term representation of neocortical memory. This representation they suggest is then associated with the episodic information stored in the CA3 cells by the modifiable CA3 to CA1 synapses. This particular suggestion is not very plausible, for the direct entorhinal cortex projection via the perforant path to CA1 is relatively weak (Levy, Colbert and Desmond, 1995), and does not appear to have the property required of an input which would force the CA1 cells into a given pattern of activity, so that the CA3 firing could become associated with the CA1 pattern. Apart from the way in which the hippocampus may lead to neocortical recall of the correct representation, the theory of McClelland *et al.* (1995) is very similar to that of Rolls (1989b–e) and Treves and Rolls (1994).

6.10 Backprojections to the neocortex, and recall from the hippocampus

Two quantitative neurocomputational issues are considered in Section 6.10. The first is how the limited capacity of the hippocampus might, according to the buffer store hypothesis, have implications for understanding retrograde amnesia which may be produced by damage to the hippocampal system. The second is how information could be retrieved from the hippocampus, as it must be if the hippocampus is to be used either as an intermediate-term buffer store, or as a long-term store of some types of information, for example episodic memories.

6.10.1 Capacity limitations of the hippocampus, and its implications for the possibility that the hippocampus acts as a buffer store, and for the gradient of retrograde amnesia

As is shown above, the number of memories that can be stored in the hippocampal CA3 network is limited by the number of synapses on the dendrites of each CA3 cell devoted to CA3 recurrent collaterals (12 000 in the rat), increased by a factor of perhaps 3–5 depending on the sparseness of activity in the representations stored in CA3. Let us assume for illustration that this number is 50 000. If more than this number were stored, then the network would become unable to retrieve memories correctly. Therefore, there should be a mechanism that prevents this capacity being exceeded, by gradually overwriting older memories. This overwriting mechanism is, in principle, distinct from the heterosynaptic long-term depression implicit in the covariance (associative) learning rule hypothesized above to be present in the hippocampus. The heterosynaptic depression is required in order to minimize interference between memories held at any one time in the store, rather than in order to gradually delete older memories. A variety of formulas, representing modifications of the basic covariance rule, may be proposed in order to explicitly model the overwriting mechanism (see Appendix A3). The rate at which this overwriting mechanism operates should be determined by the rate at which new memories are to be stored in the hippocampus, and by the need to keep the total number concurrently stored below the capacity limit. If this limit were exceeded, it would be very difficult to recall any memories reliably from the hippocampus. If older memories are overwritten in the hippocampus, the average time after which older memories are not retrievable any more from the hippocampus will not, therefore, be a characteristic time *per se*, but rather the time in which a number of new episodes, roughly of the order of the storage capacity, has been acquired for new storage in the hippocampal system.

Given the estimates for the number of memories that can be stored in the CA3 network described above, one can link the estimated number p with a measure of the time gradient of retrograde amnesia. Such measures have been tentatively produced for humans, monkeys, and rats (Squire, 1992), although there is still discussion about the evidence for a temporal gradient for retrograde amnesia (Gaffan, 1993). To the extent that there may be such a gradient as a function of the remoteness of the memory at the onset of amnesia, it is of interest to consider a possible link between the gradient and the number of memories that can be stored in the hippocampus. The notion behind the link is that the retrograde memories lost in amnesia are those not yet consolidated in longer-term storage (in the neocortex). As they are still held in the hippocampus, their number has to be less than the storage capacity of the (presumed) CA3 autoassociative memory. Therefore, the time gradient of the amnesia provides not only a measure of a characteristic time for consolidation, but also an upper bound on the rate of storage of new memories in CA3. For example, if one were to take as a measure of the time gradient in the monkey, say, 5 weeks (about 50 000 min; Squire, 1992) and as a reasonable estimate of the capacity of CA3 in the monkey, say $p = 50\,000$ (somewhat larger than a typical estimate for the rat, if C is larger and the sparseness perhaps smaller), then one would conclude that there is an upper bound on the rate of storage in CA3 of not more than one new memory per minute, on average. (This might be an average over many weeks; the fastest rate might be closer to 1 per second.)

The main point just made based on a quantitative neurocomputational approach is that the hippocampus would be a useful store for only a limited period, which might be in the order of days, weeks, or months, depending on the acquisition rate of new memories. If the animal was in a constant and limited environment, then as new information is not being added to the store, the representations in the store would remain stable and persistent. Our hypotheses have clear experimental implications, both for recordings from single neurons and for the gradient of retrograde amnesia, in stable or frequently changing environments (Treves and Rolls, 1994). For example, memories would need to become independent of the hippocampus more quickly in time particularly under the condition that many new memories are being learned, in a frequently changing environment. Other experimental implications arise when considering changes in the parameters of the circuitry, for example a decrease in the number of synapses per cell that could result from ageing. The computational approach suggests that such a change, that would result in a decrease in hippocampal storage capacity, could be evident as a shorter period to forgetting in old animals, as measured behaviourally or neurophysiologically (Barnes *et al.*, 1994; Treves, Barnes and Rolls, 1996).

6.10.2 The recall of information from the hippocampus to the cerebral cortex

What is suggested is that the hippocampus is able to recall the whole of a previously stored memory (e.g. a memory of an episode), when only a fragment of the memory is available to start the recall. This recall from a fragment of the original memory would take place particularly as a result of completion produced by the autoassociation implemented in the CA3 network. It would then be the role of the hippocampus to reinstate in the cerebral neocortex (via the entorhinal cortex) the whole of the memory. The cerebral cortex would then, with the whole of the information in the memory (of, for example, the episode) now producing firing in the correct sets of neocortical neurons, be in a position to use the information to initiate action, or on the buffer store hypothesis, to incorporate the recalled information into a long-term store in the neocortex (Rolls, 1996c).

We suggest that the modifiable connections from the CA3 neurons to the CA1 neurons allow the whole of a particular memory retrieved in CA3 to be produced in CA1. This may be assisted as described above by the direct perforant path input to CA1. This might allow details of the input key for the recall process, as well as the possibly less information-rich memory of the whole (episodic) memory recalled from the CA3 network, to contribute to the firing of CA1 neurons. The CA1 neurons would then activate directly via their termination in the deep layers of the entorhinal cortex, and via the subiculum, at least the pyramidal cells in the deep layers of the entorhinal cortex (see Fig. 6.1). These neurons would then, by virtue of their backprojections to the parts of cerebral cortex that originally provided the inputs to the hippocampus, terminate in the superficial layers of those neocortical areas, where synapses would be made onto the distal parts of the dendrites of the cortical pyramidal cells (see Rolls, 1989a–c, 1996c). The areas of cerebral neocortex in which this recall would be produced could include multimodal cortical areas (e.g. the cortex in the superior temporal sulcus which receives inputs from temporal, parietal and occipital cortical areas, and from which it is thought that cortical areas such as 39 and 40 related to language developed), and also areas of unimodal association cortex (e.g. inferior temporal visual cortex).

The architecture with which recall from the hippocampus to the neocortex would be achieved is shown in Fig. 6.1. The feedforward connections from association areas of the cerebral neocortex (solid lines in Fig. 6.1), show major convergence as information is passed to CA3, with the CA3 autoassociation network having the smallest number of neurons at any stage of the processing. The backprojections allow for divergence back to neocortical areas. The way in which we suggest that the backprojection synapses are set up to have the appropriate strengths for recall is as follows (see also Rolls, 1989a,b; Treves and Rolls, 1994). During the setting up of a new (episodic) memory, there would be strong feedforward activity progressing towards the hippocampus. During the episode, the CA3 synapses would be modified, and via the CA1 neurons and the subiculum, a pattern of activity would be produced on the backprojecting synapses to the entorhinal cortex. Here the backprojecting synapses from active backprojection axons onto pyramidal cells being activated by the forward inputs to entorhinal cortex would be associatively modified. (This process is formally pattern association, with the forward inputs to neocortical pyramidal cells the forcing stimulus, and the backprojections the to-be-learned stimulus.) A similar process would be implemented at preceding stages of neocortex, that is in the parahippocampal gyrus/

perirhinal cortex stage, and in association cortical areas, as shown in Fig. 6.1. The timing of the backprojecting activity would be sufficiently rapid for this, in that, for example, inferior temporal cortex (ITC) neurons become activated by visual stimuli with latencies of 90–110 ms and may continue firing for several hundred milliseconds (Rolls, 1992b); and hippocampal pyramidal cells are activated in visual object-and-place and conditional spatial response tasks with latencies of 120–180 ms (Rolls, Miyashita et al., 1989; Miyashita et al., 1989). Thus, backprojected activity from the hippocampus might be expected to reach association cortical areas such as the inferior temporal visual cortex within 60–100 ms of the onset of their firing, and there would be a several hundred millisecond period in which there would be conjunctive feedforward activation present with simultaneous backprojected signals in the association cortex.

During recall, the backprojection connections onto the distal synapses of cortical pyramidal cells would be helped in their efficiency in activating the pyramidal cells by virtue of two factors. The first is that with no forward input to the neocortical pyramidal cells, there would be little shunting of the effects received at the distal dendrites by the more proximal effects on the dendrite normally produced by the forward synapses. Further, without strong forward activation of the pyramidal cells, there would not be very strong feedback and feedforward inhibition via GABA cells, so that there would not be a further major loss of signal due to (shunting) inhibition on the cell body and (subtractive) inhibition on the dendrite. (The converse of this is that when forward inputs are present, as during normal processing of the environment rather than during recall, the forward inputs would, appropriately, dominate the activity of the pyramidal cells, which would be only influenced, not determined, by the backprojecting inputs (see Rolls, 1989a,b).)

The synapses receiving the backprojections would have to be Hebb-modifiable, as suggested by Rolls (1989a,b). This would solve the de-addressing problem, that is the problem of how the hippocampus is able to bring into activity during recall just those cortical pyramidal cells that were active when the memory was originally being stored. The solution hypothesized (Rolls, 1989a,b) arises because, during learning, modification occurs of the synapses from active backprojecting neurons from the hippocampal system onto the dendrites of only those neocortical pyramidal cells active at the time of learning. Without this modifiability of cortical backprojections during learning, it is difficult to see how exactly the correct cortical pyramidal cells active during the original learning experience would be activated during recall. Consistent with this hypothesis (Rolls, 1989a,b), there are NMDA receptors present especially in superficial layers of the cerebral cortex (Monaghan and Cotman, 1985), implying Hebb-like learning just where the backprojecting axons make synapses with the apical dendrites of cortical pyramidal cells.

If the backprojection synapses are associatively modifiable, we may consider the duration of the period for which their synaptic modification should persist. What follows from the operation of the system described above is that there would be no point, indeed it would be disadvantageous, if the synaptic modifications lasted for longer than the memory remained in the hippocampal buffer store. What would be optimal would be to arrange for the associative modification of the backprojecting synapses to remain for as long as the memory persists in the hippocampus. This suggests that a similar mechanism for the associative modification within the hippocampus and for that of at least one stage of the backprojecting synapses would be appropriate. It is suggested that the presence of high concentrations of NMDA synapses in the distal parts of the dendrites of neocortical pyramidal cells and within the hippocampus may reflect the similarity of the synaptic modification processes in these two regions (cf. Kirkwood *et al.*, 1993). It is noted that it would be appropriate to have this similarity of time course for at least one stage in the series of backprojecting stages from the CA3 region to the neocortex. Such stages might include the CA1 region, subiculum, entorhinal cortex, and perhaps the parahippocampal gyrus. However, from multimodal cortex (e.g. the parahippocampal gyrus) back to earlier cortical stages, it might be desirable for the backprojecting synapses to persist for a long period, so that some types of recall and top-down processing (see Rolls, 1989a,b and Section 10.2) mediated by the operation of neocortico-neocortical backprojecting synapses could be stable.

An alternative hypothesis to that above is that rapid modifiability of backprojection synapses would be required only at the beginning of the backprojecting stream. Relatively fixed associations from higher to earlier neocortical stages would serve to activate the correct neurons at earlier cortical stages during recall. For example, there might be rapid modifiability from CA3 to CA1 neurons, but relatively fixed connections from there back (McClelland *et al.*, 1995). For such a scheme to work, one would need to produce a theory not only of the formation of semantic memories in the neocortex, but also of how the operations performed according to that theory would lead to recall by setting up appropriately the backprojecting synapses, and of how neocortical long-term memories could force an adequate representation of long-term memory onto the CA1 neurons via the direct perforant path inputs to CA1.

We describe in Chapter 10 how backprojections, including cortico-cortical backprojections, and backprojections originating from structures such as the hippocampus and amygdala, may have a number of different functions (Rolls, 1989a–c, 1990a,b, 1992a). The particular function with which we have been concerned here is how memories or information stored in the hippocampus might lead to recall expressed in regions of the cerebral neocortex.

6.10.3 Quantitative constraints on the backprojection connectivity

How many backprojecting fibres does one need to synapse on any given neocortical pyramidal cell, in order to implement the mechanism outlined above? Clearly, if neural network theory were to produce a definite constraint of that sort, quantitative anatomical data could be used for verification or falsification.

Attempts to come up with an estimate of the number of synapses required have sometimes followed the simple line of reasoning presented next. (The type of argument to be described has been applied to analyse the capacity of pattern association and autoassociation memories (see for example Marr, 1971; Willshaw and Buckingham, 1990), and because there are limitations in this type of approach, we address this approach in this paragraph.) Consider first the assumption that hippocampo-cortical connections are monosynaptic and not modifiable, and that all existing synapses have the same efficacy. Consider further the assumption that a hippocampal representation across N cells of an event consists of $a_h N$ cells firing at the same elevated rate, while the remaining $(1 - a_h)N$ cells are silent. If each pyramidal cell in the association areas of neocortex receives synapses from an average C^h hippocampal axons, there will be an average probability $y = C^h/N$ of finding a synapse from any given hippocampal cell to any given neocortical one. Across neocortical cells, the number A of synapses of hippocampal origin activated by the retrieval of a particular episodic memory will follow a Poisson distribution of average $ya_hN = a_hC^h$

$$P(A) = (a_hC^h)^A \exp(-a_hC^h)/A! \tag{6.6}$$

The neocortical cells activated as a result will be those, in the tail of the distribution, receiving at least T active input lines, where T is a given threshold for activation. Requiring that $P(A > T)$ be at least equal to a_{nc}, the fraction of neocortical cells involved in the neocortical representation of the episode, results in a constraint on C^h. This simple type of calculation can be extended to the case in which hippocampo-cortical projections are taken to be poly-synaptic, and mediated by modifiable synapses. In any case, the procedure does not appear to produce a very meaningful constraint for at least three reasons. First, the resulting minimum value of C^h, being extracted by looking at the tail of an exponential distribution, varies dramatically with any variation in the assumed values of the parameters a_h, a_{nc}, and T. Second, those parameters are ill-defined in principle: a_h and a_{nc} are used having in mind the unrealistic assumption of binary distributions of activity in both the hippocampus and neocortex (although the sparseness of a representation can be defined in general, as shown above, it is the particular definition pertaining to the binary case that is invoked here); while the definition of T is based on unrealistic assumptions about neuronal dynamics (how coincident does one require the various inputs to be in time, in order to generate a single spike, or a train of spikes of given frequency, in the postsynaptic cell?). Third, the calculation assumes that neocortical cells receive no other inputs, excitatory or inhibitory. Relaxing this assumption, to include for example non-specific activation by subcortical afferents, makes the calculation extremely fragile. This argument applies in general to approaches to neurocomputation which base their calculations on what would happen in the tail of an exponential, Poisson, or binomial distribution. Such calculations must be interpreted with great care for the above reasons.

An alternative way to estimate a constraint on C^h, still based on very simple assumptions, but hopefully producing a result which is more robust with respect to relaxing those assumptions, is the following.

Consider a polysynaptic sequence of backprojecting stages, from hippocampus to neo-cortex, as a string of simple heteroassociative memories in which, at each stage, the input lines are those coming from the previous stage (closer to the hippocampus). Implicit in this framework is the assumption that the synapses at each stage are modifiable and have been indeed modified at the time of first experiencing each episode, according to some Hebbian associative plasticity rule. A plausible requirement for a successful hippocampo-directed recall operation is that the signal generated from the hippocampally retrieved pattern of activity, and carried backwards towards neocortex, remains undegraded when compared to the noise due, at each stage, to the interference effects caused by the concurrent storage of other patterns of activity on the same backprojecting synaptic systems. That requirement is equivalent to that used in deriving the storage capacity of such a series of associative memories, and it was shown in Treves and Rolls (1991) that the maximum number of independently generated activity patterns that can be retrieved is given, essentially, by the same formula as Eq. 6.2 above

$$p \approx \frac{C}{a \, \ln(1/a)} \, k'$$

(6.7)

where, however, a is now the sparseness of the representation at any given stage, and C is the average number of (back-)projections each cell of that stage receives from cells of the previous one. (k' is a similar slowly varying factor to that introduced above in Eq. 6.2.) (The interesting insight here is that multiple iterations through a recurrent autoassociative network can be treated as being formally similar to a whole series of feedforward associative networks, with each new feedforward network equivalent to another iteration in the recurrent network (Treves and Rolls, 1991). This formal similarity enables the quantitative analysis developed in general for autoassociation nets (see Treves and Rolls, 1991) to be applied directly to the series of backprojection networks (Treves and Rolls, 1994).) If p is equal to the number of memories held in the hippocampal buffer, it is limited by the retrieval capacity of the CA3 network, p_{max}. Putting together the formula for the latter with that shown here, one concludes that, roughly, the requirement implies that the number of afferents of (indirect) hippocampal origin to a given neocortical stage(C^h), must be $C^h = C^{RC} a_{nc}/a_{CA3}$, where C^{RC} is the number of recurrent collaterals to any given cell in CA3, the average sparseness of a representation is a_{nc}, and a_{CA3} is the sparseness of memory representations in CA3 (Treves and Rolls, 1994).

The above requirement is very strong: even if representations were to remain as sparse as they are in CA3, which is unlikely, to avoid degrading the signal, C^h should be as large as C^{RC}, i.e. 12 000 in the rat. Moreover, other sources of noise not considered in the present calculation would add to the severity of the constraint, and partially compensate for the relaxation in the constraint that would result from requiring that only a fraction of the p episodes would involve any given cortical area. If then C^h has to be of the same order as C^{RC}, one is led to a very definite conclusion: a mechanism of the type envisaged here could not possibly rely on a set of monosynaptic CA3-to-neocortex backprojections. This would imply that, to make a sufficient number of synapses on each of the vast number of neocortical cells, each cell in CA3 has to generate a disproportionate number of synapses (i.e. C^h times the ratio between the number of neocortical and of CA3 cells). The required divergence can be kept within reasonable limits only by assuming that the backprojecting system is polysynaptic, provided that the number of cells involved grows gradually at each stage, from CA3 back to neocortical association areas (see Fig. 6.1).

Although backprojections between any two adjacent areas in the cerebral cortex are approximately as numerous as forward projections, and much of the distal parts of the dendrites of cortical pyramidal cells are devoted to backprojections, the actual number of such connections onto each pyramidal cell may be on average only in the order of thousands. Further, not all might reflect backprojection signals originating from the hippocampus, for there are backprojections which might be considered to originate in the amygdala (see Amaral *et al.*, 1992) or in multimodal cortical areas (allowing for example for recall of a visual image by an auditory stimulus with which it has been regularly associated). In this situation, one may consider whether the backprojections from any one of these systems would be sufficiently numerous to produce recall. One factor which may help here is that when recall is being produced by the backprojections, it may be assisted by the local recurrent collaterals between nearby (within approximately 1 mm) pyramidal cells which are a feature of neocortical connectivity. These would tend to complete a partial neocortical representation being recalled by the backprojections into a complete recalled pattern. There are two alternative possibilities about how this would operate. First, if the recurrent collaterals showed slow and long-lasting synaptic modification, then they would be useful in completing the whole of long-term (e.g. semantic) memories. Second, if the neocortical recurrent collaterals showed rapid changes in synaptic modifiability with the same time course as that of hippocampal synaptic modification, then they would be useful in filling in parts of the information forming episodic memories which could be made available locally within an area of the cerebral neocortex.

6.10.4 Simulations of hippocampal operation

In order to test the operation of the whole hippocampal system for individual stages of which an analytic theory has now been developed (Treves and Rolls, 1994), Rolls (1995c) simulated the network shown in Fig. 6.7. The operation of the network was tested by presenting the entorhinal cortex with large numbers of sparse random binary patterns, allowing the hippocampal system to learn these patterns, and then, after all the learning, to present small parts of each pattern to the entorhinal cortex, and measuring how well, under different experimental conditions, the whole pattern could be retrieved in the firing of entorhinal neurons. The number of neurons in each part of the network and the number of connections from each source onto each type of neuron are shown. These numbers reflect the relative numbers of neurons and connections per neuron in the real rat hippocampus, scaled down to fit within a simulation of reasonable size, yet kept large enough so that finite size effects due to small numbers of neurons are not a limiting factor. (Some of the theory we and others have developed applies analytically to large networks (see further Treves and Rolls, 1994), but we have checked that the simulation is sufficiently large, for example by simulating CA3 alone with different numbers of neurons (1000, 2000, 4000, and 8000), and showing that the results scale down from 8000 to 1000 as expected. We have also compared the capacity of CA3 in these simulations with that calculated analytically (see Treves and Rolls, 1991), and obtained the expected values (Simmen, Treves and Rolls, 1996; Rolls, Treves, Foster and Perez-Vicente, 1997)).

The performance of the network with the standard values for the parameters, the most important of which was that the sparseness of CA3 was set to 0.05, is shown in Fig. 6.8. With the recurrent collateral effect in CA3 off, recall in CA3 was produced only by the modifiable synapses from the entorhinal cortex onto the CA3 neurons (which operate essentially as a pattern associator), and the recall was not good. (The correlation of the recalled CA3 pattern of firing with that originally stored was no better than the correlation of the entorhinal retrieval cue with the stimulus learned originally, as shown by the fact that the CA3 recall points lie approximately along a diagonal in Fig. 6.8a). In contrast, when the CA3 recurrent

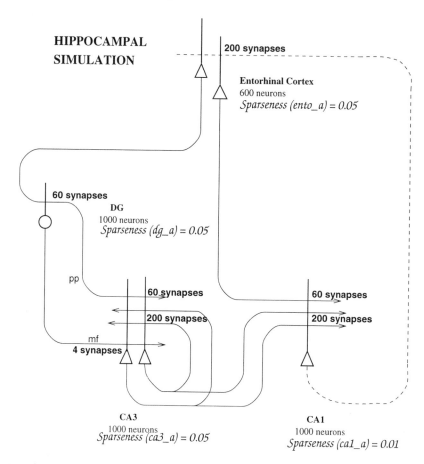

Fig. 6.7 Architectural specification of the network simulated. DG, dentate gyrus; mf, mossy fibres; pp, perforant path. The numbers of neurons, the numbers of synapses per neuron, and the sparsenesses for each stage of the network are shown.

collateral effect was allowed to operate, recall was much better in CA3. It is also shown that as the network is more heavily loaded, performance is less good, as expected. It is also shown in Fig. 6.8b that the CA1 recall is better than the CA3 recall, and the pattern recalled in the entorhinal cortex is better than that recalled in the CA1 neurons. This indicates that in this multistage system, partial recall at an intermediate stage such as CA3 can be improved upon by subsequent stages in the system, such as CA1 (which is a competitive network) and the entorhinal cortex, which improves on retrieval because it acts as a pattern associator. The maximum level of retrieval in the entorhinal cortex is limited by the number of synapses onto each entorhinal neuron: if set at a realistically scaled number of 200, the maximum correlation achieved through this pattern associator is 0.8, and if increased to 1000, as in the simulations shown, the pattern recalled in the entorhinal cortex can approach a correlation of 1.0 with that learned (see Rolls and Treves, 1990; Treves and Rolls, 1994; Rolls, 1995c).

Use of this network showed that the widespread connectivity of the CA3 recurrent collaterals is important for good retrieval from incomplete cues restricted to one part of

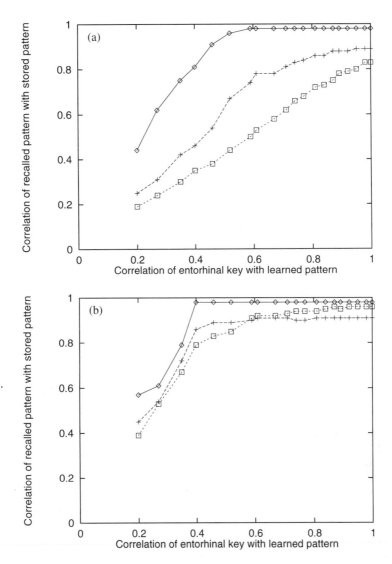

Fig. 6.8 Recall in CA3 (squares), CA1 (+) and entorhinal cortex (diamond) (expressed as the correlation between the recalled pattern and the pattern present during learning in the same set of neurons, ordinate), as a function of the correlation of the entorhinal retrieval cue with the entorhinal pattern learned (abscissa). 100 patterns learned. (a) CA3 recurrent collateral effect off in recall. (b) CA3 recurrent collateral effect on in recall. The sparseness, a, was 0.05 in the entorhinal cortex, 0.05 for the dentate granule cells; 0.05 for the CA3 cells, and 0.01 for the CA1 cells.

the topographically organized inputs (Rolls, 1995c). The simulation also showed that the network operated much better if the CA3 cells operated in binary mode (either firing or not), rather than having continuously graded firing rates (Rolls, 1995c; Rolls, Treves, Foster and Perez-Vicente, 1997). The reason for this is that given that there is a maximum amount of information that can be stored in an autoassociation system, memories that are stored as a continuously variable pattern of activity will be more corrupted at retrieval in CA3, having the impact shown on retrieval in the entorhinal cortex. The use of this simulation has thus not only enabled the theory we have developed (see Treves and Rolls, 1994; Rolls, 1996c) to be verified computationally, but has also anticipated possible future neurophysiological experiments, by showing the type of effect expected if the CA3 recurrent collaterals were made inoperative, and made it of interest to focus on the distribution of firing rates of hippocampal pyramidal cells.

6.11 Other suggestions concerning hippocampal operation and function

We have now described hypotheses on how a number of different parts of hippocampal and related circuitry might operate (see further Rolls, 1989a–c, 1990a–c; Treves and Rolls, 1992, 1994). Although these hypotheses are consistent with a theory of how the hippocampus operates, some of these hypotheses could be incorporated into other views or theories. The overall view described in this chapter is in fact close in different respects to those of a number of other investigators (Brown and Zador, 1990; Eichenbaum, Otto and Cohen, 1992; Gaffan, 1992; Gardner-Medwin, 1976; Marr, 1971; McNaughton and Nadel, 1990; McNaughton et al., 1995; Squire, 1992). In the following, we briefly review some of the different and alternative suggestions that have been made. The journal *Hippocampus* has devoted some of its issues to forums of discussion on hippocampal function (e.g. Vol. 1, No. 3, 1991, on spatial function; and Vol. 6, No. 6, 1996 on computational models).

Some theories postulate that the hippocampus performs spatial computation. The theory of O'Keefe and Nadel (1978), that the hippocampus implements a cognitive map, placed great emphasis on spatial function. It supposed that the hippocampus at least holds information about allocentric space in a form which enables rats to find their way in an environment even when novel trajectories are necessary. O'Keefe (1990) and Burgess, Recce and O'Keefe (1994) have extended this analysis and produced a computational theory of the hippocampus as a cognitive map, in which the hippocampus performs geometric spatial computations. In at least one version of this theory, the hippocampus stores the centroid and slope of the distribution of landmarks in an environment, and stores the relationships between the centroid and the individual landmarks. The hippocampus then receives as inputs information about where the rat currently is, and where the rat's target location is, and computes geometrically the body turns and movements necessary to reach the target location. In this sense, the hippocampus is taken to be a spatial computer, which produces an output which is very different from its inputs. McNaughton and colleagues have also, at various times, elaborated on the link between the hippocampus and spatial computations. McNaughton et al. (1991) have proposed a 'compass' solution to the problem of spatial navigation along novel trajectories in known environments, postulating that distances and bearings (i.e. vector quantities) from landmarks are stored, and that computation of a new trajectory involves vector subtraction by the hippocampus. The detailed mechanism envisaged includes pattern association between head direction cells and local view cells, where the local views are used to reset dead-reckoning navigation. McNaughton et al. (1996) further hypothesize that rat place fields may reflect more path integration than a constellation of sensory cues, and may be reset only, for example, when the animal hits the boundaries of the environment. The ensemble of fields expressed by hippocampal cells in one environment would constitute one of several prewired *charts*, onto which the salient cues present in any given environment would 'anchor'. This theory is again in contrast to the theory outlined in this chapter, in which the hippocampus operates as a memory utilizing autoassociation, which is able to recall what was stored in it, using a partial cue if necessary. Our theory is consistent with the presence of spatial view cells and whole body motion (e.g. vestibular) cells in the primate hippocampus (Rolls and O'Mara, 1993; Rolls, 1996d,e) (or place cells in the rat hippocampus and head direction cells in the rat presubiculum), for we note that it is often important to store and later recall where one has been (views of the environment, body turns made, etc.), and indeed such (episodic) memories are required for navigation by 'dead reckoning' in small environments.

Thus some theories postulate that the hippocampus actually performs a spatial computation, and contrast to the view presented here, and others of similar perspective, which emphasize the ability of the hippocampus to rapidly store information, and then to recall what was stored to the neocortex, by acting as an autoassociative memory. This function is often necessary for successful spatial computation, but is not itself spatial computation. Instead, we believe that spatial computation is more likely to be performed in the parietal and temporal cortex (utilizing information if necessary recalled from the hippocampus). Consistent with this view, hippocampal damage impairs the ability to learn new environments but not to perform spatial computations such as finding one's way to a place in a familiar environment. Instead, damage to neocortical areas such as the right parahippocampal cortex can produce deficits such as topographical disorientation (Habib and Sirigu, 1987), and damage to the right parieto-temporo-occipital region can produce topographical agnosia (see Grusser and Landis, 1991), both of which can be taken as requiring spatial computation. (In monkeys, there is evidence for a role of the parietal cortex in allocentric spatial computation. For example, monkeys with parietal cortex lesions are impaired at performing a landmark task, in which the object to be chosen is signified by the proximity to it of a 'landmark' (another object) (Ungerleider and Mishkin, 1982).)

The theory outlined in this chapter thus holds that spatial computation is performed by the neocortex, and that the hippocampus may be involved in spatial computation only in so far as new information may be required to be stored or recalled in order to perform a particular spatial task. We note that the quantitative arguments which have been presented here in episodic memory terms, such as the number of memories that could be stored in the system, would still apply once reformulated in terms of spatial memory, e.g. the (different) number of charts that the system could maintain, or in terms of other types of memory.

Marr (1971) had the general view that the hippocampal system operates as an intermediate-term memory. While his computational arguments were not detailed enough to identify specific functions for different parts of the hippocampal circuitry (dentate, CA3, CA1, etc.), his approach has been very influential at the systems level. Marr suggested in particular that memories might be unloaded from the hippocampus to the cerebral neocortex during sleep, a suggestion that has been taken up again by Wilson and McNaughton (1994). A comparison of the present theory to that of Marr (1971) is provided by Treves and Rolls (1994) and Rolls (1996c).

Another theory is that the hippocampus (and amygdala) are involved in recognition memory (Mishkin, 1978, 1982). It is now believed that recognition memory as tested in a visual delayed match to sample task is dependent on the perirhinal cortex, and rather less on hippocampal circuitry proper (Zola-Morgan et al., 1989, 1994; Gaffan and Murray, 1992).

Our approach to this is that we note that the hippocampal CA3 recurrent collateral system is most likely to be heavily involved in memory processing when new associations between arbitrary events which may be represented in different regions of the cerebral cortex must be linked together to form an episodic memory. Often, given the large inputs to the hippo-campus from the parietal cortex, one of these events will contain spatial information. We suppose that given the circuitry of the hippocampus, it is especially well suited for such tasks, although some mixing of inputs may occur before the hippocampus. We therefore predict that when arbitrary associations must be rapidly learned between such different events to form new episodic memories, the hippocampal circuitry is likely to become increasingly important, but we are not surprised if some memory formation of this type can occur without the hippocampus proper. In this respect we note that what is found after hippocampal damage, as well as from hippocampal recordings (Treves, Panzeri et al., 1996), will reflect *quantitatively* rather than just *qualitatively* the ability to form new (especially multimodal, with one modality space) episodic memories. Thus, while the episodic versus semantic memory distinction is very useful conceptually, one should not be surprised by findings that appear contradictory when interpreted on such a strict qualitative basis.

Sutherland and Rudy (1991) have proposed that the hippocampus is necessary for the formation of configurative memories, that is when a memory must be separated from memories of subsets of its elements. By acting as a competitive network, the dentate system may in fact enable the hippocampal CA3 system to store different memories for even somewhat similar episodes, which is a very important feature of an episodic memory system. In so far as the hippocampal system has circuitry specialized for this in the dentate granule cells, then we believe that this could allow the hippocampus to make a special contribution to the storage of such overlapping configurative memories, even though, as noted above, this is likely to be measurable as a quantitative advantage in memory tasks of having a hippocampus, rather than an all-or-none consequence of having a hippocampus or not (cf. Rudy and Sutherland, 1995).

The view proposed by McClelland, McNaughton and O'Reilly (1995) is similar to ours at the systems level, but their computational model assumes that the last set of synapses that are modified rapidly during the learning of each episode are those between the CA3 and the CA1 pyramidal cells (see Fig. 6.1). The entorhinal cortex connections via the perforant path onto the CA1 cells are taken to be non-modifiable (in the short term), and allow a representation of neocortical long-term memories to activate the CA1 cells. The new information learned in an episode by the CA3 system is then linked to existing long-term memories by the CA3 to CA1 rapidly modifiable synapses. All the connections from the CA1 back via the subiculum, entorhinal cortex, parahippocampal cortex, etc. to the association neocortex are held to be unmodifiable in the short term, during the formation of an episodic memory. Our arguments indicate, instead, that it is likely that at least for part of the way back into neocortical processing (e.g. as far as the inferior temporal cortex), the backprojecting synapses should be associatively modifiable, as rapidly as the time it takes to learn a new episodic memory. It may well be, though, that at earlier stages of cortical processing, for example from V4 to V2, the backprojections are relatively more fixed, being formed during early developmental plasticity or during the formation of new long-term semantic memory structures. Having such relatively fixed synaptic strengths in these earlier cortical backprojection systems could ensure that whatever is recalled in higher cortical areas, such as objects, will in turn recall relatively fixed and stable representations of parts of objects or features. Given that the functions of backprojections may include many top-down processing operations, including attention and priming (see Chapter 10), it may be useful to ensure that there is consistency in how higher cortical areas affect activity in earlier 'front-end' or preprocessing cortical areas.

The comparison of these different approaches to the hippocampus underlies the importance of combining multidisciplinary approaches in order to understand brain function. It is necessary to have good evidence on what function is performed by a brain region, using methods such as analysis of its connections and of the effects of lesions, single neuron neurophysiology, and brain imaging; and in order to understand how a part of the brain works, it is necessary to combine evidence from microanatomy, the biophysics and synaptic modifiability of neurons, and single and multiple single neuron neurophysiology, to produce a theory at the neuronal network level of how that part of the brain operates. The potential of current work in neuroscience is that it is leading us towards an understanding of how the brain actually works.

7 Pattern association in the brain: amygdala and orbitofrontal cortex

Networks that perform pattern association are described in Chapter 2, and analytic approaches to their performance are described in Appendix A3. The aim of this chapter is to review where in the brain pattern association networks are found, what their role is in the systems-level operation of the brain, and evidence on how they actually operate in the brain.

7.1 Pattern association involved in learning associations between sensory representations and reward or punishment

Learning about which stimuli in the environment are rewarding, punishing, or neutral is crucial for survival. For example, it takes just one trial to learn if a seen object is hot when we touch it, and associating that visual stimulus with the pain may help us to avoid serious injury in the future. Similarly, if we are given a new food which has an excellent taste, we can learn in one trial to associate the sight of it with its taste, so that we can select it in future. In our examples, the previously neutral visual stimuli become conditioned reinforcers by their association with a primary (unlearned) reinforcer such as taste or pain. To make sure that our terms are clear, we can define a reward as anything for which an animal will work, and a punishment as anything an animal will work to avoid. We can also call the reward a positive reinforcer, and the punishment a negative reinforcer, where a reinforcer is defined as an event that alters the probability of a behavioural response on which it is made contingent. (Further treatment of these concepts is provided by Rolls, 1990c.) Our examples show that learning about which stimuli are rewards and punishments is very important in the control of motivational behaviour such as feeding and drinking, and in emotional behaviour such as fear and pleasure. The type of learning involved is pattern association, between the conditioned and the unconditioned stimulus. As this type of learning provides a major example of the utility of pattern association learning by the brain, in this section we consider where in sensory processing this learning occurs, which brain structures are involved in this type of learning, and how the neuronal networks for pattern association learning may actually be implemented in these regions.

A fascinating question is the stage in information processing at which such pattern association learning occurs. Does it occur, for example, early on in the cortical processing of visual signals, or late in cortical visual processing, or even after the main stages of visual cortical processing? We shall see that at least in primates, the latter is the case, and that there are good reasons why this should be the case. Before we consider the neural processing itself, we introduce in more detail in Section 7.1.1 the brain systems involved in learning about rewards and punishments, showing how they are very important in understanding the brain mechanisms involved in emotion and motivation. Section 7.1.1 is intended to be introductory, and may be omitted if the reader is familiar with this area.

7.1.1 Emotion and motivation

Emotions can usefully be defined as states produced by instrumental reinforcing stimuli (see Rolls, 1990c, and earlier work by Millenson, 1967; Weiskrantz, 1968; Gray, 1975, 1987). (Instrumental reinforcers are stimuli which if their occurrence, termination, or omission is made contingent upon the making of a response, alter the probability of the future emission of that response.) Some stimuli are unlearned reinforcers (for example the taste of food if the animal is hungry, or pain); while others may become reinforcing by learning, because of their association with such primary reinforcers, thereby becoming 'secondary reinforcers'. This type of learning may thus be called 'stimulus-reinforcement association learning', and probably occurs via the process of classical conditioning. If a reinforcer increases the probability of emission of a response on which it is contingent, it is said to be a 'positive reinforcer' or 'reward'; if it decreases the probability of such a response it is a 'negative reinforcer' or 'punishment'. For example, fear is an emotional state which might be produced by a sound (the conditioned stimulus) that has previously been associated with an electrical shock (the primary reinforcer).

The converse reinforcement contingencies produce the opposite effects on behaviour. The omission or termination of a positive reinforcer ('extinction' and 'time out' respectively, sometimes described as 'punishing'), decrease the probability of responses. Responses followed by the omission or termination of a negative reinforcer increase in probability, this pair of negative reinforcement operations being termed 'active avoidance' and 'escape' respectively (see further Gray, 1975 and Mackintosh, 1983).

The different emotions can be described and classified according to whether the reinforcer is positive or negative, and by the reinforcement contingency. An outline of such a classification scheme, elaborated more precisely by Rolls, 1990c, is shown in Fig. 7.1.

The mechanisms described here would not be limited in the range of emotions for which they could account. Some of the factors which enable a very wide range of human emotions to be analysed with this foundation are elaborated elsewhere (Rolls, 1990c), and include the following:

1. The reinforcement contingency (see Fig. 7.1).
2. The intensity of the reinforcer (see Fig. 7.1).
3. Any environmental stimulus might have a number of different reinforcement associations. For example, a stimulus might be associated both with the presentation of a reward and of a punishment, allowing states such as conflict and guilt to arise.
4. Emotions elicited by stimuli associated with different primary reinforcers will be different.

5. Emotions elicited by different secondary reinforcing stimuli will be different from each other (even if the primary reinforcer is similar).

6. The emotion elicited can depend on whether an active or passive behavioural response is possible. For example, if an active behavioural response can occur to the omission of an S+ (where S+ signifies a positively reinforcing stimulus), then anger might be produced, but if only passive behaviour is possible, then sadness, depression, or grief might occur.

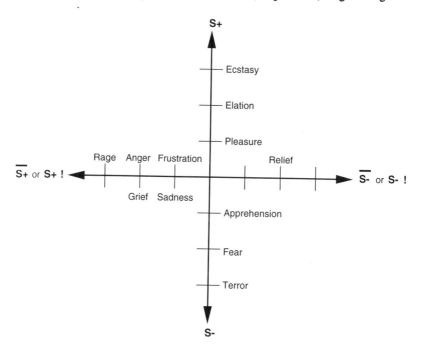

Fig. 7.1 Some of the emotions associated with different reinforcement contingencies are indicated. Intensity increases away from the centre of the diagram, on a continuous scale. The classification scheme created by the different reinforcement contingencies consists of: (1) the presentation of a positive reinforcer S+; (2) the presentation of a negative reinforcer S−; (3) the omission of a positive reinforcer $\overline{S+}$ or the termination of a positive reinforcer S+!; (4) the omission of a negative reinforcer $\overline{S-}$ or the termination of a negative reinforcer S−!.

By combining these six factors, it is possible to account for a very wide range of emotions (see Rolls, 1990c). It is also worth noting that emotions can be produced just as much by the recall of reinforcing events as by external reinforcing stimuli; that cognitive processing (whether conscious or not) is important in many emotions, for very complex cognitive processing may be required to determine whether environmental events are reinforcing or not; that emotions normally consist of cognitive processing which determines the reinforcing valence of the stimulus, and then an elicited mood change if the valence is positive or negative; and that stability of mood implies that absolute levels of reinforcement must be represented over moderately long time spans by the firing of mood-related neurons, a difficult operation which may contribute to 'spontaneous' mood swings, depression which occurs without a clear external cause, and the multiplicity of hormonal and transmitter systems which seem to be involved in the control of mood (see further Rolls, 1990c).

In terms of the neural bases of emotion, the most important point from this introduction is that in order to understand the neural bases of emotion, we need to consider brain mechanisms involved in reward and punishment, and involved in learning about which environmental stimuli are associated, or are no longer associated, with rewards and punishments.

Functions of emotion

Before considering these pattern association learning mechanisms, it is also useful to summarize the functions of emotions, because these functions are important for understanding the output systems to which brain mechanisms involved in emotion must interface. Knowing the functions of emotion is also important when considering the brain mechanisms of emotion, for the functions of emotion provide an indication of which signals (for example the sight of a face, and touch) must be associated together. The functions, described more fully elsewhere (Rolls, 1990c), can be summarized as follows:

1. The elicitation of autonomic responses (for example a change in heart rate) and endocrine responses (for example the release of adrenaline (that is epinephrine)). These prepare the body for action. There are output pathways from the amygdala and orbitofrontal cortex directly, and via the hypothalamus, to the brainstem autonomic nuclei.

2. Flexibility of behavioural responses to reinforcing stimuli. Emotional (and motivational) states allow a simple interface between sensory inputs and motor outputs, because only the valence of the stimulus to which attention is being paid need be passed to the motor system, rather than a full representation of the sensory world. In addition, when a stimulus in the environment elicits an emotional state (because it is a primary reinforcer or because of classical conditioning), we can flexibly choose any appropriate instrumental response to obtain the reward, or avoid the punishment. This is more flexible than simply learning a fixed behavioural response to a stimulus (see Gray, 1975; Rolls, 1990c). Pathways from the amygdala and orbitofrontal cortex to the striatum are implicated in these functions.

This function is based on the crucial role which rewards and punishments have on behaviour. Animals are built with neural systems that enable them to evaluate which environmental stimuli, whether learned or not, are rewarding and punishing, that is will be worked for or avoided. A crucial part of this system is that with many competing rewards, goals, and priorities, there must be a selection system for enabling the most important of these goals to become the object of behaviour at any one time. This selection process must be capable of responding to many different types of reward decoded in different brain systems that have evolved at different times, even including the use in humans of a language system to enable long-term plans to be made (see Rolls, 1997a). These many different brain systems, some involving implicit evaluation of rewards, and others explicit, verbal, conscious, evaluation of rewards and planned long-term goals (see Rolls, 1997a, b), must all enter into the selector of behaviour. This selector, although itself poorly understood, might include a process of competition between all the competing calls on output, and might involve the basal ganglia (see Chapter 9; Rolls and Johnstone, 1992; Rolls, 1994d).

3. Emotion is motivating. For example, fear learned by stimulus-reinforcement association formation provides the motivation for actions performed to avoid noxious stimuli.

4. Communication. For example, monkeys may communicate their emotional state to others, by making an open-mouth threat to indicate the extent to which they are willing to compete for resources, and this may influence the behaviour of other animals. There are neural systems in the amygdala and overlying temporal cortical visual areas, and the orbitofrontal cortex, which are specialized for the face-related aspects of this processing (see Sections 7.1.3, 7.1.4, and Chapter 8).

5. Social bonding. Examples of this are the emotions associated with the attachment of the parents to their young, and the attachment of the young to their parents (see Dawkins, 1989).

6. The current mood state can affect the cognitive evaluation of events or memories (see Blaney, 1986), and this may have the function of facilitating continuity in the interpretation of the reinforcing value of events in the environment. A hypothesis on the neural pathways which implement this is presented in Section 7.2.

7. Emotion may facilitate the storage of memories. One way in which this occurs is that episodic memory (that is one's memory of particular episodes) is facilitated by emotional states. This may be advantageous in that storing many details of the prevailing situation when a strong reinforcer is delivered may be useful in generating appropriate behaviour in situations with some similarities in the future. This function may be implemented by the relatively non-specific projecting systems to the cerebral cortex and hippocampus, including the cholinergic pathways in the basal forebrain and medial septum, and the ascending noradrenergic pathways (see Section 7.1.5; Treves and Rolls, 1994; Wilson and Rolls, 1990a,b). A second way in which emotion may affect the storage of memories is that the current emotional state may be stored with episodic memories, providing a mechanism for the current emotional state to affect which memories are recalled. A third way in which emotion may affect the storage of memories is by guiding the cerebral cortex in the representations of the world which are set up. For example, in the visual system, it may be useful to build perceptual representations or analysers which are different from each other if they are associated with different reinforcers, and to be less likely to build them if they have no association with reinforcement. Ways in which backprojections from parts of the brain important in emotion (such as the amygdala) to parts of the cerebral cortex could perform this function are discussed in Chapter 4 (Section 4.5 on the guidance of competitive learning), in Chapter 10, and by Rolls (1989b,e, 1990d).

8. Another function of emotion is that by enduring for minutes or longer after a reinforcing stimulus has occurred, it may help to produce persistent motivation and direction of behaviour.

9. Emotion may trigger the recall of memories stored in neocortical representations. Amygdala backprojections to the cortex could perform this for emotion in a way analogous to that in which the hippocampus could implement the retrieval in the neocortex of recent (episodic) memories (Rolls, 1990c; Treves and Rolls, 1994).

For emotional behaviour, rapid learning is clearly important. For some types of emotional behaviour, rapid *relearning* of the reinforcement value of a stimulus may be useful. For example, in social behaviour, reinforcers are constantly being exchanged (for example a positive glance to indicate continuing cooperation), and the current reinforcement association therefore of a person must correspondingly continually be updated. It appears that, particularly in primates, brain systems have evolved to perform this very rapid stimulus-reinforcement relearning. The relearning is often tested by reversal, in which one stimulus is paired with reward (for example a sweet taste) and another with punishment (for example a salt taste), and then the reinforcement contingency is reversed.

Motivational behaviour, such as eating and drinking, also requires reinforcement-related learning, so that we can, for example, learn to associate the sight of a food with its taste (which is innately rewarding or aversive, and is described as a primary reinforcer).

In conclusion, a theme of this section that is closely related to learning in neuronal networks is the importance of learning associations between previously neutral visual stimuli such as the sight of an object (an example of a secondary reinforcer) and innately rewarding stimuli (or primary reinforcers) such as the taste of food, or a pleasant or painful somatosensory input.

7.1.2 Processing in the cortical visual areas

We now consider whether associations between visual stimuli and reinforcement are learned, and stored, in the visual cortical areas which proceed from the primary visual cortex, V1, through V2, V4, and the inferior temporal visual cortex (see Figs 1.8–1.10, and 7.2 for schematic diagrams of the organization in the brain of some of the systems being considered). Is the emotional or motivational valence of visual stimuli represented in these regions? A schematic diagram summarizing some of the conclusions that will be reached is shown in Fig. 7.2b.

(a)

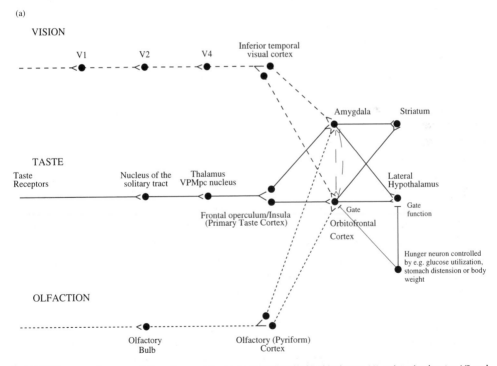

Fig. 7.2 (a) Diagrammatic representation of some of the connections described in this chapter. V1, striate visual cortex. V2 and V4, cortical visual areas. In primates, sensory analysis proceeds in the visual system as far as the inferior temporal cortex and the primary gustatory cortex; beyond these areas, in for example the amygdala and orbitofrontal cortex, the hedonic value of the stimuli, and whether they are reinforcing or are associated with reinforcement, is represented (see text).

(b)   Brain Mechanisms of Emotion

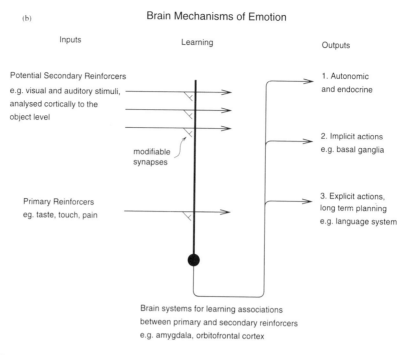

Fig. 7.2 (b) Schematic diagram showing the organization of brain networks involved in learning reinforcement associations of visual and auditory stimuli.

One way to answer this is to test monkeys in a learning paradigm in which one visual stimulus is associated with reward (for example glucose taste, or fruit juice taste), and another visual stimulus is associated with an aversive taste, such as saline. Rolls, Judge, and Sanghera did just such an experiment in 1977, and found that single neurons in the inferior temporal visual cortex did not respond differently to objects based on their reward association. To test whether a neuron might be influenced by the reward association, the monkey performed a visual discrimination task in which the reinforcement contingency could be reversed during the experiment. (That is, the visual stimulus, for example a triangle, to which the monkey had to lick to obtain a taste of fruit juice, was after the reversal associated with saline: if the monkey licked to the triangle after the reversal, he obtained mildly aversive salt solution.) An example of such an experiment is shown in Fig. 7.3. The neuron responded more to the triangle, both before reversal when it was associated with fruit juice, and after reversal, when the triangle was associated with saline. Thus the reinforcement association of the visual stimuli did not alter the response to the visual stimuli, which was based on the physical properties of the stimuli (for example their shape, colour, or texture). The same was true for the other neurons recorded in this study. This independence from reward association seems to be characteristic of neurons right through the temporal visual cortical areas, and must be true in earlier cortical areas too, in that they provide the inputs to the inferior temporal visual cortex.

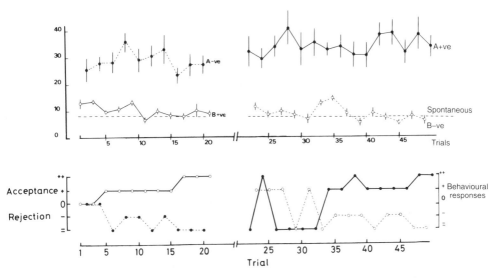

Fig. 7.3 Example of the responses of a neuron in the inferior temporal visual cortex, showing that its responses (firing rate in spikes/s, upper panel) do not reverse when the reward association of the visual stimuli reverses. For the first 21 trials of the visual discrimination task, visual stimulus A was aversive (–ve, because if the monkey licked he obtained saline), and visual stimulus B was associated with reward (+ve, because if the monkey licked when he saw this stimulus, he obtained fruit juice). The neuron responded more to stimulus A than to stimulus B. After trial 21, the contingencies reversed (so that A was now +ve, and B –ve). The monkey learned the reversal correctly by about trial 35 (lower panel). However, the inferior temporal cortex neuron did not reverse when the reinforcement contingency reversed—it continued to respond to stimulus A after the reversal, even though the stimulus was now +ve. Thus this, and other inferior temporal cortex neurons, respond to the physical aspects of visual stimuli, and not to the stimuli based on their reinforcement contingency. (From Rolls, Judge and Sanghera, 1977.)

Why reward and punishment associations of stimuli are not represented early in information processing in the primate brain

The processing stream that has just been considered is that concerned with objects, that is with *what* is being looked at. Two fundamental points about pattern association networks for stimulus-reinforcement association learning can be made from what we have considered. The first point is that sensory processing in the primate brain proceeds as far as the invariant representation of objects (invariant with respect to, for example, size, position on the retina, and even view), independently of reward versus punishment association. Why should this be, in terms of systems level brain organization? The suggestion that is made is that the visual properties of the world about which reward associations must be learned are generally objects (for example the sight of a banana, or of an orange), and are not just raw pixels or edges, with no invariant properties, which is what is represented in the retina and V1. The implication is that the sensory processing must proceed to the stage of the invariant representation of objects before it is appropriate to learn reinforcement associations. The invariance aspect is important too, for if we had different representations for an object at different places in our visual field, then if we learned when an object was at one point on the retina that it was rewarding, we would not generalize correctly to it when presented at another position on the retina. If it had previously been punishing at that retinal position, we might find the same object rewarding when at one point on the retina, and punishing when at another. This is inappropriate given the world in which we live, and in which our brain evolved, in that the most appropriate assumption is that objects have the same reinforcement association wherever they are on the retina.

The same systems-level principle of brain organization is also likely to be true in other sensory systems, such as those for touch and hearing. For example, we do not generally want to learn that a particular pure tone is associated with reward or punishment. Instead, it might be a particular complex pattern of sounds such as a vocalization that carries a reinforcement signal, and this may be independent of the exact pitch at which it is uttered. Thus, cases in which some modulation of neuronal responses to pure tones in parts of the brain such as the medial geniculate (the thalamic relay for hearing) where tonotopic tuning is found (LeDoux, 1994) may be rather special *model* systems (that is simplified systems on which to perform experiments), and not reflect the way in which auditory-to-reinforcement pattern associations are normally learned. The same may be true for touch in so far as one considers associations between objects identified by somatosensory input, and primary reinforcers. An example might be selecting a food object from a whole collection of objects in the dark.

So far we have been considering where the reward association of objects is represented in the systems that process information about *what* object is being looked at. It is also of importance in considering where pattern associations between representations of objects and their reward or punishment association are formed, to know where in the information processing of signals such as taste and touch the reward value of these primary reinforcers is decoded and represented. In the case of taste in primates, there is evidence that processing continues beyond the primary taste cortex (see Fig. 7.2a) before the reward value of the taste is decoded. The evidence for this is that hunger does affect the responses of taste neurons in the secondary taste cortex (see Fig. 7.4), but not in the primary taste cortex (Rolls, Sienkiewicz and Yaxley, 1989; Rolls, Scott, Sienkiewicz and Yaxley, 1988; Rolls, 1989a, 1995a, 1997b; Yaxley, Rolls and Sienkiewicz, 1988). (Hunger modulates the reward value of taste, in that if hunger is present, animals work to obtain the taste of a food, and if satiated, do not work to obtain the taste of a food.) The systems-level principle here is that identification of what the taste is should ideally be performed independently of how rewarding or pleasant the taste is. The adaptive value of this is that even when neurons that reflect whether the taste (or the sight or smell of the food) is still rewarding have ceased responding because of feeding to satiety, it may still be important to have a representation of the sight/smell/taste of food, for then we can still (with other systems) learn, for example, where food is in the environment, even when we do not want to eat it. It would not be adaptive, for example, to become blind to the sight of food after we have eaten it to satiety, and the same holds for taste. On the other hand, when associations must be made explicitly with primary reinforcers such as taste, the stage in processing where the taste representation is appropriate may be where its reward value is represented, and this appears to be in primates beyond the primary taste cortex, in regions such as the secondary taste cortex (Rolls, 1989a, 1995a, 1997c). In the case of punishing somatosensory input, the situation may be different, for there pain may be decoded early in information processing, by being represented in a special set of sensory afferents, the C fibres. From a systems-level perspective, the primary reinforcer input to pattern associators for learning about which (for example visual or auditory) stimuli are associated with pain could in principle be derived early on in pain processing.

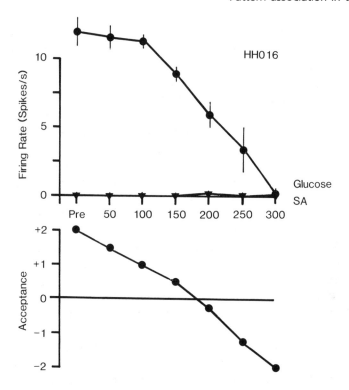

Fig. 7.4 Secondary taste cortex: a taste neuron stopped responding to the taste of glucose when the monkey was fed to satiety with 20% glucose. At the start of the experiment (Pre the feeding to satiety), the neuron increased its firing rate to approximately 12 spikes/s (closed circles, mean and standard error of the mean). The monkey was fed 50 ml aliquots of 20% glucose (see abscissa) until he was satiated (lower panel, which indicates the behaviour). The neuron did not increase its firing above the spontaneous rate (SA, inverted triangles) to the taste of glucose at the end of the experiment. Thus the neuron only responded to the taste of glucose when the monkey was hungry, and the taste was rewarding. (From Rolls, Sienkiewicz and Yaxley, 1989.)

The second point, which complements the first, is that the visual system is not provided with the appropriate primary reinforcers for such pattern association learning, in that visual processing in the primate brain is mainly unimodal to and through the inferior temporal visual cortex (see Fig. 7.2a). It is only after the inferior temporal visual cortex, when it projects to structures such as the amygdala and orbitofrontal cortex, that the appropriate convergence between visual processing pathways and pathways conveying information about primary reinforcers such as taste and touch/pain occurs (Fig. 7.2). We now, therefore, turn our attention to the amygdala and orbitofrontal cortex, to consider whether they might be the brain regions that contain the neuronal networks for pattern associations involving primary reinforcers. We note at this stage that in order to make the results as relevant as possible to brain function and its disorders in humans, the system being described is that present in primates such as monkeys. In rats, although the organization of the amygdala may be similar, the areas which may correspond to the primate inferior temporal visual cortex and orbitofrontal cortex are hardly developed.

7.1.3 The amygdala

Bilateral damage to the amygdala produces a deficit in learning to associate visual and other stimuli with a primary (that is unlearned) reward or punishment. For example, monkeys with damage to the amygdala when shown foods and non-foods pick up both and place them in their mouths. When such visual discrimination performance and learning is tested more formally, it is found that monkeys have difficulty in associating the sight of a stimulus with whether it produces a taste reward or is noxious and should be avoided (see Rolls, 1990c, 1992a). Similar changes in behaviour have been seen in humans with extensive damage to the temporal lobe.

Connections

The amygdala is a subcortical region in the anterior part of the temporal lobe. It receives massive projections in the primate from the overlying visual and auditory temporal lobe cortex (see Van Hoesen, 1981; chapters in Aggleton, 1992) (see Fig. 7.5). These come in the monkey to overlapping but partly separate regions of the lateral and basal amygdala from the inferior temporal visual cortex, the superior temporal auditory cortex, the cortex of the temporal pole, and the cortex in the superior temporal sulcus. Thus the amygdala receives inputs from the inferior temporal visual cortex, but not from earlier stages of cortical visual information processing. Via these inputs, the amygdala receives inputs about objects that could become secondary reinforcers, as a result of pattern association in the amygdala with primary reinforcers. The amygdala also receives inputs that are potentially about primary reinforcers, for example taste inputs (from the secondary taste cortex, via connections from the orbitofrontal cortex to the amygdala), and somatosensory inputs, potentially about the rewarding or painful aspects of touch (from the somatosensory cortex via the insula) (Mesulam and Mufson, 1982). The outputs of the amygdala include projections to the hypothalamus: from the lateral amygdala via the ventral amygdalofugal pathway to the lateral hypothalamus, potentially allowing autonomic outputs; and from the medial amygdala, which is relatively small in the primate, via the stria terminalis to the medial hypothalamus, potentially allowing influences on endocrine systems. The ventral amygdalofugal pathway is now known to contain some long descending fibres that project directly to the autonomic centres in the medulla oblongata (for example the dorsal motor nucleus of the vagus), and provide a route for cortically processed signals to influence autonomic responses. (Such autonomic responses include salivation and insulin release to visual stimuli associated with the taste of food, and heart rate changes to visual stimuli associated with anxiety.) A further interesting output of the amygdala is to the ventral striatum including the nucleus accumbens, for via this route information processed in the amygdala could gain access to the basal ganglia and thus influence behavioural output, such as whether we approach or avoid an object. In addition, the amygdala has direct projections back to many areas of the temporal, orbitofrontal, and insular cortices from which it receives inputs.

These anatomical connections of the amygdala indicate that it is placed to receive highly processed information from the cortex and to influence motor systems, autonomic systems, some of the cortical areas from which it receives inputs, and other limbic areas. The functions

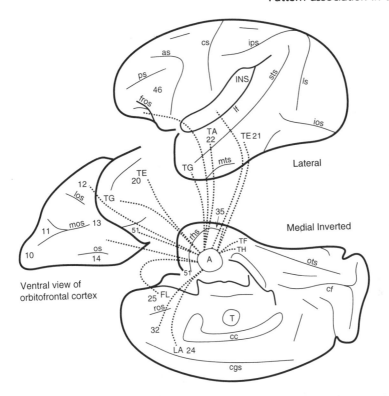

Fig. 7.5 Connections of the amygdala shown on lateral, ventral, and medial views of the monkey brain (after Van Hoesen, 1981). Abbreviations: as, arcuate sulcus; cc, corpus callosum; cf, calcarine fissure; cgs, cingulate sulcus; cs, central sulcus; ls, lunate sulcus; ios, inferior occipital sulcus; mos, medial orbital sulcus; os, orbital sulcus; ots, occipito-temporal sulcus; ps, principal sulcus; rhs, rhinal sulcus; sts, superior temporal sulcus; lf, Lateral (or Sylvian) fissure (which has been opened to reveal the insula); A, amygdala; INS, insula; T, thalamus; TE (21), inferior temporal visual cortex; TA (22), superior temporal auditory association cortex; TF and TH, parahippocampal cortex; TG, temporal pole cortex; 12, 13, 11, orbitofrontal cortex; 35, perirhinal cortex; 51, olfactory (prepyriform and periamygdaloid) cortex. The cortical connections shown provide afferents to the amygdala, but are reciprocated.

mediated through these connections will now be considered, using information available from the effects of damage to the amygdala and from the activity of neurons in the amygdala.

Effects of amygdala lesions

Bilateral removal of the amygdala in monkeys produces tameness, a lack of emotional responsiveness, excessive examination of objects, often with the mouth, and eating of previously rejected items such as meat (Weiskrantz, 1956). These behavioural changes comprise much of the Kluver–Bucy syndrome which is produced in monkeys by bilateral anterior temporal lobectomy (Kluver and Bucy, 1939). In analyses of the bases of these behavioural changes, it has been observed that there are deficits in learning to associate stimuli with primary reinforcement, including both punishments (Weiskrantz, 1956) and rewards (Jones and Mishkin, 1972; Gaffan, 1992; Aggleton, 1993). The association learning deficit is present when the associations must be learned from a previously neutral stimulus (for example the sight of an object) to a primary reinforcing stimulus (such as the taste of food). The impairment is not found when the association learning is between a visual stimulus and an auditory stimulus which is already a secondary reinforcer (because of prior pairing with food). Thus the amygdala is involved in learning associations between neutral stimuli and primary (but not secondary) reinforcers (see Gaffan, 1992; Gaffan and Harrison, 1987; Gaffan, Gaffan and Harrison, 1988, 1989; Baylis and Gaffan, 1991). Further evidence linking the amygdala to reinforcement mechanisms is that monkeys will work in order to obtain electrical stimulation of the amygdala, and that single neurons in the amygdala are activated by brain-stimulation reward of a number of different sites (Rolls, 1974, 1975, 1976, 1979; Rolls, Burton and Mora, 1980).

The symptoms of the Kluver–Bucy syndrome, including the emotional changes, could be a result of this type of deficit in learning stimulus-reinforcement associations (Jones and Mishkin, 1972; Aggleton and Passingham, 1981; Mishkin and Aggleton, 1981; Rolls, 1986, 1990c, 1992a). For example, the tameness, the hypo-emotionality, the increased orality, and the altered responses to food would arise because of damage to the normal mechanism by which stimuli become associated with reward or punishment.

The amygdala is well placed anatomically for learning associations between objects and primary reinforcers, for it receives inputs from the higher parts of the visual system, and from systems processing primary reinforcers such as taste, smell, and touch (see Fig. 7.2a). The association learning in the amygdala may be implemented by Hebb-modifiable synapses from visual and auditory neurons onto neurons receiving inputs from taste, olfactory or somato-sensory primary reinforcers. Consistent with this, at least one type of associative learning in the amygdala (and not its retention) can be blocked by local application to the amygdala of an NMDA receptor blocker (see Davis, 1992; Davis *et al.*, 1994). Once the association has been learned, the outputs from the amygdala could be driven by the conditioned as well as by the unconditioned stimuli. In line with this, LeDoux, Iwata, Cichetti and Reis (1988) were able to show that lesions of the lateral hypothalamus (which receives from the central nucleus of the amygdala) blocked conditioned heart rate (autonomic) responses. Lesions of the central grey of the midbrain (which also receives from the central nucleus of the amygdala) blocked the conditioned freezing but not the conditioned autonomic response to the aversive conditioned stimulus. Further, Cador, Robbins and Everitt (1989) obtained evidence consistent with the hypothesis that the learned incentive (conditioned reinforcing) effects of previously neutral stimuli paired with rewards are mediated by the amygdala acting through the ventral striatum, in that amphetamine injections into the ventral striatum enhanced the effects of a conditioned reinforcing stimulus only if the amygdala was intact (see further Everitt and Robbins, 1992).

There is, thus, much evidence that the amygdala is involved in responses made to stimuli associated with primary reinforcement. There is evidence that it may also be involved in whether novel stimuli are approached, for monkeys with amygdala lesions place novel foods and non-food objects in their mouths, and rats with amygdala lesions have decreased neophobia, in that they more quickly accept new foods (Rolls and Rolls, 1973; see also Dunn and Everitt, 1988; Rolls, 1992a; Wilson and Rolls, 1993).

Neuronal activity in the primate amygdala to reinforcing stimuli

Recordings from single neurons in the amygdala of the monkey have shown that some neurons do respond to visual stimuli, consistent with the inputs from the temporal lobe visual cortex (Sanghera, Rolls and Roper-Hall, 1979). Other neurons responded to auditory, gustatory, olfactory, or somatosensory stimuli, or in relation to movements. In tests of whether the neurons responded on the basis of the association of stimuli with reinforcement, it was found that approximately 20% of the neurons with visual responses had responses which occurred primarily to stimuli associated with reinforcement, for example to food and to a range of stimuli which the monkey had learned signified food in a visual discrimination task (Sanghera et al., 1979; Rolls, 1981a,b, 1992a; Wilson and Rolls, 1993). However, none of these neurons (in contrast to some neurons in the hypothalamus and orbitofrontal cortex described below) responded exclusively to rewarded stimuli, in that all responded at least partly to one or more neutral, novel, or aversive stimuli. Neurons with responses which are probably similar to these have also been described by Ono et al. (1980), and by Nishijo, Ono and Nishino (1988) (see Ono and Nishijo, 1992).

The degree to which the responses of these amygdala neurons are associated with reinforcement has also been assessed in learning tasks. When the association between a visual stimulus and reinforcement was altered by reversal (so that the visual stimulus formerly associated with juice reward became associated with aversive saline and vice versa), it was found that 10 of 11 neurons did not reverse their responses (and for the other neuron the evidence was not clear) (Sanghera, Rolls and Roper-Hall, 1979; Wilson and Rolls, 1993; see Rolls, 1992a). On the other hand, in a rather simpler relearning situation in which salt was added to a piece of food such as a water melon, the responses of 4 amygdala neurons to the sight of the water melon diminished (Nishijo, Ono and Nishino, 1988). More investigations are needed to show the extent to which amygdala neurons do alter their activity flexibly and rapidly in relearning tests such as these (for discussion, see Rolls, 1992a). What has been found in contrast is that neurons in the orbitofrontal cortex do show *very rapid reversal* of their responses in visual discrimination reversal, and it therefore seems likely that the orbitofrontal cortex is especially involved when repeated relearning and re-assessment of stimulus-reinforcement associations is required, as described below, rather than initial learning, in which the amygdala may be involved.

Responses of these amygdala neurons to novel stimuli which are reinforcing

As described above, some of the amygdala neurons that responded to rewarding visual stimuli also responded to some other stimuli that were not associated with reward. Wilson and Rolls (1993, 1997) discovered a possible reason for this. They showed that these neurons with reward-related responses also responded to relatively novel visual stimuli in, for example, visual recognition memory tasks. When monkeys are given such relatively novel stimuli outside the task, they will reach out for and explore the objects, and in this respect the novel stimuli are reinforcing. Repeated presentation of the stimuli results in habituation of the responses of these amygdala neurons and of behavioural approach, if the stimuli are not associated with primary reinforcement. It is thus suggested that the amygdala neurons described provide an output if a stimulus is associated with a positive reinforcer, or is positively reinforcing because of relative novelty. The functions of this output may be to influence the interest shown in a stimulus; whether a stimulus is approached or avoided; whether an affective response occurs to a stimulus; and whether a representation of the stimulus is made or maintained via an action mediated through either the basal forebrain nucleus of Meynert (see Section 7.1.5) or the backprojections to the cerebral cortex (see Sections 4.5 and 10.2) (Rolls, 1987, 1989b, 1990c, 1992a).

It is an important adaptation to the environment to explore relatively novel objects or situations, for in this way advantage due to gene inheritance can become expressed and selected for. This function appears to be implemented in the amygdala in this way. Lesions of the amygdala impair the operation of this mechanism, in that objects are approached and explored indiscriminately, relatively independently of whether they are associated with positive or negative reinforcement, or are novel or familiar.

The details of the neuronal mechanisms that implement this in the amygdala are not currently known, but could be as follows. Cortical visual signals which do not show major habituation with repeated visual stimuli, as shown by recordings in the temporal cortical visual areas (see Rolls, 1994a, 1997d; Rolls, Judge and Sanghera, 1977), reach the amygdala. In the amygdala, neurons respond to these at first, and have the property that they gradually habituate unless the pattern association mechanism in the amygdala detect co-occurrence of these stimuli with a primary reinforcer, in which case it strengthens the active synapses for that object, so that it continues to produce an output from amygdala neurons that respond to either rewarding or punishing visual stimuli. Neurophysiologically, the habituation condition would correspond in a pattern associator to long-term depression (LTD) of synapses with high presynaptic activity but low postsynaptic activity, that is to homosynaptic LTD (see Fig. 7.6).

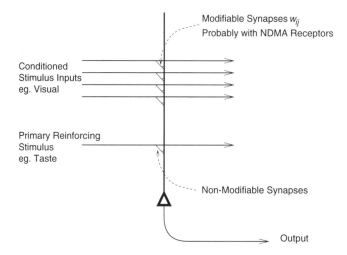

Fig. 7.6 Pattern association between a primary reinforcer, such as the taste of food, which activates neurons through non-modifiable synapses, and a potential secondary reinforcer, such as the sight of food, which has modifiable synapses onto the same neurons. Such a mechanism appears to be implemented in the amygdala and orbitofrontal cortex. (Homosynaptic) long-term depression in a pattern associator in the amygdala could account for the habituating responses to novel visual stimuli which are not associated with primary reinforcers.

We thus know that some amygdala neurons can be driven by primary reinforcers such as taste or touch, that some can be activated by visual stimuli associated with primary reinforcers, and that in the rat associative conditioning may require NMDA receptor activation for the learning, though not necessarily for the later expression of the learning. We also know that autonomic responses learned to such stimuli can depend on outputs from the amygdala to the hypothalamus, and that the effects that learned incentives have on behaviour may involve outputs from the amygdala to the ventral striatum. We also know that there are similar neurons in the ventral striatum to some of those described in the amygdala (Williams *et al.*, 1993). All this is consistent with the hypothesis that there are neuronal networks in the amygdala that perform the required pattern association. Interestingly, there is somewhat of a gap in our knowledge here, for the microcircuitry of the amygdala has been remarkably little studied. It is known from Golgi studies in young rats in which sufficiently few amygdala cells are stained that it is possible to see them individually, that there are pyramidal cells in the amygdala with large dendrites and many synapses (Millhouse and DeOlmos, 1983; McDonald, 1992; Millhouse, 1986). What these studies have not yet defined is whether visual and taste inputs converge anatomically onto some cells, and whether (as might be predicted) the taste inputs are likely to be strong (for example large synapses close to the cell body), whereas the visual inputs are more numerous, and on a part of the dendrite with NMDA receptors. Clearly, to bring our understanding fully to the network level, such evidence is required, together with evidence that the appropriate synapses are modifiable by a Hebb-like rule (such as might be implemented using the NMDA receptors), in a network of the type shown in Fig. 7.6.

7.1.4 Orbitofrontal cortex

The orbitofrontal cortex is strongly connected to the amygdala, and is involved in a related type of learning. Damage to the orbitofrontal cortex produces deficits in tasks in which associations learned to reinforcing stimuli must be extinguished or reversed. The orbitofrontal cortex receives inputs about primary reinforcers, such as the taste of food, and information about objects in the world, for example about what objects or faces are being seen. In that it receives information from the ventral visual system (see Chapters 1 and 8), and also from the taste, smell, and somatosensory systems, it is a major site in the brain of information about what stimuli are present in the world, and potentially for forming associations between these different 'what' representations. This part of the cortex appears to be involved in the rapid learning and relearning of which objects in the world are associated with reinforcement, and probably does this using pattern association learning between a visual stimulus and a reward or punishment (unconditioned stimulus). The orbitofrontal cortex develops very greatly in monkeys and humans compared to rats (in which the sulcal prefrontal cortex, which may be a homologue, is very little developed). The orbitofrontal cortex may be a pattern associator which is specialized to allow the very rapid re-evaluation of the current reinforcement value of many different objects or other animals or humans, which continually alter in, for example, social situations. Consistent with this, as described next, the impairments produced in humans by damage to the orbitofrontal cortex include emotional and social changes.

Effects of orbitofrontal cortex lesions

The learning tasks impaired by orbitofrontal cortex damage include tasks in which associations learned to reinforcing stimuli must be extinguished or reversed. For example, in work with non-human primates, it has been found that if a reward (a positive reinforcer) is no longer delivered when a behavioural response is made, then that response normally gradually stops, and is said to show extinction. If the orbitofrontal cortex is damaged, then such behavioural responses continue for long periods even when no reward is given. In another example, in a visual discrimination task, a monkey learns that choice of one of two visual stimuli leads to a food reward and of the other to no reward. If the reinforcement contingency is then reversed, normal monkeys can learn to reverse their choices. However, if the orbitofrontal cortex is damaged, then the monkey keeps choosing the previously rewarded visual stimulus, even though it is no longer rewarded. This produces a visual discrimination reversal deficit (see Butter, 1969; Jones and Mishkin, 1972; Rolls, 1990c, 1996b). The visual discrimination learning deficit shown by monkeys with orbitofrontal cortex damage (Jones and Mishkin, 1972; Baylis and Gaffan, 1991), may be due to the tendency of these monkeys not to withhold responses to non-rewarded stimuli (Jones and Mishkin, 1972). Some of these effects of lesions of the orbitofrontal cortex are summarized in Fig. 7.7.

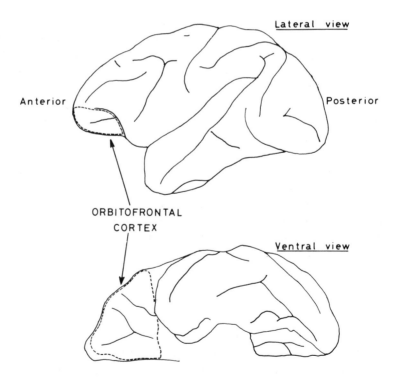

Fig. 7.7 The orbitofrontal cortex of the monkey. The effects of orbitofrontal lesions include: (a) changes in emotional behaviour; (b) changes in food-selection behaviour; (c) impaired extinction; (d) difficulty in reversing responses during visual discrimination reversal.

In humans, analogous deficits have been found after damage to the ventral part of the frontal lobe, in a region that includes the orbitofrontal cortex. It has long been known that in patients with frontal lobe damage there is an impairment in the Wisconsin Card Sorting Task, in which a special set of cards must be sorted into piles based on criteria such as the colour of the items on a card, their form, or the number of items on a card. After the subject has sorted correctly according to colour, the sorting rule is shifted to another dimension (form) without warning. Patients with frontal lobe damage often have problems shifting to the new sorting dimension, and continue sorting according to the previously reinforced sorting dimension. Because this task involves complex cognitive strategies including shifting attention from one stimulus dimension to another, Rolls, Hornak, Wade and McGrath (1994) designed a much simpler visual discrimination reversal task, to test much more directly the hypothesis that damage to the orbitofrontal cortex impairs the rapid re-learning of which visual stimulus is associated with reward. Their patients touched one of two presented stimuli on a video monitor in order to obtain points, and had not to touch the other stimulus, otherwise they would lose points. After learning this, when the rewarded stimulus was reversed, the patients tended to keep choosing the previously rewarded stimulus, and were thus impaired at visual discrimination reversal (Rolls et al., 1994). This deficit is a perseveration to the previously rewarded stimulus, and not a response perseveration which may be produced by damage to other parts of the frontal lobe (see Rolls, 1996b).

Damage to the caudal orbitofrontal cortex in the monkey produces emotional changes. These include decreased aggression to humans and to stimuli such as a snake and a doll, and a reduced tendency to reject foods such as meat (Butter, Snyder and McDonald, 1970; Butter and Snyder, 1972; Butter, McDonald and Snyder, 1969). In the human, euphoria, irresponsibility, and lack of affect can follow frontal lobe damage (see Kolb and Whishaw, 1996). These changes in emotional and social behaviour may be accounted for, at least in part, by a failure to respond appropriately to stimuli on the basis of their previous and often rapidly changing association with reinforcement (see further Rolls, 1990c, 1996b, 1997c; Rolls, Hornak, Wade and McGrath, 1994). Interestingly, in a study prompted by our finding that there are some neurons in the orbitofrontal cortex that respond to faces, we have found that some of these patients are impaired at discriminating facial expression correctly (though face recognition is not impaired) (Hornak, Rolls and Wade, 1996). Face expression is used as a reinforcer in social situations (for example a smile to reinforce), and the hypothesis is that this is part of the input, equivalent to the unconditioned stimulus, to a pattern associator involved in learning associations between individual people or objects and their current reward status or value.

Connections

The inputs to the orbitofrontal cortex include many of those required to determine whether a visual or auditory stimulus is associated with a primary reinforcer such as taste or smell (see Fig. 7.8). As a result of quite recent work, it is now known that the orbitofrontal cortex contains the secondary taste cortex (in that it receives direct inputs from the primary taste cortex—Baylis, Rolls and Baylis, 1994), and that information about the reward value of taste, a primary or unlearned reinforcer, is represented here (in that neurons here only respond to,

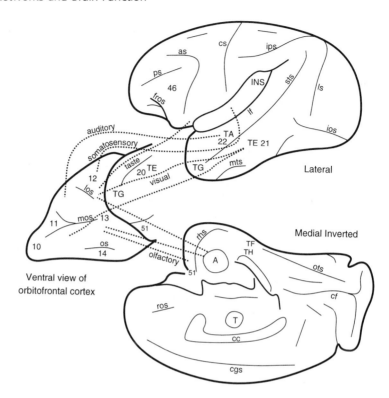

Fig. 7.8 Connections of the orbitofrontal cortex shown on lateral, ventral and medial views of the monkey brain. Abbreviations: as, arcuate sulcus; cc, corpus callosum; cf, calcarine fissure; cgs, cingulate sulcus; cs, central sulcus; ls, lunate sulcus; ios, inferior occipital sulcus; mos, medial orbital sulcus; os, orbital sulcus; ots, occipito-temporal sulcus; ps, principal sulcus; rhs, rhinal sulcus; sts, superior temporal sulcus; lf, Lateral (or Sylvian) fissure (which has been opened to reveal the insula); A, amygdala; INS, insula; T, thalamus; TE (21), inferior temporal visual cortex; TA (22), superior temporal auditory association cortex; TF and TH, parahippocampal cortex; TG, temporal pole cortex; 12, 13, 11, orbitofrontal cortex; 35, perirhinal cortex; 51, olfactory (prepyriform and periamygdaloid) cortex. The cortical connections shown provide afferents to the orbitofrontal cortex, but are reciprocated.

for example, the taste of glucose if the monkey is hungry—Rolls, Sienkiewicz and Yaxley, 1989; see Rolls, 1995a). The orbitofrontal cortex also receives somatosensory input (via the insula—see Mesulam and Mufson, 1982), and this could convey primary reinforcing information, for example about the reward value of touch (cf. Taira and Rolls, 1996) or about aversive, painful, stimulation. Consistent with the latter, patients with frontal damage may say that the pain is still present, but that it no longer has its full emotional and motivational value. In a recent functional magnetic resonance imaging (fMRI) investigation, Rolls, Francis, *et al.* (1997) have shown that the lateral orbitofrontal cortex of humans is activated much more by a pleasant than a neutral touch (relative to the primary somatosensory cortex), thus producing direct evidence that the orbitofrontal cortex is involved in the affective, reward-related, representation of touch. The orbitofrontal cortex also receives inputs about visual and auditory stimuli, both directly from the inferior temporal visual cortex, the cortex in the superior temporal sulcus (which contains visual, auditory, and somatosensory areas, see Baylis, Rolls and Leonard, 1987), the temporal pole (another multimodal cortical area), and the amygdala (Jones and Powell, 1970; Seltzer and Pandya, 1989; Barbas, 1988); and indirectly via the thalamic projection nucleus of the orbitofrontal cortex, the medial, magnocellular, part of the mediodorsal nucleus (Krettek and Price, 1974, 1977). Olfactory inputs are received from the pyriform (primary olfactory) cortex, first to a caudal part of the orbitofrontal cortex designated area 13a, and then more widespread in the caudal orbitofrontal cortex (Price *et al.*, 1991; Carmichael *et al.*, 1994). There are also modulatory influences, which could modulate learning, in the form of a dopaminergic input. The orbitofrontal cortex has outputs, through which it can influence behaviour, to the striatum (see Kemp and Powell, 1970) (including the ventral striatum), and also projections back to temporal lobe areas such as the inferior temporal cortex, and, in addition, to the entorhinal cortex and cingulate cortex (Van Hoesen *et al.*, 1975). The orbitofrontal cortex also projects to the preoptic region and lateral hypothalamus (through which learning could influence autonomic responses to conditioned stimuli) (Nauta, 1964).

The orbitofrontal cortex thus has inputs which provide it with information about primary reinforcing inputs such as taste and touch, inputs which provide it with information about objects in the world (for example visual inputs from the inferior temporal cortex), and outputs which could influence behaviour (via the striatum) and autonomic responses (via the preoptic area and lateral hypothalamus).

Neuronal activity in the primate orbitofrontal cortex to reinforcing stimuli

Orbitofrontal cortex neurons with taste responses were described by Thorpe, Rolls and Maddison (1983) and have now been analysed further (Rolls, 1989b, 1995a, 1997c). They are in a mainly caudal and lateral part of the orbitofrontal cortex, and can be tuned quite finely to gustatory stimuli such as a sweet taste (Rolls, Yaxley and Sienkiewicz, 1990; Rolls, 1989b). Moreover, their activity is related to reward, in that those which respond to the taste of food do so only if the monkey is hungry (Rolls, Sienkiewicz and Yaxley, 1989). These orbitofrontal neurons receive their input from the primary gustatory cortex, in the frontal operculum (Scott, Yaxley, Sienkiewicz and Rolls, 1986a; Rolls, 1989a, 1995a, 1997c; Baylis, Rolls and Baylis, 1994). There is also direct neurophysiological evidence that somatosensory inputs affect orbitofrontal cortex neurons, for the taste responses of some are affected by the texture of the food (see Rolls, 1997c).

Visual inputs also influence some orbitofrontal cortex neurons, and in some cases convergence of visual and gustatory inputs onto the same neuron is found (Thorpe *et al.*, 1983; Rolls, Critchley, Mason and Wakeman, 1996). Moreover, in at least some cases the visual stimuli to which these neurons respond correspond to the taste to which they respond (Thorpe *et al.*, 1983; Rolls, Critchley, Mason and Wakeman, 1996). This suggests that learning of associations between visual stimuli and the taste with which they are associated influences the responses of orbitofrontal cortex neurons. This has been shown to be the case in formal investigations of the activity of orbitofrontal cortex visual neurons, which in many cases reverse their responses to visual stimuli when the taste with which the visual stimulus is associated is reversed by the experimenter (Thorpe *et al.*, 1983; Rolls, Critchley, Mason and Wakeman, 1996). An example of the responses of an orbitofrontal cortex cell that reverses the stimulus to which it responds during reward reversal is shown in Fig. 7.9. This reversal by orbitofrontal visual neurons can be very fast, in as little as one trial, that is a few seconds (see for example Fig. 7.10). Another class of neurons responds only in certain non-reward situations (Thorpe *et al.*, 1983). For example, some neurons responded in extinction, immediately after a lick had been made to a visual stimulus which had previously been associated with fruit juice reward, and other neurons responded in a reversal task, immediately after the monkey had responded to the previously rewarded visual stimulus, but had obtained punishment rather than reward (Thorpe *et al.*, 1983).

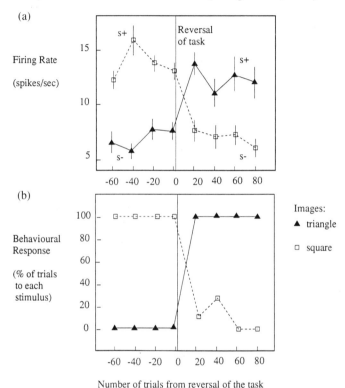

Fig. 7.9 Orbitofrontal cortex: visual discrimination reversal. The activity of an orbitofrontal visual neuron during performance of a visual discrimination task and its reversal. The stimuli were a triangle and a square presented on a video monitor. (a) Each point represents the mean poststimulus activity in a 500 ms period of the neuron to approximately 10 trials of the different visual stimuli. The standard errors of these responses are shown. After 60 trials of the task the reward associations of the visual stimuli were reversed. (+ indicates that a lick response to that visual stimulus produces fruit juice reward; – indicates that a lick response to that visual stimulus results in a small drop of aversive tasting saline. This neuron reversed its responses to the odorants following the task reversal. (b) The behavioural response of the monkey to the task. It is shown that the monkey performs well, in that he rapidly learns to lick only to the visual stimulus associated with fruit juice reward. (Reprinted from Rolls, Critchley, Mason and Wakeman, 1996.)

These cells thus reflect the information about which stimulus to make behavioural responses to during reversals of visual discrimination tasks. If a reversal occurs, then the taste cells provide the information that an unexpected taste reinforcer has been obtained, another group of cells shows a vigorous discharge which could signal that reversal is in progress, and the visual cells with reinforcement association-related responses reverse the

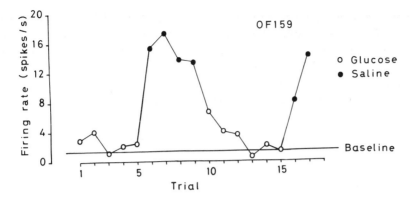

Fig. 7.10 Orbitofrontal cortex: one trial visual discrimination reversal by a neuron. On trials 1–5, no response of the neuron occurred to the sight of a 2 ml syringe from which the monkey had been given orally glucose solution to drink on the previous trial. On trials 6–9, the neuron responded to the sight of the same syringe from which he had been given aversive hypertonic saline to drink on the previous trial. Two more reversal trials (10–15, and 16–17) were performed. The reversal of the neuron's response when the significance of the same visual stimulus was reversed shows that the responses of the neuron only occurred to the sight of the visual stimulus when it was associated with a positively reinforcing and not with a negatively reinforcing taste. Moreover, it is shown that the neuronal reversal took only one trial. (Reprinted from Thorpe, Rolls and Maddison, 1983.)

stimulus to which they are responsive. These neurophysiological changes take place rapidly, in as little as 5 s, and are presumed to be part of the neuronal learning mechanism that enables primates to alter their knowledge of the reinforcement association of visual stimuli so rapidly. This capacity is important whenever behaviour must be corrected when expected reinforcements are not obtained, in for example feeding, emotional, and social situations (see Rolls, 1996b).

The details of the neuronal network architecture which underlies this reversal learning have not been finalized for the orbitofrontal cortex. On the basis of the evidence described, it can be proposed that it is as shown in Fig. 7.6. It is likely that NMDA receptors help to implement the modifiability of the visual inputs onto taste neurons. This could be tested by using NMDA receptor blockers applied locally to individual orbitofrontal cortex visual neurons during visual discrimination reversal, or by perfusion of the orbitofrontal cortex with an NMDA receptor blocker to investigate whether this interferes with behavioural visual discrimination reversal. The difference predicted for this system from that in the amygdala is that the orbitofrontal associativity would be more rapidly modifiable (in one trial) than that in the amygdala (which might take very many trials). Indeed, the cortical reversal mechanism in the orbitofrontal cortex may be effectively a fast version of what is implemented in the amygdala, that has evolved particularly to enable rapid updating by received reinforcers in social and other situations in primates. This hypothesis, that the orbitofrontal cortex, as a rapid learning mechanism, effectively provides an additional route for some of the functions performed by the amygdala, and is very important when this stimulus-reinforcement learning must be rapidly readjusted, has been developed elsewhere (Rolls, 1990c, 1992a, 1996b). Although the mechanism has been described so far for visual to taste association learning, this is because experiments on this are most direct. It is likely, given the evidence from the effects of lesions, that taste is only one type of primary reinforcer about which such learning occurs in the orbitofrontal cortex, and is likely to be an example of a much more general type of stimulus-reinforcement learning system. Some of the evidence for this is that humans with orbitofrontal cortex damage are impaired at visual discrimination reversal when working for a reward that consists of points (Rolls, Hornak, Wade and McGrath, 1994; Rolls, 1996b). Moreover, as described above, there is now evidence that the representation of the affective aspects of touch is represented in the human orbitofrontal cortex (Rolls, Francis *et al.*, 1997), and learning about what stimuli are associated with this class of primary reinforcer is also likely to be an important aspect of the stimulus-reinforcement association learning performed by the orbitofrontal cortex.

Further evidence on other types of stimulus-reinforcement association learning in which neurons in the orbitofrontal cortex are involved is starting to become available. One such example comes form learning associations between olfactory stimuli and tastes. It has been shown that some neurons in the orbitofrontal cortex respond to taste and to olfactory stimuli (Rolls and Baylis, 1994). It is probably here in the orbitofrontal cortex that the representation of flavour in primates is built. To investigate whether these bimodal cells are built by olfactory to taste association learning, we measured the responses of these olfactory cells during olfactory to taste discrimination reversal (Rolls, Critchley, Mason and Wakeman, 1996). We found that some of these cells did reverse (see for example Fig. 7.11), others during reversal stopped discriminating between the odours, while others still did not reverse (see Table 7.1). When reversal did occur, it was quite slow, often needing 20–50 trials. This evidence thus suggests that although there is some pattern association learning between olfactory stimuli and taste implemented in the orbitofrontal cortex, the mechanism involves quite slow learning, perhaps because in general it is important to maintain rather stable representations of flavours. Moreover, olfactory to taste association learning occurs for only some of the olfactory neurons. The olfactory neurons which do not reverse may be carrying information which is in some cases independent of reinforcement association (i.e. is about olfactory identity). In other cases, the olfactory representation in the orbitofrontal cortex may reflect associations of odours with other primary reinforcers (for example whether sickness has occurred in association with some smells), or may reflect primary reinforcement value provided by some olfactory stimuli. (For example, the smell of flowers may be innately pleasant and attractive, and some other odours may be innately unpleasant.) In this situation, the olfactory input to some orbitofrontal cortex neurons may represent an unconditioned stimulus input, with which other (for example visual) inputs may become associated.

Part of the importance of investigating what functions are performed by the orbitofrontal cortex, and how they are performed, is that these have implications for understanding the effects of damage to this part of the brain in patients. The patients we investigated (Rolls, Hornak, Wade and McGrath, 1994) with damage to the ventral part of the frontal lobe had high scores on a behaviour questionnaire which reflected the degree of disinhibited and socially inappropriate behaviour exhibited by the patients. They also were impaired at stimulus-reinforcement association reversal, and had impairments in identifying facial expression. Their altered social and emotional behaviour appear to be at least partly related to the impairment in learning correctly to alter their behaviour in response to changing reinforcing contingencies in the environment (see further Rolls, 1990c, 1996b; Rolls *et al.*, 1994; Hornak, Rolls and Wade, 1996).

In conclusion, some of the functions of the orbitofrontal cortex and amygdala are related to stimulus-reinforcement pattern association learning, involving systems-level connections of the form summarized in Fig. 7.2, and involving neuronal networks of the type shown in Fig. 7.6.

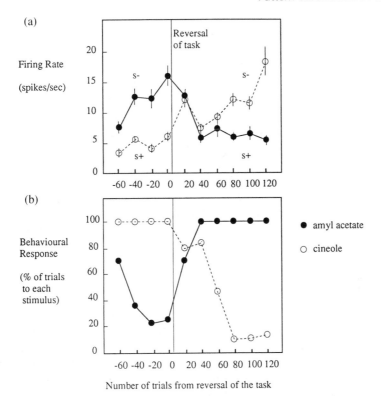

(a)

Firing Rate

(spikes/sec)

(b)

Behavioural Response

(% of trials to each stimulus)

● amyl acetate

○ cineole

Number of trials from reversal of the task

Fig. 7.11 Orbitofrontal cortex: olfactory to taste association reversal. (a) The activity of a single orbitofrontal olfactory neuron during the performance of a two-odour olfactory discrimination task and its reversal is shown. Each point represents the mean poststimulus activity of the neuron in a 500 ms period to approximately 10 trials of the different odorants. The standard errors of these responses are shown. The odorants were amyl acetate (●) (initially S–) and cineole (o) (initially S+). After 80 trials of the task the reward associations of the stimuli were reversed. This neuron reversed its responses to the odorants following the task reversal. (b) The behavioural responses of the monkey during the performance of the olfactory discrimination task. The number of lick responses to each odorant is plotted as a percentage of the number of trials to that odorant in a block of 20 trials of the task. (Reprinted from Rolls, Critchley, Mason and Wakeman, 1996.)

Table 7.1 Proportion of neurons in the primate orbitofrontal cortex showing reversal, or extinction (ceasing to discriminate after the reversal), or no change of responses, during visual or olfactory discrimination reversal (from Rolls, Critchley, Mason and Wakeman, 1996).

	Olfactory cells	%	Visual cells	%
Reversal	7	25.0	12	70.6
Extinction	12	42.9	4	23.5
No change	9	32.1	1	5.9
Total	28	100	17	100

7.1.5 Basal forebrain cholinergic systems, and the noradrenergic system

Basal forebrain cholinergic neurons

Before leaving pattern association learning systems in the amygdala and orbitofrontal cortex, it is useful to consider the role in memory of one of the systems to which they project, the basal forebrain magnocellular nuclei of Meynert. These cells are just lateral to the lateral hypothalamus in the substantia innominata, and extend forward through the preoptic area into the diagonal band of Broca (see Mesulam, 1990). These cells, many of which are cholinergic, project directly to the cerebral cortex (Kievit and Kuypers, 1975; Divac, 1975; Mesulam, 1990). These cells provide the major cholinergic input to the cerebral cortex, in that if they are lesioned, the cortex is depleted of acetylcholine (Mesulam, 1990). Loss of these cells does occur in Alzheimer's disease, and there is consequently a reduction in cortical acetylcholine in Alzheimer's disease (Mesulam, 1990). This loss of cortical acetylcholine may contribute to the memory loss in Alzheimer's disease, although it may well not be the primary factor in the etiology.

In order to investigate the role of the basal forebrain nuclei in memory, Aigner *et al.* (1991) made neurotoxic lesions of them in monkeys. Some impairments on a simple test of recognition memory, delayed non-match-to-sample, were found. Analysis of the effects of similar lesions in rats showed that performance on memory tasks was impaired, perhaps because of failure to attend properly (Muir *et al.*, 1994). There are quite limited numbers of these basal forebrain neurons (in the order of thousands). Given that there are relatively few of these neurons, it is not likely that they carry the information to be stored in cortical memory circuits, for the number of different patterns that could be represented and stored is so small. (The number of different patterns that could be stored is dependent in a leading way on the number of input connections onto each neuron in the pattern associator, see Chapter 2). With this few neurons distributed throughout the cerebral cortex, the memory capacity of the whole system would be impractically small. This argument alone indicates that they are unlikely to carry the information to be stored in cortical memory systems. Instead, they could modulate storage in the cortex of information derived from what provides the numerically major input to cortical cells, other cortical cells. This modulation may operate by setting thresholds for cortical cells to the appropriate value, or by more directly influencing the cascade of processes involved in long-term potentiation (see Chapter 1). There is indeed evidence that acetylcholine is necessary for cortical synaptic modifiability, as shown by studies in which depletion of acetylcholine and noradrenaline impaired cortical LTP/synaptic modifiability (Bear and Singer, 1986).

The question then arises of whether the basal forebrain cholinergic neurons tonically release acetylcholine, or whether they release it particularly in response to some external influence. To examine this, recordings have been made from basal forebrain neurons, at least some of which will have been the cholinergic neurons just described. It has been found that some of these neurons respond to visual stimuli associated with rewards such as food (Rolls, 1975, 1981b, 1986, 1990c, 1993, 1994b; Rolls, Burton and Mora, 1976; Burton, Rolls and Mora, 1976; Mora, Rolls and Burton, 1976; Wilson and Rolls, 1990b,c), or with punishment (Rolls, Sanghera and Roper-Hall, 1979), that others respond to novel visual stimuli (Wilson and Rolls, 1990a), and that others respond to a range of visual stimuli. For example, in one set of recordings, one group of these neurons (1.5%) responded to novel visual stimuli while monkeys performed recognition or visual discrimination tasks (Wilson and Rolls, 1990a). A complementary group of neurons more anteriorly responded to familiar visual stimuli in the same tasks (Rolls et al., 1982; Wilson and Rolls, 1990a). A third group of neurons (5.7%) responded to positively reinforcing visual stimuli in visual discrimination and in recognition memory tasks (Wilson and Rolls, 1990b,c). In addition, a considerable proportion of these neurons (21.8%) responded to any visual stimuli shown in the tasks, and some (13.1%) responded to the tone cue which preceded the presentation of the visual stimuli in the task, and was provided to enable the monkey to alert to the visual stimuli (Wilson and Rolls, 1990a). These neurons did not respond to touch to the leg which induced arousal, so their responses did not simply reflect arousal. Neurons in this region receive inputs from the amygdala (see Mesulam, 1990; Aggleton, 1992), and it is probably via the amygdala that the information described here reaches the basal forebrain neurons, for neurons with similar response properties have been found in the amygdala, and the amygdala appears to be involved in decoding visual stimuli that are associated with reinforcement, or are novel (Rolls, 1990c, 1992a; Wilson and Rolls, 1993, 1998).

It is therefore suggested that the normal physiological function of these basal forebrain neurons is to send a general activation signal to the cortex when certain classes of environmental stimuli occur. These stimuli are often stimuli to which behavioural activation is appropriate or required, such as positively or negatively reinforcing visual stimuli, or novel visual stimuli. The effect of the firing of these neurons on the cortex is excitatory, and in this way produces activation. This cortical activation may produce behavioural arousal, and may thus facilitate concentration and attention, which are both impaired in Alzheimer's disease. The reduced arousal and concentration may themselves contribute to the memory disorders. But the acetylcholine released from these basal magnocellular neurons may in addition be more directly necessary for memory formation, for Bear and Singer (1986) showed that long-term potentiation, used as an indicator of the synaptic modification which underlies learning, requires the presence in the cortex of acetylcholine as well as noradrenaline. For comparison, acetylcholine in the hippocampus makes it more likely that LTP will occur, probably through activation of an inositol phosphate second messenger cascade (Markram and Siegel, 1992; see Siegel and Auerbach, 1996; see also Hasselmo et al., 1995; Hasselmo and Bower, 1993). The adaptive value of the cortical strobe provided by the basal magnocellular neurons may thus be that it facilitates memory storage especially when significant (for example reinforcing) environmental stimuli are detected. This means that memory storage is likely to be conserved (new memories are less likely to be laid down) when significant environmental stimuli are not present. In that the basal forebrain projection spreads widely to many areas of the cerebral cortex, and in that there are relatively few basal forebrain neurons (in the order of thousands), the basal forebrain neurons do not determine the actual memories that are stored. Instead the actual memories stored are determined by the active subset of the thousands of cortical afferents onto a strongly activated cortical neuron (cf. Treves and Rolls, 1994). The basal forebrain magnocellular neurons would then according to this analysis simply when activated increase the probability that a memory would be stored. Impairment of the normal operation of the basal forebrain magnocellular neurons would be expected to interfere with normal memory by interfering with this function, and this interference could contribute in this way to the memory disorder in Alzheimer's disease.

Another property of cortical neurons emphasized recently (Markram and Tsodyks, 1996; Abbott *et al.*, 1996) is that they tend to adapt with repeated input. However, this adaptation is most marked in slices, in which there is no acetylcholine. One effect of acetylcholine is to reduce this adaptation. When recordings are made from single neurons operating in physiological conditions in the awake behaving monkey, peristimulus time histograms of inferior temporal cortex neurons to visual stimuli show only some adaptation. There is typically an onset of the neuronal response at 80–100 ms after the stimulus, followed within 50 ms by the highest firing rate. There is after that some reduction in the firing rate, but the firing rate is still typically more than half maximal 500 ms later (see example in Fig. 10.12). Thus under normal physiological conditions, firing rate adaptation can occur, but does not involved a major adaptation, even when cells are responding fast (at for example 100 spikes/s) to a visual stimulus. One of the factors that keeps the response relatively maintained may, however, be the presence of acetylcholine. Its depletion in some disease states could lead to less sustained neuronal responses (that is more adaptation), and this may contribute to the symptoms found.

Noradrenergic neurons

The source of the noradrenergic projection to the neocortex is the locus coeruleus (noradrenergic cell group A6) in the pons (see Green and Costain, 1981). There are a few thousand of these neurons that innervate the whole of the cerebral cortex, and the amygdala, so it is unlikely that the noradrenergic neurons convey the specific information stored in synapses that specifies each memory. Instead, to the extent that the noradrenergic neurons are involved in memory (including pattern association), it is likely that they would have a modulatory role on cell excitability, which would influence the extent to which the voltage-dependent NMDA receptors are activated, and thus the likelihood that information carried on specific afferents would be stored (cf. Siegel and Auerbach, 1996). Evidence that this may be the case comes from a study in which it was shown that neocortical LTP is impaired if noradrenergic and simultaneously cholinergic inputs to cortical cells are blocked pharmacologically (Bear and Singer, 1986). Further, in a study designed to show whether the noradrenergic modulation is necessary for memory, Borsini and Rolls (1984) showed that intra-amygdaloid injections of noradrenergic receptor blockers did impair the type of learning in which rats gradually learned to accept novel foods. The function implemented by this noradrenergic input may be general activation, rather than a signal that carries information about whether reward versus punishment has been given, for noradrenergic neurons in rats respond to both rewarding and punishing stimuli, and one of the more effective stimuli for producing release of noradrenaline is placing the feet in cool water (see McGinty and Szymusiak, 1988).

7.2 Backprojections in the cerebral cortex—the role of pattern association

There are as many backward connections between adjacent areas in the cerebral cortex as there are forward connections, and there are also backprojections from the amygdala and hippocampus to the areas of the association cortex which project forward towards these structures. It is thought that one of the functions of these backprojections is to implement recall of neuronal activity in earlier cortical stages, and that this recall operates by pattern association (Rolls, 1989b; Treves and Rolls, 1994; Chapters 6 and 10). Evidence that during recall neural activity does occur in cortical areas involved in the original processing comes, for example, from investigations which show that when humans are asked to recall visual scenes in the dark, blood flow is increased in visual cortical areas, stretching back from association cortical areas as far as early (possibly even primary) visual cortical areas (Roland and Friberg, 1985; Kosslyn, 1994). The way in which the backprojections could operate is described in Section 10.2. The important point here, in relation to pattern association, is that it is hypothesized that the backprojections in terminating on the apical dendrites of cortical pyramidal cells form associatively modifiable synapses. Consistent with this, there is a high density of NMDA receptors on the apical dendrites of cortical pyramidal cells. Functions such as recall, and top-down processing involved in attention, constraint satisfaction, and priming, are thought to be implemented through this pattern association system (see Section 10.2).

In relation to the amygdala, it is important to note that in addition to the axons and their terminals in layer 1 from the succeeding cortical stage, there are also axons and terminals in layer 1 in many stages of the cortical hierarchy from the amygdala (Turner, 1981; Amaral and Price, 1984; Amaral, 1986, 1987; Amaral et al., 1992) (see Fig. 7.12). In the context of the brain systems involved in emotion, it is of interest that the amygdala has widespread

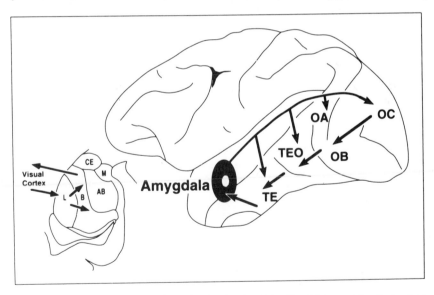

Fig. 7.12 Amygdala backprojections. Schematic illustration showing the relationship of the amygdala with the visual related cortices of the temporal and occipital lobes. The ventral visual pathway is shown projecting hierarchically from the striate visual cortex, V1 (OC in the diagram) via prestriate cortex (OB) and TEO (posterior inferotemporal cortex) to TE (inferior temporal cortex), which in turn projects into the amygdala. The backprojections from the amygdala to the visual pathways spread beyond TE to earlier visual areas. The inset diagram shows in a coronal section through the amygdala that the visual cortex inputs reach the lateral nucleus (L) of the amygdala, which projects to the basal nucleus (B), from which the backprojections to the cortex arise. AB is the basal accessory nucleus of the amygdala; and CE is the central nucleus of the amygdala.

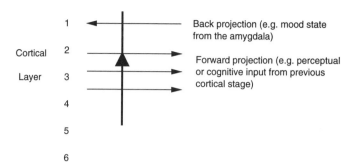

Fig. 7.13 Pyramidal cells in for example layers 2 and 3 of the temporal lobe association cortex receive forward inputs from preceding cortical stages of processing, and also backprojections from the amygdala. It is suggested that the backprojections from the amygdala make synapses on the apical dendrites of cortical pyramidal cells which are modifiable during learning when amygdala neurons are active in relation to a mood state; and that the backprojections from the amygdala via these modified synapses allow mood state to influence later cognitive processing, for example by facilitating some perceptual representations.

backprojections to association cortical areas involved in the sensory analysis of what visual, auditory, etc. stimuli are present (for example inferior temporal visual cortex and V4). Associative modification in these backprojection pathways from the amygdala (and orbitofrontal cortex) may be a mechanism by which emotional or mood states represented in the amygdala and orbitofrontal cortex can influence or bias cognitive processing in neocortical association areas (see Fig. 7.13). For example, cortical memories that are retrieved are influenced by the ongoing mood state (see for example Blaney, 1986), and such backprojections could be the mechanism by which this is achieved.

7.3 Pattern association within the hippocampus

Although autoassociation in the CA3 recurrent collateral system and competitive learning in the dentate gyrus and CA1 regions are aspects of hippocampal function discussed in Chapter 6, pattern association in some parts of the hippocampus is part of the model described in Chapter 6. One region in which pattern association is part of the principle of operation of the model is in the direct perforant path input to CA3. It is suggested that pattern association between the activity of the perforant path activity and the firing of CA3 neurons (itself produced by the mossy fibre input) occurs during learning. Later, a fragment of the original episode reaching the entorhinal cortex produces through the modified perforant path to CA3 synapses a retrieval cue for the CA3 neurons. Retrieval would then be completed by the CA3 recurrent collateral connections. Another part of the hippocampus within which pattern association may occur is in the direct perforant path input to the CA1 neurons, which may be useful during recall, and would, if so, need to be associatively modified during learning. In addition, the CA1 to entorhinal cortex synapses would operate as a pattern associator allowing correct recall of entorhinal activity. In doing this, they would be an example of how backprojection systems may operate (see Section 6.10). The hippocampal system thus provides a good example of how different classes of network may be combined together in a brain system. Each brain system probably combines the principles of several types of network described in Chapters 2–5.

8 Cortical networks for invariant pattern recognition

8.1 Introduction

One of the major problems that is solved by the visual system in the cerebral cortex is the building of a representation of visual information which allows recognition to occur relatively independently of size, contrast, spatial frequency, position on the retina, angle of view, etc. It is important to realize that this type of generalization is not a simple property of one-layer neural networks. Although neural networks do generalize well, the type of generalization they show naturally is to vectors which have a high dot product or correlation with what they have already learned. To make this clear, Fig. 8.1 is a reminder that the activation h_i of each neuron is computed as

$$h_i = \Sigma_j r'_j w_{ij} \qquad (8.1)$$

where the sum is over the C input axons, indexed by j. Now consider translation (or shift) of the input pattern vector by one position. The dot product will now drop to a low level, and the neuron will not respond, even though it is the same pattern, just shifted by one location. This makes the point that special processes are needed to compute invariant representations. Network approaches to such invariant pattern recognition are described in this chapter. Once an invariant representation has been computed by a sensory system, it is in a form which is suitable for presentation to a pattern association or autoassociation neural network (see Chapters 2 and 3).

Before considering networks that can perform invariant pattern recognition, it is useful to consider the preprocessing that the visual system performs in the stages as far as the primary visual cortex. (The visual system of the monkey is referred to in the descriptions that follow, because in primates there is a highly developed temporal lobe that contains large cortical areas involved in invariant pattern recognition; because there have been many studies of this part of the visual system in monkeys; and because knowledge of the operation of the visual system of the non-human primate is as relevant as possible to understanding visual processing in humans.)

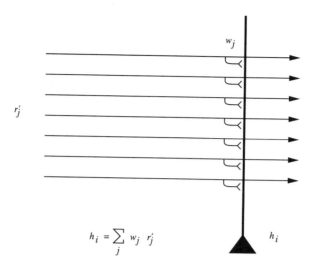

Fig. 8.1 A neuron which computes a dot product of the input pattern with its synaptic weight vector generalises well to other patterns based on their similarity measured in terms of dot product or correlation, but shows no translation (or size, etc.) invariance.

8.2 Preprocessing in the visual pathways as far as the primary visual cortex

The visual pathways from the cones in the retina to the primary visual cortex are shown schematically in Fig. 8.2. The responses of the neural elements to visual stimuli alter through these pathways. The cone receptors respond optimally to spots of light. The ganglion cells have centre-surround receptive fields (see Fig. 8.3). They respond, for example, to a spot of light in the 'on' centre, are switched off if there is also light in the surround, and respond best if light covers the centre but not (much of) the surround. They thus respond best if there is spatial discontinuity in the image on the retina over their receptive field. The lateral geniculate cells also have centre-surround organization. In the primary visual cortex (V1), some cells, for example pyramidal cells in layers 2 and 3, and 5 and 6, have receptive fields that are elongated, responding best, for example, to a bar or edge in the receptive field (see Fig. 8.3). Cells with responses of this type which have definite on and off regions of the elongated receptive field, and which do not respond to uniform illumination which covers the whole receptive field, are called simple cells (Hubel and Wiesel, 1962). They are orientation sensitive, with approximately four orientations being sufficient to account for their orientation sensitivity. The goals of this early visual processing include making the information explicit about the location of characteristic features of the world, such as elongated edges and their orientation; and removing redundancy from the visual input by not responding to areas of continuous brightness (Marr, 1982; Barlow, 1989).

An approach to how such preprocessing might self-organize in a multilayer neuronal network has been described by Linsker (1986, 1988). In the model architecture he describes, neurons with simple cell-like receptive fields develop. They develop even in the absence of visual input, and require only random firing of the neurons in the system. Visual input would

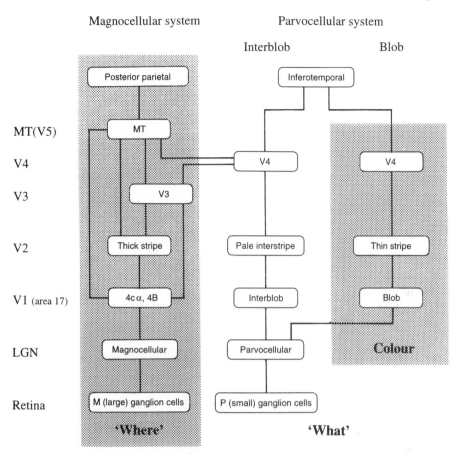

Fig. 8.2 A schematic diagram of the visual pathways from the retina to visual cortical areas. V1, primary visual cortex; V2, V3, V4 etc., other visual cortical areas; M, magnocellular; P, parvocellular.

help the network to learn by providing evidence on the statistics of the visual world. The network and its properties are described next, not because this is necessarily a sufficient approach to understanding how the preprocessing stages of the visual system develop (see Rose and Dobson, 1985 for alternative models), but because it is very helpful to understand what can develop in even quite a simple multilayer processing system which uses local Hebb-like synaptic modification rules.

The architecture of Linsker's net is shown in Fig. 8.4. It is a multilayer feedforward network. A neuron in one layer receives from a small neighbourhood of the preceding layer, with the density of connections received falling off with a Gaussian profile within that neighbourhood. These limited receptive fields are crucial, for they enable units to respond to spatial correlations in the previous layer. Layer A receives the input, and is analogous to the receptors of the retina, for example cones. The neurons are linear (that is have linear activation functions). This means that the multilayer network could be replaced by a single-layer network, for a set of linear transforms can be replaced by a single linear transform, the product of the set of transforms. This does not of course imply that any local unsupervised

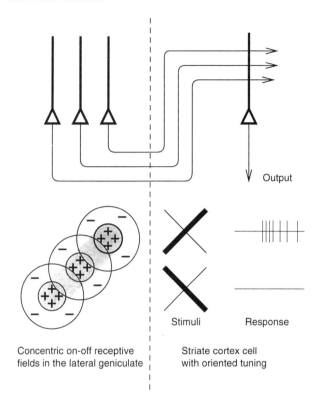

Fig. 8.3 Receptive fields in the lateral geniculate have concentric on-centre off-surround (or vice versa) receptive fields (left). Neurons in the striate cortex, such as the simple cell illustrated on the right, respond to elongated lines or edges at certain orientations. A suggestion that the lateral geniculate neurons might combine their responses in the way shown to produce orientation-selective simple cells was made by Hubel and Wiesel (1962).

learning rule would be able in a one-layer net to produce the equivalent transform. What is of interest in this network is the type of receptive fields that self-organize with the architecture and a local Hebb-like learning rule, and this is what Linsker (1986, 1988) has investigated. The modified Hebb rule he used (for the jth synaptic weight w_j from an axon with firing rate r'_j on a neuron with firing rate r) is

$$\delta w_j = k \ (r r'_j + b r'_j + c r + d) \tag{8.2}$$

where the first product inside the brackets is the Hebbian term, and k is the learning rate. The weights in Linsker's network are clipped at maximum and minimum values. The second to fourth terms in the brackets alter how the network behaves within these constraints, with separate constants b, c and d for each of these terms. The rule acts, as does Oja's (see Chapter 4) to maximize the output variance of the neuron's outputs given a set of inputs, that is it finds the first principal component of the inputs (Linsker, 1988; Hertz *et al.*, 1991).

Random firing of the input neurons was used to train the network. (A parallel might be the conditions when the visual system is set up before birth, so that at birth there is some organization of the receptive fields already present.) For a range of parameters, it was found

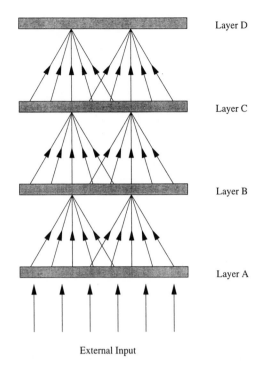

Layer D

Layer C

Layer B

Layer A

External Input

Fig. 8.4 Architecture of Linsker's multilayer Hebbian learning network, showing the receptive fields of a few neurons.

that neurons in layer B simply averaged the input activity over their receptive fields. Because the receptive fields of neighbouring neurons in layer B overlapped considerably and they therefore received correlated inputs, the activity of neighbouring neurons in layer B was highly correlated. Correspondingly, the activity of neurons further apart in layer B was not highly correlated. This led to the emergence of a new type of receptive field in layer C, with centre-surround organization, in which a neuron might respond to inputs in the centre of its receptive field, and be inhibited by inputs towards the edge of its receptive field, responding maximally, for example, to a bright spot with a dark surround (Fig. 8.5a). Other cells in layer C developed off-centre on-surround receptive fields.

Within layer C, nearby units were positively correlated while further away there was a ring of negative correlation. (This is again related to the geometry of the connections, with nearby neurons receiving from overlapping and therefore correlated regions of layer B.) The result of the neighbourhood correlations in layer C, which had a spatial profile like that of a Mexican hat (see Fig. 4.7), was that later layers of the simulation (D-F) also had centre-surround receptive fields, which had sharper and sharper Mexican hat correlations.

For layer G the parameters were altered, and it was found that many of the neurons were no longer circularly symmetric, but had elongated receptive fields, which responded best to bars or edges of a particular orientation. Thus, symmetry had been broken, and cells with similarities to simple cells in V1 with oriented bar or edge selectivity had developed.

Cells in layer G developed the orientation to which they responded best independently of their neighbours. In the architecture described so far, there were no lateral connections. If

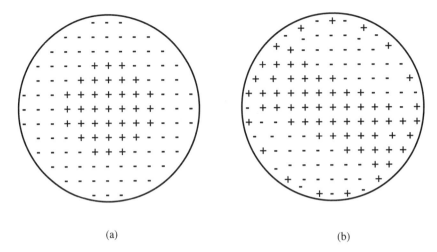

Fig. 8.5 Diagram of positive and negative connection strengths within the receptive fields of units in Linsker's network: (a) a centre-surround cell in layer C; (b) an orientation-selective cell in layer G, which would respond well to a light bar. (After Hertz, Krogh and Palmer, 1991, Fig. 8.7.)

lateral connections were introduced (for example short-range excitatory connections) in layer G, then neighbouring cells developed nearby orientation preferences, and there was a transition of orientation-sensitivity across the cell layer, reminiscent of orientation columns in V1 (Linsker, 1986). This is an example of the formation of topological maps described in Chapter 4.

The exact way in which different types of receptive fields develop in this architecture depending on the parameters in the learning rule have been analysed in detail by Kammen and Yuille (1988), Yuille *et al.* (1989), and MacKay and Miller (1990). Perhaps the most important point from these results in relation to the network just described is that a multilayer network with local feedforward connectivity, and using a local, modified Hebb, learning rule, can even without structured inputs, develop quite complicated receptive fields which would be biologically useful as preprocessors, and which have similarities to the receptive fields of neurons in the mammalian visual system.

There are a number of issues that are not addressed by Linsker's network which are of fundamental importance, and which we now go on to address. One is that in Linsker's network neighbouring neurons develop within a layer independently of any information from their neighbours about how they are responding, so that the layer as a whole cannot learn to remove redundancy from the representations it forms, and in this way produce efficient encoding (cf. Barlow, 1989). (The lateral connections added as an option to layer G of Linsker's network assist in the formation of maps, but not directly in redundancy removal.) A second issue is that Linsker's net was trained on random input firings. What now needs exploration is how networks can learn the statistics of the visual world they are shown, to set up analysers appropriate for efficiently encoding that world, and removing the redundancies present in the visual scenes we see. If the visual world consisted of random pixels, then this would not be an issue. But because our visual world is far from random, with, for example, many elongated edges (at least over the short range), this implies that it is useful to extract these elements as features. A third issue is that higher-order representations are not explicitly addressed by the network just described. A higher- order feature might consist of a horizontal line or bar above a vertical line or bar, which if occurring in combination might signify a letter 'T'. It is the co-occurrence, that is the pairwise correlation, of these two features which is significant in detecting the T, and which is a specific higher-order (in this case second-order) combination of the two line elements. Such higher-order representations are crucial for distinguishing real objects in the world. A fourth issue is that nothing of what has been considered so far deals with or solves the problem of invariant representation, that is how objects can be identified irrespective of their position, size, etc. on the retina, and even independently of view. A fifth issue is that Linsker's net is relevant to early visual processing, as far as V1, but does not consider how processing beyond that might occur, including object identification. A sixth issue is that, for reasons of mathematical tractability, Linsker's network was linear. We know that many mappings cannot be performed in a single-layer network (to which Linsker's is equivalent, in terms of how it maps inputs to outputs), and it is therefore appropriate to consider how non-linearities in the network may help in the solution of some of the issues just raised.

These issues are all addressed in Section 8.4 on visual processing in a multilayer network which models some of the aspects of visual processing in the primate visual system between V1 and the anterior inferior temporal visual cortex (IT). Before describing the model, VisNet, a short summary is given of some of what is known about processing in this hierarchy, as this is what is to be modelled. Knowing about processing in the brain may also help to lead to models which can solve difficult computational problems, such as invariant object recognition. There is at least an existence proof in the brain that the problem can be solved, and by studying the brain, many indications about how it is solved can be obtained.

8.3 Processing to the inferior temporal cortex in the primate visual system

A schematic diagram to indicate some aspects of the processing involved in object identification from the primary visual cortex, V1, through V2 and V4 to the posterior inferior temporal cortex (TEO) and the anterior inferior temporal cortex (TE) is shown in Fig. 8.6 (see also Fig. 1.10). Their approximate location on the brain of a macaque monkey is shown in Fig. 8.7, which also shows that TE has a number of different subdivisions. The different TE areas all contain visually responsive neurons, as do many of the areas within the cortex in the superior temporal sulcus (Baylis, Rolls and Leonard, 1987). For the purposes of this summary, these areas will be grouped together as anterior inferior temporal (IT), except where otherwise stated. Some of the information processing that takes place through these pathways that must be addressed by computational models is as follows. A fuller account is provided by Rolls (1994a, 1995b, 1997d). Many of the studies on neurons in the inferior temporal cortex and cortex in the superior temporal sulcus have been performed with neurons which respond particularly to faces, because such neurons can be found regularly in recordings in this region, and therefore provide a good population for systematic studies (see Rolls, 1994a, 1997d).

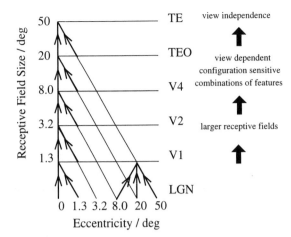

Fig. 8.6 Schematic diagram showing convergence achieved by the forward projections in the visual system, and the types of representation that may be built by competitive networks operating at each stage of the system from the primary visual cortex (V1) to the inferior temporal visual cortex (area TE) (see text). LGN, lateral geniculate nucleus. Area TEO forms the posterior inferior temporal cortex. The receptive fields in the inferior temporal visual cortex (for example in the TE areas) cross the vertical midline (not shown).

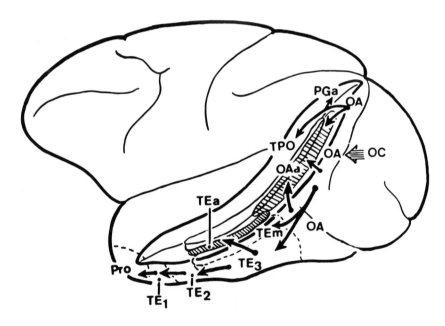

Fig. 8.7 Lateral view of the macaque brain showing the different architectonic areas (for example TEm, TPO) in and bordering the anterior part of the superior temporal sulcus (STS) of the macaque (see text). The borders of the opened superior temporal sulcus are indicated by bold outline. Terminology of Seltzer and Pandya (1978).

8.3.1 Hierarchical feedforward organization

Cortical visual processing for object recognition is considered to be organized as a set of hierarchically connected cortical regions consisting at least of V1, V2, V4, posterior inferior temporal cortex (TEO), and anterior inferior temporal cortex. (This stream of processing has many connections with a set of cortical areas in the anterior part of the superior temporal sulcus, including area TPO, which receive motion information.) There is a definite hierarchy of processing as shown by the following. First, each cortical area receives information particularly in layers 4, 2 and 3 from the preceding cortical area (see Fig. 10.7). The superficial pyramidal cells, in layer 2 and 3, project forwards to the next cortical area. The deep pyramidal cells in layer 5 project backwards to end in layer 1 of the preceding cortical area (on the apical dendrites of pyramidal cells). This asymmetry of the connectivity allows one to identify a definite forward and backward direction of the connectivity, starting from the primary visual cortex (V1), which in turn receives its inputs from the retina via the lateral geniculate. Although there are backprojections in this system, there is a definite forward and backward direction to the connectivity, and, as will be shown, object identification is possible when there is insufficient time for full processing in the backprojecting pathways. Second, the latencies of activation of neurons in this pathway increase from shortest values in V1 of about 40–50 ms, to values of 80–100 ms in the anterior inferior temporal cortical areas. Response latencies between V1 and the inferior temporal cortex increase from stage to stage (Thorpe and Imbert, 1989), with time for approximately 15 ms to be added at each stage (Rolls, 1994a, 1995b, 1997d). Third, evidence on a gradual change in the response properties such as the receptive field size and response complexity of cells along this pathway (see below) is also indicative of a processing hierarchy.

8.3.2 Receptive field size and translation invariance

There is convergence from each small part of a region to the succeeding region (or layer in the hierarchy) in such a way that the receptive field sizes of neurons (for example 1 near the fovea in V1) become larger by a factor of approximately 2.5 with each succeeding stage (and the typical parafoveal receptive field sizes found would not be inconsistent with the calculated approximations of, for example, 8° in V4, 20° in TEO, and 50° in inferior temporal cortex, Boussaoud et al., 1991) (see Fig. 8.6). Such zones of convergence would overlap continuously with each other (see Fig. 8.6). This connectivity provides part of the basis for the fact that many neurons in the temporal cortical visual areas respond to a stimulus relatively independently of where it is in their receptive field, and moreover maintain their stimulus selectivity when the stimulus appears in different parts of the visual field (Gross et al., 1985; Tovee, Rolls and Azzopardi, 1994). This is called translation or shift invariance. In addition to having topologically appropriate connections, it is necessary for the connections to have the appropriate synaptic weights to perform the mapping of each set of features, or object, to the same set of neurons in IT. How this could be achieved is addressed in the neural network model that follows.

8.3.3 Size and spatial frequency invariance

Some neurons in IT/STS respond relatively independently of the size of an effective face stimulus, with a mean size invariance (to a half maximal response) of 12 times (3.5 octaves) (Rolls and Baylis, 1986). This is not a property of a simple single-layer network (see Fig. 8.1), nor of neurons in V1, which respond best to small stimuli, with a typical size invariance of 1.5 octaves. (Some neurons in IT/STS also respond to face stimuli which are blurred, or which are line drawn, showing that they can also map the different spatial frequencies with which objects can be represented to the same representation in IT/STS, see Rolls, Baylis and Leonard, 1985.)

8.3.4 Combinations of features in the correct spatial configuration

Many cells in this processing stream respond to combinations of features (including objects), but not to single features presented alone, and the features must have the correct spatial arrangement. This has been shown, for example, with faces, for which it has been shown by masking out or presenting parts of the face (for example eyes, mouth, or hair) in isolation, or by jumbling the features in faces, that some cells in the cortex in IT/STS respond only if two or more features are present, and are in the correct spatial arrangement (Perrett *et al.*, 1982; Rolls, Tovee, Purcell *et al.*, 1994). Corresponding evidence has been found for non-face cells. For example, Tanaka *et al.* (1990) showed that some posterior inferior temporal cortex neurons might only respond to the combination of an edge and a small circle if they were in the correct spatial relation to each other. Evidence consistent with this suggestion, that neurons are responding to combinations of a few variables represented at the preceding stage of cortical processing, is that some neurons in V2 and V4 respond to end-stopped lines, to tongues flanked by inhibitory subregions, or to combinations of colours (see references cited by Rolls, 1991c). Neurons that respond to combinations of features but not to single features indicate that the system is non-linear.

8.3.5 A view-independent representation

For recognizing and learning about objects (including faces), it is important that an output of the visual system should be not only translation and size invariant, but also relatively view invariant. In an investigation of whether there are such neurons, we found that some temporal cortical neurons reliably responded differently to the faces of two different individuals independently of viewing angle (Hasselmo, Rolls, Baylis and Nalwa, 1989), although in most cases (16/18 neurons) the response was not perfectly view-independent. Mixed together in the same cortical regions there are neurons with view-dependent responses (for example Hasselmo *et al.*, 1989; Rolls and Tovee, 1995a). Such neurons might respond, for example, to a view of a profile of a monkey but not to a full-face view of the same monkey (Perrett *et al.*, 1985). These findings, of view-dependent, partially view-independent, and view-independent representations in the same cortical regions are consistent with the hypothesis discussed below that view-independent representations are being built in these regions by associating together the outputs of neurons that respond to different views of the same individual. These findings also provide evidence that one output of the visual system includes representations of what is being seen, in a view-independent way that would be useful for object recognition and for learning associations about objects; and that another output is a view-based representation that would be useful in social interactions to determine whether another individual is looking at one, and for selecting details of motor responses, for which the orientation of the object with respect to the viewer is required. Further evidence that some neurons in the temporal cortical visual areas have object-based responses comes from a population of neurons that responds to moving faces, for example to a head undergoing ventral flexion, irrespective of whether the view of the head was full face, of either profile, or even of the back of the head (Hasselmo, Rolls, Baylis and Nalwa, 1989).

8.3.6 Distributed encoding

An important question for understanding brain function is whether a particular object (or face) is represented in the brain by the firing of one or a few gnostic (or 'grandmother') cells (Barlow, 1972), or whether instead the firing of a group or ensemble of cells each with somewhat different responsiveness provides the representation. Advantages of distributed codes (see Chapter 2) include generalization and graceful degradation (fault tolerance), and a potentially very high capacity in the number of stimuli that can be represented (that is exponential growth of capacity with the number of neurons in the representation). If the ensemble encoding is sparse, this provides a good input to an associative memory, for then large numbers of stimuli can be stored (see Chapters 2 and 3). We have shown that in the IT/STS, responses of a group of neurons, but not of a single neuron, provide evidence on which face was shown. We showed, for example, that these neurons typically respond with a graded set of firing to different faces, with firing rates from 120 spikes/s to the most effective face, to no response at all to a number of the least effective faces (Baylis et al., 1985; Rolls and Tovee, 1995a). The sparseness a of the activity of the neurons was 0.65 over a set of 68 stimuli including 23 faces and 45 non-face natural scenes, and a measure called the response sparseness a_r of the representation, in which the spontaneous rate was subtracted from the firing rate to each stimulus so that the responses of the neuron were being assessed (cf. Eqn. 1.4), was 0.38 across the same set of stimuli (Rolls and Tovee, 1995a).

It has been possible to apply information theory to show that each neuron conveys on average approximately 0.4 bits of information about which face in a set of 20 faces has been seen (Tovee and Rolls, 1994; cf. Tovee et al., 1993). If a neuron responded to only one of the faces in the set of 20, then it could convey (if noiseless) 4.6 bits of information about one of the faces (when that face was shown). If, at the other extreme, it responded to half the faces in the set, it would convey 1 bit of information about which face had been seen on any one trial. In fact, the average maximum information about the best stimulus was 1.8 bits of information. This provides good evidence not only that the representation is distributed, but also that it is a sufficiently reliable representation that useful information can be obtained from it.

The most impressive result obtained so far is that when the information available from a population of neurons about which of 20 faces has been seen is considered, the information increases approximately linearly as the number of cells in the population increases from 1 to 14 (Abbott, Rolls and Tovee, 1996; Rolls, Treves and Tovee, 1997; see Fig. 8.8). Remembering

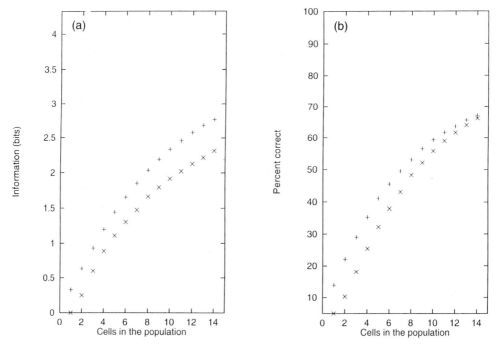

Fig. 8.8 (a) The values for the average information available in the responses of different numbers of neurons in the temporal cortical visual areas on each trial in a 500 ms period, about which of a set of 20 face stimuli has been shown. The decoding method was Dot Product (×) or Probability Estimation (+). (b) The percentage correct for the corresponding data to those shown in Fig. 8.8a. (Fig. 4 of Rolls, Treves and Tovee, 1997.)

that the information in bits is a logarithmic measure, this shows that the representational capacity of this population of cells increases exponentially. This is the case both when an optimal, probability estimation, form of decoding of the activity of the neuronal population is used, and also when the neurally plausible dot product type of decoding is used (Fig. 8.8a). (The dot product decoding assumes that what reads out the information from the population activity vector is a neuron or a set of neurons which operates just by forming the dot product of the input population vector and its synaptic weight vector—see Rolls *et al.*, 1997.) By simulation of further neurons and further stimuli, we have shown that the capacity grows very impressively, approximately as shown in Fig. 1.6 (Abbott, Rolls and Tovee, 1996). This result is exactly what would be hoped for from a distributed representation. This result is not what would be expected for local encoding, for which the number of stimuli that could be encoded would increase linearly with the number of cells. (Even if the grandmother cells were noisy, adding more replicates to increase reliability would not lead to more than a linear increase in the number of stimuli that can be encoded as a function of the number of cells.) These findings provide very firm evidence that the encoding built at the end of the visual system *is* distributed, and that part of the power of this representation is that by receiving inputs from relatively small numbers of such neurons, neurons at the next stage of processing (for example in memory structures such as the hippocampus, amygdala and orbitofrontal cortex) would obtain information about which of a very great number of stimuli had been shown. This representational capacity of neuronal populations has fundamental implications for the connectivity of the brain, for it shows that neurons need not have hundreds of thousands or millions of inputs to have available to them information what is represented in another population of cells, but that instead the real numbers of perhaps 8 000–10 000 synapses per neuron would be adequate for them to receive considerable information from the several different sources between which this set of synapses is allocated.

It may be noted that it is unlikely that there are further processing areas beyond those described where ensemble coding changes into grandmother cell encoding. Anatomically, there does not appear to be a whole further set of visual processing areas present in the brain; and outputs from the temporal lobe visual areas such as those described are taken to limbic and related regions such as the amygdala and via the entorhinal cortex the hippocampus (see Rolls, 1994a, 1997d). Indeed, tracing this pathway onwards, we have found a population of neurons with face-selective responses in the amygdala, and in the majority of these neurons, different responses occur to different faces, with ensemble (not local) coding still being present (Leonard et al., 1985; Rolls, 1992a). The amygdala, in turn, projects to another structure which may be important in other behavioural responses to faces, the ventral striatum, and comparable neurons have also been found in the ventral striatum (Williams et al., 1993). We have also recorded from face-responding neurons in the part of the orbitofrontal cortex that receives from the IT/STS cortex, and have found that the encoding there is also not local (Rolls and Critchley, in preparation; Rolls, 1996b).

8.3.7 Speed of processing

There is considerable evidence that the processing time required in each cortical area for useful computation is in the order of 20–30 ms only. This has important implications for the nature of the computation that could be performed in each cortical stage. It would be unlikely to be sufficiently long for a stochastic iterative process, or for temporal encoding and synchronization of multiple different populations of neurons.

The first type of evidence is that in terms of response latency, there is sufficient time for only 15 or perhaps 20 ms per stage of processing in the visual system (see above). Given that the first few spikes of a neuron in an area far removed from the input, for example IT/STS, are not poorly tuned, and that considerable information can be extracted from them about which face was seen (Tovee et al., 1993; Tovee and Rolls, 1995), there must be useful information emanating from the first 15–20 ms of the processing in *each* cortical area in the hierarchy.

The second type of evidence comes from analysis of the amount of information available in short temporal epochs of the responses of temporal cortical face-selective neurons about which face had been seen. We found that if a period of the firing rate of 50 ms was taken, then this contained 84.4% of the information available in a much longer period of 400 ms about which of four faces had been seen. If the epoch was as little as 20 ms, the information was 65% of that available from the firing rate in the 400 ms period (Tovee et al., 1993; Tovee and Rolls, 1995). These high information yields were obtained with the short epochs taken near the start of the neuronal response, for example in the post-stimulus period 100–120 ms (Tovee and Rolls, 1995). These investigations thus showed that there was considerable information about which stimulus had been seen actually available from short time epochs near the start of the response of temporal cortex neurons.

The third approach was to measure how short a period of activity is sufficient in a cortical area for successful object recognition to occur. This approach used a visual backward masking paradigm. In this paradigm there is a brief (20 ms) presentation of a test stimulus which is rapidly followed (within 1–100 ms) by the presentation of a second stimulus (the mask), which impairs or masks the perception of the test stimulus (Rolls and Tovee, 1994). When there was no mask the cell responded to a 16 ms presentation of the test stimulus for 200–300 ms, far longer than the presentation time. It is suggested that this reflects the operation of a short term memory system implemented in local cortico-cortical circuitry, in the form of an attractor or autoassociation implemented by recurrent collaterals between cortical pyramidal cells. If the mask was a stimulus which did not stimulate the cell (either a non-face pattern mask consisting of black and white letters N and O, or a face which was a non-effective stimulus for that cell), then as the interval between the onset of the test stimulus and the onset of the mask stimulus (the stimulus onset asynchrony, SOA) was reduced, the length of time for which the cell fired in response to the test stimulus was reduced. This reflected an abrupt interruption of neuronal activity produced by the effective face stimulus. When the SOA was 20 ms, face-selective neurons in the inferior temporal cortex of macaques responded for a period of 20–30 ms before their firing was interrupted by the mask (Rolls and Tovee, 1994). We went on to show that under these conditions (a test-mask stimulus onset asynchrony of 20 ms), human observers looking at the same displays could just identify which of six faces was shown (Rolls, Tovee, Purcell, Stewart and Azzopardi, 1994). (Consistent psychophysical evidence, that multiple images can be recognized when they are shown with rapid sequential visual presentation, was described by Thorpe and Imbert, 1989).

These results provide evidence that a cortical area can perform the computation necessary for the recognition of a visual stimulus in 20–30 ms, and provide a fundamental constraint which must be accounted for in any theory of cortical computation.

8.3.8 Rapid learning

Learning to identify new objects or images never seen before can occur rapidly. Just a very few seconds of seeing a new face or picture will enable us to recognize it later. Although at least verbal (declarative) statements about whether an object is recognized may involve later processing stages such as the perirhinal cortex or hippocampus (see Chapter 6), there is now neurophysiological evidence that new representations can be built in the IT/STS cortex (the macaque inferior temporal cortex and the cortex in the superior temporal sulcus) in a few seconds of visual experience. Part of the importance of this for computational models of processing in neural networks in the visual system is that this rapid learning rules out some algorithms, such as backpropagation, which require thousands of training trials. Instead, the process must be very fast, within 15 trials.

Part of the evidence for new representations being set up rapidly during visual learning in the temporal cortical visual areas, when a set of new faces was shown in random order (with 1 s for each presentation) to face-responsive neurons, is that some altered the relative degree to which they responded to the different members of the set of novel faces over the first few (12) presentations of the set (Rolls, Baylis, Hasselmo and Nalwa, 1989). If in a different experiment a single novel face was introduced when the responses of a neuron to a set of familiar faces were being recorded, it was found that the responses to the set of familiar faces were not disrupted, while the responses to the novel face became stable within a few presentations.

The second type of evidence that these neurons can learn new representations very rapidly comes from an experiment in which binarized black and white images of faces which blended with the background were used. These did not strongly activate face-selective neurons. Full grey-scale images of the same photographs were then shown for ten 0.5 s presentations. It was found that in a number of cases, if the neuron happened to be responsive to that face, when the binarized version of the same face was shown next, the neurons responded to it (Tovee, Rolls and Ramachandran, 1996). This is a direct parallel to the same phenomenon which we observed psychophysically, and provides dramatic evidence that these neurons are influenced by only a very few seconds (in this case 5 s) of experience with a visual stimulus. Following up this study in humans, we have been able to show in a positron emission tomography (PET) study that corresponding areas in the human temporal lobe alter their activity during rapid visual learning (Dolan *et al.*, 1997).

These experiments indicate that candidate neural networks for processing and forming representations of visual stimuli should ideally be capable of performing both these functions rapidly. A visual system which took thousands of trials to learn about a new object would not be very useful biologically.

Learning has also been demonstrated to affect the responses of inferior temporal cortex neurons, though often many training trials were used. For example, a tendency of some temporal cortical neurons to associate together visual stimuli when they have been shown over many repetitions separated by several seconds has also been described by Miyashita and Chang (1988) (see also Miyashita, 1993; Naya, Sakai and Miyashita, 1996; Higuchi and Miyashita, 1996). In addition, Logothetis *et al.* (1994) using extensive training (600 training trials) showed that neurons could alter their responses to different views of computer-generated objects.

It is also the case that there is a much shorter-term form of memory in which some of these neurons are involved, for whether a particular visual stimulus (such as a face) has been seen recently, in that some of these neurons respond differently to recently seen stimuli in short-term visual memory tasks (Baylis and Rolls, 1987; Miller and Desimone, 1994).

8.4 VisNet—an approach to biologically plausible visual object identification

The neurophysiological investigations just described provide indications, when taken with evidence on the connections of the visual system, about how visual information processing for object identification might take place. They have led to hypotheses about how this part of the visual system might operate (Rolls, 1992b, 1994a, 1995b, 1996c), which in turn have been investigated by a neuronal net simulation which we call VisNet (Rolls, 1994a, 1995b, 1997d; Wallis, Rolls and Foldiak, 1993; Wallis and Rolls, 1997). The hypotheses are described here in the context of the neuronal net simulation.

VisNet is a four-layer network, with successive layers corresponding approximately to V2, V4, the posterior temporal cortex, and the anterior temporal cortex (cf. Figs. 8.6 and 8.9). The forward connections to a cell in one layer are derived from a topologically corresponding region of the preceding layer, using a Gaussian distribution of connection probabilities to determine the exact neurons in the preceding layer to which connections are made. This schema is constrained to preclude the repeated connection of any cells. Each cell in the simulation receives 100 connections from the 32 × 32 cells of the preceding layer, with a 67% probability that a connection comes from within 4 cells of the distribution centre. One underlying computational hypothesis in the design of this part of the architecture is that it should provide the basis for the connectivity for neurons in layer 4 (the top layer) to respond in a translation invariant way, that is, independently of where stimuli are in the visual field. Another is that the connectivity of any neuron in the architecture is limited to a small part of the preceding layer, in order to reduce the combinatorial explosion if a neuron in layer 1 had to search for all possible combinations of features over the whole visual field. The hypothesis instead is that the particular combinations of inputs that do occur in the redundant world on which the network is trained will have neurons allocated to them in whatever input space is seen by each neuron in the network. By layer 4, neurons will have the possibility (given appropriate training) to respond to the actual feature combinations (such as objects) found in the visual world, independently of where they are in the visual field.

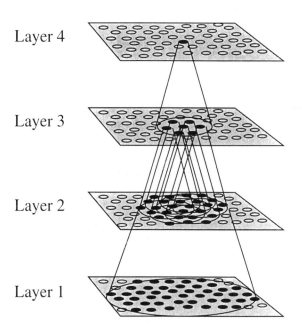

Fig. 8.9 Hierarchical network structure of VisNet.

Within each layer, lateral inhibition between neurons has a radius of effect just greater than the radius of feedforward convergence just defined. The lateral inhibition is to enable competitive learning (with a modification soon to be described) to operate locally within each region of a layer. (This corresponds directly to the inhibitory neurons for feedback inhibition present within the cortex, with typical dendritic and axonal domains of approximately 12 mm. The domain is only that of an inhibitory interneuron, and the 'domains' overlap continuously in each layer.) The hypotheses underlying the 'soft' competitive learning (with graded responses, not with a single winner within a domain) are as follows. First, competitive networks operate to detect correlations between the activity of the input neurons, and to allocate output neurons to respond to each cluster or 'feature' of such correlated inputs. These networks thus act as categorizers. In relation to visual information processing, they would remove redundancy from the input representation, and would develop low entropy ('information rich') representations of the information (cf. Barlow, 1985; Barlow et al., 1989). Second, such competitive nets are biologically plausible, in that they utilize forward excitatory connections that are modifiable by a local Hebb-like learning rule, with competitive inhibition mediated by cortical inhibitory neurons. Third, the network is effectively unsupervised, and does not need a teacher for every output neuron with backpropagation of errors to earlier layers of the network, nor a single global reinforcing teacher. Fourth, the soft or graded competition, which would not result in the formation of 'winner-take-all' or 'grandmother' cells, but would instead result in a small ensemble of active neurons representing each input (Rolls, 1989a–c), has the advantages that the output neurons learn better to distribute themselves between the input patterns (see Chapter 4); that the sparse representations formed (which provide 'coarse coding') have utility in maximizing the number of memories that can be stored when, towards the end of the visual system, the visual representation of objects is interfaced to associative memory (Rolls, 1989a,b; Rolls and Treves, 1990); and that the finite width of the response region of each neuron which tapers from a maximum at the centre is important for enabling the system to generalize smoothly from the examples with which it has learned. Fifth, the competitive learning in a multilayer network enables higher-order features to be detected and represented. First-order statistical properties include the probability of a variable, such as a feature, being present in the visual scene. These are not sufficient for visual pattern recognition, in that if the features in the scene were rearranged, new objects would be constructed, but the first-order probabilities of the features would remain unchanged. Second-order statistical properties refer to the probability that pairs of variables or features will occur together, such as a horizontal and vertical bar to form a 'T'. Competitive networks can learn such higher-order statistics, and given features such as bars and edges mapped in visual space as they are in V1, could form higher-order feature analysers which included the spatial arrangement of the bars or edges. It is an important part of this suggestion that some local spatial information would be inherent in the features which were being combined. For example, cells might not respond to the combination of an edge and a small circle unless they were in the correct spatial relation to each other. (This is in fact consistent with the neurophysiological data of Tanaka et al. (1990), and with our data on face neurons, in that some face neurons require the face features to be in the correct spatial configuration, and not jumbled: Rolls, Tovee, Purcell et al., 1994.) The local spatial information in the features being combined would ensure that the representation at the next level would contain some information about the (local) arrangement of features. Further low-order combinations of such neurons (that is including inputs from a few neurons) at the next stage would include sufficient local spatial information so that another arbitrary spatial arrangement of the same features would not activate the same neuron. (This is the proposed, and limited, solution which this mechanism would provide for the feature binding problem described later in relation to the work of von der Malsburg, 1990). By several such hierarchical stages of processing, neurons could be formed that would respond to complex spatial arrangements of features on which the network had been trained. (In VisNet, the lateral inhibition is simulated via a local contrast-enhancing filter active on each neuron. The cell activation is then passed through a non-linear cell activation function, which also produces contrast enhancement of the firing rates for the competition. Details are provided by Wallis and Rolls, 1997.)

To make the results of the simulation particularly relevant to understanding processing in higher cortical visual areas, the inputs to layer 1 of VisNet come from a separate input layer which provides an approximation to the encoding found in visual area 1 (V1) of the primate visual system. These response characteristics of neurons in the input layer are provided by a series of spatially tuned filters with image contrast sensitivities chosen to accord with the general tuning profiles observed in the simple cells of V1. Currently, even-symmetric (bar detecting) filter shapes are used. The precise filter shapes were computed by weighting the difference of two Gaussians by a third orthogonal Gaussian (see Wallis *et al.*, 1993; Wallis and Rolls, 1997). Four filter spatial frequencies (in the range 0.0625 to 0.5 pixels^{-1} over four octaves), each with one of four orientations (0° to 135°) were implemented. Cells of layer 1 receive a topologically consistent, localized, random selection of the filter responses in the input layer, under the constraint that each cell samples every filter spatial frequency and receives a constant number of inputs. Use of such filters also allows the network to be presented with real images.

The crucial part of the functional architecture for learning invariant responses is the learning rule used. This is essentially the same rule used in a competitive network, but includes a short memory trace of preceding neuronal activity (in the postsynaptic term). The hypothesis is as follows (see further Foldiak, 1991; Rolls, 1992b; Wallis, Foldiak and Rolls, 1993; Rolls, 1994a, 1995b; Wallis and Rolls, 1997). The hypothesis is that because real objects viewed normally in the real world have continuous properties in space and time, an object at one place on the retina might activate feature analysers at the next stage of cortical processing, and when the object was translated to a nearby position (or transformed in some other way), because this would occur in a short period (for example 0.5 s) of the previous activation, the membrane of the postsynaptic neuron would still be in its 'Hebb-modifiable' state (caused, for example, by calcium entry as a result of the voltage-dependent activation of NMDA receptors), and the presynaptic afferents activated with the object in its new position would thus become strengthened on the still-activated postsynaptic neuron. It is suggested that the short temporal window (for example 0.5 s) of Hebb-modifiability in this system helps neurons to learn the statistics of objects moving in the physical world, and at the same time to form different representations of different feature combinations or objects, as these are physically discontinuous and present less regular correlations to the visual system. (In the real world, objects are normally viewed for periods of a few hundred milliseconds, with perhaps then saccades to another part of the same object, and finally a saccade to another object in the scene.) It is this temporal pattern with which the world is viewed that enables the network to learn using just a Hebb-rule with a short memory trace. Foldiak (1991) has proposed computing an average activation of the postsynaptic neuron to implement the trace in a simulation. One idea here is that the temporal properties of the biologically implemented learning mechanism are such that it is well suited to detecting the relevant continuities in the world of real objects. Another suggestion is that a memory trace for what has been seen in the last 300 ms appears to be implemented by a mechanism as simple as continued firing of inferior temporal neurons after the stimulus has disappeared, as was found in the masking experiments described above (see also Rolls and Tovee, 1994; Rolls, Tovee *et al.*, 1994). Rolls (1992b, 1994a, 1995b) also suggests that other invariances, for example size, spatial frequency, and rotation invariance, could be learned by a comparable process. (Early processing in V1 which enables different neurons to represent inputs at different spatial scales would allow combinations of the outputs of such neurons to be formed at later stages. Scale invariance would then result from detecting at a later stage which neurons are almost conjunctively active as the size of an object alters.) It is suggested that this process takes place at most stages of the multiple-layer cortical processing hierarchy, so that invariances are learned first over small regions of space, about quite simple feature combinations, and then over successively larger regions. This limits the size of the connection space within which correlations must be sought. In layer 1, Hebbian learning without a trace is implemented in the network, so that simple feature combinations are learned here.

In this functional network architecture, another hypothesis is that view-independent representations could be formed by the same type of trace-rule learning, operating to combine a limited set of views of objects. The plausibility of providing view-independent recognition of objects by combining a set of different views of objects has been proposed by a number of investigators (Koenderink and Van Doorn, 1979; Poggio and Edelman, 1990; Logothetis *et al.*, 1994; Ullman, 1996). Consistent with the suggestion that the view-independent representations are formed by combining view-dependent representations in the primate visual system, is the fact that in the temporal cortical visual areas, neurons with view-independent representations of faces are present in the same cortical areas as neurons with view-dependent representations (from which the view-independent neurons could receive inputs) (Hasselmo, Rolls, Baylis and Nalwa, 1989; Perrett *et al.*, 1987).

The actual synaptic learning rule used in VisNet can be summarized as follows:

$$\delta w_{ij} = k\, m_i r'_j \tag{8.3}$$

and

$$m_i^{(t)} = (1-\eta)r_i^{(t)} + \eta m_i^{(t-1)} \tag{8.4}$$

where r'_j is the jth input to the ith neuron, r_i is the output of the ith neuron, w_{ij} is the jth weight on the ith neuron, η governs the relative influence of the trace and the new input (it takes values in the range 0–1, and is typically 0.4–0.6), and $m_i^{(t)}$ represents the value of the ith cell's memory trace at time t. In the simulation the neuronal learning was bounded by normalization of each cell's dendritic weight vector, as in standard competitive learning (see Chapter 4). An alternative, more biologically relevant implementation, using a local weight bounding operation, has in part been explored using a version of the Oja update rule (Oja, 1982; Kohonen, 1984).

To train the network to produce a translation invariant representation, one stimulus was placed successively in a sequence of nine positions across the input (or 'retina'), then the next stimulus was placed successively in the same sequence of nine positions across the input, and so on through the set of stimuli. The idea was to enable the network to learn whatever was common at each stage of the network about a stimulus shown in different positions. To train on view invariance, different views of the same object were shown in succession, then different views of the next object were shown in succession, and so on.

One test of the network used a set of three non-orthogonal stimuli, based upon probable edge cues (such as 'T, L, and +' shapes). During training these stimuli were chosen in random sequence to be swept across the 'retina' of the network, a total of 1000 times. In order to assess the characteristics of the cells within the net, a two-way analysis of variance was performed on the set of responses of each cell, with one factor being the stimulus type and the other the position of the stimulus on the 'retina'. A high F ratio for stimulus type (F_s) (indicating large differences in the responses to the different stimuli), and a low F ratio for stimulus position (F_p) would imply that a cell had learned a position invariant representation of the stimuli. The discrimination factor of a particular cell was then simply defined as the ratio F_s/F_p (a factor useful for ranking at least the most invariant cells). To assess the utility of the trace learning rule, nets trained with the trace rule were compared with nets trained with standard Hebbian learning without a trace, and with untrained nets (with the initial random weights). The results of the simulations, illustrated in Fig. 8.10, show that networks trained with the trace learning rule do have neurons with much higher values of the discrimination factor than networks trained with the Hebb rule or (as a control) untrained networks. Examples of the responses of two cells in layer 4 of such a simulation are illustrated in Fig. 8.11. Similar position invariant encoding has been demonstrated for a stimulus set consisting of eight faces. View invariant coding has also been demonstrated for a set of three faces each shown in seven views (for full details see Wallis and Rolls, 1997). For example, seven different views of three different faces presented centrally on the retina were used in one simulation. During 800 epochs of learning, each face was chosen at random, and a sequence of preset views of it was shown, sweeping the face either clockwise or counterclockwise. Examples of invariant layer 4 neuron response profiles are shown in Fig. 8.12.

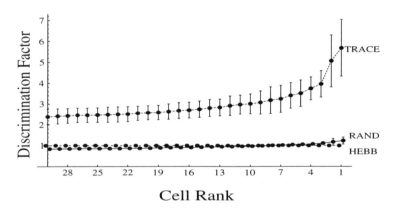

Fig. 8.10 Comparison of network discrimination when trained with the trace learning rule, with a Hebb rule (no trace), and when not trained (random). The discrimination factor (see text) for the 32 most invariant neurons is shown. The network was trained with three shapes (L, T and +) in each of nine different positions on the retina.

These results show that the proposed learning mechanism and neural architecture can produce cells with responses selective for stimulus type with considerable position or view invariance. Use of the network enables issues which cannot yet be treated analytically to be

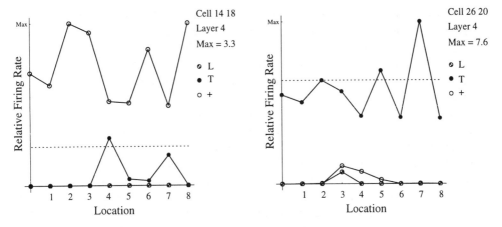

Fig. 8.11 The responses of two layer 4 cells in the simulation. The cell on the right had a translation invariant response to stimulus 1, the letter T, relatively independently of which of the nine locations it was presented on the retina. The cell on the left had a translation invariant response to stimulus 3, the +, relatively independently of which of the nine locations it was presented on the retina.

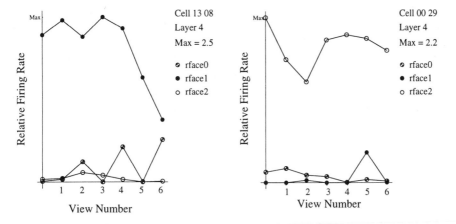

Fig. 8.12 Response profiles for two cells in layer 4 of VisNet when trained on three faces each with seven different views.

investigated. One is the capacity of the network. This depends on the types of redundancy present in the visual scenes on which the network is trained. It is expected that the same set of feature analysers with limited invariant properties early in the network will be appropriate for and used for many different objects. Only at the later stages of the network will neurons become in any way specialized in the actual objects to which they respond. The degree to which the network can take advantage of such redundancies with real visual scenes, and therefore have the capacity to learn invariant responses to many different objects, is then an issue which for the present is most easily investigated by neural network simulation, in this case using VisNet. The ability of the network to be trained with natural scenes may be one way in which it can help to advance our understanding of encoding in the real visual system.

The trace rule used in VisNet is local and hence biologically plausible, in that the signals required to alter the synaptic strength during learning are the presynaptic firing and the postsynaptic activation, both available locally at the synapse. The use of such a learning rule sets this proposal apart from most other proposals for how invariant representations might be formed. The system also operates by self-organizing competitive learning, which is also biologically plausible, in that the learning can be driven by the actual inputs received with no external teacher needed, and in that lateral inhibition, which implements competition, is a well- known property of cortical architecture. Other models have typically combined various less attractive elements such as supervised or non-local learning (Poggio and Edelman, 1990; Fukushima, 1980), extremely idealized or simplified stimuli (Foldiak, 1991; Hinton, 1981), prohibitive object by object matching processes (Olshausen *et al.*, 1993; Buhmann *et al.*, 1990); or non-localized connectivity (Hummel and Biederman, 1992). On the other hand, some of these models have some advantages over the model described here. For example, the model of Olshausen *et al.* (1993, 1995) more explicitly addresses the issue of locating and attending to objects throughout the visual field. VisNet only addresses object recognition within the high acuity centre of the visual field, and would require some other mechanism for locating and fixating stimuli. It is possible that in the brain this process is not performed by the ventral visual system, but is instead performed by the dorsal visual system (see Chapters 1 and 10).

8.5 The cognitron and neocognitron

Fukushima (1975, 1980, 1989, 1991) developed an early biologically inspired multilayer learning network. His approach used two types of cell within each layer to approach the problem of invariant representations. In each layer, a set of simple cells, with defined position, orientation, etc. sensitivity for the stimuli to which they responded, was followed by a set of complex cells, which generalized a little over position, orientation, etc. This simple cell-complex cell pairing within each layer provided some invariance. When a neuron in the network using competitive learning with its stimulus set, which was typically letters on a 16×16 pixel array, learned that a particular feature combination had occurred, that type of feature analyser was replicated in a non-local manner throughout the layer, to provide further translation invariance. Invariant representations were thus learned in a different way from VisNet. Up to eight layers were used. The network could learn to differentiate letters, even with some translation, scaling, or distortion. Although internally it is organised and learns very differently to VisNet, it is an independent example of the fact that useful invariant pattern recognition can be performed by multilayer hierarchical networks.

8.6 An alternative approach for object-based representations

The solution to 'object-based' representations provided by VisNet is very different from that traditionally proposed for artificial vision systems, in which the relative coordinates in 3D object-based space of parts of objects are stored in a database, and general-purpose algorithms operate on the inputs to perform transforms such as translation, rotation, and scale change in 3D space to see if there is any match to a stored 3D representation (for example Marr, 1982). One problem with implementing such a scheme in the brain is that a detailed syntactical description of the relations between the parts of the 3D object is required, for example body > thigh > shin > foot > toes. Such syntactical relations are difficult to implement in neuronal networks, because if the representations of all the features just mentioned were active simultaneously, how would the spatial relations between the features also be encoded? (How would it be apparent just from the firing of neurons that the toes were linked to the rest of foot but not to the body?) In VisNet, which is a much more limited but more biologically plausible scheme, this problem is solved by allocating neurons to representing low-order combinations of features in the particular topological arrangement, based on the simultaneously active neurons in the spatially mapped input neurons. Low-order combinations are used in this scheme to limit the number of 'combination-responding' neurons required, to limit the combinatorial explosion. (Low-order combinations are combinations of only a few variables.) The type of representation developed in VisNet (and present in IT/STS of the brain) would be suitable for recognition of an object, and for linking associative memories to objects, but would be less good for making actions in 3D space to particular parts of, or inside, objects, as the 3D coordinates of each part of the object would not be explicitly available. It is therefore proposed that visual fixation is used to locate in foveal vision part of an object to which movements must be made, and that local disparity and other measurements of depth then provide sufficient information for the motor system to make actions relative to the small part of space in which a local, *view-dependent*, representation of depth would be provided (cf. Ballard, 1990).

8.7 Syntactic binding of separate neuronal ensembles by synchronization

The problem of syntactic binding of neuronal representations, in which some features must be bound together to form one object, and other simultaneously active features must be bound together to represent another object, has been addressed by von der Malsburg (see 1990). He has proposed that this could be performed by temporal synchronization of those neurons that were temporarily part of one representation in a different time slot from other neurons that were temporarily part of another representation. The idea is attractive in allowing arbitrary relinking of features in different combinations. Singer, Engel, Konig, and colleagues (see Engel *et al.*, 1992) have obtained some evidence that when features must be bound, synchronization of neuronal populations can occur. The extent to which it is required for visual object recognition is an open issue, and evidence on the possible role of synchronization in cortical computation is considered further in Section 10.4.7.

Synchronization to implement syntactic binding has a number of disadvantages and limitations, as described in Section 10.4.7. In the context of VisNet, and how the real visual system may operate to implement object recognition, the use of synchronization does not appear to match the way in which the visual system is organized. For example, using only a two-layer network, von der Malsburg has shown that synchronization could provide the necessary feature linking to perform object recognition with relatively few neurons, because they can be reused again and again, linked differently for different objects. In contrast, the primate uses a considerable part of its brain, perhaps 50% in monkeys, for visual processing, with therefore what must be in the order of 10^9 neurons and 10^{13} synapses involved, so that the solution adopted by the real visual system may be one which relies on many neurons with simpler processing than arbitrary syntax implemented by synchronous firing of separate assemblies suggests. On the other hand, a solution such as that investigated by VisNet which forms low- order combinations of what is represented in previous layers, is very demanding in terms of the number of neurons required. It will be fascinating to see how research on these different approaches to processing in the primate visual system develops. For the development of both approaches, the use of well-defined neuronal network models is proving to be very helpful.

8.8 Different processes involved in different types of object identification

To conclude this chapter, it is proposed that there are three different types of process that could be involved in object identification. The first is the simple situation where different objects can be distinguished by different non-overlapping sets of features. An example might be a banana and an orange, where the list of features of the banana might include yellow, elongated, and smooth surface; and of the orange its orange colour, round shape, and dimpled surface. Such objects could be distinguished just on the basis of a list of the properties, which could be processed appropriately by a competitive network, pattern associator, etc. No special mechanism is needed for view invariance, because the list of properties is very similar from most viewing angles. Object recognition of this type may be extremely common in animals, especially those with visual systems less developed than those of primates.

A second type of process might involve the ability to generalize across a small range of views of an object, where cues of the first type cannot be used to solve the problem. An example might be generalization across a range of views of a cup when looking into the cup, from just above the near lip until the bottom inside of the cup comes into view. Such generalization would work because the neurons are tuned as filters to accept a range of variation of the input within parameters such as relative size and orientation of the components of the features. Generalization of this type would not be expected to work when there is a catastrophic change in the features visible, as for example occurs when the cup is rotated so that one can suddenly no longer see inside it, and the outside bottom of the cup comes into view.

The third type of process is one that can deal with the sudden catastrophic change in the features visible when an object is rotated to a completely different view, as in the cup example just given (cf. Koenderink, 1990). Another example, quite extreme to illustrate the point, might be when a card with different images on its two sides is rotated so that one face and then the other is in view. This makes the point that this third type of process may involve arbitrary pairwise association learning, to learn which features and views are different aspects of the same object. Another example occurs when only some parts of an object are visible. For example, a red-handled screwdriver may be recognized either from its round red handle, or from its elongated, silver coloured blade.

The full view-invariant recognition of objects which occurs even when the objects share the same features, such as colour, texture, etc. is an especially computationally demanding task which the primate visual system is able to perform with its highly developed temporal cortical visual areas. The neurophysiological evidence and the neuronal networks described in this chapter suggest how the primate visual system may perform this task.

9   Motor systems: cerebellum and basal ganglia

In this chapter we consider two non-neocortical brain systems involved in different aspects of the control of movement. The cerebellum has a remarkably regular neural network architecture which, together with evidence on its synaptic modification rules, has stimulated great interest in its theory of operation. The basal ganglia are of great interest because they are implicated in motor habit learning, might play a role in interfacing many cerebral cortical areas to systems for behavioural output, or at least in allowing some interaction between different cortical systems competing for output, and because they have considerable regularity in their structure, so that a theoretical approach is starting to appear attractive.

9.1 The cerebellum

The cerebellum is involved in the accurate control of movements. If the cerebellum is damaged, movements can still be initiated, but the movements are not directed accurately at the target, and frequently oscillate on the way to the target (see Ito, 1984). There is insufficient time during rapid movements for feedback control to operate, and the hypothesis is that the cerebellum performs feedforward control by learning to control the motor commands to the limbs and body in such a way that movements are smooth and precise. The cerebellum is thus described as a system for adaptive feedforward motor control. Network theories of cerebellar function are directed at showing how it learns to take the context for a movement, which might consist of a desired movement from the cerebral neocortex and the starting position of the limbs and body, and produce the appropriate output signals. The appropriate output signals are learned as a result of the errors which have occurred on previous trials.

Much of the fundamental anatomy and physiology of the cerebellum is described by Eccles, Ito and Szentagothai (1967), Albus (1971), Ito (1984, 1993a), and Raymond, Lisberger and Mauk (1996) and reference is given here therefore only to selected additional papers.

9.1.1 Architecture of the cerebellum

The overall architecture of the cerebellum and the main hypothesis about the functions of the different inputs it receives are shown schematically in Fig. 9.1. The mossy fibres convey the context or to-be-modified input which is applied, after expansion recoding implemented via the granule cells, via modifiable parallel fibre synapses onto the Purkinje cells, which provide the output of the network. There is a climbing fibre input to each Purkinje cell, and this is thought to carry the teacher or error signal that is used during learning to modify the strength of the parallel fibre synapses onto each Purkinje cell. This network architecture is implemented beautifully in the very regular cerebellar cortex. As cortex, it is a layered system. The cerebellar cortex has great regularity and precision, and this has helped analysis of its network connectivity.

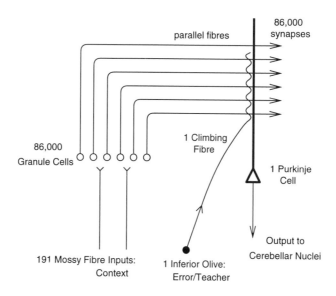

Fig. 9.1 Overall architecture of the cerebellum. Inputs relayed from the pontine nuclei form the mossy fibres which synapse onto dentate granule cells. The dentate granule cells via their parallel fibres form modifiable synapses on the Purkinje cells, from which outputs leave the cerebellar cortex. Each Purkinje cell receives one climbing fibre input. The numbers indicate the approximate numbers of cells of different types, relative to one Purkinje cell (see Ito, 1984). The numbers indicate expansion recoding of the mossy fibre inputs to the parallel fibres. A working hypothesis is that the context for a movement (for example limb state and motor command) reaches the Purkinje cells via the parallel fibres, and that the effect that this input has on the Purkinje cell output is learned through synaptic modification taught by the climbing fibre, so that the cerebellar output produces smooth error-free movements.

9.1.1.1 The connections of the parallel fibres onto the Purkinje cells

A small part of the anatomy of the cerebellum is shown schematically in Fig. 9.2. Large numbers of parallel fibres course at right angles over the dendritic trees of Purkinje cells. The dendritic tree of each Purkinje cell has the shape of a flat fan approximately 250 m by 250 m by 6 m thick. The dendrites of different Purkinje cells are lined up as a series of flat plates (see Fig. 9.3). As the parallel fibres cross the Purkinje cell dendrites, each fibre makes one synapse with approximately every fifth Purkinje cell (see Ito, 1984). Each parallel fibre runs for approximately 2 mm, in the course of which it makes synapses with approximately 45 Purkinje cells. Each Purkinje cell receives approximately 80 000 synapses, each from a different parallel fibre. The great regularity of this cerebellar anatomy thus beautifully provides for a simple connectivity matrix, with even sampling by the Purkinje cells of the parallel fibre input.

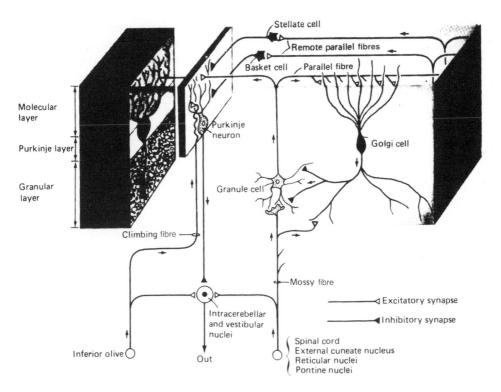

Fig. 9.2 Schematic diagram providing a 3D impression of the connectivity of a small part of the cerebellar cortex. Some of the inhibitory interneurons (Golgi, stellate, and basket cells) are shown.

9.1.1.2 The climbing fibre input to the Purkinje cell

The other main input to the Purkinje cell is the climbing fibre. There is one climbing fibre to each Purkinje cell, and this climbing fibre spreads to reach every part of the dendritic tree of the Purkinje cell. (Although the climbing fibre does not reach quite to the end of every dendrite, the effects of it on the Purkinje cell will reach to the apical extremity of every dendrite.) Although each Purkinje cell has an input from only one climbing fibre, each climbing fibre does branch and innervate approximately 10–15 different Purkinje cells in different parts of the cerebellum. The climbing fibres have their cell bodies in the inferior olive.

9.1.1.3 The mossy fibre to granule cell connectivity

The parallel fibres do not arise directly from the neurons in the pontine and vestibular nuclei (see Fig. 9.1). Instead, axons of the pontine and vestibular cells become the mossy fibres of the cerebellum, which synapse onto the granule cells, the axons of which form the parallel fibres (see Fig. 9.2). There are approximately 450 times as many granule cells (and thus parallel fibres) as mossy fibres (see Ito, 1984, p. 116). In the human cerebellum, it is estimated that there are 10^{10}–10^{11} granule cells (see Ito, 1984, p. 76), which makes them the most numerous cell type in the brain. What is the significance of this architecture? It was suggested by Marr (1969), who was the pioneer of quantitative theories of cerebellar function, that the expansion recoding achieved by the remapping onto granule cells is to decorrelate or orthogonalize the representation before it is applied to the modifiable synapses onto the Purkinje cells. It was shown in Chapter 2 how such expansion recoding can maximize capacity, reduce interference, and allow arbitrary mappings through a simple associative or similar network.

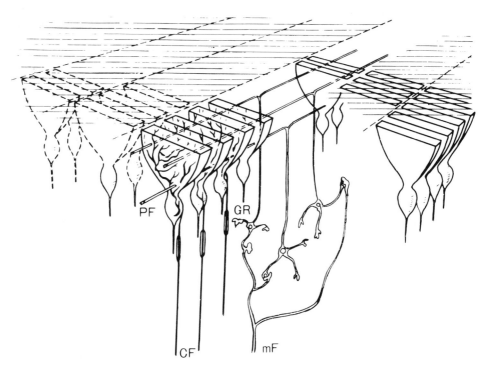

Fig. 9.3 The connections of the parallel fibres (PF) onto the Purkinje cells. Five rows of Purkinje cells with their fan-shaped dendritic trees are shown. CF, climbing fibre; mF, mossy fibre. (Reprinted with permission from Ito, 1984, Fig. 19, after Szentagothai, 1968.)

As implied by the numbers given above, and as summarized in Fig. 9.1 in which estimates of the relative numbers of the cells discussed so far in the cerebellum are given, there is a massive expansion in the number of cells used to code the same amount of information, in that the ratio of granule cells to mossy fibres is approximately 450. This provides a system comparable to a binary decoder. Consider an 8-bit number representation system, which can be used to represent the numbers 0–255 (decimal) (that is 2^8 numbers). Now, the bits which represent these numbers are often highly correlated with each other. For example, the bits that are set in the binary representations of the numbers 127 and 255 are identical apart from the highest order bit. (127 is 01111111 in binary and 255 is 11111111.) These two numbers would thus appear very similar to an associative memory system, such as that described in Chapter 2, so there would be great interference between the two numbers in such a memory system. However, if the numbers 0–255 were each represented by a different active axon synapsing onto the output cell, then there would be no interference between the different numbers, that is the representations would be orthogonalized by binary decoding. The 450-fold increase in the number of fibres as information passes from mossy to parallel fibres potentially allows this expansion recoding to take place.

This architecture of the brain provides an interesting contrast with that of conventional digital computers. The type of information storage device which has evolved in the brain uses a distributed representation over a large number of input connections to a neuron, and gains from this the useful properties of generalization, graceful degradation, and the capacity to function with low information capacity stores (that is synapses), but pays the penalty that some decorrelation of the inputs is required. The brain apparently has evolved a number of specialized mechanisms to do this. One is exemplified by the granule cell system of the cerebellum, with its enormous expansion recoding. In effect, the brain provides an enormous number of what may be low resolution storage locations, which in the cerebellum are the vast number of parallel fibre to Purkinje cell synapses. It is astounding that the number of storage locations in the (human) cerebellum would thus be approximately 3×10^{10} parallel fibres with every parallel fibre making approximately 300 synapses, or approximately 10^{13} modifiable storage locations.

The recoding which could be performed by the mossy fibre to granule cell system can be understood more precisely by considering the quantitative aspects of the recoding. Each mossy fibre ends in several hundred rosettes, where excitatory contact is made with the dendrites of granule cells. Each granule cell has dendrites in 1–7 mossy rosettes, with an average of four. Thus, on average, each granule cell receives from four mossy fibres. Due to the spacing of the mossy fibres, in almost all cases the four inputs to a granule cell will be from different mossy fibres. Approximately 28 granule cell dendrites contact each rosette. There are approximately 450 times as many granule cells as mossy fibres. The result of this architecture is that the probability that any given granule cell has all its C inputs active, and is therefore likely to fire, is low. It is in fact P^C, where P is the probability that any mossy fibre is active. Thus if the probability that a mossy fibre is active is 0.1, the probability that all of the average four inputs to a granule cell are active is 0.1^4 or 0.0001. The probability that three of the four inputs to a given granule cell are active would be 0.001. Given that each mossy fibre forms rosettes over a rather wide area of the cerebellum, that is over several folia, cells with three or four active inputs would be dotted randomly, with a low probability, within any one area.

One further aspect of this recoding must be considered. The output of the parallel fibres is sampled by the Golgi cells of the cerebellum via synapses with parallel fibres. Each Golgi cell feeds back inhibitory influences to about 5000–10 000 granule cells. Neighbouring Golgi cells overlap somewhat in their dendritic fields, which are approximately 300 μm across, and in their axon arborization. Every granule cell is inhibited by at least one Golgi cell. The Golgi cells terminate directly on the mossy rosettes, inhibiting the granule cells at this point. This very broad inhibitory feedback system suggests the function of an automatic gain control. Thus Albus (1971) argued that the Golgi cells serve to maintain granule cell, and hence parallel fibre, activity at a constant rate. If few parallel fibres are active, Golgi inhibitory feedback decreases, allowing granule cells with lower numbers (for example 3 instead of 4) of excitatory inputs to fire. If many parallel fibres become active, Golgi feedback increases, allowing only those few granule cells with many (for example 5 or 6) active mossy inputs to fire. In addition to this feedback inhibitory control of granule cell activity, there is also a feedforward inhibitory control implemented by endings of the mossy fibres directly on Golgi cells. This feedforward inhibition probably serves a speed-up function, to make the overall inhibitory effect produced by the Golgi cells more smooth in time. The net effect of this inhibitory control is probably thus to maintain a relatively low proportion, perhaps 1%, of parallel fibres active at any one time. This relatively low proportion of active input fibres to the Purkinje cell learning mechanism optimizes information storage in the synaptic storage matrix, in that the sparse representation enables many different patterns to be learned by the parallel fibre to Purkinje cell system.

9.1.2 Modifiable synapses of parallel fibres onto Purkinje cell dendrites

Evidence that the climbing fibres take part in determining the modifiability of the parallel fibre to Purkinje cell synapses is described next.

An experiment to investigate the modifiability of the parallel fibre to Purkinje cell synapses was performed by Ekerot and Kano (1983), using the experimental arrangement shown in Fig. 9.4A. First, they showed that stimulation through a microelectrode of either a bundle Pf1 or a bundle Pf2 of parallel fibres produced firing of a Purkinje cell PC from which they were recording (see peristimulus histograms in Figs 9.4Ba and 9.4Bb respectively). Then they paired stimulation of parallel fibre bundle Pf1 with climbing fibre activation. The details of this conjunctive stimulation were that the beam of parallel fibres was stimulated with 0.2 ms pulses of 20 to 60 μA, and the inferior olive, from which the climbing fibres originate, was stimulated 6 to 12 ms later. This conjunctive stimulation occurred at a rate of 4 Hz for one or two minutes. After this, the magnitude of the response of the Purkinje cell to stimulation of the parallel fibres was measured. It was found that the number of action potentials elicited by stimulation of bundle Pf1 had decreased, whereas the number of action potentials produced by stimulation of bundle Pf2 was unchanged (Fig. 9.4B). This experiment thus showed that if activity on a climbing fibre occurred when a parallel fibre was active, then the synaptic weights from the parallel fibre to the Purkinje cell *decreased*. The modification could last for many minutes (see Figs 9.4C and 9.4D), with the effect of the learning often evident after one hour. This type of synaptic modification is called long-term depression (LTD) (see Ito, 1989).

This modification required conjunctive activity in the climbing fibre and the parallel fibres, as shown by the demonstration that stimulation of fibre bundle Pf2 without conjunctive stimulation of the climbing fibre input did not decrease the activation of the Purkinje cell produced by stimulation of bundle Pf2 (see Fig. 9.4C, stimulation indicated by an open arrow).

The observation just noted, on stimulation of bundle Pf2 without conjunctive stimulation of the climbing fibre, is of great interest in relation to Hebb's (1949) hypothesis, for learning failed even though the presynaptic activity of Pf2 could produce action potentials in the postsynaptic neuron (Ito, 1984). Hebb's hypothesis is as follows: 'When an axon of cell A is near enough to excite a cell B and repeatedly or persistently takes part in firing it, some growth or metabolic change takes place in one or both cells such that A's efficiency, as one of

of the mechanism of LTD because LTD occurs best when the parallel fibre stimulation followed climbing fibre stimulation within intervals of approximately this duration. For example, after conjunctive stimulation of climbing fibres and parallel fibres at 2 Hz for 8 min a significant depression of parallel fibre responses occurred in 67% of the Purkinje cells tested at an interval of 20 ms, in 50% of cells tested at intervals of 125 and 250 ms, and only in 29% of the cells tested at a 375 ms interval (Ekerot and Oscarsson, 1981). It is possible that the prolonged depolarization produced by a climbing fibre impulse is produced by a rise in calcium conductance, and would give rise to an increased intradendritic calcium concentration. The calcium ions might activate a second messenger responsible for reducing the sensitivity of conjunctively activated synaptic receptors for the transmitter in the parallel fibres (see Ekerot, 1985; Ito, 1989). It may also be noted that the fact that the climbing fibre does not produce many spikes in a Purkinje cell, and only usually fires at 1–2 Hz, also suggests that the Hebb rule does not operate in cerebellar learning. That is, it is not fast firing of the postsynaptic neuron (which often occurs without climbing fibre activity, because of parallel fibre activity), in conjunction with parallel fibre activity which produces modification, but rather it is climbing fibre activity in conjunction with parallel fibre activity that modifies the parallel fibre to Purkinje cell synaptic strength.

9.1.3 The cerebellum as a perceptron

The cerebellum does not operate as a pattern associator of the type described in Chapter 2. In particular, the climbing fibre inputs to the Purkinje cells do not act as unconditioned or forcing stimuli, which produce a level of firing of the Purkinje cells which can then be associated with the pattern of activity on the parallel fibres. In fact, the climbing fibres do not make a large contribution to the overall number of action potentials produced by the Purkinje cells. Part of the evidence for this is that complex spikes, which are produced by the climbing fibre inputs, represent a relatively small proportion of the firing of Purkinje cells. Moreover, the firing rates of the climbing fibres are typically 0–4 spikes/s. Instead, the climbing fibres appear to set the modifiability of the parallel fibre to Purkinje cell synapses, as described above, without making a large direct contribution to the firing rates of the Purkinje cells. The climbing fibres can thus be seen as setting the efficacy of each parallel fibre to Purkinje cell synapse, based on the activity of the presynaptic terminal at the time the climbing fibre is active. In effect, the climbing fibre teaches the parallel fibre synapses the pattern of activity on the parallel fibres to which the Purkinje cell should respond. To be even more correct, the climbing fibres teach the synapses that when a particular pattern appears on the parallel fibres, the Purkinje cell should not be activated by that pattern. This formulation arises from the synaptic modification rule implemented in the cerebellum, which operates to weaken parallel to Purkinje cell synapses when the climbing fibre fires conjunctively with parallel fibre input (see above). The operation of the climbing fibres as teachers in this way means that the type of learning implemented in the cerebellum is very similar to that of the perceptron, in which the climbing fibre would convey the error signal. In addition, this type of learning may be quite slow, based on many learning trials, and is thus an interesting contrast to the rapid, one-shot type of learning of episodic events in which the hippocampus is implicated (see Chapter 6).

The overall concept that the work of Marr (1969), Albus (1971), and Ito (1979, 1984) has led towards is that the cerebellum acts as an adaptive feedforward controller with the general form shown in Fig. 9.1. The command for the movement and its context given by the current state of the limbs, trunk, etc., signalled by proprioceptive input would be fed via the mossy fibres to the parallel fibres. There the appropriate firing of the Purkinje cells to enable the movement to be controlled smoothly would be produced via the parallel fibre to Purkinje cell synapses. These synapses would be set to the appropriate values by error signals received on many previous trials and projected via the climbing fibres to provide the teaching input for modifying the synapses utilizing LTD. The proposal described by Ito (1984) is that the cerebellum would be in a sideloop, with the overall gain or output of the system resulting from a direct, unmodifiable output and the variable gain output contributed by the cerebellum (see Fig. 9.5).

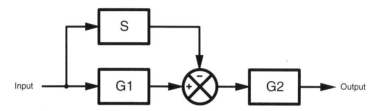

Fig. 9.5 Generic sideloop controller. There is a direct route to the output with a gain G1, and a sideloop with a variable, adaptive gain S. In this example the sideloop gain subtracts from the effect produced directly (indicated by the summing node, indicated by the circle), and the overall gain of this part of the pathway is G1–S. The cerebellum is hypothesized to be an adaptive gain sideloop controller, with its gain represented by S. The direct path is in the brainstem.

This is a generic model of cerebellar function. Different parts of the cerebellum are concerned with different types of motor output (see Raymond *et al.*, 1996). One part, the flocculus, is involved in the control of eye movements, and in particular with the way in which eye movements compensate for head movements, helping the images to remain steady during head movement. This is called the vestibulo-ocular reflex (VOR). Another part of the cerebellum is concerned with the accurate and smooth control of limb movements. Another part is implicated in the classical conditioning of skeletal responses, in which a sensory input such as a tone comes by learning to produce learned reflex responses. In order to assess how the generic circuitry of the cerebellum contributes to these types of adaptive control of motor output, we now turn to consider evidence on how the cerebellum contributes to each of these types of motor learning. It is the goal of studying networks in the brain to understand not only how each network operates, but also how each network performs its contribution as part of a connected set of networks, and for this goal to be achieved it is necessary to study the operation of the networks at the brain systems-level.

9.1.4 Systems-level analysis of cerebellar function

9.1.4.1 The vestibulo-ocular reflex

The cerebellum is connected to a number of different motor systems. The operation of the cerebellum in one such system, that concerned with the vestibulo-ocular reflex (VOR), has been intensively studied, and we shall start with this.

The VOR produces eye movements which compensate for head movements. For example, when the head is rotated to the left, the eyes move to the right. This prevents images of the visual world from slipping across the retina during head movements, and allows an animal to see its environment while its head is moving. The VOR is driven by signals from the semicircular canals and otolith organs, and is effected by coordinated contraction and relaxation of six extraocular eye muscles. Of the three types of head rotation (horizontal, vertical, and rotatory), the horizontal VOR has been investigated most, and is considered here. The gain of the VOR is specified by the ratio of the amplitude of head movement to that of eye movement. (The ratio of the velocities can also be used.) The gain (tested by measuring eye movements in the dark produced by head rotation) is near 1.0 in monkeys and humans, and the phase shift is close to 0 for frequencies of head rotation between about 0.05 and 1 Hz. This means that the eye movements almost exactly compensate for head movements, over a considerable range of velocities of head movement.

The anatomy of the VOR system is shown in Fig. 9.6. The vestibular system senses head movement, and sends this information to the vestibular nuclei, and through mossy fibres to the floccular part of the cerebellum. The vestibular nuclei connect to the oculomotor nuclei (cranial nerve nuclei III and VI), which control the medial and lateral rectus muscles of the eye. In addition to this main reflex arc (vestibular organs–vestibular nuclei–oculomotor nuclei), there is also a pathway from the vestibular organ via the mossy fibres to the granule cells of the cerebellum, which synapse with the Purkinje cells, which send output connections to the vestibular nuclei. Visual information from the retina reaches the pretectal area, from which there are connections to the inferior olive, with the result that visual signals are sent up the climbing fibres (see Fig. 9.6).

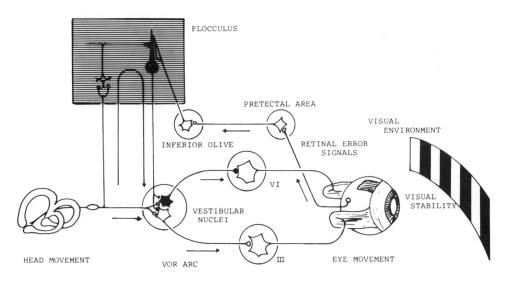

Fig. 9.6 Circuitry of the vestibulo-ocular reflex (VOR). III and VI: oculomotor nerve nuclei. (Reprinted with permission from Ito, 1984, Fig. 159.)

The functional significance of this anatomy has been suggested by Ito (1970, 1972, 1974; see Ito, 1984), and is consistent with the perceptron theory of cerebellar function (Albus, 1971, see above). According to this floccular hypothesis of the VOR, the cerebellum acts as a sidepath (with a gain S in Fig. 9.5) to the main path (with gain G1 from the vestibular organ to the vestibular nuclei. The transfer function of the system shown in Fig. 9.5 is $T = (G1–S)G2$ (where G2 is the gain of the motoneuronal part of the system). Thus, if the transfer function of the sidepath is modified, the overall transfer function of the system will change. According to the hypothesis, the gain (or with more generality the transfer function) of sidepath S is altered by the action of climbing fibres on the parallel fibre to Purkinje cell synapses. In this way the overall transfer function of the system can be modified by experience so that it is close to 1.0, that is so that eye movements compensate exactly for head movements. The error correction signal which is applied to the climbing fibres is then retinal slippage (which is zero if the gain is correct).

There are many findings which are consistent with this hypothesis. Many are based on adaptation of the gain of the VOR, which can be produced by altered experience. For example, if human subjects wear left–right inverting prisms, then at first the eye movements are in the incorrect direction for maintaining stability of the visual scene when the head is moved, but gradually, over 2 days, the gain of the VOR decreases, so that the errors of eye movement are less, and over the next 5–10 days the phase of the VOR reverses so that the eye movements are in the same direction as the head movement, and thus (partially) compensate for the change produced by the reversing prisms (Gonshor and Melvill-Jones, 1976). (After 10 days the gain was 0.45, and the phase was 135° lag, at a frequency of 1/6 Hz with a sinusoidal head oscillation with 60°/s velocity amplitude.)

A first example of evidence consistent with the hypothesis described above is that lesions of the floccular part of the cerebellum abolish adaptation of the VOR in the rabbit, cat and monkey (Ito, 1974; Robinson, 1976; Ito, Jastreboff and Miyashita, 1982). It may be noted that, as might be expected from the hypothesis, the effects of the lesions on the gain of the VOR were rather unpredictable across species, in line with the point that the cerebellum is conceived as an adaptive sidepath, the contribution of which to the overall gain might be different in different species. The significant aspect of the findings for the hypothesis was that whatever change of gain was produced by the lesions, no adaptation of the gain as a result of further experience took place after the lesion. Between the different studies, it was shown that it was cell damage in the flocculus which impaired adaptation (rather than retrograde degeneration of neurons projecting into the cerebellum), and that it was the flocculus rather than another part of the cerebellum which was important for adaptation. Further, lesions which damage the inferior olive or its inputs, through which the retinal error signal is thought to reach the cerebellum, abolish VOR adaptation (Ito and Miyashita, 1975; Haddad et al., 1980; Demer and Robinson, 1982).

Second, alterations in the responses of neurons in the flocculus are found during adaptation of the VOR, and the changes found are generally consistent with those predicted by the flocculus hypothesis of VOR control. Simple spikes in Purkinje cells reflect primarily mossy fibre inputs, and in the flocculus are produced by vestibular input, as shown by the observation that in the monkey Purkinje cells fire in relation to head rotation in the darkness (Watanabe, 1984). The vestibular input can also be measured in the light by the firing of simple spikes of the Purkinje cells which occurs in phase with head rotation when the visual field is moved with the head so that it remains stationary on the retina precluding visual input signals. It is found that this modulation of the simple spike activity of Purkinje cells produced by vestibular input shifts in amplitude and phase during adaptation in such a way as to produce the adaptation. For example, in an experiment of Watanabe (1984) the visual field was moved in phase with head movements producing a decrease in the gain of the VOR of 0.2, and it was found that the responses of Purkinje cells shifted so that they tended to occur more in-phase with the head velocity, that is towards 0° in Fig. 9.7. Because the output of Purkinje cells onto vestibular nucleus neurons is inhibitory, the firing of these neurons produced by the vestibular afferents, which are active in phase with head velocity and are excitatory, will be reduced (see Ito, 1984, pp. 363–4). Thus the control signals from the vestibular nucleus neurons which control the eye movements for the VOR are reduced, and the gain diminishes. The opposite result occurs if the gain of the VOR is increased by 0.2 by adaptation using visual stimuli which rotate out of phase with head movements. The simple spike discharges of Purkinje cells shift their phase so that they fire more out of phase (towards 180° in Fig. 9.7 left) with head movements. The effect of this is to produce extra modulation of the firing of vestibular nucleus neurons by enhancing inhibition when their firing rate would normally be low (out of phase with head movement), thus increasing the gain of the VOR.

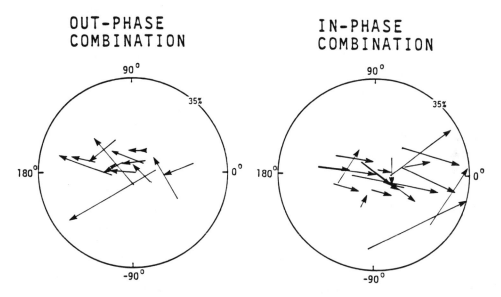

Fig. 9.7 Phase changes of the responses of Purkinje cells in the flocculus during adaptation of the VOR (see text). (From Watanabe, 1984; reprinted with permission from Ito, 1984, Fig. 180.)

The complex spike discharges of Purkinje cells are produced by the climbing fibre input, and appear to have some of the properties appropriate for an error-correction signal. Thus in the rabbit (Ghelarducci *et al.*, 1975) and monkey (Watanabe, 1984), during combined visual and vestibular movements, if the visual field moves in phase with the head, then complex spikes are modulated out of phase with the head velocity. In contrast, if the visual field is stationary or moves out of phase with the head rotation, then complex spikes of Purkinje cells are modulated in phase with head velocity. In the latter case, the modulation of Purkinje cells by in-phase vestibular input would be decreased (by the climbing fibre learning rule), there would be less inhibition on vestibular nucleus neurons, and the gain of the system would thus increase. Another way of describing this is that in-phase climbing fibre activity, which these complex spikes reflect, thus (because the climbing fibre learning rule produces a sign reversal due to climbing fibre activity weakening mossy fibre efficacy in influencing Purkinje cells) enhances out-of-phase simple spike activity of Purkinje cells and, as above, the extra modulation enhances the gain of the VOR. The increase in VOR gain of course means that the eyes make the appropriate overcompensatory movements when head rotation occurs. For example, if the head rotates 10° to the right, but simultaneously the visual field does not stay stationary but moves 2° to the left, then the eyes must overcompensate by moving 12° to the left, and this is achieved by the increase in the gain of the VOR.

In conclusion, studies of the VOR and its modification by experience have provided detailed insight into how one part of the cerebellum, the flocculus, utilizes the modifiability of the cerebellum for a particular function. The analyses are consistent with the hypothesis that at least this part of the cerebellar modifiability is put to use in a perceptron-like form of learning which utilizes an error signal (retinal slip) presented to the climbing fibres to alter the effect which the vestibular organ has on eye rotation. More recent studies have supported the evidence that the cerebellar cortex is involved in VOR learning in this way, although its role in timing has been emphasized, and it has been suggested that, in addition, the cerebellar cortex may guide learning in the deep cerebellar nuclei (Raymond, Lisberger and Mauk, 1996).

9.1.4.2 The control of limb movement

It is only into one part of the cerebellum, a part of the flocculus, that the mossy fibres carry a vestibular signal, and the climbing fibres a visual signal that could be a retinal slip error signal. Mossy fibres to some other parts of the cerebellum convey somaesthetic information, which is often proprioceptive from a single muscle. This information is conveyed from cell bodies in the spinal cord through the spinocerebellar tracts. In this way the cerebellum is provided with information about, for example, limb position and loading. Another major mossy fibre input arises from the pontine nuclei, which receive information from the cerebral cortex. All the climbing fibres of the cerebellum originate in the inferior olive which, in addition to the vestibular inputs described above, also receives inputs from the spinal cord and from the cerebral cortex. There are a number of ways in which the cerebellar modifiability may be used for these other systems.

One is to use the cerebellum as a sideloop control system for cerebral motor commands, which would be routed both directly to the red nucleus and thus via the rubrospinal tracts to the spinal cord, and indirectly to the red nuclei via the pontine nuclei, intermediate part of the cerebellum (IM), and the interpositus cerebellar nuclei (see Fig. 9.8A). For this system the cerebellum would receive error feedback from the periphery so that the gain of the system could be adaptively set.

Another possible sideloop mode of operation is for one cerebral cortical area such as the premotor cortex to project directly to the motor cortex, area 4, and to project indirectly to it via the pontine nuclei, mossy fibres to the cerebellar cortex, and thus through Purkinje cells to the cerebellar nuclei, and the ventrolateral nucleus of the thalamus to the primary motor cortex (Fig. 9.8B). This would enable the influence which the premotor cortex has on the motor cortex to be subject to a major influence of learning. The error or learning signal presented to the climbing fibres in this system might include information from the periphery, so that premotor influences might be regulated adaptively by the effects they produce.

Fig. 9.8 Sideloop variable gain cerebellar subsystems suggested by Ito (1984). A–E are different brain subsystems. FL, flocculus; VN, vestibular nucleus; VO, vestibular organ; OM, motoneurons innervating eye muscles; VE, vermis; RST, reticulospinal tract; FA, fastigial nucleus; SG, segmental neuron pool; IP, interpositus; RN, red nucleus; CB (CBa, CBb, CBc), cerebral cortex; CBm, motor cortex; PON, pontine nucleus. (Reprinted with permission from Ito, 1984, Fig. 152.)

Another possible mode of operation is for the mossy fibres to present the context for a movement, including, for example, the position the limb has now reached, to the Purkinje cells, which influence motor output via the cerebellar nuclei and then systems which descend to the spinal cord or systems which ascend to the motor cortex. The mossy fibre to Purkinje cell synapses in this system would learn according to an error signal, which might reflect the difference between a cortical command for a movement and what was being achieved, presented to the climbing fibres.

There is some neurophysiological evidence relevant to the adaptive changes which occur as part of the systems function of other parts of the cerebellum. For example, Gilbert and Thach (1977) recorded from Purkinje cells of monkeys performing a task in which they had to keep the arm within limits indicated by a visual stimulus. The arm was perturbed by fixed flexor or extensor loads, and the monkey learned to bring the arm quickly back to within the prescribed window in order to obtain reward. Then the magnitude of one of the flexor or extensor loads was altered, and over 12 to 100 trials the monkey learned not to overshoot or undershoot when the perturbations occurred, but instead to return to the central position quickly and smoothly. During this learning, Purkinje cells in lobules III through V of the intermediate cortex of the cerebellum showed an increased occurrence of complex spikes, which occurred 50 to 150 ms after the perturbation. Over the same number of trials, during which the monkeys' performance improved, the incidence of simple spikes from these neurons decreased (Fig. 9.9). These results are consistent with the proposal that the climbing fibres carry the error signal required to recalibrate the arm movements against the load, and that the learning involves a decreased strength in these particular neurons of the mossy fibre to Purkinje cell synapses, evident as fewer simple spikes during the perturbations after the learning (Gilbert and Thach, 1977; Ito, 1984, p. 440).

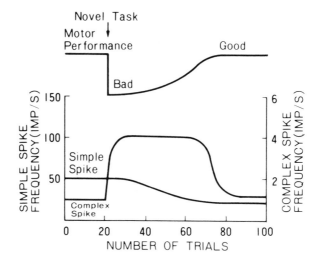

Fig. 9.9 Alterations in complex spike frequency of cerebellar Purkinje cells (reflecting climbing fibre inputs) during adaptation of arm movements to an increased load (after Gilbert and Thach, 1977). The load was altered at the arrow labelled 'Novel Task'.

9.1.4.3 Classical conditioning of skeletal muscle responses

Although classical conditioning is usually of autonomic or endocrine responses (for example conditioned salivation to the sound of a bell), and the circuitry implicated is in the amygdala and connected structures (see Chapter 7), it is possible to obtain classical conditioning of skeletal muscle responses. Classical conditioning of skeletal muscle responses involved in eyelid conditioning has been used as a model (or simple) system for studying how different parts of cerebellar circuitry are involved in learning (Thompson, 1986; Thompson and Krupa, 1994). The unconditioned stimulus (US) is a corneal airpuff, the response is closure of the nictitating membrane, and the conditioned stimulus (CS) is a tone. The circuitry underlying this conditioning is shown in Fig. 9.10. The US pathway seems to consist of somatosensory projections to the dorsal accessory portion of the inferior olive and its climbing fibre projections to the cerebellum. This corresponds to the teaching or error signal in the perceptron model. The tone CS pathway seems to consist of auditory projections to the pontine nuclei and their mossy fibre projections to the cerebellum. The efferent (eyelid closure) conditioned response pathway projects from the interpositus nucleus (Int) of the

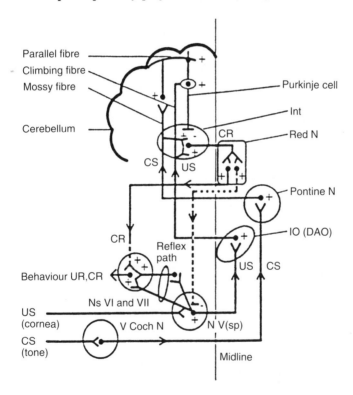

Fig. 9.10 Schematic diagram of pathways involved in classical conditioning of skeletal muscle responses using the model system of conditioned eyelid closure, which is the unconditioned and conditioned response, UR and CR. The unconditioned stimulus (US) is a corneal air puff, and the conditioned stimulus (CS) a tone (see text) (reprinted with permission from Thompson, 1986). Pluses indicate excitatory and minuses inhibitory synaptic action. DAO, dorsal accessory portion of the inferior olive; Int, interpositus nucleus; IO, inferior olive; N V (sp), spinal fifth cranial nucleus; N VI, sixth cranial nucleus; N VII, seventh cranial nucleus; Pontine n, pontine nucleus; Red N, red nucleus; V Coch N, ventral cochlear nucleus.

cerebellum (which receives from Purkinje cells) to the red nucleus (Red N) and via the descending rubral pathway to act ultimately on motor neurons. The hypothesis (Thompson, 1986) is that the parallel fibre synapses onto the Purkinje cells convey the CS, and that these synapses modify appropriately to produce the conditioned response when the CS is paired with a US (or teacher) reaching the Purkinje cells via the climbing fibres. This hypothesis is supported by evidence showing that the conditioned response (but not the unconditioned response) is abolished by lesions or inactivation of the interpositus nucleus (see Thompson, 1986; Hardiman, Ramnani and Yeo, 1996) and probably of the appropriate part of the cerebellar cortex (Yeo *et al.*, 1985; Yeo and Hardiman, 1992), that the tone CS can be substituted by electrical stimulation of the mossy fibre system, and that the US can be substituted by electrical stimulation of the dorsal accessory olive (see Thompson, 1986). Raymond *et al.* (1996) have emphasized the similarity of the functions of the cerebellar cortex in learning eyeblink conditioning and in learning VOR.

9.2 The basal ganglia

The basal ganglia are parts of the brain that include the striatum, globus pallidus, substantia nigra, and subthalamic nucleus and that are necessary for the normal initiation of movement. For example, depletion of the dopaminergic input to the striatum leads to the lack in the initiation of voluntary movement that occurs in Parkinson's disease. The basal ganglia receive inputs from all parts of the cerebral cortex, including the motor cortex, and have outputs directed strongly towards the premotor and prefrontal cortex via which they could influence movement initiation. There is an interesting organization of the dendrites of the neurons in the basal ganglia which has potentially important implications for understanding how the neuronal network architecture of the basal ganglia enables it to perform its functions. Although the operation of the basal ganglia is by no means understood yet, many pieces of the jigsaw puzzle of evidence necessary to form a theory of the operation of the basal ganglia are becoming available, and are considered here. Important issues to be considered are where the different parts of the input stage of the basal ganglia, the striatum, receive their inputs from, and what information is received via these inputs. These, and the effects of damage to different parts of the basal ganglia, are systems-level questions, necessary for understanding *what* computational role may be performed by the basal ganglia. Next, we must consider the internal neuronal network organization of the basal ganglia, to provide evidence on whether the different inputs received by the basal ganglia can interact with each other, and to lead towards a theory of *how* the basal ganglia perform their function.

9.2.1 Systems-level architecture of the basal ganglia

The point-to-point connectivity of the basal ganglia as shown by experimental anterograde and retrograde neuroanatomical path tracing techniques in the primate is indicated in Figs 9.11 and 9.12. The general connectivity is for cortical or limbic inputs to reach the striatum, which then projects to the globus pallidus and substantia nigra, which in turn project via the thalamus back to the cerebral cortex. Within this overall scheme, there are a set of at least partially segregated parallel processing streams, as illustrated in Figs 9.11 and 9.12 (see

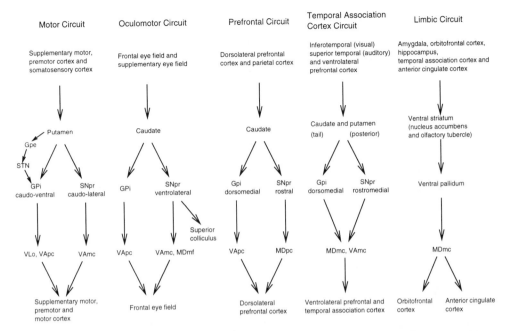

Fig. 9.11 A synthesis of some of the anatomical studies (see text) of the connections of the basal ganglia. GPe, Globus Pallidus, external segment; GPi, Globus Pallidus, internal segment; MD, nucleus medialis dorsalis; SNpr, Substantia Nigra, pars reticulata; VAmc, n. ventralis anterior pars magnocellularis of the thalamus; VApc, n. ventralis anterior pars compacta; VLo, n. ventralis lateralis pars oralis; VLm, n. ventralis pars medialis. An indirect pathway from the striatum via the external segment of the globus pallidus and the subthalamic nucleus (STN) to the internal segment of the globus pallidus is present for the first four circuits (left to right in Fig. 9.11) of the basal ganglia.

reviews by Strick *et al.*, 1995; Rolls and Johnstone, 1992; De Long *et al.*, 1984; Middleton and Strick, 1996a,b; Alexander *et al.*, 1990). First, the motor cortex (area 4) and somatosensory cortex (areas 3, 1, and 2) project somatotopically to the putamen, which has connections through the globus pallidus and substantia nigra to the ventral anterior thalamic nuclei and thus to the supplementary motor cortex. Recent experiments with a virus transneuronal pathway tracing technique have shown that there may be at least partial segregation within this stream, with different parts of the globus pallidus projecting via different parts of the ventrolateral (VL) thalamic nuclei to the supplementary motor area, the primary motor cortex (area 4), and to the ventral premotor area on the lateral surface of the hemisphere (Middleton and Strick, 1996b). Second, there is an oculomotor circuit (see Fig. 9.11). Third, the dorsolateral prefrontal and the parietal cortices project to the head and body of the caudate nucleus, which has connections through parts of the globus pallidus and substantia nigra to the ventral anterior group of thalamic nuclei and thus to the dorsolateral prefrontal cortex. Fourth, the inferior temporal visual cortex and the ventrolateral (inferior convexity) prefrontal cortex to which it is connected project to the posterior and ventral parts of the putamen and the tail of the caudate nucleus (Kemp and Powell, 1970; Saint-Cyr *et al.*, 1990; Graybiel and Kimura, 1995). Moreover, part of the globus pallidus, perhaps the part influenced by the temporal lobe visual cortex, area TE, may project back (via the thalamus) to area TE (Middleton and Strick, 1996a). Fifth, limbic and related structures such as the amygdala, orbitofrontal cortex and hippocampus project to the ventral striatum,

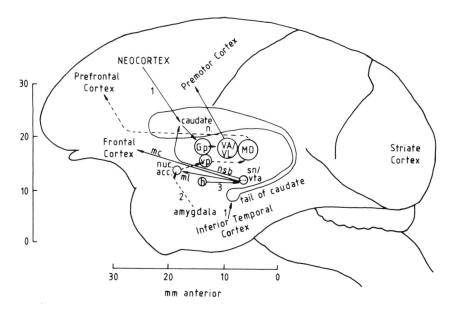

Fig. 9.12 Some of the striatal and connected regions in which the activity of single neurons is described shown on a lateral view of the brain of the macaque monkey. Gp, globus pallidus; h, hypothalamus; sn, substantia nigra, pars compacta (A9 cell group), which gives rise to the nigrostriatal dopaminergic pathway, or nigrostriatal bundle (nsb); vta, ventral tegmental area, containing the A10 cell group, which gives rise to the mesocortical dopamine pathway (mc) projecting to the frontal and cingulate cortices and to the mesolimbic dopamine pathway (ml), which projects to the nucleus accumbens (nuc acc).

which has connections through the ventral pallidum to the mediodorsal nucleus of the thalamus and thus to the prefrontal and cingulate cortices (Strick *et al.*, 1995). These same regions may also project to the striosomes or patches (in for example the head of the caudate nucleus), which are set in the matrix formed by the other cortico-striatal systems (Graybiel and Kimura, 1995). Thus, anatomical studies provide evidence for at least partial segregation of a number of major systems through the basal ganglia, with each system finally connecting to different cortical areas.

A crucial question for understanding basal ganglia function has become the extent to which any convergence or interaction between these at least partially segregated basal ganglia loops occurs within the basal ganglia. One possibility is that when afferents reach a part of the striatum, there is convergence from different cortical regions. Another possibility is that there is convergence from different parts of the striatum into the same part of the pallidum. There is some evidence that this type of convergence does occur to at least a limited extent (see Graybiel and Kimura, 1995). For example, projections from representations of the same body part in different somatosensory cortical and even motor cortical areas do converge into the same part of the striatum. Also, inputs to one part of the pallidum may converge from different parts of the striatum (see Graybiel and Kimura, 1995). Another possibility is that the large flat striatal dendritic trees and even larger pallidal and nigral dendritic trees (for example Percheron *et al.*, 1984a,b, 1994, and described in more detail below) allow convergence of signals from widely different parts of the cerebral neocortex. However, these anatomical studies were not able to show whether there was convergence onto exactly the same neurons in a given zone of convergence or of dendritic overlap. The results with the virus tracing techniques (see Strick *et al.*, 1995; Middleton and Strick, 1996a,b) suggest that there may not be convergence onto individual neurons, in that in many cases even when much of the basal ganglia circuit has been traversed by the virus, anatomical segregation of the outputs is still apparent, which would not be predicted if some mixing had occurred. However, it is possible that the virus technique is not sufficiently sensitive to reveal the mixing that might be present between the loops. Neurophysiological evidence in this situation might be useful in revealing whether or not convergence occurs. Some evidence that some convergence does occur is provided below. Another crucial line of evidence would be on whether as one moves through the basal ganglia (for example to the globus pallidus and substantia nigra), the majority of the neurons reflect the type of processing that occurs in motor structures, or whether instead all the non-motor-related neuronal types found in, for example, the caudate nucleus and ventral striatum (as described below) are equally represented in the (appropriate parts of the) globus pallidus. Although many more studies of this type need to be performed, the majority of neurons recorded when one is in the globus pallidus and substantia nigra do seem to be movement-related or at least to reflect striatal convergence (for example Rolls and Johnstone, 1992; Johnstone and Rolls, 1990; Mora, Mogenson and Rolls, 1977; Fabre *et al.*, 1983; Rolls, Burton and Mora, 1976). Unless there are as many of the more cognitive neurons in the nigra and pallidum as one would expect from the relative number of cognitive neurons to movement-related neurons in the striatum (see below), then this would imply that there is some convergence.

9.2.2 Systems-level analysis of the basal ganglia: effects of striatal lesions

Damage to the striatum produces effects which suggest that it is involved in orientation to stimuli and in the initiation and control of movement. Lesions of the dopamine pathways in animals which deplete the striatum of dopamine lead to a failure to orient to stimuli, a failure to initiate movements which is associated with catalepsy, and to a failure to eat and drink (Marshall *et al.*, 1974). In humans, depletion of dopamine in the striatum is found in Parkinson's disease, in which there is akinesia, that is a lack of voluntary movement, bradykinesia, rigidity and tremor (Hornykiewicz, 1973). However, consistent with the anatomical evidence, the effects of damage to different regions of the striatum also suggest that there is functional specialization within the striatum (Divac and Oberg, 1979; Oberg and Divac, 1979; Iversen, 1984). The selective effects may be related to the function of the cortex or limbic structure from which a region of the striatum receives inputs. For example, in the monkey, lesions of the anterodorsal part of the head of the caudate nucleus disrupted delayed spatial alternation performance, a task which requires spatial short-term memory, which is also impaired by lesions of the corresponding cortical region, the dorsolateral prefrontal cortex. Lesions of the ventrolateral part of the head of the caudate nucleus (as of the orbitofrontal cortex which projects to it), impaired object reversal performance, which measures the ability to reverse stimulus-reinforcement associations (see Chapter 7). Lastly, lesions of the tail of the caudate nucleus (as of the inferior temporal visual cortex which projects to this part of the caudate) produced a visual pattern discrimination deficit (Divac *et al.*, 1967; Iversen, 1979).

Analogously, in the rat, lesions of the anteromedial head of the caudate nucleus (or of the medial prefrontal cortex, which projects to it) impaired spatial habit reversal, while lesions of the ventrolateral part of the head of the caudate nucleus (or of the orbital prefrontal cortex from which it receives) impaired the withholding of responses in a go/no-go task or in extinction (Dunnett and Iversen, 1981; Iversen, 1984). Further, in the rat a sensori-motor orientation deficit was produced by damage to a part of the dorsal striatum which receives inputs from lateral cortical areas (Dunnett and Iversen, 1982b; Iversen, 1984). Similar deficits are produced by selective depletion of dopamine in each of these areas using 6-hydroxydopamine (Dunnett and Iversen, 1982a,b; Iversen, 1984). In addition, there is evidence linking the ventral striatum and its dopamine input to reward, for manipulations of this system alter the incentive effects which learned rewarding stimuli have on behaviour (Everitt and Robbins, 1992; Robbins *et al.*, 1989). Further evidence linking the ventral striatum to some types of reward is that rats will self-administer amphetamine into the nucleus accumbens (a part of the ventral striatum), and lesions of the nucleus accumbens attenuate the intravenous self-administration of cocaine (Phillips and Fibiger, 1990). In addition to this role in reward, related probably to inputs to the ventral striatum from the amygdala and orbitofrontal cortex (Rolls, 1990b, see Chapter 7), the ventral striatum is also implicated in effects which could be mediated by the hippocampal inputs to the ventral striatum. For example, spatial learning (Schacter *et al.*, 1989) and locomotor activity elicited by novel stimuli (see Iversen, 1984) are influenced by manipulations of the nucleus accumbens.

Monkeys with lesions of the hippocampus and amygdala can perform 'habit' learning. For example they can learn a visual discrimination task when the trials are well spaced (e.g. with one trial every 24 h). Learning of this type is slow, and inflexible in that it does not reverse. This type of learning is impaired by lesions of the inferior temporal visual cortex, and it has been suggested that the striatum is involved in this type of habit learning (Phillips *et al.*, 1988; Petri and Mishkin, 1994).

9.2.3 Systems-level analysis of the basal ganglia: neuronal activity in different parts of the striatum

The evidence from the connections of the striatum and from the effects of damage to the striatum thus suggests that there may be functional segregation within the striatum. To investigate this more directly, the activity of single neurons has been recorded in different parts of the striatum of behaving macaque monkeys, usually in tasks known to be impaired by damage to the striatum. These investigations are described next.

9.2.3.1 Tail of the caudate nucleus, and posteroventral putamen

The projections from the inferior temporal cortex and the prestriate cortex to the striatum arrive mainly, although not exclusively, in the tail (and genu) of the caudate nucleus and in the posteroventral portions of the putamen (Kemp and Powell, 1970; Saint-Cyr *et al.*, 1990). Since these regions of the caudate nucleus and putamen are adjacent and have a common anatomical input, we refer to them together as the caudal neostriatum. Although there is this visual projection directly into the caudal neostriatum, there have been few studies of the functions of these visual pathways and their importance in visually controlled behaviour.

Divac *et al.* (1967) reported that stereotaxic lesions placed in the tail of the caudate nucleus in the region which receives input from the inferior temporal visual cortex produced a deficit in visual discrimination learning. The lesion did not produce impairment in an object reversal task, though in two out of four monkeys it did disturb delayed alternation performance. Buerger *et al.* (1974) found that lesions of the ventral putamen in the monkey produced a deficit in the retention of a pre-operatively learnt visual discrimination problem but the lesion did not disturb retention of auditory discrimination or delayed alternation tasks. The deficit produced by both neostriatal lesions seems therefore to reflect predominantly a loss of visual functions rather than a general loss of cognitive functions.

Since so little is known about the nature of visual processing within the caudal neostriatum and the fate of the visual input from inferior temporal cortex to this area, the activity of single neurons was recorded in the tail of the caudate nucleus and adjoining part of the ventral putamen (Caan *et al.*, 1984). Of 195 neurons analysed in two macaque monkeys, 109 (56%) responded to visual stimuli, with latencies of 90–150 ms for the majority of the neurons. The neurons responded to a limited range of complex visual stimuli, and in some cases responded to simpler stimuli such as bars and edges. Typically (for 75% of neurons tested) the neurons habituated rapidly, within 1–8 exposures, to each visual stimulus, but remained responsive to other visual stimuli with a different pattern. This habituation was orientation specific, in that the neurons responded to the same pattern shown at an orthogonal orientation. The habituation was also relatively short-term, in that at least partial dishabituation to one stimulus could be produced by a single intervening presentation of a different visual stimulus. These neurons were relatively unresponsive in a visual discrimination task, having habituated to the stimuli which had been presented in the task on many previous trials.

The main characteristics of the responses of these neurons in the tail of the caudate nucleus and adjoining part of the putamen were rapid habituation to specific visual patterns, their sensitivity to changes in visual pattern, and the relatively short-term nature of their habituation to a particular pattern, with dishabituation occurring to a stimulus by even one intervening trial with another stimulus. Given these responses, it may be suggested that these neurons are involved in short-term pattern-specific habituation to visual stimuli. This system would be distinguishable from other habituation systems (involved, for example, in habituation to spots of light) in that it is specialized for patterned visual stimuli which have been highly processed through visual cortical analysis mechanisms, as shown not only by the nature of the neuronal responses, but also by the fact that this system receives inputs from the inferior temporal visual cortex. It may also be suggested that this sensitivity to visual pattern change may have a role in alerting the monkey's attention to new stimuli. This suggestion is consistent with the changes in attention and orientation to stimuli produced by damage to the striatum. Thus, damage to the dopaminergic nigrostriatal bundle produces an inability to orient to visual and other stimuli in the rat (Marshall *et al.*, 1974). In view of these neurophysiological findings, and the finding that in a visual discrimination task neurons which reflected the reinforcement contingencies of the stimuli were not found, we (Caan *et al.*, 1984) suggested that the tail of the caudate nucleus is not directly involved in the development and maintenance of reward associations to stimuli, but may aid visual discrimination performance by its sensitivity to change in visual stimuli. Neurons in some other parts of the striatum may, however, be involved in connecting visual stimuli to appropriate motor responses. For example, in the putamen some neurons have early movement-related firing during the performance of a visual discrimination task (Rolls *et al.*, 1984); and some neurons in the head of the caudate nucleus respond to environmental cues which signal that reward may be obtained (Rolls, Thorpe and Maddison, 1983).

9.2.3.2 Posterior putamen

Following these investigations on the caudal striatum which implicated it in visual functions related to a short term habituation or memory process, a further study has been performed to investigate the role of the posterior putamen in visual short-term memory tasks (Johnstone and Rolls, 1990; Rolls and Johnstone, 1992). Both the inferior temporal visual cortex and the prefrontal cortex project to the posterior ventral parts of the putamen (e.g. Goldman and Nauta, 1977; Van Hoesen *et al.*, 1981) and these cortical areas are known to subserve a variety of complex functions, including functions related to memory. For example, cells in both areas respond in a variety of short-term memory tasks (Fuster, 1973, 1989; Fuster and Jervey, 1982; Baylis and Rolls, 1987; Miyashita and Chang, 1988). Since striatal activity often is related to the area from which it receives (Rolls, 1984a; Rolls and Williams, 1987a), we asked whether neurons in the posterior putamen responded during short-term memory tasks in a manner related to the neocortical areas from which it receives. By comparing striatal responses to neocortical responses, we could examine the progression of information processing in relation to the question of integration. Second, the convergence of inputs from the striatum to the pallidum suggests that further integration of information could occur beyond the striatum in the pallidum. We addressed this question by recording in both the posterior putamen and the pallidum and comparing the responses in both areas. The prediction was that pallidal activity should represent a further stage of information processing from the striatum.

Two main groups of neuron with memory-related activity were found in delayed match to sample (DMS) tasks, in which the monkey was shown a sample stimulus, and had to remember it during a 2–5 s delay period, after which if a matching stimulus was shown he could make one response, but if a non-matching stimulus was shown he had to make no response. First, 11% of the 621 neurons studied responded to the test stimulus which followed the sample stimulus, but did not respond to the sample stimulus. Of these neurons, 43% responded only on non-match trials (test different from sample), 16% only on match trials (test same as the sample), and 41% to the test stimulus irrespective of whether it was the same or different from the sample. These neuronal responses were not related to the licking since (i) the units did not respond in other tasks in which a lick response was required (for example, in an auditory delayed match to sample task which was identical to the visual delayed match to sample task except that auditory short-term memory rather than visual short term memory was required, in a serial recognition task, or in a visual discrimination task), and (ii) a periresponse time spike density function indicated that the stimulus onset better predicted neuronal activity.

Second, 9.5% of the neurons responded in the delay period after the sample stimulus, during which the sample was being remembered. These neurons did not respond in the auditory version of the task, indicating that the responses were visual modality-specific (as were the responses of all other neurons in this part of the putamen with activity related to the delayed match to sample task). Given that the visual and auditory tasks were very similar apart from the modality of the input stimuli, this suggests that the activity of the neurons was not related to movements, or to rewards obtained in the tasks, but instead to modality-specific short-term memory-related processing.

The different neuron types described were located throughout the posterior 5 mm of the striatum (see Rolls and Johnstone, 1992). Within this posterior region there was a tendency for the units to be located medially, thus overlapping the termination zones of the inferior temporal and prefrontal projections, but the neurons were found in an area wider than might have been predicted from the termination sites alone. This could be related to the spread of the dendrites of the cells from which the recordings were made (cf. Percheron *et al.*, 1984b). Consistent with an influence on the striatal cells recorded from, among the cortical cells which project into this region, neurons which respond during the delay of a DMS task and other neurons which discriminate recently seen visual stimuli from novel stimuli are found in the frontal (Fuster, 1973, 1989) and temporal cortices (Fuster and Jervey, 1982; Baylis and Rolls, 1987; Miyashita and Chang, 1988).

In recordings made from pallidal neurons it was found that some responded in both visual and auditory versions of the task (Johnstone and Rolls, 1990; Rolls and Johnstone, 1992). Of 37 units responsive in the visual DMS task which were also tested in the auditory version, seven (19%) responded also in the auditory DMS task. The finding that some of the pallidal neurons active in the DMS task were not modality-specific, whereas only visual modality-specific DMS units were located in the posterior part of the striatum, suggests that the pallidum may represent a further stage in information processing in which information from different parts of the striatum can converge.

9.2.3.3 Head of the caudate nucleus

The activity of 394 neurons in the head of the caudate nucleus and most anterior part of the putamen was analysed in three behaving rhesus monkeys (Rolls, Thorpe and Maddison, 1983). 64.2% of these neurons had responses related to environmental stimuli, movements, the performance of a visual discrimination task, or eating. However, only relatively small proportions of these neurons had responses that were unconditionally related to visual (9.6%), auditory (3.5%), or gustatory (0.5%) stimuli, or to movements (4.1%). Instead, the majority of the neurons had responses that occurred conditionally in relation to stimuli or movements, in that the responses occurred in only some test situations, and were often dependent on the performance of a task by the monkeys. Thus, it was found that in a visual discrimination task 14.5% of the neurons responded during a 0.5 s tone/light cue which signalled the start of each trial; 31.1% responded in the period in which the discriminative visual stimuli were shown, with 24.3% of these responding more either to the visual stimulus which signified food reward or to that which signified punishment; and 6.2% responded in relation to lick responses. Yet these neurons typically did not respond in relation to the cue stimuli, to the visual stimuli, or to movements, when these occurred independently of the task or performance of the task was prevented. Similarly, although of the neurons tested during feeding, 25.8% responded when the food was seen by the monkey, 6.2% when he tasted it, and 22.4% during a cue given by the experimenter that a food or non-food object was about to be presented, only few of these neurons had responses to the same stimuli presented in different situations. Further evidence on the nature of these neuronal responses was that many of the neurons with cue-related responses only responded to the tone/light cue stimuli when they were cues for the performance of the task or the presentation of food, and some responded to the different cues used in these two situations.

The finding that such neurons may respond to environmental stimuli only when they are significant (Rolls *et al.*, 1979b, 1983) was confirmed by Evarts and his colleagues. They showed that some neurons in the putamen only responded to the click of a solenoid when it indicated that a fruit juice reward could be obtained (see Evarts and Wise, 1984). We have found that this decoding of the significance of environmental events which are signals for the preparation for or initiation of a response is represented in the firing of a population of neurons in the dorsolateral prefrontal cortex, which projects into the head of the caudate nucleus (E.T.Rolls and G.C.Baylis, unpublished observations, 1984). These neurons respond to the tone cue only if it signals the start of a trial of the visual discrimination task, just as do the corresponding population of neurons in the head of the caudate nucleus. The indication that the decoding of significance is performed by the cortex, and that the striatum receives only the result of the cortical computation, is considered below and elsewhere (Rolls and Williams, 1987a).

These findings indicate that the head of the caudate nucleus and most anterior part of the putamen contain populations of neurons which respond to cues which enable preparation for the performance of tasks such as feeding and tasks in which movements must be initiated, and others which respond during the performance of such tasks in relation to the stimuli used and the responses made, yet that the majority of these neurons do not have unconditional sensory or motor responses. It has therefore been suggested (Rolls *et al.*, 1983) that the anterior neostriatum contains neurons which are important for the utilization of environmental cues for the preparation for behavioural responses, and for particular behavioural responses made in particular situations to particular environmental stimuli, that is in stimulus–motor response habit formation. Different neurons in the cue-related group often respond to different subsets of environmentally significant events, and thus convey some information which would be useful in switching behaviour, in preparing to make responses, and in connecting inputs to particular responses (Rolls *et al.*, 1979b, 1983; Rolls, 1984a). Striatal neurons with similar types of response have also been recorded by Schultz and colleagues (see for example Schultz, Apicella, Romo and Scarnati, 1995), and striatal (tonically active) interneurons whose activity probably reflects the responses of such neurons have been described by Graybiel and Kimura (1995).

It may be suggested that deficits in the initiation of movements following damage to striatal pathways may arise in part because of interference with these functions of the anterior neostriatum. Thus the akinesia or lack of voluntary movement produced by damage to the dopaminergic nigrostriatal bundle in animals and present in Parkinson's disease in man (Hornykiewicz, 1973) may arise at least in part because of dysfunction of a system which normally is involved in utilizing environmental stimuli which are used as cues in the preparation for the initiation of movements. Such preparation may include for example postural adjustments. The movement disorder may also be due in part to the dysfunction of the system of neurons in the head of the caudate nucleus which appears to be involved in the generation of particular responses to particular environmental events (see also Rolls *et al.*, 1979b; Rolls *et al.*, 1983; Rolls and Williams, 1987a; Rolls, 1990a).

9.2.3.4 Anterior putamen

The anterior putamen receives inputs from the sensorimotor cortex, areas 3, 2, 1, 4 and 6 (Kunzle, 1975, 1977, 1978; Kunzle and Akert, 1977; Jones *et al.*, 1977; DeLong *et al.*, 1983). It is clear that the activity of many neurons in the putamen is related to movements (Anderson, 1978; DeLong *et al.*, 1984; Crutcher and DeLong, 1984a,b). There is a somatotopic organization of neurons in the putamen, with separate areas containing neurons responding to arm, leg, or orofacial movements. Some of these neurons respond only to active movements, and others to active and to passive movements. Some of these neurons respond to somatosensory stimulation, with multiple clusters of neurons responding, for example, to the movement of each joint (see DeLong *et al.*, 1984; Crutcher and DeLong, 1984a). Some neurons in the putamen have been shown in experiments in which the arm has been given assisting and opposing loads to respond in relation to the direction of an intended movement, rather than in relation to the muscle forces required to execute the movement (Crutcher and DeLong, 1984b). Also, the firing rate of neurons in the putamen tends to be linearly related to the amplitude of movements (Crutcher and DeLong, 1984b), and this is of potential clinical relevance, since patients with basal ganglia disease frequently have difficulty in controlling the amplitude of their limb movements.

In order to obtain further evidence on specialization of function within the striatum, the activity of neurons in the putamen has been compared with the activity of neurons recorded in different parts of the striatum in the same tasks (Rolls *et al.*, 1984). Of 234 neurons recorded in the putamen of two macaque monkeys during the performance of a visual discrimination task and the other tests in which other striatal neurons have been shown to respond (Rolls *et al.*, 1983; Caan *et al.*, 1984), 68 (29%) had activity that was phasically related to movements (Rolls *et al.*, 1984). Many of these responded in relation to mouth movements such as licking. The neurons did not have activity related to taste, in that they responded, for example, during tongue protrusion made to a food or non-food object. Some of these neurons responded in relation to the licking mouth movements made in the visual discrimination task, and always also responded when mouth movements were made during clinical testing when a food or non-food object was brought close to the mouth. Their responses were thus unconditionally related to movements, in that they responded in whichever testing situation was used, and were therefore different from the responses of neurons in the head of the caudate nucleus (Rolls *et al.*, 1983). Of the 68 neurons in the putamen with movement-related activity in these tests, 61 had activity related to mouth movements, and 7 had activity related to movements of the body. Of the remaining neurons, 24 (10%) had activity which was task related in that some change of firing rate associated with the presentation of the tone cue or the opening of the shutter occurred on each trial (see Rolls *et al.*, 1984), 4 had auditory responses, 1 responded to environmental stimuli (see Rolls *et al.*, 1983), and 137 were not responsive in these test situations.

These findings (Rolls *et al.*, 1984) provide further evidence that differences between neuronal activity in different regions of the striatum are found even in the same testing situations, and also that the inputs which activate these neurons are derived functionally from the cortex which projects into a particular region of the striatum (in this case sensori-motor cortex, areas 3, 1, 2, 4 and 6).

9.2.3.5 Ventral striatum

To analyse the functions of the ventral striatum, the responses of more than 1000 single neurons were recorded in a region which included the nucleus accumbens and olfactory tubercle in five macaque monkeys in test situations in which lesions of the amygdala, hippocampus, and inferior temporal cortex produce deficits, and in which neurons in these structures respond (Rolls and Williams, 1987b; Williams *et al.*, 1993; Rolls, 1990b–e; 1992a,b; Rolls, Cahusac *et al.*, 1993). While the monkeys performed visual discrimination and related feeding tasks, the different populations of neurons found included neurons which responded to novel visual stimuli; to reinforcement-related visual stimuli such as (for different neurons) food-related stimuli, aversive stimuli, or faces; to other visual stimuli; in relation to somatosensory stimulation and movement; or to cues which signalled the start of a task. The neurons with responses to reinforcing or novel visual stimuli may reflect the inputs to the ventral striatum from the amygdala and hippocampus, and are consistent with the hypothesis that the ventral striatum provides a route for learned reinforcing and novel visual stimuli to influence behaviour.

9.2.4 What computations are performed by the basal ganglia?

The neurophysiological investigations described in Section 9.2.3 indicate first that there are differences between neuronal activity in different regions of the striatum, so that there is segregation of function in the striatum; and second that the signals which activate the neurons in a part of the striatum are derived functionally from the cortical region or limbic structure which projects into that region of the striatum (for a more complete analysis, see Rolls and Johnstone, 1992). Given that in some of these regions, neurons have movement-related activity (the dorsal putamen), but in other regions have activity related to signals that prepare the animal for movement (head of caudate), to novel visual stimuli (tail of caudate nucleus), to visual short-term memory (ventral putamen), and to reinforcers (ventral striatum), we may ask whether some of these other regions might not have sensory or cognitive functions. It is possible to ask whether these other parts of the striatum are actually performing a sensory computation, or a cognitive computation (for the present purposes one in which neither the inputs nor the outputs are directly related to sensory or motor function, see also Oberg and Divac, 1979), or whether these parts of the striatum provide an essential output route for a cortical area with a sensory or cognitive function, but do not themselves have sensory or cognitive functions.

One way to obtain evidence on this is to analyse neurophysiologically the computation being performed by a part of the striatum, and relate this to the computation being performed in its input and output regions. For example, the taste and visual information necessary for the computation that a visual stimulus is no longer associated with taste reward reaches the orbitofrontal cortex, and the putative output of such a computation, namely neurons which respond in this non-reward situation, are found in the orbitofrontal cortex (Thorpe *et al.*, 1983; Rolls, 1989a, 1996b). However, such neurons which represent the necessary sensory information for this computation, and neurons which respond to the non-reward, were not found in the head of the caudate nucleus (Rolls, Thorpe and Maddison, 1983). Instead, in the head of the caudate nucleus, neurons in the same test situation responded in relation to whether the monkey had to make a response on a particular trial, that is many of them responded more on Go than on No-Go trials. This could reflect the *output* of a cognitive computation performed by the orbitofrontal cortex, indicating whether on the basis of the available sensory information, the current trial should be a Go trial, or a No-Go trial because a visual stimulus previously associated with punishment had been shown.

A similar comparison can be made for the tail of the caudate nucleus. Here the visual responses shown by neurons typically habituated to zero within a few trials, whereas such marked habituation was less common in neurons in the inferior temporal visual cortex, which projects to the tail of the caudate nucleus (see Caan *et al.*, 1984). In this case, the signal being processed by the striatum thus occurred when a patterned visual stimulus changed, and this could be of use in switching attention or orienting to the changed stimulus.

A comparison can also be made for the region of the posterior putamen containing neurons which respond in a visual delayed match to sample short-term memory task. These neurons did not respond differentially to the *particular* visual stimuli being remembered, but instead responded on, for example, all match trials, or on all non-match trials, or during all delay periods. Thus the activity of these neurons did not reflect the information necessary to solve the memory task, but instead appeared to reflect the output of such a (cortical) mechanism, producing in the striatum a signal which would be useful in preparing, initiating, or preventing a movement as appropriate on that trial (see above). The signal clearly did not simply reflect a movement, as shown (amongst other criteria) by the finding that the neurons were modality-specific.

Another comparison can be made between neurons in the head of the caudate that respond to cues that signal preparation for a trial of a task. Neurons similar to those in the head of the caudate nucleus described in Section 9.2.3.3 that respond to a 0.5 s tone or light cue that precedes a trial of a visual discrimination task are also found in the overlying dorsolateral prefrontal cortex that projects into the head of the caudate (observations of E.T. Rolls and G. Baylis, see Rolls, Thorpe and Maddison, 1983). These prefrontal cortex neurons are also influenced by learning of the type reflected in the responses of head of caudate neurons, in that the cortical neurons come in a few trials to respond to whichever cue (for example tone or light) precedes the start of a trial, and stop responding to the tone or light when it no longer signals the start of a trial. These findings suggest that the site of this type of learning is in or before the prefrontal cortex, and that the responses of head of caudate neurons just reflect this learning, and the output of the prefrontal cortex neurons. Although it is tempting to ascribe this type of learning to mechanisms within the striatum itself (Graybiel and Kimura, 1995), it is always necessary to show that the site of the learned modification is not represented at a prestriatal stage, as it is in this case.

In these four parts of the striatum in which a comparison can be made of processing in the striatum with that in the cortical area which projects to that part of the striatum, it appears that the full information represented in the cortex does not reach the striatum, but that rather the striatum receives the output of the computation being performed by a cortical area, and could use this to switch or alter behaviour. For example, the processing being performed in the ventral striatum is in many cases not just sensory, in that many of the neurons which respond to visual inputs do so preferentially on the basis of whether the stimuli are recognized, or are associated with reinforcement (Williams *et al.*, 1993). Much of the sensory and memory-related processing required to determine whether a stimulus is a face, is recognized, or is associated with reinforcement has been performed in and is evident in neuronal responses in structures such as the amygdala (Leonard *et al.*, 1985; Rolls, 1992a,b), orbitofrontal cortex (Thorpe *et al.*, 1983; Rolls, 1996b) and hippocampal system (see Chapter 6). Comparable evidence is available for the head of the caudate nucleus, the tail of the caudate nucleus, and the ventral putamen (see Rolls and Johnstone, 1992).

The hypothesis arises from these findings that some parts of the striatum, particularly the caudate nucleus, ventral striatum, and posterior putamen, receive the output of these memory-related and cognitive computations, but do not themselves perform them. Instead, on receiving the cortical and limbic outputs, the striatum may be involved in switching behaviour as appropriate as determined by the different, sometimes conflicting, information received from these cortical and limbic areas. On this view, the striatum would be particularly involved in the selection of behavioural responses, and in producing one coherent stream of behavioural output, with the possibility to switch if a higher priority input was received. This process may be achieved by a laterally spreading competitive interaction between striatal or pallidal neurons, which might be implemented by direct connections between nearby neurons in the striatum and globus pallidus. In addition, the inhibitory interneurons within the striatum, the dendrites of which in the striatum may cross the boundary between the matrix and striosomes, may play a part in this interaction between striatal processing streams (Groves, 1983; Groves *et al.*, 1995; Graybiel and Kimura, 1995). Dopamine could play an important role in setting the sensitivity of this response selection function, as discussed elsewhere (Rolls and Williams, 1987a; Rolls *et al.*, 1984). In addition to this response selection function by competition, the basal ganglia may, by the convergence discussed, enable signals originating from non-motor parts of the cerebral cortex to be mapped into motor signals to produce behavioural output. The ways in which these computations might be performed are considered next.

9.2.5 How do the basal ganglia perform their computations?

On the hypothesis just raised, different regions of the striatum, or at least the outputs of such regions, would need to interact. Is there within the striatum the possibility for different regions to interact, and is the partial functional segregation seen within the striatum maintained in processing beyond the striatum? For example, is the segregation maintained throughout the globus pallidus and thalamus with projections to different premotor and even prefrontal regions reached by different regions of the striatum, or is there convergence or the possibility for interaction at some stage during this post-striatal processing?

Given the anatomy of the basal ganglia, interactions between signals reaching the basal ganglia could happen in a number of different ways. One would be for each part of the striatum to receive at least some input from a number of different cortical regions. As discussed above, there is evidence for patches of input from different sources to be brought adjacent to each other in the striatum (Seleman and Goldman-Rakic, 1985; Van Hoesen *et al.*, 1981; Graybiel and Kimura, 1995). For example, in the caudate nucleus, different regions of association cortex project to adjacent longitudinal strips (Seleman and Goldman-Rakic, 1985). Now, the dendrites of striatal neurons have the shape of large plates which lie at right angles to the incoming cortico-striate fibres (Percheron *et al.*, 1984a,b, 1994) (see Figs 9.13 and 9.14). Thus one way in which interaction may start in the basal ganglia is by virtue of the same striatal neuron receiving inputs on its dendrites from more than just a limited area of the cerebral cortex. This convergence may provide a first level of integration over limited sets of cortico-striatal fibres. The large number of cortical inputs received by each striatal neuron, in the order of 10,000 (Wilson, 1995), is consistent with the hypothesis that convergence of inputs carrying different signals is an important aspect of the function of the basal ganglia. The computation which could be performed by this architecture is discussed below for the inputs to the globus pallidus, where the connectivity pattern is comparable (Percheron *et al.*, 1984a,b, 1994).

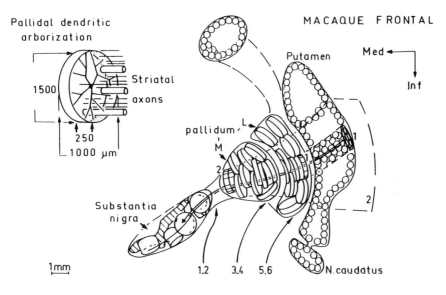

Fig. 9.13 Semi-schematic spatial diagram of the striato-pallido-nigral system (see text). The numbers represent the numbers of non-overlapping arborizations of dendrites in the plane shown. (Reprinted with permission from Percheron, Yelnik and Francois, 1984b.)

9.2.5.1 Interaction between neurons and selection of output

The regional segregation of neuronal response types in the striatum described above is consistent with mainly local integration over limited, adjacent, sets of cortico-striatal inputs, as suggested by this anatomy. Short-range integration or interactions within the striatum may also be produced by the short length (for example 0.5 mm) of the intrastriatal axons of striatal

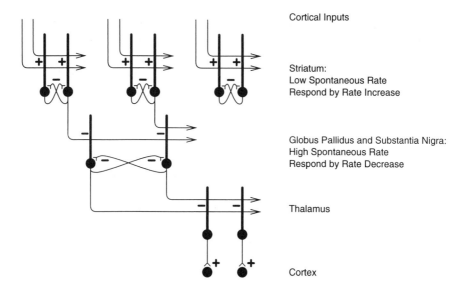

Fig. 9.14 Simple hypothesis of basal ganglia network architecture. A key aspect is that in both the striatum and globus pallidus there are direct inhibitory connections (–) between the principal neurons, as shown. These synapses use GABA as a transmitter. Excitatory inputs to the striatum are shown as +.

neurons. These could produce a more widespread influence if the effect of a strong input to one part of the striatum spread like a lateral competition signal (cf Groves, 1983; Groves *et al.*, 1995). Such a mechanism could contribute to behavioural response selection in the face of different competing input signals to the striatum. The lateral inhibition could operate, for example, between the striatal principal (that is spiny) neurons by direct connections (they receive excitatory connections from the cortex, respond by increasing their firing rates, and could inhibit each other by their local axonal arborizations, which spread in an area as large as their dendritic trees, and which utilize GABA as their inhibitory transmitter). Further lateral inhibition could operate in the pallidum and substantia nigra (see Fig. 9.14). Here again there are local axon collaterals, as large as the very large pallidal and nigral dendritic fields. The lateral competition could again operate by direct connections between the cells. (Note that pallidal and nigral cells have high spontaneous firing rates (often 25–50 spikes/s), and respond (to their inhibitory striatal inputs) by decreasing their firing rates below this high spontaneous rate. Such a decrease in the firing rate of one neuron would release inhibition on nearby neurons, causing them to increase their firing rates, equivalent to responding less.) A selection function of this type between processing streams in the basal ganglia, even without any convergence anatomically between the processing streams, might provide an important computational *raison d'être* for the basal ganglia. The direct inhibitory local connectivity between the principal neurons within the striatum and globus pallidus would seem to provide a simple, and perhaps evolutionarily old, way in which to implement competition between neurons and processing streams. This might even be a primitive design principle that characterizes the basal ganglia. This design principle contrasts interestingly with that of the cerebral cortex (including the hippocampal and pyriform cortex), in which the principal neurons are excitatory, and direct local connections between the neurons are useful for providing attractor-based short-term memories, for maintaining low average firing rates and thus energy consumption (in contrast to the globus pallidus and substantia nigra (pars reticulata)), etc. (see Chapter 10). The cerebral cortex has also developed its learning mechanism using NMDA receptors, which may facilitate associative operations because of its voltage-dependent properties (see Chapter 1). Another feature of directly coupled excitatory networks such as cortex is that they are inherently unstable, so requiring the evolution of very efficient (inhibitory) interneurons to maintain stability, whereas a system such as the basal ganglia with direct inhibitory recurrent collaterals may have evolved more easily, and been especially appropriate in motor systems in which instability could produce movement and co-ordination difficulties.

This hypothesis of lateral competition between the neurons of the basal ganglia can be sketched simply (see also Fig. 9.14). The inputs from the cortex to the striatum are excitatory, and competition between striatal neurons is implemented by the use of an inhibitory transmitter (GABA), and direct connections between striatal neurons, within an area which is approximately co-extensive with the dendritic arborization. Given that the lateral connections between the striatal neurons are collaterals of the output axons, the output must be inhibitory onto pallidal and nigral neurons. This means that to transmit signals usefully, and in contrast to striatal neurons, the neurons in the globus pallidus and substantia nigra (pars reticulata) must have high spontaneous firing rates, and respond by decreasing their firing rates. These pallidal and nigral neurons then repeat the simple scheme for lateral competition between output neurons by having direct lateral inhibitory connections to the other pallidal and nigral neurons. When nigral and pallidal neurons respond by decreasing their firing rates, the reduced inhibition through the recurrent collaterals allows the connected pallidal and nigral neurons to fire faster, and also at the same time the main output of the pallidal and nigral neurons allows the thalamic neurons to fire faster. The thalamic neurons then have the standard excitatory influence on their cortical targets. The simple, and perhaps evolutionarily early, aspect of this basal ganglia architecture is that the striatal, pallidal, and nigral neurons implement competition (for selection) by direct inhibitory recurrent lateral connections of the main output neurons onto other output neurons, with the inputs to a stage synapsing directly onto the output neurons (see Fig. 9.14).

Another possible mechanism for interaction within the striatum is provided by the dopaminergic pathway, through which a signal which has descended from, for example, the ventral striatum might influence other parts of the striatum (Nauta and Domesick, 1978). Because of the slow conduction speed of the dopaminergic neurons, this latter system would probably not be suitable for rapid switching of behaviour, but only for more tonic, long-term adjustments of sensitivity.

Further levels for integration within the basal ganglia are provided by the striato-pallidal and striato-nigral projections (Percheron et al., 1984a,b, 1994). The afferent fibres from the striatum again cross at right angles a flat plate or disc formed by the dendrites of the pallidal or nigral neurons (see Fig. 9.13). The discs are approximately 1.5 mm in diameter, and are stacked up one upon the next at right angles to the incoming striatal fibres. The dendritic discs are so large that in the monkey there is room for only perhaps 50 such discs not to overlap in the external pallidal segment, for 10 non-overlapping discs in the medial pallidal segment, and for one overlapping disc in the most medial part of the medial segment of the globus pallidus and in the substantia nigra.

One result of this convergence achieved by this stage of the medial pallidum/substantia nigra is that even if inputs from different cortical regions were kept segregated by specific wiring rules onto different neurons, nevertheless there might well be the possibility for mutual competition between different pallidal neurons, implemented by interneurons. Given the relatively small number of neurons into which the cortical signals had now been compressed, it would be feasible to have competition (the same effect as lateral inhibition would achieve elsewhere) implemented between the relatively small population of neurons, now all collected into a relatively restricted space, so that the competition could spread widely within these nuclei. This could allow selection by competition between these pathways, that is effectively between information processing in different cortical areas. This could be important in allowing each cortical area to control output when appropriate (depending on the task being performed). Even if full segregation was maintained in the return paths to the cerebral cortex, the return paths could influence each cortical area, allowing it to continue processing if it had the strongest 'call'. Each cortical area on a fully segregated hypothesis might thus have its own non-basal ganglia output routes, but might according to the current suggestion utilize the basal ganglia as a system to select a cortical area or set of areas, depending on how strongly each cortical area is calling for output. The thalamic outputs from the basal ganglia (areas VA and VLo of the thalamus) might according to this hypothesis have to some extent an activity or gain controlling function on an area (such as might be mediated by diffuse terminals in superficial cortical layers), rather than the strong and selective inputs implemented by a specific thalamic nucleus such as the lateral geniculate.

9.2.5.2 Convergent mapping within the basal ganglia

In addition to this selection function, it is also attractive to at least consider the further hypothesis that there is some convergent mapping achieved by the basal ganglia. This hypothesis is now considered in more detail. The anatomical arrangement just described does provide a possibility for some convergence onto single striatal neurons of cortical input, and onto single pallidal and nigral neurons of signals from relatively different parts of the striatum. For what computation might such anatomy provide a structural basis? Within the pallidum, each dendritic disc is flat, is orthogonal to the input fibres which pierce it, but is not filled with dendritic arborizations. Instead, each dendrite typically consists of 4-5 branches which are spread out to occupy only a small part of the surface area of the dendritic disc (see Fig. 9.13). There are thousands of such sparsely populated plates stacked on top of one another. Each pallidal neuron is contacted by a number of the mass of fibres from the striatum which pass it, and given the relatively small collecting area of each pallidal or nigral neuron (4 or 5 dendritic branches in a plane), each such neuron is thus likely to receive a random combination of inputs from different striatal neurons within its collection field. The thinness of the dendritic sheet may help to ensure that each axon does not make more than a few synapses with each dendrite, and that the combinations of inputs received by each dendrite are approximately random. This architecture thus appears to be appropriate for bringing together at random onto single pallidal and nigral neurons inputs which originate from quite diverse parts of the cerebral cortex. (This is a two-stage process, cortex to striatum, and striatum to pallidum and substantia nigra). By the stage of the medial pallidum and substantia nigra, there is the opportunity for the input field of a single neuron to effectively become very wide, although whether in practice this covers very different cortical areas, or is instead limited to a rather segregated cortex–basal ganglia–cortex loop, remains to be confirmed.

Given then that this architecture could allow individual pallidal and nigral neurons to receive random combinations of inputs from different striatal neurons, the following functional implications arise. Simple Hebbian learning in the striatum would enable strongly firing striatal neurons to increase the strength of the synapses from the active cortical inputs. In the pallidum, such conjunctive learning of coactive inputs would be more complex, requiring for example a strongly inhibited pallidal neuron to show synaptic strengthening from strongly firing but inhibitory inputs from the striatum. Then, if a particular pallidal or nigral neuron received inputs by chance from striatal neurons which responded to an environmental cue signal that something significant was about to happen, and from striatal neurons which fired because the monkey was making a postural adjustment, this conjunction of events might make that pallidal or nigral neuron become inhibited (that is respond to) either input alone. Then, in the future, the occurrence of only one of the inputs, for example only the environmental cue, would result in a decrease of firing of that pallidal or nigral neuron, and thus in the appropriate postural adjustment being made by virtue of the output connections of that pallidal or nigral neuron.

This is a proposal that the basal ganglia are able to detect combinations of conjunctively active inputs from quite widespread regions of the cerebral cortex using their combinatorial architecture and a property of synaptic modifiability. In this way it would be possible to trigger any complex pattern of behavioural responses by any complex pattern of environmental inputs, using what is effectively a pattern associator. According to this possibility the unconditioned input would be the motor input, and the to-be-associated inputs the other inputs to the basal ganglia.

It may be noted that the input events need not include only those from environmental stimuli represented in the caudate nucleus and ventral striatum, but also, if the overlapping properties of the dendrites described above provide sufficient opportunity for convergence, of the context of the movement, provided by inputs via the putamen from sensorimotor cortex. This would then make a system appropriate for triggering an appropriate motor response (learned by trial and error, with the final solution becoming associated with the triggering input events) to any environmental input state. As such, this hypothesis provides a suggested neural basis for 'habit' learning in which the basal ganglia have been implicated (Phillips *et al.*, 1988; Petri and Mishkin, 1994). The hypothesis could be said to provide a basis for the storage of motor plans in the basal ganglia, which would be instantiated as a series of look-ups of the appropriate motor output pattern to an evolving sequence of input information. An interesting aspect of this hypothesis is that other parts of the motor system, such as the cortico-cortical pathways, may mediate the control of action in a voluntary, often slow way in the early stages of learning. The input context for the movement and the appropriate motor signals (originating during learning from motor cortical areas) could then be learned by the basal ganglia, until after many trials the basal ganglia can perform the required look-up of the correct motor output in an automated, 'habit', mode. In this sense, the cortico-cortical pathways would set up the conditions, which because of their continuing repetition would be learned by the basal ganglia. The hypothesis introduced above also may provide a basis for the switching between different types of behaviour proposed as a function of the basal ganglia, for if a strong new pattern of inputs was received by the basal ganglia, this would result in a different pattern of outputs being 'looked up' than that currently in progress.

9.2.5.3 Discussion

According to this hypothesis, the striatum followed by the globus pallidus/substantia nigra/ ventral pallidum form a two-stage system which is involved in the selection of behavioural responses when inputs compete, and in eliciting learned responses appropriate for the context. The first stage, the cortico-striatal projection, would find regularities in the inputs it receives from a small part of the cerebral cortex. The low contact probability, given the few radial dendrites of each striatal neuron, would help each striatal neuron to learn to respond to a different combination of inputs to other striatal neurons. Competition may contribute to this process. The second stage, implemented by the striato-pallidal and striato-nigral systems would then detect regularities originating effectively over larger areas of the cerebral cortex, for example from sensory and motor areas. The advantage of the two-stage system is suggested to be that this enables pallidal neurons to respond to inputs originating from different parts of the cerebral cortex, without needing to have the enormous numbers of inputs that would be needed if the inputs were received directly from the cerebral cortex. The architecture is thus reminiscent of the multistage convergence seen in the visual cortical pathways with, however, the difference that there is a more obvious funnelling numerically as one progresses through the striatum to the basal ganglia. There are many fewer striatal neurons than cortical neurons, and many fewer pallidal and nigral than striatal neurons. It has been suggested above that a first purpose of this funnelling is to bring the neurons sufficiently close and small in number so that competition can be implemented over a large part of the population of medial pallidal/nigral neurons by lateral connections, helping the basal ganglia to produce one output at a time. The limitation in the number of neurons would be useful for this purpose, because the neurons in the pallidum and nigra have inputs and outputs which are limited in number (in the order of 10 000), so could not possibly perform the same function if they were connected to receive inputs directly from the vast number of cortical neurons. A second purpose of the funnelling might be to allow neurons in the medial part of the pallidum, and the substantia nigra, to have dendrites large enough that they can cover a large part of the information space, and thus potentially respond to combinations of inputs received from very different areas of the cerebral cortex.

The outputs of the globus pallidus and substantia nigra directed via the thalamus to motor regions such as the supplementary motor cortex and the premotor cortex potentially provide important output routes for the basal ganglia to produce actions (see Fig. 9.11). However, a possible objection to this hypothesis is that there are also outputs of the basal ganglia to structures which may not be primarily motor, such as the dorsolateral prefrontal cortex of primates (Middleton and Strick, 1994, 1996b), and the inferior temporal visual cortex (Middleton and Strick, 1996a). One possible function of basal ganglia influences on non-motor cortical areas, that of selection of whether a cortical area can be allowed to dominate output (even if its output is not mediated by pathways through the basal ganglia), has been described above. Another possible reason for an output which might be motor to be projected to the prefrontal cortex might be the following. It is well accepted that the dorsolateral prefrontal cortex is involved in working memory concerned with responses, for example of where spatial responses should be or have been made, whether by the hand or by the eye (Goldman-Rakic, 1987, 1996). (In the macaque this is a bilateral function, in humans a right dorsolateral prefrontal function; see Milner and Petrides, 1984.) If this area is involved in remembering where responses have been made, or should be made (for example in delayed spatial response tasks), then it would be essential for the prefrontal cortex to be provided with information about which response has been selected or has just been performed, so that the prefrontal cortex can implement a short-term memory for the body response. It is this information about actions which it is suggested is conveyed by the basal ganglia outputs which reach the prefrontal cortex. It is appropriate for the prefrontal cortex to implement the response short-term memory, for it is a feature of cortical (but not of basal ganglia) architecture that there are excitatory connections between the cells which could implement a short-term autoassociative memory, in which the neurons would by their firing hold the memory as one of the attractor states of the network (cf. Treves and Rolls, 1994; see Chapter 10).

The hypothesis of basal ganglia function just described incorporates associative learning of coactive inputs onto neurons, at both the cortico-striatal stages, and the striato-pallidal and nigral stages. Consistent with this hypothesis (Rolls, 1979, 1987, 1984b, 1994d; Rolls and Johnstone, 1992), it has now been possible to demonstrate long-term potentiation in at least some parts of the basal ganglia. For example, Pennartz et al. (1993) demonstrated LTP of limbic inputs to the nucleus accumbens, and were able to show that such LTP is facilitated by dopamine (see also Calabresi et al., 1992; Wickens and Kotter, 1995; Wickens, Begg and Arbuthnott, 1996). This reminds us that the dopamine pathways project to the striatum. Given that there are relatively few dopaminergic neurons, it is likely that the information conveyed by the dopamine pathway is relatively general or modulatory, rather than conveying the specific information that must be learned and mapped to the output of the basal ganglia, or the full details of the primary reinforcer obtained (for example which taste, which touch, etc). One possible function is setting the threshold or gain of basal ganglia neurons (Rolls et al., 1984). In this way, or perhaps by a different mechanism, the dopamine might also modulate learning in the basal ganglia. A very simple hypothesis is that the modulation of learning is likely to be quite general, making it more likely that information will be stored, that is that conjunctive activity at the inputs to the basal ganglia will become associated. An alternative possibility is that activity in the dopamine pathways carries a teaching signal, which might operate as a reinforcer in the type of reinforcement learning network described in Chapter 5. Schultz et al. (1995b) have argued from their recordings from dopamine neurons that this may be the case. For example, dopamine neurons can respond to the taste of a liquid reward in an operant task. However, these neurons may stop responding to such a primary (unlearned) reinforcer quite rapidly as the task is learned, and instead respond only to the earliest indication that a trial of the task is about to begin (Schultz et al., 1995b). Thus they could not convey information about a primary reward obtained if the trial is successful, but instead appear to convey information which would be much better suited to a preparation or 'Go' role for dopamine release in the striatum. Further evidence that the dopamine projection does not convey a specific 'reward' signal is that dopamine release occurs not only to rewards (such as food or brain-stimulation reward, or later in training to an indication that a reward might be given later), but also to aversive stimuli such as aversive stimulation of the medial hypothalamus (see Hoebel *et al.*, 1996; Hoebel, personal communication, 1996). These findings are much more consistent with the hypothesis that instead of acting as the reinforce or error signal in a reinforcement learning system, as suggested by Houk, Adams and Barto (1995), the dopamine projection to the striatum may act as a 'Go' or 'preparation' signal to set the thresholds of neurons in the striatum, and/or as a general modulatory signal that could help to strengthen synapses of conjunctively active pre- and postsynaptic neurons. In such a system, what is learned would be dependent on the presynaptic and postsynaptic terms, and would not be explicitly guided by a reinforce/ teacher signal that would provide feedback *after* each trial on the degree of success of each trial as in the reinforcement learning algorithm (see Chapter 5).

Evidence relevant to determining the utility of the hypothesis described above of basal ganglia function will include further information on the extent to which convergence from widely separate parts of the cerebral cortex is achieved in the basal ganglia, and if not, if it is achieved in the thalamic or cortical structures to which the basal ganglia project (see above); and evidence on whether pallidal and nigral neurons show learning of conjunctions of their inputs. Some evidence for convergence in the globus pallidus from the striatum is that some pallidal neurons were found to respond in both visual and auditory delayed match to sample tasks, whereas in the posterior putamen the neurons responded only in the visual delayed match to sample task, as described above. It may be noted that one reason for having two stages of processing in the basal ganglia, from the cortex to striatal neurons, and from striatal neurons to pallidal or nigral neurons, may be to ensure that the proportion of responding input neurons to any stage of the memory is kept relatively low, in order to reduce interference in the association memory (Rolls, 1987; Rolls and Treves, 1990). Another reason may be that with the very high degree of convergence, from an enormous number of corticostriatal afferents, to relatively few pallidal and nigral neurons, a one-stage system would not suffice, given that the limit on the number of inputs to any one neuron is in the order of 10,000.

An alternative view of striatal function is that the striatum might be organized as a set of segregated and independent transmission routes, each one of which would receive from a given region of the cortex, and project finally to separate premotor or prefrontal regions (Strick *et al.*, 1995; Middleton and Strick, 1996a,b) (see Fig. 9.11). Even if this is correct, the detection of combinations of conjunctively active inputs, but in this case from limited populations of input axons to the basal ganglia, might still be an important aspect of the function of the basal ganglia. More investigations are needed to lead to further understanding of these concepts on the function of the basal ganglia.

10 Cerebral neocortex

The fine structure of the neocortex is complicated and the connections between the different neuron types are still poorly understood, so that it is not possible now to produce complete computational theories of its operation. However, because an understanding of neocortical function is crucial to understanding the higher functions of the brain, it is important to consider what is known of its structure and connectivity, in order to provide a basis for the development of computational theories of its function. Indeed, a step towards a computational theory of the neocortex will be made in this chapter, by proposing a theory of the remarkable property of the neocortex that, in addition to forward connections from one area to the next, there are also as many projections backwards to the preceding cortical area.

10.1 The fine structure and connectivity of the neocortex

The neocortex consists of many areas which can be distinguished by the appearance of the cells (cytoarchitecture) and fibres or axons (myeloarchitecture), but nevertheless, the basic organization of the different neocortical areas has many similarities, and it is this basic organization which is considered here. Differences between areas, and the ways in which different cortical areas are connected together at the systems level, are in part considered in Chapter 1. Useful sources for descriptions of neocortical structure and function are the book 'Cerebral Cortex' edited by Jones and Peters (1984; Peters and Jones, 1984); and Douglas and Martin (1990). Approaches to quantitative aspects of the connectivity are provided by Braitenberg and Schuz (1991) and by Abeles (1991). Some of the connections described in Sections 10.1.1 and 10.1.2 are shown schematically in Fig. 10.6.

10.1.1 Excitatory cells and connections

Some of the cell types found in the neocortex are shown in Fig. 10.1. Cells A–D are pyramidal cells. The dendrites (shown thick in Fig. 10.1) are covered in spines, which receive the excitatory synaptic input to the cell. Pyramidal cells with cell bodies in different laminae of the cortex (shown in Fig. 10.1 as I–VI) not only have different distributions of their dendrites, but also different distributions of their axons (shown thin in Fig. 10.1), which connect both within that cortical area and to other brain regions outside that cortical area (see labelling at the bottom of Fig. 10.1).

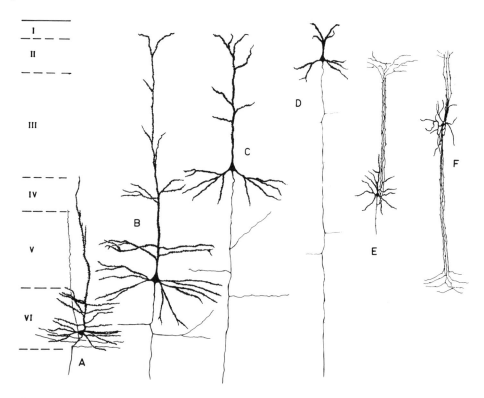

Fig. 10.1 Cell types in the cerebral neocortex. The different laminae of the cortex are designated I–VI, with I at the surface. Cells A–E are pyramidal cells in the different layers. Cell E is a spiny stellate cell, and F is a double bouquet cell. (After Jones, 1981; see Jones and Peters, 1984, p. 7.)

The main information-bearing afferents to a cortical area have many terminals in layer IV. (By these afferents, we mean primarily those from the thalamus or from the preceding cortical area. We do not mean the cortico-cortical backprojections, nor the subcortical cholinergic, noradrenergic, dopaminergic, and serotonergic inputs, which are numerically minor, although they are important in setting cortical cell thresholds, excitability, and adaptation, see for example Douglas and Martin, 1990.) In primary sensory cortical areas only there are **spiny stellate** cells in a rather expanded layer IV, and the thalamic terminals synapse onto these cells (Lund, 1984; Martin, 1984; Douglas and Martin, 1990; Levitt, Lund and Yoshioka, 1996). (Primary sensory cortical areas receive their inputs from the primary sensory thalamic nucleus for a sensory modality. An example is the primate striate cortex which receives inputs from the lateral geniculate nucleus, which in turn receives from the retinal ganglion cells. Spiny stellate cells are so-called because they have radially arranged, star-like, dendrites. Their axons usually terminate within the cortical area in which they are located.) Each thalamic axon makes 1,000–10,000 synapses, not more than several (or at most 10) of which are onto any one spiny stellate cell. In addition to these afferent terminals, there are some terminals of the thalamic afferents onto pyramidal cells with cell bodies in layers 6 and 3 (Martin, 1984), (and terminals onto inhibitory interneurons such as basket cells, which thus provide for a feedforward inhibition) (see Fig. 10.4). Even in layer 4, the thalamic axons provide less than 20% of the synapses. The spiny stellate neurons in layer 4 have axons which terminate in layers 3 and 2, as illustrated in Fig. 10.2, at least partly on dendrites of pyramidal cells with cell bodies in layers 3 and 2. (These synapses are of Type I, that is are asymmetrical and are on spines, so that they are probably excitatory. Their transmitter is probably glutamate.) These layer 3 and 2 pyramidal cells provide the onward cortico-cortical projection, with axons which project into layer 4 of the next cortical area. For example, layer 3 and 2 pyramidal cells in the primary visual (striate) cortex of the macaque monkey project into the second visual area (V2), layer 4.

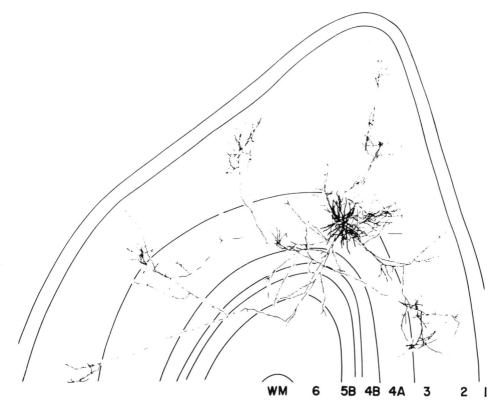

Fig. 10.2 Spiny stellate neurons. The spiny stellate neurons in layer 4 have axons which terminate in layers 3 and 2, at least partly on dendrites of pyramidal cells with cell bodies in layers 3 and 2. In this and subsequent figures, the thick processes are dendrites, and the thin processes are axons. The dendrites of spiny stellate cells are in the region of 500 μm in diameter, and their axons can distribute in patches 200–300 μm across separated by distances of up to 1 mm. (Reproduced with permission from Martin, 1984.)

In non-primary sensory areas, important information-bearing afferents from a preceding cortical area terminate in layer 4, but there are no or few spiny stellate cells in this layer (Lund, 1984; Levitt *et al.*, 1996). Layer 4 still looks 'granular' (due to the presence of many small cells), but these cells are typically small pyramidal cells (Lund, 1984). (It may be noted here that spiny stellate cells and small pyramidal cells are similar in many ways, with a few main differences including the absence of a major apical dendrite in a spiny stellate which accounts for its non-pyramidal, star-shaped, appearance; and for many spiny stellate cells, the absence of an axon which projects outside its cortical area.) The terminals presumably make synapses with these small pyramidal cells, and also presumably with the dendrites of cells from other layers, including the basal dendrites of deep layer 3 pyramidal cells (see Fig. 10.1).

The axons of the **superficial (layer 2 and 3) pyramidal cells** have collaterals and terminals in layer 5 (see Fig. 10.1), and synapses are made with the dendrites of the layer 5 pyramidal cells (Martin, 1984). The axons also typically project out of that cortical area, and on to the next cortical area in sequence, where they terminate in layer 4, forming the forward cortico-cortical projection (see Fig. 10.3). It is also from these pyramidal cells that projections to the amygdala arise in some sensory areas which are high in the hierarchy (Amaral *et al.*, 1992).

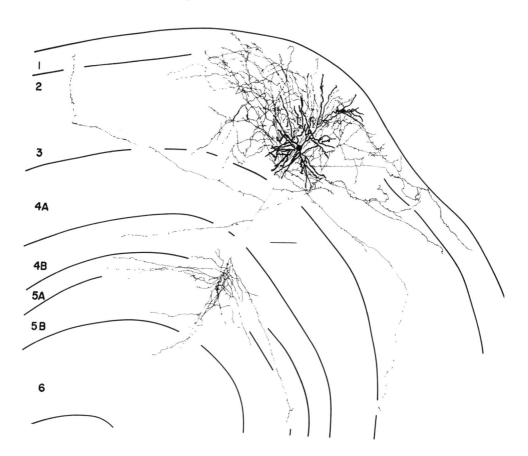

Fig. 10.3 The axons of the superficial (layer 2 and 3) pyramidal cells have collaterals and terminals in layer 5 as well as layer 3, and synapses are made with the dendrites of the layer 5 pyramidal cells. The axons also typically project out of that cortical area, and on to the next cortical area in sequence, where they terminate in layer 4, forming the forward cortico-cortical projection. The dendrites of layer 2 and 3 pyramidal cells can be approximately 300 μm in diameter, but after this the relatively narrow column appears to become less important, for the axons of the superficial pyramidal cells can distribute over 1 mm or more, both in layers 2 and 3, and in layer 5. (Reproduced with permission from Martin, 1984.)

The axons of the **layer 5 pyramidal cells** have many collaterals in layer 6 (see Fig. 10.1), where synapses could be made with the layer 6 pyramidal cells (based on indirect evidence, see Martin, 1984, Fig. 13), and axons of these cells typically leave the cortex to project to subcortical sites (such as the striatum), or back to the preceding cortical area to terminate in layer 1 (Fig. 10.3). It is remarkable that there are as many of these backprojections as there are forward connections between two sequential cortical areas. The possible computational significance of this connectivity is considered below.

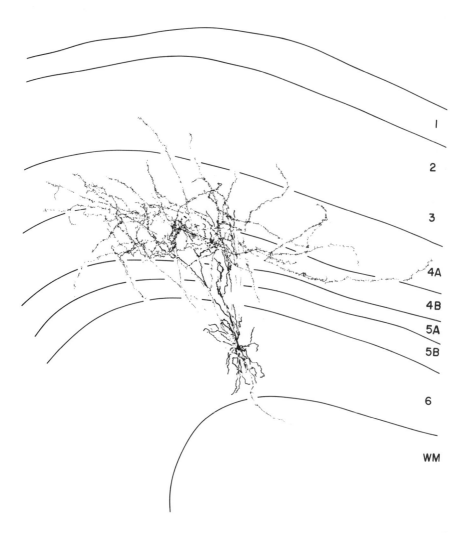

Fig. 10.4 The layer 6 pyramidal cells have prolific dendritic arborizations in layer 4, and receive synapses from thalamic afferents, and also presumably from pyramidal cells in other cortical layers. (Reproduced with permission from Martin, 1984). The axons of the layer 6 pyramidal cells form backprojections to the thalamic nucleus which projects into that cortical area, and also axons of cells in layer 6 contribute to the backprojections to layer 1 of the preceding cortical area.

The **layer 6 pyramidal cells** have prolific dendritic arborizations in layer 4 (see Figs. 10.1 and 10.4), and receive synapses from thalamic afferents (Martin, 1984), and also presumably from pyramidal cells in other cortical layers (see Fig. 10.4). The axons of these cells form backprojections to the thalamic nucleus which projects into that cortical area, and also axons of cells in layer 6 contribute to the backprojections to layer 1 of the preceding cortical area (see Jones and Peters, 1984; Peters and Jones, 1984; see Figs. 10.1 and 10.3).

Although the pyramidal and spiny stellate cells form the great majority of neocortical neurons with excitatory outputs, there are in addition several further cell types (see Peters and Jones, 1984, Ch. 4). Bipolar cells are found in layers 3 and 5, and are characterised by having two dendritic systems, one ascending and the other descending, which, together with the axon distribution, is confined to a narrow vertical column often less than 50 µm in diameter (Peters, 1984b). Bipolar cells form asymmetrical (presumed excitatory) synapses with pyramidal cells, and may serve to emphasize activity within a narrow vertical column.

10.1.2 Inhibitory cells and connections

There are a number of types of neocortical inhibitory neurons. All are described as smooth in that they have no spines, and use GABA as a transmitter. (In older terminology they were called Type II.) A number of types of inhibitory neuron can be distinguished, best by their axonal distributions (see Szentagothai, 1978; Peters and Regidor, 1981; Douglas and Martin, 1990). One type is the **basket cell**, present in layers 3–6, which has few spines on its dendrites so that it is described as smooth, and has an axon which participates in the formation of weaves of preterminal axons which surround the cell bodies of pyramidal cells and form synapses directly onto the cell body, but also onto the dendrite spines (Somogyi *et al.*, 1983) (Fig. 10.5). Basket cells comprise 5–7% of the total cortical cell population, compared to approximately 72% for pyramidal cells (Sloper and Powell, 1979a,b). Basket cells receive synapses from the main extrinsic afferents to the neocortex, including thalamic afferents (Fig. 10.5), so that they must contribute to a feedforward type of inhibition of pyramidal cells. The inhibition is feedforward in that the input signal activates the basket cells and the pyramidal cells by independent routes, so that the basket cells can produce inhibition of pyramidal cells which does not depend on whether the pyramidal cells have already fired. Feedforward inhibition of this type not only enhances stability of the system by damping the responsiveness of the pyramidal cell simultaneously with a large new input, but can be conceived of as a mechanism which normalizes the magnitude of the input vector received by each small region of neocortex (see further Chapters 4 and 8). In fact, the feedforward mechanism allows the pyramidal cells to be set at the appropriate sensitivity for the input they are about to receive. Basket cells can also be polysynaptically activated by an afferent volley in the thalamo-cortical projection (Martin, 1984), so that they may receive inputs from pyramidal cells, and thus participate in feedback inhibition of pyramidal cells.

The transmitter used by the basket cells is gamma-aminobutyric acid (GABA), which opens chloride channels in the postsynaptic membrane. Because the reversal potential for Cl⁻ is approximately −10mV relative to rest, opening the Cl⁻ channels does produce an inhibitory postsynaptic potential (IPSP), which results in some hyperpolarization, especially in the dendrites. This is a subtractive effect, hence it is a linear type of inhibition (see Douglas and

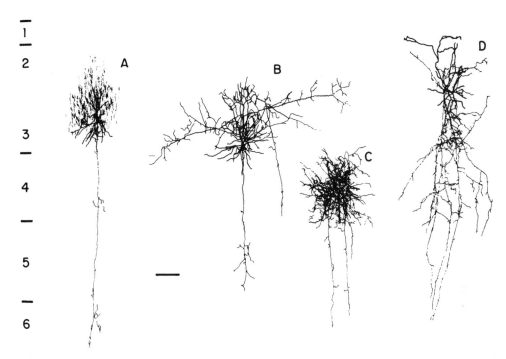

Fig. 10.5 Smooth cells from cat visual cortex. (A) Chandelier or axoaxonic cell. (B) Large basket cell of layer 3. Basket cells, present in layers 3–6, have few spines on their dendrites so that they are described as smooth, and have an axon which participates in the formation of weaves of preterminal axons which surround the cell bodies of pyramidal cells and form synapses directly onto the cell body. (C) Small basket or clutch cell of layer 3. The major portion of the axonal arbor is confined to layer 4. (D) Double bouquet cell. The axon collaterals run vertically. The cortical layers are as indicated. Bar = 100 μm. (Reproduced with permission from Douglas and Martin, 1990.)

Martin, 1990). However, a major effect of the opening of the Cl⁻ channels in the cell body is that this decreases the membrane resistance, thus producing a shunting effect. The importance of shunting is that it decreases the magnitude of excitatory postsynaptic potentials (EPSPs) (cf. Andersen *et al.*, 1980 for hippocampal pyramidal cells), so that the effect of shunting is to produce division (i.e. a multiplicative reduction) of the excitatory inputs received by the cell, and not just to act by subtraction (see further Bloomfield, 1974; Martin, 1984; Douglas and Martin, 1990). Thus, when modelling the normalization of the activity of cortical pyramidal cells, it is common to include division in the normalization function (cp. Chapter 4 and Appendix A5). It is notable that the dendrites of basket cells can extend laterally 0.5 mm or more (primarily within the layer in which the cell body is located), and that the axons can also extend laterally from the cell body 0.5–1.5 mm. Thus the basket cells produce a form of lateral inhibition which is quite spatially extensive. There is some evidence that each basket cell may make 4–5 synapses with a given pyramidal cell, that each pyramidal cell may receive from 10–30 basket cells, and that each basket cell may inhibit approximately 300 pyramidal cells (Martin, 1984; Douglas and Martin, 1990). The basket cells are sometimes called *clutch* cells.

A second type of GABA-containing inhibitory interneuron is the **axoaxonic** (or '**chandelier**') cell, named because it synapses onto the initial segment of the axon of pyramidal cells. The pyramidal cells receiving this type of inhibition are almost all in layers 2 and 3, and much less in the deep cortical layers. One effect which axoaxonic cells probably produce is thus prevention of outputs from layer 2 and 3 pyramidal cells reaching the pyramidal cells in the deep layers, or from reaching the next cortical area. Up to five axoaxonic cells converge onto a pyramidal cell, and each axoaxonic cell may project to several hundred pyramidal cells scattered in a region which may be several hundred microns in length (see Martin, 1984; Peters, 1984a). This implies that axoaxonic cells provide a rather simple device for preventing runaway overactivity of pyramidal cells, but little is known yet about the afferents to axoaxonic cells, so that the functions of these neurons are very incompletely understood.

A third type of (usually smooth and inhibitory) cell is the **double bouquet** cell, which has primarily vertically organized axons. These cells have their cell bodies in layer 2 or 3, and have an axon traversing layers 2–5, usually in a tight bundle consisting of varicose, radially oriented collaterals often confined to a narrow vertical column 50 μm in diameter (Somogyi and Cowey, 1984). Double bouquet cells receive symmetrical, type II (presumed inhibitory) synapses, and also make type II synapses, perhaps onto the apical dendrites of pyramidal cells, so that these neurons may serve by this double inhibitory effect to emphasize activity within a narrow vertical column.

Another type of GABA-containing inhibitory interneuron is the smooth and sparsely spinous non-pyramidal (multipolar) neuron with local axonal plexuses (Peters and Saint Marie, 1984). In addition to extrinsic afferents, these neurons receive many type I (presumed excitatory) terminals from pyramidal cells, and have inhibitory terminals on pyramidal cells, so that they may provide for the very important function of feedback or recurrent lateral inhibition (see Chapters 3 and 4).

10.1.3 Quantitative aspects of cortical architecture

Some quantitative aspects of cortical architecture are described, because, although only preliminary data are available, they are crucial for developing an understanding of how the neocortex could work. Further evidence is provided by Braitenberg and Schuz (1991), and by Abeles (1991). Typical values, many of them after Abeles (1991) are shown in Table 10.1. The figures given are for a rather generalized case, and indicate the order of magnitude. The number of synapses per neuron (20 000) is an estimate for monkeys; those for humans may be closer to 40 000, and for the mouse, closer to 8000. The value of 18 000 excitatory synapses made by a pyramidal cell is set to match the number of excitatory synapses received by pyramidal cells, for the great majority of cortical excitatory synapses are made from axons of cortical, principally pyramidal, cells.

Microanatomical studies show that pyramidal cells rarely make more than one connection with any other pyramidal cell, even when they are adjacent in the same area of the cerebral cortex. An interesting calculation takes the number of local connections made by a pyramidal cell within the approximately 1 mm of its local axonal arborization (say 9000), and the number of pyramidal cells with dendrites in the same region, and suggests that the probability that a pyramidal cell makes a synapse with its neighbour is low, approximately 0.1 (Braitenberg and Schuz, 1991; Abeles, 1991). This fits with the estimate from simultaneous recording of nearby pyramidal cells using spike-triggered averaging to monitor time-locked EPSPs (see Abeles, 1991; Thomson and Deuchars, 1994).

Table 10.1 Typical quantitative estimates for neocortex (partly after Abeles (1991) and reflecting estimates in macaques)

Neuronal density	20 000–40 000/mm^3
Neuronal composition:	
Pyramidal	75%
Spiny stellate	10%
Inhibitory neurons, for example smooth stellate, chandelier	15%
Synaptic density	8×10^8/mm^3
Numbers of synapses on pyramidal cells:	
Excitatory synapses from remote sources per neuron	9,000
Excitatory synapses from local sources per neuron	9,000
Inhibitory synapses per neuron	2,000
Pyramidal cell dendritic length	10 mm
Number of synapses made by axons of pyramidal cells	18,000
Number of synapses on inhibitory neurons	2,000
Number of synapses made by inhibitory neurons	300
Dendritic length density	400 m/mm^3
Axonal length density	3200 m/mm^3
Typical cortical thickness	2 mm
Cortical area	
human (assuming 3 mm for cortical thickness)	300 000 mm^2
macaque (assuming 2 mm for cortical thickness)	30 000 mm^2
rat (assuming 2 mm for cortical thickness)	300 mm^2

Now the implication of the pyramidal cell to pyramidal cell connectivity just described is that within a cortical area of perhaps 1 mm^2, the region within which typical pyramidal cells have dendritic trees and their local axonal arborization, there is a probability of excitatory-to-excitatory cell connection of 0.1. Moreover, this population of mutually interconnected neurons is served by 'its own' population of inhibitory interneurons (which have a spatial receiving and sending zone in the order of 1 mm^2), enabling local threshold setting and optimization of the set of neurons with 'high' (0.1) connection probability in that region. Such an architecture is effectively recurrent or re-entrant. It may be expected to show some of the properties of recurrent networks, including the fast dynamics described in Appendix A5. Such fast dynamics may be facilitated by the fact that cortical neurons in the awake behaving monkey generally have a low spontaneous rate of firing (personal observations; see for example Rolls, and Tovee, 1995a; Rolls, Treves, Tovee and Panzeri, 1997), which means that even any small additional input may produce some spikes sooner than would otherwise have occurred, because some of the neurons may be very close to a threshold for firing. It might also show some of the autoassociative retrieval of information typical of autoassociation networks, if the synapses between the nearby pyramidal cells have the appropriate (Hebbian) modifiability. In this context, the value of 0.1 for the probability of a connection between nearby neocortical pyramidal cells is of interest, for the connection probability between hippocampal CA3 pyramidal is approximately 0.02–0.04, and this is thought to be sufficient to sustain associative retrieval (see Chapter 6 and Appendix A4). As Abeles (1991) has pointed out, the recurrent connectivity typical of local cortical circuits is unlike that of two other major types of cortex, the cerebellar cortex and the hippocampal cortex. Within the cerebellum, there is no re-entrant path (see Chapter 9). In the hippocampus, the connectivity is primarily feedforward (dentate to CA3 to CA1). However, and not pointed out by Abeles, one part of the hippocampus, the CA3 system, does form a re-entrant circuit. Indeed, the CA3 system might be directly compared with a local region of the cerebral neocortex in terms of the recurrent connectivity. However, whereas in the CA3 cells a single network may be implemented (on each side of the brain, see Chapter 6), in the neocortex, each 1 mm^2 region probably overlaps somewhat continuously with the next.

This raises the issue of modules in the cortex, described by many authors as regions of the order of 1 mm^2 (with different authors giving different sizes), in which there are vertically oriented columns of neurons which may share some property (for example responding to the same orientation of visual stimulus), and which may be anatomically marked (for example Powell, 1981; Mountcastle, 1984; see Douglas *et al.*, 1996). The anatomy just described, with the local connections between nearby (1 mm) pyramidal cells, and the local inhibitory neurons, may provide a network basis for starting to understand the columnar architecture of the neocortex, for it implies that local recurrent connectivity on this scale implementing local re-entrancy is a feature of cortical computation. We can note that the neocortex could not be a single, global, autoassociation network, because the number of memories that could be stored in an autoassociation network rather than increasing with the size of the network is limited by the number of recurrent connections per neuron, which is in the order of ten thousand (see Table 10.1), or less, depending on the species, as pointed out by O'Kane and Treves (1992). This would be an impossibly small capacity for the whole cortex. It is suggested that instead a principle of cortical design is that it does have in part local connectivity, so that each part can have its own processing and storage, which may be triggered by other modules, but is a distinct operation from what occurs simultaneously in other modules.

An additional remarkable parallel between the hippocampus and any small patch of neocortex is the allocation of an array of many small excitatory (usually non-pyramidal, spiny stellate or granular) cells at the input side. In neocortex this is layer 4, in the hippocampus the dentate gyrus. In both cases, these cells receive the feedforward inputs and relay them to a population of pyramidal cells (in layers 2–3 of the neocortex and in the CA3 field of the hippocampus) with extensive recurrent collateral connections. In both cases, the pyramidal cells receive inputs both as relayed by the front-end preprocessing array and directly (see Chapter 6 and below). Such analogies might indicate that the functional roles of neocortical layer 4 cells and of dentate granule cells could be partially the same, although one should note also perspicuous differences, such as the lack of the analogue of the rather peculiar mossy fibre projections in the neocortex. More detailed analyses of the architecture and physiology of layer 4 circuits (particularly in higher association cortices) are required in order to clarify this issue.

The short-range high density of connectivity may also contribute to the formation of cortical topographic maps, as described in Chapter 4, which may help to ensure that different parameters of the input space are represented in a nearly continuous fashion across the cortex, to the extent that the reduction in dimensionality allows it, or by the clustering of cells with similar response properties, when preserving strict continuity is not possible, as illustrated for example by colour 'blobs' in striate cortex.

If an excitatory recurrent network utilizes an anti-Hebbian rule, decreasing the synaptic strength between two pyramidal cells if their firing is correlated, this can implement a form of optimal coding by neurons, by decorrelating their activity (Foldiak, 1990) (see Section 10.6). This is another possibility for the local cortico-cortical connectivity, although there is no evidence that an anti-Hebbian rule is implemented.

10.1.4 Functional pathways through the cortical layers

Because of the complexity of the circuitry of the cerebral cortex, some of which is summarized in Fig. 10.6, there are only preliminary indications available now of how information is processed by the cortex. In primary sensory cortical areas, the main extrinsic 'forward' input is from the thalamus, and ends in layer 4, where synapses are formed onto spiny stellate cells. These in turn project heavily onto pyramidal cells in layers 3 and 2, which in turn send projections forward to the next cortical area. The situation is made more complex than this by the fact that the thalamic afferents synapse also onto the basal dendrites in or close to the layer 2 pyramidal cells, as well as onto layer 6 pyramidal cells and inhibitory interneurons (see Fig. 10.4). Given that the functional implications of this particular architecture are not fully clear, it would be of interest to examine the strength of the functional links between thalamic afferents and different classes of cortical cell using cross-correlation techniques, to determine which neurons are strongly activated by thalamic afferents with monosynaptic or polysynaptic delays. Given that this is technically difficult, an alternative approach has been to use electrical stimulation of the thalamic afferents to classify cortical neurons as mono- or polysynaptically driven, then to examine the response properties of the neuron to physiological (visual) inputs, and finally to fill the cell with horseradish peroxidase so that its full structure can be studied (see for example Martin, 1984). Using these techniques, it has been shown in the cat visual cortex that spiny stellate cells can indeed be driven monosynaptically by thalamic afferents to the cortex. Further, many of these neurons have S-type receptive fields, that is they have distinct on and off regions of the receptive field, and respond with orientation tuning to elongated visual stimuli (see Martin, 1984). Further, consistent with the anatomy just described, pyramidal cells in the deep part of layer 3, and in layer 6, could also be monosynaptically activated by thalamic afferents, and had S-type receptive fields (see Fig. 10.4; Martin, 1984). Also consistent with the anatomy just described, pyramidal cells in layer 2 were di- (or poly-) synaptically activated by stimulation of the afferents from the thalamus, but also had S-type receptive fields.

Inputs could reach the layer 5 pyramidal cells from the pyramidal cells in layers 2 and 3, the axons of which ramify extensively in layer 5, in which the layer 5 pyramidal cells have widespread basal dendrites (see Figs 10.1 and 10.4), and also perhaps from thalamic afferents. Many layer 5 pyramidal cells are di- or trisynaptically activated by stimulation of the thalamic afferents, consistent with them receiving inputs from monosynaptically activated deep layer 3 pyramidal cells, or from disynaptically activated pyramidal cells in layer 2 and upper layer 3 (Martin, 1984). Interestingly, many of the layer 5 pyramidal cells had C-type receptive fields, that is they did not have distinct on and off regions, but did respond with orientation tuning to elongated visual stimuli (Martin, 1984).

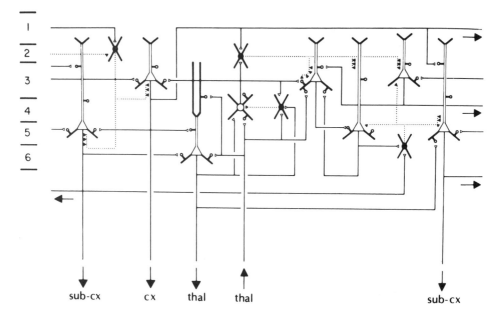

Fig. 10.6 Basic circuit for visual cortex. Excitatory neurons, which are spiny and use glutamate as a transmitter, and include the pyramidal and spiny stellate cells, are indicated by open somata; their axons are indicated by solid lines, and their synaptic boutons by open symbols. Inhibitory (smooth, GABAergic) neurons are indicated by black (filled) somata; their axons are indicated by dotted lines, and their synaptic boutons by solid symbols. thal, thalamus; cx, cortex; sub-cx, subcortical. Cortical layers I–VI are as indicated. (Reproduced with permission from Douglas and Martin, 1990.)

Studies on the function of inhibitory pathways in the cortex are also beginning. The fact that basket cells often receive strong thalamic inputs, and that they terminate on pyramidal cell bodies where part of their action is to shunt the membrane, suggests that they act in part as a feedforward inhibitory system which normalizes the thalamic influence on pyramidal cells by dividing their response in proportion to the average of the thalamic input received (see Chapter 4). The smaller and numerous smooth (or sparsely spiny) non-pyramidal cells which are inhibitory may receive inputs from pyramidal cells as well as inhibit them, so that these neurons could perform the very important function of recurrent or feedback inhibition (see Chapter 3). It is only feedback inhibition which can take into account not only the inputs received by an area of cortex, but also the effects which these inputs have, once multiplied by the synaptic weight vector on each neuron, so that recurrent inhibition is necessary for competition and contrast enhancement (see Chapter 4).

Another way in which the role of inhibition in the cortex can be analysed is by applying a drug such as bicuculline using iontophoresis (which blocks GABA receptors to a single neuron), while examining the response properties of the neuron (see Sillito, 1984). With this technique, it has been shown that in the visual cortex of the cat, layer 4 simple cells lose their orientation and directional selectivity. Similar effects are observed in some complex cells, but the selectivity of other complex cells may be less affected by blocking the effect of endogenously released GABA in this way (Sillito, 1984). One possible reason for this is that the inputs to complex cells must often synapse onto the dendrites far from the cell body (see Fig. 10.4), and distant synapses will probably be unaffected by the GABA receptor blocker released near the cell body. The experiments reveal that inhibition is very important for the normal selectivity of many visual cortex neurons for orientation and the direction of movement. Many of the cells displayed almost no orientation selectivity without inhibition. This implies that not only is the inhibition important for maintaining the neuron on an appropriate part of its activation function (see Chapter 2), but also that lateral inhibition between neurons is important because it allows the responses of a single neuron (which need not be markedly biased by its excitatory input) to have its responsiveness set by the activity of neighbouring neurons (see Chapter 4).

10.1.5 The scale of lateral excitatory and inhibitory effects, and the concept of modules

The forward cortico-cortical afferents to a cortical area sometimes have a columnar pattern to their distribution, with the column width 200–300 µm in diameter (see Eccles, 1984). Similarly, individual thalamo-cortical axons often end in patches in layer 4 which are 200–300 µm in diameter (Martin, 1984). The dendrites of spiny stellate cells are in the region of 500 µm in diameter, and their axons can distribute in patches 200–300 µm across separated by distances of up to 1 mm (for example Fig. 10.2; Martin, 1984). The dendrites of layer 2 and 3 pyramidal cells can be approximately 300 µm in diameter, but after this the relatively narrow column appears to become less important, for the axons of the superficial pyramidal cells can distribute over 1 mm or more, both in layers 2 and 3, and in layer 5 (for example Fig. 10.3; Martin, 1984). Other neurons which may contribute to the maintenance of processing in relatively narrow columns are the double bouquet cells, which because they receive inhibitory inputs, and themselves produce inhibition, all within a column perhaps 50 µm across (see above), would tend to enhance local excitation. The bipolar cells, which form excitatory synapses with pyramidal cells, may also serve to emphasize activity within a narrow vertical column approximately 50 µm across. These two mechanisms for enhancing local excitation operate against a much broader ranging set of lateral inhibitory processes, and could it is suggested have the effect of increasing contrast between the firing rates of pyramidal cells 50 µm apart, and thus be very important in competitive interactions between pyramidal cells. Indeed, the lateral inhibitory effects are broader than the excitatory effects described so far, in that for example the axons of basket cells spread laterally 500 µm or more (see above) (although those of the small, smooth non-pyramidal cells are closer to 300 µm, see Peters and Sainte-Marie, 1984). Such short-range local excitatory interactions with longer range inhibition not only provide for contrast enhancement and for competitive interactions, but also can result in the formation of maps in which neurons with similar responses are grouped together and neurons with dissimilar response are more widely separated (see Chapter 4). Thus these local interactions are consistent with the possibilities that cortical pyramidal cells form a competitive net (see Chapter 4 and below), and that cortical maps are formed at least partly as a result of local interactions of this kind in a competitive net (see Chapter 4).

In contrast to the relatively localized terminal distributions of forward cortico-cortical and thalamo-cortical afferents, the cortico-cortical backward projections which end in layer 1 have a much wider horizontal distribution, of up to several mm. It is suggested below that this enables the backward projecting neurons to search over a larger number of pyramidal cells in the preceding cortical area for activity which is conjunctive with their own (see below).

10.2 Theoretical significance of backprojections in the neocortex

In Chapter 8 a possible way in which processing could operate through a hierarchy of cortical stages was described. Convergence and competition were key aspects of the system suggested. This processing could act in a feedforward manner, and indeed, in the experiments on backward masking described in Chapter 8, there was insufficient time for top-down processing to occur. (Neurons in each cortical stage respond for 20–30 ms when an object can just be seen, and given that the time from V1 to inferior temporal cortex takes about 50 ms, there is insufficient time for a return projection from IT to reach V1, influence processing there, and in turn for V1 to project up to IT to alter processing there.) Nevertheless, backprojections are a major feature of cortical connectivity, and we next consider hypotheses about their possible function.

10.2.1 Architecture

The forward and backward projections in the neocortex that will be considered are shown in Fig. 10.7 (for further anatomical information see Jones and Peters, 1984; Peters and Jones, 1984). As described above, in primary sensory cortical areas, the main extrinsic 'forward' input is from the thalamus and ends in layer 4, where synapses are formed onto spiny stellate cells. These in turn project heavily onto pyramidal cells in layers 3 and 2, which in turn send projections forwards to terminate strongly in layer 4 of the next cortical layer (on small pyramidal cells in layer 4 or on the basal dendrites of the layer 2 and 3 (superficial) pyramidal cells, and also onto layer-6 pyramidal cells and inhibitory interneurons). Inputs reach the layer 5 (deep) pyramidal cells from the pyramidal cells in layers 2 and 3 (Martin, 1984), and it is the deep pyramidal cells that send backprojections to end in layer 1 of the preceding cortical area (see Fig. 10.7), where there are apical dendrites of pyramidal cells. It is important to note that in addition to the axons and their terminals in layer 1 from the succeeding cortical stage, there are also axons and terminals in layer 1 in many stages of the cortical hierarchy from the amygdala and (via the subiculum, entorhinal cortex, and parahippocampal cortex) from the hippocampal formation (see Fig. 10.7) (Van Hoesen, 1981; Turner, 1981; Amaral and Price, 1984; Amaral, 1986, 1987; Amaral et al., 1992).

One point made in Figs. 1.8 and 6.1 is that the amygdala and hippocampus are stages of information processing at which the different sensory modalities (such as vision, hearing, touch, taste, and smell for the amygdala) are brought together, so that correlations between inputs in different modalities can be detected in these regions, but not at prior cortical processing stages in each modality, as these cortical processing stages are mainly unimodal. Now, as a result of bringing together any two modalities, significant correspondences between the two modalities can be detected. One example might be that a particular visual stimulus is associated with the taste of food. Another example might be that another visual stimulus is associated with painful touch. Thus at these limbic (and orbitofrontal cortex, see Chapter 7) stages of processing, but not before, the significance of, for example, visual and auditory stimuli can be detected and signalled. Sending this information back to the neocortex could thus provide a signal which indicates to the cortex that information should be stored. Even more than this, the backprojection pathways could provide patterns of firing

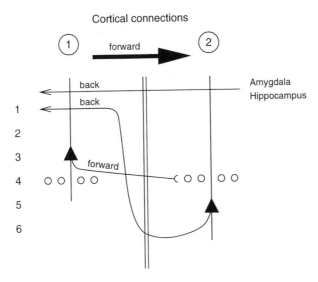

Fig. 10.7 Schematic diagram of forward and backward connections between adjacent neocortical areas. (Area 1 projects forwards to area 2 in the diagram. Area 1 would in sensory pathways be closest to the sense organs.) The superficial pyramidal cells (triangles) in layers 2 and 3 project forwards (in the direction of the arrow) to the next cortical area. The deep pyramidal cells in layer 5 project backwards to end in layer 1 of the preceding cortical area (on the apical dendrites of pyramidal cells). The hippocampus and amygdala are also the source of backprojections that end mainly in layer 1 of the higher association cortical areas. Spiny stellate cells are represented by small circles in layer 4. See text for further details.

that could help the neocortex to store the information efficiently, one of the possible functions of backprojections in and to the neocortex considered next.

10.2.2 Learning

The way in which the backprojections could assist learning in the cortex can be considered using the architecture shown in Figs. 10.7 and 4.5 (see also Section 4.5). The input stimulus occurs as a vector applied to (layer 3) cortical pyramidal cells through modifiable synapses in the standard way for a competitive net (input A in the schematic Fig. 4.5). If it is a primary cortical area, the input stimulus is at least partly relayed through spiny stellate cells, which may help to normalize and orthogonalize the input patterns in a preliminary way before the patterns are applied to the layer 3 pyramidal cells. If it is a non-primary cortical area, the cortico-cortical forward axons may end more strongly on the basal dendrites of neurons in the superficial cortical layers (3 and possibly 2). The lower set of synapses on the pyramidal cells would then start by competitive learning to set up representations on these neurons which would represent correlations in the input information space, and could be said to correspond to features in the input information space, where a feature is defined simply as the representation of a correlation in the input information space (see Chapter 4).

Consider now the conjunctive application of a pattern vector via the backprojection axons with terminals in layer 1 (B in Fig. 4.5) with the application of one of the input stimulus vectors. If, to start with, all synapses can be taken to have random weights, some of the pyramidal cells will by chance be strongly influenced both by the input stimulus and by the backprojecting vector. These strongly activated neurons will then compete with each other as in a standard competitive net, to produce contrast enhancement of their firing patterns. The relatively short-range (50 μm) excitatory operations produced by the bipolar and double bouquet cells, together with more widespread (300–500 μm) recurrent lateral inhibition produced by the smooth non-pyramidal cells and perhaps the basket cells, may be part of the mechanism of this competitive interaction. Next, Hebbian learning takes place as in a competitive net (see Chapter 4), with the addition that not only the synapses between forward projecting axons and active neurons are modified, but also the synapses between the backward projecting axons and the active neurons, which are in layer 1, are associatively modified.

This functional architecture has the following properties (see also Section 4.5). First, orthogonal backprojecting inputs can help the neurons to separate input stimuli (on the forward projection lines, A in Fig. 4.5) even when the input stimuli are very similar. This is achieved by pairing two somewhat different input stimuli A with very different (for example orthogonal) backprojection stimuli B. This is easily demonstrated in simulations (for example Rolls, 1989b,e). Conversely, if two somewhat different input stimuli A are paired with the same backprojection stimulus B, then the outputs of the network to the two input stimuli are more similar than they would otherwise be (see Section 4.5). This is also easily demonstrated in simulations.

In the neocortex, the backprojecting 'tutors' (see Rolls, 1989b,e) can be of two types. One originates from the amygdala and hippocampus, and by benefiting from cross-modal comparison, can for example provide orthogonal backprojected vectors. This backprojection, moreover, may only be activated if the multimodal areas detect that the visual stimulus is significant, because for example it is associated with a pleasant taste. This provides one way in which guidance can be provided for a competitive learning system as to what it should learn, so that it does not attempt to lay down representations of all incoming sensory information. The type of guidance is to influence which categories are formed by the competitive network.

The second type of backprojection 'tutor' is that from the next cortical area in the hierarchy. The next cortical area could operate in the same manner, and because it is a competitive system, is able to further categorize or orthogonalize the stimuli it receives, benefiting also from additional convergence (see for example Chapter 8). It then projects back these more orthogonal representations as tutors to the preceding stage, to effectively build better filters for the categories it is finding (cf. Fig. 4.6). These categories might be higher order, for example for two parallel lines on the retina, and even though the receptive fields of neurons at the preceding area might never receive inputs about both lines because the receptive fields are too small, the backprojections could still help to build feature analyzers at the earlier stage that would be tuned to the components of what can be detected as higher order feature at the next stage (see Chapter 8 (and Fig. 4.6)).

Another way for what is laid down in neocortical networks to be influenced is by neurons which 'strobe' the cortex when new or significant stimuli are shown. The cholinergic system originating in the basal forebrain, and the noradrenergic input to layer 1 of the cortex from the locus coeruleus, may contribute to this function (see Section 7.1.5). By modulating whether storage occurs according to the arousal or activation being produced by the environment, the storage of new information can be promoted at important times only, thus making good use of inevitably limited storage capacity. This influence on neocortical storage is not explicit guidance about the categories formed, as could be produced by backprojections, but instead consists of influencing which patterns of neuronal activity in the cortex are stored.

10.2.3 Recall

Evidence that during recall neural activity does occur in cortical areas involved in the original processing comes, for example, from investigations which show that when humans are asked to recall visual scenes in the dark, blood flow is increased in visual cortical areas, stretching back from association cortical areas as far as early (possibly even primary) visual cortical areas (Roland and Friberg, 1985; Kosslyn, 1994). Recall is a function that could be produced by cortical backprojections (see also Section 7.2).

If in Figs. 10.7 or 4.5 only the backprojection input (B in Fig. 4.5) is presented after the type of learning just described, then the neurons originally activated by the forward projecting input stimuli (A in Fig. 4.5) are activated. This occurs because the synapses from the backprojecting axons onto the pyramidal cells have been associatively modified only where there was conjunctive forward and backprojected activity during learning. This thus provides a mechanism for recall. The crucial requirement for recall to operate in this way is that in the backprojection pathways, the backprojection synapses would need to be associatively modifiable, so that the backprojection input could operate when presented alone effectively as a pattern associator, to produce recall. Some aspects of neocortical architecture consistent with this hypothesis (Rolls, 1989b,e) are as follows. First, there are many NMDA receptors on the apical dendrites of cortical pyramidal cells, where the backprojection synapses terminate. These receptors are implicated in associative modifiability of synapses (see Chapter 1), and indeed plasticity is very evident in the superficial layers of the cerebral cortex (Diamond et al., 1994). Second, the backprojection synapses in ending on the apical dendrite, quite far from the cell body, might be expected to be sufficient to dominate the cell firing when there is no forward input close to the cell body. In contrast, when there is forward input to the neuron, activating synapses closer to the cell body than the backprojecting inputs, this would tend to electrically shunt the effects received on the apical dendrite. This could be beneficial during the original learning, in that during the original learning the forward input would have the stronger effect on the activation of the cell, with mild guidance then being provided by the backprojections.

An example of how this recall could operate is provided next. Consider the situation when in the visual system the sight of food is forward projected onto pyramidal cells in higher order visual cortex, and conjunctively there is a backprojected representation of the taste of the food from, for example, the amygdala or orbitofrontal cortex. Neurons which have conjunctive inputs from these two stimuli set up representations of both, so that later if only the taste representation is backprojected, then the visual neurons originally activated by the sight of that food will be activated. In this way many of the low-level details of the original visual stimulus might be recalled. Evidence that during recall relatively early cortical processing stages are activated comes from cortical blood flow studies in humans, in which it has been found, for example, that quite posterior visual areas are activated during recall of visual (but not auditory) scenes (Kosslyn, 1994). The backprojections are probably in this situation acting as pattern associators.

The quantitative analysis of the recall that could be implemented through the hippocampal backprojection synapses to the neocortex, and then via multiple stages of cortico-cortical backprojections, made it clear that the most important quantitative factor influencing the number of memories p that can be recalled is the number C of backprojecting synapses onto each cortical neuron in the backprojecting pathways (see Section 6.10.3). Indeed, it was shown in Treves and Rolls (1991) that the maximum number of independently generated activity patterns p that can be retrieved in such a multilayer associative system is given by Eq. 6.7

$$ p \approx \frac{C}{a \ln(1/a)} k' \qquad (6.7) $$

where a is the sparseness of the representation at any given stage, and C is the average number of (back)projections each cell of that stage receives from cells of the previous one (in the backprojection direction). (k' is a similar slowly varying factor to that introduced in Eq. 6.2.) This provides an interpretation of why there are in general as many backprojecting synapses between two adjacent cortical areas as forward connections. The number C needs to be large to recall as many memories as possible, but need not be larger than the number of forward inputs to each neuron, which influences the number of possible classifications that the neuron can perform with its forward inputs (see Chapters 2, 4, and 5).

An implication of these ideas is that if the backprojections are used for recall, as seems likely as just discussed, then this would place severe constraints on their use for functions such as error backpropagation (see Chapter 5). It would be difficult to use the backprojections in cortical architecture to convey an appropriate error signal from the output layer back to the earlier, hidden, layers if the backprojection synapses are also to be set up associatively to implement recall.

10.2.4 Semantic priming

A third property of this backprojection architecture is that it could implement semantic priming, by using the backprojecting neurons to provide a small activation of just those neurons which are appropriate for responding to that semantic category of input stimulus.

10.2.5 Attention

In the same way, attention could operate from higher to lower levels, to selectively facilitate only certain pyramidal cells by using the backprojections. Indeed, the backprojections described could produce many of the 'top-down' influences that are common in perception (cf. Fig. 4.6).

10.2.6 Autoassociative storage, and constraint satisfaction

If the forward connections from one cortical area to the next, and the return backprojections, are both associatively modifiable, then the coupled networks could be regarded as, effectively, an autoassociative network. A pattern of activity in one cortical area would be associated with a pattern in the next which occurred regularly with it. This could enable higher cortical areas to influence the state of earlier cortical areas, and could be especially influential in the type of situation shown in Fig. 4.6 in which some convergence occurs at the higher area. For example, if one of the earlier stages (for example the olfactory stage in Fig. 4.6) had a noisy input on a particular occasion, its representation could be cleaned up if a taste input normally associated with it was present. The higher cortical area would be forced into the correct pattern of firing by the taste input, and this would feed back as a constraint to affect the state into which the olfactory area settled. This could be a useful general effect in the cerebral cortex, in that constraints arising only after information has converged from different sources at a higher level could feed back to influence the representations that earlier parts of the network settle into. This is a way in which top-down processing could be implemented.

The autoassociative effect between two forward and backward connected cortical areas could also be used in short term memory functions, to implement the types of short term memory effect described in Chapter 3 and Section 10.3. Such connections could also be used to implement a trace learning rule as described in Chapter 8 and Section 10.6.

10.2.7 Backprojections from the primary visual cortex, V1, to the lateral geniculate nucleus

One aspect of backprojection architecture that has led to much speculation is the enormous number (perhaps 15–30 million) of backprojecting fibres from V1 to the lateral geniculate nucleus. This is much greater than the number of fibres from the lateral geniculate to V1 (approximately 1 million, figures for Old World monkeys). Many functions have been suggested. One very simple possible explanation for the much larger number of back-projecting axons in this V1 to geniculate architecture is that deep cortical pyramidal cells have a genetic specification to have backprojecting axons, and because there are many more V1 cells than lateral geniculate cells (15–30 million in layer 6, the origin of the backprojecting axons), there are more backprojecting than forward projecting fibres in this particular part of the system. The suggestion is that this simple genetic specification works well for most cortico-cortical systems, producing approximately equal numbers of forward and back-projecting fibres because of the approximately similar numbers of cells in connected cortical regions that are adjacent in the hierarchy. Use of this simple rule economizes on special genetic specification for the particular case of backprojections from V1 to the lateral geniculate.

10.3 Cortical short term memory systems and attractor networks

There are a number of different short term memory systems, each implemented in a different cortical area. The particular systems considered here each implement short term memory by subpopulations of neurons which show maintained activity in a delay period, while a stimulus or event is being remembered. These memories may operate as autoassociative attractor networks, as described next. The actual autoassociation could be implemented by associatively modifiable synapses between connected pyramidal cells within an area, or between adjacent cortical areas in a hierarchy as just described.

One short term memory system is in the dorso-lateral prefrontal cortex, area 46. This is involved in remembering the locations of spatial responses, in for example delayed spatial response tasks (see Goldman-Rakic, 1996). In such a task performed by a monkey, a light beside one of two response keys illuminates briefly, there is then a delay of several seconds, and then the monkey must touch the appropriate key in order to obtain a food reward. The monkey must not initiate the response until the end of the delay period, and must hold a central key continuously in the delay period. Lesions of the prefrontal cortex in the region of the principal sulcus impair the performance of this task if there is a delay, but not if there is no delay. Some neurons in this region fire in the delay period, while the response is being remembered (see Fuster, 1973, 1989; Goldman-Rakic, 1996). Different neurons fire for the two different responses.

There is an analogous system in a more ventral and posterior part of the prefrontal cortex involved in remembering the position in visual space to which an eye movement (a saccade) should be made (Funahashi, Bruce and Goldman-Rakic, 1989; Goldman-Rakic, 1996). In this case, the monkey may be asked to remember which of eight lights appeared, and after the delay to move his eyes to the light that was briefly illuminated. The short term memory function is topographically organized, in that lesions in small parts of the system impair remembered eye movements only to that eye position. Moreover, neurons in the appropriate part of the topographic map respond to eye movements in one but not in other directions (see Fig. 10.8). Such a memory system could be easily implemented in such a topographically organized system by having local cortical connections between nearby pyramidal cells which implement an attractor network (Fig. 10.9). Then triggering activity in one part of the topographically organized system would lead to sustained activity in that part of the map, thus implementing a short term or working memory for eye movements to that position in space.

Another short term memory system is implemented in the inferior temporal visual cortex, especially more ventrally towards the parahippocampal areas. This memory is for whether a particular visual stimulus (such as a face) has been seen recently. This is implemented in two ways. One is that some neurons respond more to a novel than to a familiar visual stimulus in such tasks, or in other cases respond to the familiar, selected, stimulus (Baylis and Rolls, 1987; Miller and Desimone, 1994). The other is that some neurons, especially more ventrally, continue to fire in the delay period of a delayed match to sample task (see Fahy et al., 1993; Miyashita, 1993). These neurons can be considered to reflect the implementation of an attractor network between the pyramidal cells in this region (Amit, 1995). A cortical syndrome that may reflect loss of such a short-term visual memory system is simultanagnosia, in which more than one visual stimulus cannot be remembered for more than a few seconds (Warrington and Weiskrantz, 1973).

Continuing firing of neurons in short term memory tasks in the delay period is also found in other cortical areas. For example, it is found in a delayed match to sample short term

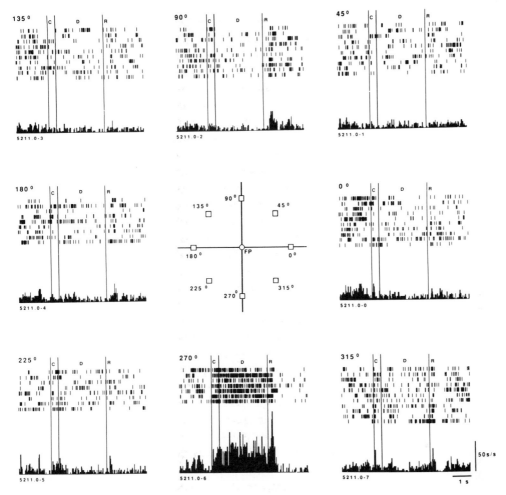

Fig. 10.8 The activity of a single neuron in the ventrolateral prefrontal cortical area involved in remembered saccades. Each row is a single trial, with each spike shown by a vertical line. A cue is shown in the cue (C) period, there is then a delay (D) period without the cue in which the cue position must be remembered, then there is a response (R) period. The monkey fixates the central fixation point (FP) during the cue and delay periods, and saccades to the position where the cue was shown, in one of the eight positions indicated, in the response period. The neuron increased it activity primarily for saccades to position 270. The increase of activity was in the cue, delay, and response period while the response was made. The time calibration is 1 s. (Reproduced with permission from Funahashi, Bruce and Goldman-Rakic, 1989.)

memory task in the inferior frontal convexity cortex, in a region connected to the ventral temporal cortex (see Fuster, 1989; Wilson, O'Sclaidhe and Goldman-Rakic, 1993). Delay-related neuronal firing is also found in the parietal cortex when monkeys are remembering a target to which a saccade should be made (see Andersen, 1995); and in the motor cortex when a monkey is remembering a direction in which to reach with the arm (see Georgopoulos, 1995).

Another short term memory system is human auditory–verbal short term memory, which appears to be implemented in the left hemisphere at the junction of the temporal, parietal, and occipital lobes. Patients with damage to this system are described clinically as showing

Local autoassociation networks in the prefrontal cortex
for delayed spatial responses, eg. saccades

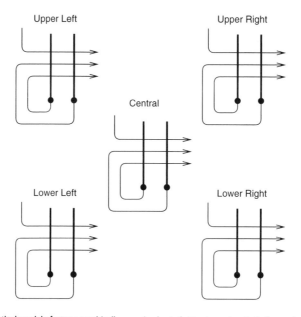

Fig. 10.9 A possible cortical model of a topographically organized set of attractor networks in the prefrontal cortex that could be used to remember the position to which saccades should be made. Excitatory local recurrent collateral Hebb-modifiable connections would enable a set of separate attractors to operate. The input that would trigger one of the attractors into continuing activity in a memory delay period would come from the parietal cortex, and topology of the inputs would result in separate attractors for remembering different positions in space. (For example inputs for the Upper Left of space would trigger an attractor that would remember the upper left of space.) Neurons in different parts of this cortical area would have activity related to remembering one part of space; and damage to a part of this cortical area concerned with one part of space would result in impairments in remembering targets to which to saccade for only one part of space.

conduction aphasia, in that they cannot repeat a heard string of words (cannot conduct the input to the output) (Warrington and Weiskrantz, 1973).

10.4 Coding in the cortex

How is information encoded in the cerebral cortex? Can we read the code being used by the cortex? What are the advantages of the encoding scheme used for the neuronal network computations being performed in different areas of the cortex? These are some of the key issues considered in this section. Because information is exchanged between the computing elements of the cortex, the neurons, by their spiking activity, which is conveyed by their axon to synapses onto other neurons, the appropriate level of analysis is how single neurons, and populations of single neurons, encode information in their firing. More global measures which reflect the averaged activity of large numbers of neurons (for example PET (positron emission tomography) and fMRI (functional magnetic resonance imaging), EEG (electro-encephalographic recording), and ERPs (event-related potentials)) cannot reveal *how* the information is represented, or *how* the computation is being performed.

We summarise some of the types of representation that might be found at the neuronal level next (cf. Chapter 1). A **local** representation is one in which all the information that a particular stimulus or event occurred is provided by the activity of one of the neurons. This is sometimes called a grandmother cell representation, because in a famous example, a single neuron might be active only if one's grandmother was being seen (see Barlow, 1995). A **fully distributed** representation is one in which all the information that a particular stimulus or event occurred is provided by the activity of the full set of neurons. If the neurons are binary (for example either active or not), the most distributed encoding is when half the neurons are active for any one stimulus or event. A **sparse distributed** representation is a distributed representation in which a small proportion of the neurons is active at any one time. These definitions are introduced in Chapter 1, and Eq. 1.4 defines a measure of the sparseness, a. For a binary representation, a is 0.5 for a fully distributed representation, and $1/S$ if a neuron responds to one of a set of S stimuli. Another measure of the sparseness which has been used is the breadth of tuning index of Smith and Travers (1979), which takes the value 1 for a fully distributed representation, and 0.0 if the neuron responds to one of a set of S stimuli[1]. Another measure of sparseness is the kurtosis of the distribution, which is the fourth moment of the distribution. It reflects the length of the tail of the distribution. (An actual distribution of the firing rates of a neuron to a set of 65 stimuli is shown in Fig. 10.11. The sparseness a for this neuron was 0.69, see Rolls, Treves, Tovee and Panzeri, 1997.)

10.4.1 Distributed representations evident in the firing rate distributions

Barlow (1972) proposed a single neuron doctrine for perceptual psychology. He proposed that sensory systems are organized to achieve as complete a representation as possible with the minimum number of active neurons. He suggested that at progressively higher levels of sensory processing, fewer and fewer cells are active, and that each represents a more and more specific happening in the sensory environment. He suggested that 1000 active neurons (which he called cardinal cells) might represent the whole of a visual scene. An important principle involved in forming such a representation was the reduction of redundancy (see Appendix A2 for an introduction to information theory and redundancy). The implication of Barlow's (1972) approach was that when an object is being recognized, there are towards the end of the visual system one or a few neurons (the cardinal cells) that are so specifically tuned that the activity of these neurons encodes the information that one particular object is being seen. (He thought that an active neuron conveys something of the order of complexity of a word.) The encoding of information in such a system is described as local, in that knowing the activity of just one neuron provides evidence that a particular stimulus (or, more exactly, a given 'trigger feature') was present. Barlow (1972) eschewed 'combinatorial rules of usage of nerve cells', and believed that the subtlety and sensitivity of perception results from the mechanisms determining when a single cell becomes active. In contrast, with distributed or ensemble

---

[1] Smith and Travers (1979) used the proportion of a neuron's total response that is devoted to each of the stimuli to calculate its coefficient of entropy (H). The measure of entropy is derived from information theory, and is calculated as

$$H = -k\,\Sigma_i\, p_i \log p_i$$

where H = breadth of responsiveness, k = scaling constant (set so that $H = 1.0$ when the neuron responds equally well to all stimuli in the set of size S), p_i = the response to stimulus i expressed as a proportion of the total response to all the S stimuli in the set.

encoding, the activity of several or many neurons must be known in order to identify which stimulus was present, that is to read the code. It is the relative firing of the different neurons in the ensemble that provides the information about which object is present.

At the time Barlow (1972) wrote, there was little actual evidence on the activity of neurons in the higher parts of the visual and other sensory systems. There is now considerable evidence, which is now described.

In the higher parts of the visual system in the temporal lobe visual cortical areas, there are neurons that are tuned to respond to faces (see Rolls, 1994a, 1995b, 1997d) or to objects (Rolls, Booth and Treves, 1996). (Both classes of neurons are described as being tuned to provide information about faces or objects, in that their responses can be view-invariant; see Chapter 8.) Neurons that respond to faces can regularly be found on tracks into the temporal cortical visual areas, and they therefore provide a useful set of cells for systematic studies about how information about a large set of different visual stimuli, in this case different faces, is represented (Rolls, 1992b, 1994a). First, it has been shown that the representation of which particular object (face) is present is rather distributed. Baylis, Rolls and Leonard (1985) showed this with the responses of temporal cortical neurons that typically responded to several members of a set of five faces, with each neuron having a different profile of responses to each face (see examples in Fig. 10.10). It would be difficult for most of these single cells to tell which of even five faces, let alone which of hundreds of faces, had been seen. (At the same time, the neurons discriminated between the faces reliably, as shown by the values of d', taken in the case of the neurons to be the number of standard deviations of the neuronal responses which separated the response to the best face in the set from that to the least effective face in the set. The values of d' were typically in the range 1–3.) The breadth of tuning (Smith and Travers, 1979) indices for these cells had values that were for the majority of neurons in the range 0.7–0.95.

In a recent study, the responses of another set of temporal cortical neurons to 23 faces and 42 non-face natural images were measured, and again a distributed representation was found (Rolls and Tovee, 1995a). The tuning was typically graded, with a range of different firing rates to the set of faces, and very little response to the non-face stimuli (see Fig. 10.11). The spontaneous firing rate of the neuron in Fig. 10.11 was 20 spikes/s, and the histogram bars indicate the change of firing rate from the spontaneous value, produced by each stimulus. Stimuli which are faces are marked F, or P if they are in profile. B refers to images of scenes which included either a small face within the scene, sometimes as part of an image which included a whole person, or other body parts, such as hands (H) or legs. The non-face stimuli are unlabelled. The neuron responded best to three of the faces (profile views), had some response to some of the other faces, and had little or no response, and sometimes had a small decrease of firing rate below the spontaneous firing rate, to the non-face stimuli. The sparseness value a for this cell across all 68 stimuli was 0.69, and the response sparseness a_r (based on the evoked responses minus the spontaneous firing of the neuron) was 0.19. It was found that the sparseness of the representation of the 68 stimuli by each neuron had an average across all neurons of 0.65 (Rolls and Tovee, 1995a). This indicates a rather distributed representation. (If neurons had a continuum of responses equally distributed between zero and maximum rate, a would be 0.75, while if the probability of each response decreased linearly, to reach zero at the maximum rate, a would be 0.67). If the spontaneous firing rate was subtracted from the firing

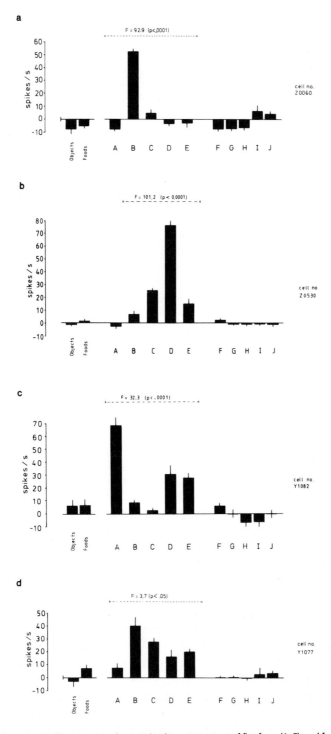

Fig. 10.10 Responses of four different temporal cortex visual neurons to a set of five faces (A–E), and for comparison to a wide range of non-face objects and foods. F–J are non-face stimuli. The means and standard errors of the responses computed over 8–10 trials are shown. (From Baylis, Rolls and Leonard, 1985.)

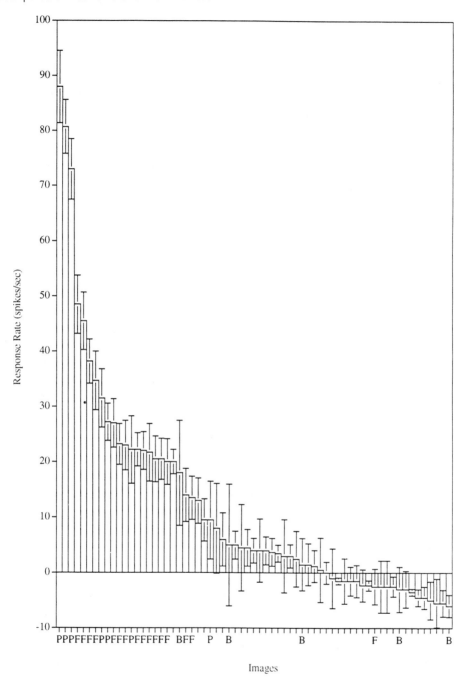

Fig. 10.11 The firing rates of a temporal visual cortex neuron to a set of 23 face stimuli and 42 non-face stimuli. Neuron am242. The firing rate of the neuron is shown on the ordinate, the spontaneous firing rate of the neuron was 20 spikes/s, and the histogram bars are drawn to show changes of firing rate from the spontaneous rate (i.e. neuronal responses) produced by each stimulus. Stimuli which are faces are marked F, or P if they are in profile. B refers to images of scenes which included either a small face within the scene, sometimes as part of an image which included a whole person, or other body parts, such as legs; and H is used if hands were a prominent part of such images. The non-face stimuli are unlabelled. (From Rolls, Treves, Tovee and Panzeri, 1997.)

rate of the neuron to each stimulus, so that the changes of firing rate, that is the *active responses* of the neurons, were used in the sparseness calculation, then the 'response sparseness' had a lower value, with a mean of 0.33 for the population of neurons, or 0.60 if calculated over the set of faces rather than over all the face and non-face stimuli. Thus the representation was rather distributed. (It is, of course, important to remember the relative nature of sparseness measures, which (like the information measures to be discussed below) depend strongly on the stimulus set used.) Thus we can reject a cardinal cell representation. As shown below, the readout of information from these cells is actually much better in any case than would be obtained from a local representation, and this makes it unlikely that there is a further population of neurons with very specific tuning that use local encoding.

10.4.2 The representation of information in the responses of single neurons to a set of stimuli

The use of an information theoretic approach (Appendix A2) makes it clear that there is considerable information in this distributed encoding of these temporal visual cortex neurons about which face was seen. Figure 10.12 shows typical firing rate changes on different trials to each of several different faces. This makes it clear that from the firing rate on any one trial, information is available about which stimulus was shown. In order to clarify the representation of individual stimuli by individual cells, Fig. 10.13 shows the information $I(s,R)$ available in the neuronal response about each of 20 face stimuli calculated for the neuron (am242) whose firing rate response profile to the set of 65 stimuli is shown in Fig. 10.11. Unless otherwise stated, the information measures given are for the information available on a single trial from the firing rate of the neuron in a 500 ms period starting 100 ms after the onset of the stimuli. It is shown in Fig. 10.13 that 2.2, 2.0, and 1.5 bits of information were present about the three face stimuli to which the neuron had the highest firing rate responses. The neuron conveyed some but smaller amounts of information about the remaining face stimuli. The average information $I(S,R)$ about this set (S) of 20 faces for this neuron was 0.55 bits. The average firing rate of this neuron to these 20 face stimuli was 54 spikes/s. It is clear from Fig. 10.13 that little information was available from the responses of the neuron to a particular face stimulus if that response was close to the average response of the neuron across all stimuli. At the same time, it is clear from Fig. 10.13 that information was present depending on how far the firing rate to a particular stimulus was from the average response of the neuron to the stimuli. Of particular interest, it is evident that information is present from the neuronal response about which face was shown if that neuronal response was below the average response, as well as when the response was greater than the average response.

One intuitive way to understand the data shown in Fig. 10.13 is to appreciate that low probability firing rate responses, whether they are greater than or less than the mean response rate, convey much information about which stimulus was seen. This is of course close to the definition of information (see Appendix A2). Given that the firing rates of neurons are always positive, and follow an asymmetric distribution about their mean, it is clear that deviations above the mean have a different probability to occur than deviations by the same amount, below the mean. One may attempt to capture the relative likelihood of different firing rates above and below the mean by computing a z score obtained by dividing the difference between the mean response to each stimulus and the overall mean response by the standard

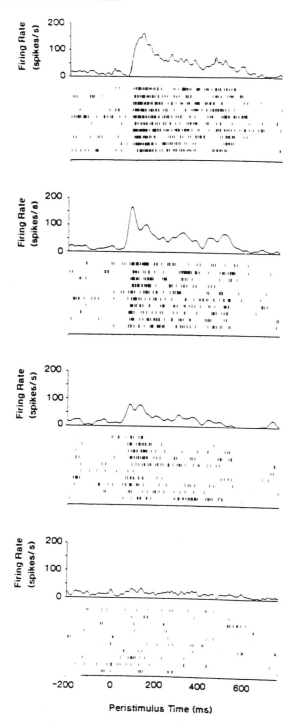

Fig. 10.12 Peristimulus time histograms and rastergrams showing the responses on different trials (originally in random order) of a face-selective neuron to four different faces. (In the rastergrams each vertical line represents one spike from the neuron, and each row is a separate trial.)

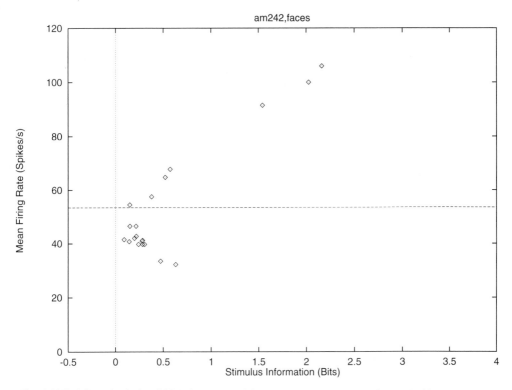

Fig. 10.13 The information $I(s,R)$ available in the response of the same neuron as in Fig. 10.11 about each of the stimuli in the set of 20 face stimuli (abscissa), with the firing rate of the neuron to the corresponding stimulus plotted as a function of this on the ordinate. (From Rolls, Treves, Tovee and Panzeri, 1997.)

deviation of the response to that stimulus. The greater the number of standard deviations (i.e. the greater the z score) from the mean response value, the greater the information might be expected to be. We therefore show in Fig. 10.14 the relation between the z score and $I(s,R)$. (The z score was calculated by obtaining the mean and standard deviation of the response of a neuron to a particular stimulus s, and dividing the difference of this response from the mean response to all stimuli by the calculated standard deviation for that stimulus.) This results in a C-shaped curve in Figs. 10.13 and 10.14, with more information being provided by the cell the further its response to a stimulus is in spikes per second or in z scores either above *or below* the mean response to all stimuli (which was 54 spikes/s). The specific C-shape is discussed further in Appendix 2.

We show in Fig. 10.15 the information $I(s,R)$ about each stimulus in the set of 65 stimuli for the same neuron, am242. The 23 face stimuli in the set are indicated by a diamond, and the 42 non-face stimuli by a cross. Using this much larger and more varied stimulus set, which is more representative of stimuli in the real world, a C-shaped function again describes the relation between the information conveyed by the cell about a stimulus and its firing rate to that stimulus. In particular, this neuron reflected information about most, but not all, of the faces in the set, that is those faces that produced a higher firing rate than the overall mean firing rate to all the 65 stimuli, which was 31 spikes/s. In addition, it conveyed information about the majority of the 42 non-face stimuli by responding at a rate below the overall mean response of the neuron to the 65 stimuli. This analysis usefully makes the point that the information available in the neuronal responses about which stimulus was shown is relative to (dependent upon) the nature and range of stimuli in the test set of stimuli.

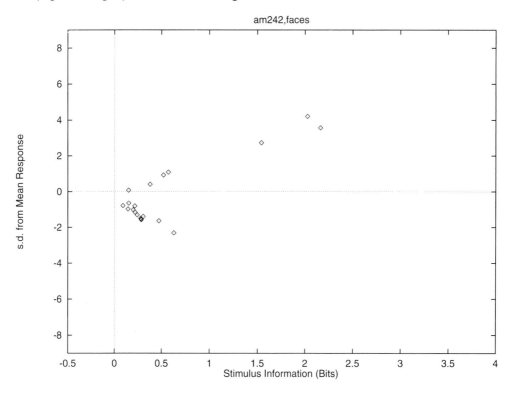

Fig. 10.14 The relation between the number of standard deviations the response to a stimulus was from the average response to all stimuli (see text, z score) plotted as a function of $I(s,R)$, the information available about the corresponding stimulus, s. (From Rolls, Treves, Tovee and Panzeri, 1997, Fig. 2c.)

This evidence makes it clear that a single cortical visual neuron tuned to faces conveys information not just about one face, but about a whole set of faces, with the information conveyed on a single trial related to the difference in the firing rate response to a particular stimulus compared to the average response to all stimuli.

The actual distribution of the firing rates to a wide set of natural stimuli is of interest, because it has a rather stereotypical shape, typically following a graded unimodal distribution with a long tail extending to high rates (see for example Fig. 10.16). The mode of the distribution is close to the spontaneous firing rate, and sometimes it is at zero firing. If the number of spikes recorded in a fixed time window is taken to be constrained by a fixed maximum rate, one can try to interpret the distribution observed in terms of optimal information transmission (Shannon, 1948), by making the additional assumption that the coding is noiseless. An exponential distribution, which maximizes entropy (and hence information transmission for noiseless codes) is the most efficient in terms of energy

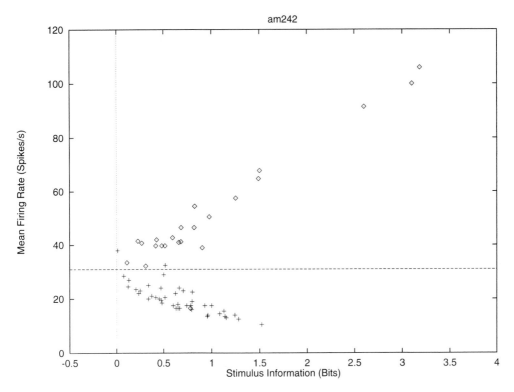

Fig. 10.15 The information $I(s,R)$ available in the response of the same neuron about each of the stimuli in the set of 23 face and 42 non-face stimuli (abscissa), with the firing rate of the neuron to the corresponding stimulus plotted as a function of this on the ordinate. The 23 face stimuli in the set are indicated by a diamond, and the 42 non-face stimuli by a cross. (From Rolls, Treves, Tovee and Panzeri, 1997, Fig. 2d.)

consumption if its mean takes an optimal value which is a decreasing function of the relative metabolic cost of emitting a spike (Levy and Baxter, 1996). This argument would favour sparser coding schemes the more energy expensive neuronal firing is (relative to rest). Although the tail of actual firing rate distributions is often approximately exponential (see for example Fig. 10.16; Baddeley *et al.*, 1998; Rolls, Treves, Tovee and Panzeri, 1997), the maximum entropy argument cannot apply as such, because noise is present and the noise level varies as a function of the rate, which makes entropy maximization different from information maximization. Moreover, a mode at low but non-zero rate, which is often observed, is inconsistent with the energy efficiency theorem. A simpler explanation for the stereotypical distribution arises by considering that the value of the activation of a neuron across stimuli, reflecting a multitude of contributing factors, will typically have a Gaussian distribution; and by considering a physiological input–output transform and realistic noise levels. In fact, an input–output transform which is supralinear in a range above threshold results from a fundamentally linear transform and fluctuations in the activation, and produces a variance in the output rate, across repeated trials, that increases with the rate itself, consistent with common observations. At the same time, such a supralinear transform tends to convert the Gaussian tail of the activation distribution into

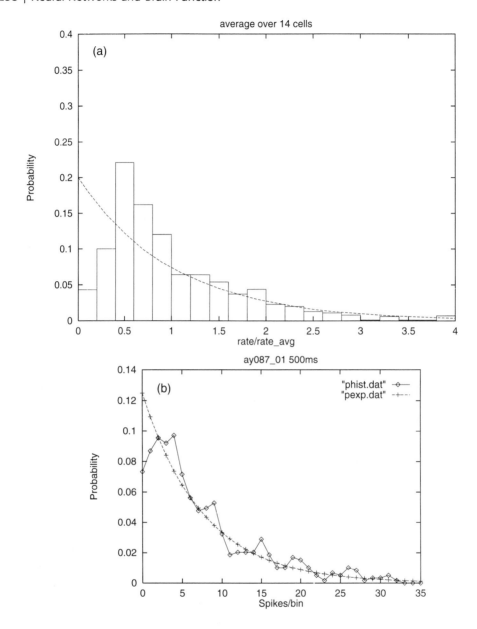

Fig. 10.16 (a) The distribution of the firing rates of a temporal visual cortex neuron to a set of visual stimuli. The frequency or probability of the different firing rates being produced is shown. The neuron responded to some faces, and the stimulus set consisted of 23 face stimuli and 42 non-face stimuli. An exponential curve is fitted to the actual distribution. (After Rolls, Treves, Tovee and Panzeri, 1997.) (b) The probability of different firing rates of a temporal cortex neuron calculated over a 5 min video showing natural scenes, including faces.

an approximately exponential tail, without implying a fully exponential distribution with the mode at zero. Such basic assumptions yield excellent fits with observed distributions (Panzeri *et al.*, 1996a; Treves *et al.*, 1998).

The observed distributions often differ from exponential in that there are too few very low rates observed, and too many low rates (Rolls, Treves, Tovee and Panzeri, 1997). This peak at low but non-zero rates may be related to the low firing rate spontaneous activity which is typical of many cortical neurons. Keeping the neurons close to threshold in this way may maximize the speed with which a network can respond to new inputs (because time is not required to bring the neurons from a strongly hyperpolarized state up to threshold). The advantage of having low spontaneous firing rates may be a further reason why a curve such as an exponential cannot sometimes be fitted to the experimental data.

10.4.3 The representation of information in the responses of a population of cortical visual neurons

Complementary evidence comes from applying information theory to analyse how information is represented by a population of these neurons. Figure 10.17 shows that if we know the average firing rate of each cell in a population to each stimulus, then on any single trial we can guess the stimulus that was present by taking into account the response of all the cells. We can expect that the more cells in the sample, the more accurate may be the estimate of the stimulus. If the encoding was local, the number of stimuli encoded by a population of neurons would be expected to rise approximately linearly with the number of neurons in the population. In contrast, with distributed encoding, provided that the neuronal responses are sufficiently independent, and are sufficiently reliable (not too noisy), information from the ensemble would be expected to rise linearly with the number of cells in the ensemble, and (as information is a log measure) the number of stimuli encodable by the population of neurons might be expected to rise exponentially as the number of neurons in the sample of the population was increased. The information available about which of 20 equiprobable faces had been shown that was available from the responses of different numbers of these neurons is shown in Fig. 8.8 (Rolls, Treves and Tovee, 1997; Abbott, Rolls and Tovee, 1996). First, it is clear that some information is available from the responses of just one neuron: on average approximately 0.34 bits. Thus, knowing the activity of just one neuron in the population does provide some evidence about which stimulus was present. This evidence that information is available in the responses of individual neurons in this way, without having to know the state of all the other neurons in the population, indicates that information is made explicit in the firing of individual neurons in a way that will allow neurally plausible decoding, involving computing a sum of input activities each weighted by synaptic strength, to work (see below, Section 10.4.4.2). Second, it is clear (Fig. 8.8) that the information rises approximately linearly, and the number of stimuli encoded thus rises approximately exponentially, as the number of cells in the sample increases (Rolls, Treves and Tovee, 1997). It must be mentioned that the data described were obtained without simultaneous recording of neuronal activity. It is likely that simultaneous recordings would allow more information to be extracted (because, for example, enhanced attention might increase the responses of all neurons for a few trials, and this would be invisible to dot product decoding), but it is just possible that there would be less information with simultaneous recordings. The experiment should be performed. The

How well can one predict which stimulus was
shown on a single trial from the mean responses
of different neurons to each stimulus?

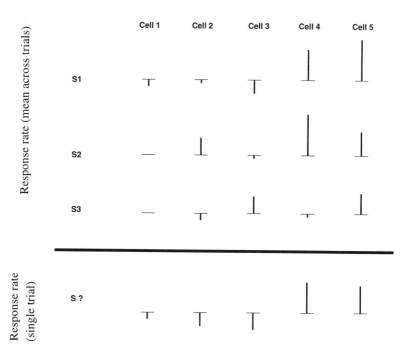

Fig. 10.17 This diagram shows the average response for each of several cells (Cell 1 etc.) to each of several stimuli (S1 etc.). The change of firing rate from the spontaneous rate is indicated by the vertical line above or below the horizontal line which represents the spontaneous rate. We can imagine guessing from such a table the stimulus S? that was present on any one trial (see text).

preliminary indication, from simultaneously recorded pairs of temporal cortex visual neurons, is that the information they convey is relatively independent (Gawne and Richmond, 1993).

This direct neurophysiological evidence thus demonstrates that the encoding is distributed, and the responses are sufficiently independent and reliable that the representational capacity increases exponentially. The consequence of this is that large numbers of stimuli, and fine discriminations between them, can be represented without having to measure the activity of an enormous number of neurons. Although the information rises approximately linearly with the number of neurons when this number is small, gradually each additional neuron does not contribute as much as the first. In the sample analysed by Rolls, Treves and Tovee, 1997, the first neuron contributed 0.34 bits, on average, with 3.23 bits available from the 14 neurons analysed. This reduction is, however, exactly what could be expected to derive from a simple ceiling effect, in which the ceiling is just the information in the stimulus set, or $\log_2 20 = 4.32$ bits, as discussed further in Appendix A2 (cf Fig. 8.8). This indicates that, on the one hand, each neuron does not contribute independently to the sum, and there is some overlap or redundancy in what is contributed by each neuron; and that, on the other hand, the degree of redundancy is not a property of the neuronal representation, but just a contingent feature dependent on the particular set of stimuli used in probing that representation. The data available is consistent with the hypothesis, explored by Abbott, Rolls and Tovee (1996) through simulations, that if the ceiling provided by the limited number of stimuli that could be presented were at much higher levels, each neuron would continue to contribute as much as the first few, up to much larger neuronal populations, so that the number of stimuli that could be encoded would still continue to increase exponentially even with larger numbers of neurons (Fig. 1.6; Abbott, Rolls and Tovee, 1996). The redundancy observed could be characterized as flexible, in that it is the task that determines the degree to which large neuronal populations need to be sampled. If the task requires discriminations with very fine resolution in between many different stimuli (i.e. in a high-dimensional space), then the responses of many neurons must be taken into account. If very simple discriminations are required (requiring little information), small subsets of neurons or even single neurons may be sufficient. The importance of this type of flexible redundancy in the representation is discussed below.

We believe that the same type of ensemble encoding of what stimulus is present (i.e. stimulus identity) is likely to be used in other sensory systems, and have evidence that this is the case for the primate taste and olfactory systems, in particular for the cortical taste and olfactory areas in the orbitofrontal cortex (Rolls, Critchley and Treves, 1997). This type of ensemble encoding is also used in the primate hippocampus, in that the information about which spatial view is being seen rises approximately linearly with the number of hippocampal neurons in the sample (Rolls, Treves, Robertson, Georges-François and Panzeri, 1997).

10.4.4 Advantages of the distributed representations found of objects for brain processing

Three key types of evidence that the visual representation provided by neurons in the temporal cortical areas, and the olfactory and taste representations in the orbitofrontal cortex, are distributed have been provided, and reviewed above. One is that the coding is not sparse (Baylis, Rolls and Leonard, 1985; Rolls and Tovee, 1995; Rolls, 1989a, 1995a). The second is that different neurons have different response profiles to a set of stimuli, and thus have at least partly independent responses (Baylis, Rolls and Leonard, 1985; Rolls and Tovee, 1995a; Rolls, 1989a, 1995a, 1997c). The third is that the capacity of the representations rises exponentially with the number of neurons (Rolls, Treves and Tovee, 1997). The advantages of such distributed encoding are now considered, and apply to both fully distributed and to more sparse (but not to local) encoding schemes.

10.4.4.1. Exponentially high coding capacity

This property arises from a combination of the encoding being sufficiently close to independent by the different neurons (i.e. factorial), and sufficiently distributed. The independence or factorial requirement is simply to ensure that the information $I(n)$ from the population of neurons rises linearly with the number of neurons in the population. (The requirement for nearly factorial codes is evident when considering the expressions in

Appendix A2 for the information carried by neuronal responses: for the number of stimuli coded to rise exponentially, the information must rise linearly, or at least there must be a linear term in the rise of information, with the number n of cells sampled, that is:

$$I\ (n) = \text{average information per neuron} \cdot n + \text{subleading terms.})$$

The requirement for distributed codes is equivalent to demanding that the average information provided by each neuron (on any one trial) is not negligibly small. For example, if a binary unit is active only once every million stimuli, $P_i = 0.000001$, it could provide, even when contributing to a fully factorial coding, only 0.00002 bits of information, and the exponential rise in the number of coded stimuli would theoretically still hold, on average, but at such a slow pace (1 full bit, or the encoding of 2 stimuli, requiring 50,000 units!) as to be irrelevant. We note that if local encoding were used, the information would increase in proportion to the logarithm of the number of cells, which is not what has been found.

Part of the biological significance of such exponential encoding capacity is that a receiving neuron or neurons can obtain information about which one of a very large number of stimuli is present by receiving the rate of firing of relatively small numbers of inputs from each of the neuronal populations from which it receives. In particular, if neurons received from something in the order of 100 inputs from the population described here, they would have a great deal of information about which stimulus was in the environment. In particular, the characteristics of the actual visual cells described here indicate that the activity of 15 would be able to encode 192 face stimuli (at 50% accuracy), of 20 neurons 768 stimuli, of 25 neurons 3072 stimuli, of 30 neurons 12288 stimuli, and of 35 neurons 49152 stimuli (Abbott, Rolls and Tovee, 1996; the values are for the optimal decoding case). Given that most neurons receive a limited number of synaptic contacts, in the order of several thousand, this type of encoding is ideal. It would enable, for example, neurons in the amygdala and orbitofrontal cortex to form pattern associations of visual stimuli with reinforcers such as the taste of food when each neuron received a reasonable number, perhaps in the order of hundreds, of randomly assigned inputs from the visually responsive neurons in the temporal cortical visual areas which specify which visual stimulus or object is being seen (see Rolls, 1990a, 1992b, 1994a, 1995b). Such a representation would also be appropriate for interfacing to the hippocampus, to allow an episodic memory to be formed, so that for example a particular visual object was seen in a particular place in the environment (Rolls, 1989b–e; Treves and Rolls, 1994). Here we should emphasize that although the sensory representation may have exponential encoding capacity, this does not mean that the associative networks that receive the information can store such large numbers of different patterns. Indeed, there are strict limitations on the number of memories that associative networks can store (Chapters 2, 3 and Appendices A3 and A4). The particular value of the exponential encoding capacity of sensory representations is that very fine discriminations can be made, as there is much information in the representation, and that the representation can be decoded if the activity of even a limited number of neurons in the representation is known.

One of the underlying themes here is the neural representation of objects. How would one know that one has found a neuronal representation of objects in the brain? The criterion we suggest that arises from this research (Rolls, 1994a) is that when one can identify the object or stimulus that is present (from a large set of stimuli, perhaps thousands or more) with a realistic number of neurons, say in the order of 100, then one has a representation of the object. This criterion appears to imply exponential encoding, for only then could such a large number of stimuli be represented with a relatively small number of units, at least for units with the response characteristics of actual neurons. (In artificial systems a few multilevel errorless units could represent a much larger set of objects, but in a non-neural-like way.) Equivalently, we can say that there is a representation of the object when the information required to specify which of many stimuli or objects is present can be decoded from the responses of a limited number of neurons.

We may note at this point that an additional criterion for an object representation is that the representation of the stimulus or object readable from the ensemble of neurons should show at least reasonable invariance with respect to a number of transforms which do not affect the identity of the object. In the case of the visual representation these invariances include translation (shift), size, and even view invariance. These are transforms to which the responses of some neurons in the temporal cortical visual areas are robust or invariant (see Rolls, 1994a, 1995b, 1997d; Tovee, Rolls and Azzopardi, 1994; Rolls, Booth and Treves, 1996). To complete the example, we can make it clear that although information about visual stimuli passes through the optic nerve from the retina, the representation at this level of the visual system is not of objects, for no decoding of a small set of neurons in the optic nerve would provide information in an invariant way about which of many objects was present on the retina.

The properties of the representation of faces, and of olfactory and taste stimuli, have been evident when the readout of the information was by measuring the firing rate of the neurons, typically over a 500 ms period. Thus, at least where objects are represented in the visual, olfactory, and taste systems (for example individual faces, objects, odours, and tastes), information can be read out without taking into account any aspects of the possible temporal synchronization between neurons (Engel et al., 1992), or temporal encoding within a spike train (Tovee et al., 1993; Rolls, Treves and Tovee, 1997).

One question which arises as a result of this demonstration of exponential encoding capacity is that of why there are so many neurons in the temporal cortical visual areas, if so few can provide such a high capacity representation of stimuli. One answer to this question is that high information capacity is needed for fine discrimination. Another point is that the 14 cells analysed to provide the data shown in Figs. 8.8, 10.18, and 1.6, or the 38 olfactory cells analysed by Rolls, Treves and Critchley (1997), were a selected subset of the cells in the relevant brain regions. The subsets were selected on the basis of the cells individually providing significant information about the stimuli in the set of visual, olfactory, or taste stimuli presented. If a random sample of say temporal cortical visual neurons had been taken, then that sample would have needed to be one to two orders of magnitude larger to include the subset of neurons in the sample. It is likely that the ensemble of neurons that projects to any particular cell in a receiving area is closer to a random sample than to our selected sample.

10.4.4.2. Ease with which the code can be read by receiving neurons: the compactness of the distributed representation

For brain plausibility, it would also be a requirement that the decoding process should itself not demand more than neurons are likely to be able to perform. This is why when we have estimated the information from populations of neurons, we have used in addition to a probability estimation (optimal, in the Bayesian sense) method, a dot product measure, which is a way of specifying that all that is required of decoding neurons would be the property of adding up postsynaptic potentials produced through each synapse as a result of the activity of each incoming axon (Rolls, Treves and Tovee, 1997). More formally, the way in which the activation h of a neuron would be produced is by the following principle:

$$h = \Sigma_j \, r'_j \, w_j$$

where r'_j is the firing of the jth axon, and w_j is the strength of its synapse. The firing r of the neuron is a function of the activation

$$r = f\,(h).$$

This output activation function f may be linear, sigmoid, binary threshold, etc.

It was found that with such a neurally plausible algorithm (the dot product, DP, algorithm), which calculates which average response vector the neuronal response vector on a single test trial was closest to by performing a normalized dot product (equivalent to measuring the angle between the test and the average vector), the same generic results were obtained, with only at most a 40% reduction of information compared to the more efficient (optimal) algorithm. This is an indication that the brain could utilize the exponentially increasing capacity for encoding stimuli as the number of neurons in the population increases. For example, by using the representation provided by the neurons described here as the input to an associative or autoassociative memory, which computes effectively the dot product on each neuron between the input vector and the synaptic weight vector, most of the information available would in fact be extracted (see Rolls and Treves, 1990; Treves and Rolls, 1991).

Such a dot product decoding algorithm requires that the representation be formed of linearly separable patterns. (Such a decoder might consist of neurons in associative networks in the next brain region such as the orbitofrontal cortex, amygdala, or hippocampus.) Moreover, the representation should be in a form in which approximately equal numbers of neurons are active for each pattern (i.e. a constant sparseness of the representation), and similar patterns should be represented by similar subsets of active neurons. This implies that the representation must not be too compact. Consider, for example, the binary encoding of numbers (0 = 0000; 1 = 0001; 2 = 0010; 3 = 0011; 4 = 0100 . . . 7 = 0111; 8 = 1000 etc.). Now take the encoding of the numbers 7 (0111) and 8 (1000). Although these numbers are very similar, the neurons that represent these numbers are completely different, and the sparseness of the representation fluctuates wildly. So although such binary encoding is optimal in terms of the number of bits used, and has exponential encoding capacity, dot product decoding could not be used to decode it, and the number of neurons active for any stimulus fluctuates wildly. Also, for such binary encoding, a receiving neuron would need to receive from all the neurons in the representation; and generalization and graceful degradation would not occur. This indicates that a neuronal code must not be too compact in order to possess all these properties of generalization, constant sparseness, dot product decoding, and exponential capacity. The use of this type of neuronal code by the brain is one of the factors that enables single neuron recording to be so useful in understanding brain function—a correlation can frequently be found between the activity of even a single neuron and a subset of the stimuli being shown, of the motor responses being made, etc.

10.4.4.3. Higher resistance to noise

This, like the next few properties, is in general an advantage of distributed over local representations, which applies to artificial systems as well, but is presumably of particular value in biological systems in which some of the elements have an intrinsic variability in their operation. An example might be the stochastic release of transmitter at central synapses, where the probability of release after each action potential is thought to be rather low (Hessler, Shirke and Malinow, 1993). Because the decoding of a distributed representation involves assessing the activity of a whole population of neurons, and computing a dot product or correlation, a distributed representation provides more resistance to variation in individual components than does a local encoding scheme.

10.4.4.4. Generalization

Generalization to similar stimuli is again a property that arises in neuronal networks if distributed but not if local encoding is used, and when dot product operation is involved (e.g. see Chapters 2 and 3). The distributed encoding found in the cerebral cortex allows this generalization to occur.

10.4.4.5. Completion

Completion occurs in associative memory networks by a similar process (e.g. see Chapters 2 and 3). The distributed encoding found in the cerebral cortex is appropriate for allowing such completion to occur.

10.4.4.6. Graceful degradation or fault tolerance

This also arises only if the input patterns have distributed representations, and not if they are local (e.g. see Chapters 2 and 3). The distributed encoding found in the cerebral cortex is appropriate for allowing such graceful degradation including tolerance to missing connections to occur.

10.4.4.7. Speed of readout of the information

The information available in a distributed representation can be decoded by an analyzer more quickly than can the information from a local representation, given comparable firing rates.

Within a fraction of an interspike interval, with a distributed representation, much information can be extracted (Treves, 1993; Rolls, Treves, Tovee, 1997; see Appendix A2). An example of how rapidly information can be made available in short time periods from an ensemble of neurons is shown in Fig. 10.18. In effect, spikes from many different neurons can contribute to calculating the angle between a neuronal population and a synaptic weight vector within an interspike interval. With local encoding, the speed of information readout depends on the exact model used, but if the rate of firing needs to be taken into account, this will necessarily take time, because of the time needed for several spikes to accumulate in order to estimate the firing rate. It is likely with local encoding that the firing rate of a neuron would need to be measured to some degree of accuracy, for it seems implausible to suppose that a single spike from a single neuron would be sufficient to provide a noise-free representation for the next stage of processing.

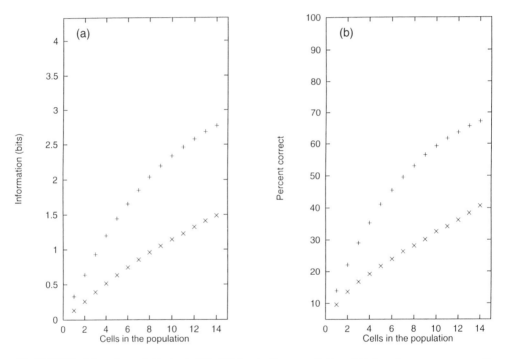

Fig. 10.18 (a) The information available about which of 20 faces had been seen that is available from the responses measured by the firing rates in a time period of 500 ms (+) or a shorter time period of 50 ms (×) of different numbers of temporal cortex cells. (b) The corresponding percentage correct from different numbers of cells. (From Rolls, Treves and Tovee, 1997.)

We may note that with integrate-and-fire neurons (see Appendix A5), there will not be a great delay in reading the information, provided that the neurons are kept close to threshold, so that even a small additional input will cause them to fire earlier than otherwise. The fact that many cortical neurons show spontaneous activity may reflect the fact that cortical neurons are indeed kept close to threshold, so that at least some of them can respond quickly when an input is received.

10.4.5 Should one neuron be as discriminative as the whole organism, in object encoding systems?

In the analysis of random dot motion with a given level of correlation among the moving dots, single neurons in area MT in the dorsal visual system of the primate can be approximately as sensitive or discriminative as the psychophysical performance of the whole animal (Zohary *et al.*, 1994). The arguments and evidence presented here suggest that this is not the case for the ventral visual system, concerned with object identification. Why should there be this difference? We suggest that the dimensionality of what is being computed may account for the difference. In the case of visual motion (at least in the study referred to), the problem was effectively one-dimensional, in that the direction of motion of the stimulus along a line in 2D space was extracted from the activity of the neurons. In this low-dimensional stimulus space, the neurons may each perform one of a few similar computations on a particular (local) portion of 2D space, with the side effect that by averaging one can extract a signal of a global nature. Indeed, the average of the neuronal activity can be used to obtain the average direction of motion, but it is clear that most of the parallel computational power in this system has been devoted to sampling visual space and analysing local motion, rather than being geared to an effective solution of the particular problem of detecting the global average of random dot motion.

In contrast, in the higher dimensional space of objects, in which there are very many different objects to represent as being different from each other, and in a system which is not concerned with location in visual space but on the contrary tends to be relatively invariant with respect to location, the goal of the representation is to reflect the many aspects of the input information in a way that enables many different objects to be represented, in what is effectively a very high dimensional space. This is achieved by allocating cells, each with an intrinsically limited discriminative power, to sample as thoroughly as possible the many dimensions of the space. Thus the system is geared to use efficiently the parallel computations of all its units precisely for tasks such as that of face discrimination, which was used as an experimental probe. Moreover, object representation must be kept higher dimensional, in that it may have to be decoded by dot-product decoders in associative memories, in which the input patterns must be in as high a dimensional space as possible (i.e. the activity on different input axons should not be too highly correlated). In this situation, each neuron should act somewhat independently of its neighbours, so that each provides its own separate contribution which adds together with that of the other neurons (in a linear manner, see above and Figs. 8.8, 10.18 and 1.6) to provide *in toto* sufficient information to specify which out of perhaps several thousand visual stimuli was seen. The computation involves in this case not an average of neuronal activity, but instead comparing the dot product of the activity of the population of neurons with a previously learned vector, stored in, for example, associative memories as the weight vector on a receiving neuron or neurons.

Zohary, Shadlen and Newsome (1994) put another argument which suggested to them that the brain could hardly benefit from taking into account the activity of more than a very limited number of neurons. The argument was based on their measurement of a small (0.12) correlation between the activity of simultaneously recorded neurons in area MT. They suggested that there would because of this be decreasing signal-to-noise ratio advantages as more neurons were included in the population, and that this would limit the number of neurons that it would be useful to decode to approximately 100. However, a measure of correlations in the activity of different neurons depends entirely on the way the space of neuronal activity is sampled, that is on the task chosen to probe the system. Among face cells in the temporal cortex, for example, much higher correlations would be observed when the task is a simple two-way discrimination between a face and a non-face, than when the task involves finer identification of several different faces. (It is also entirely possible that some face cells could be found that perform as well in a given particular face–non-face discrimination as the whole animal.) Moreover, their argument depends on the type of decoding of the activity of the population that is envisaged. It implies that the average of the neuronal activity must be estimated accurately. If a set of neurons uses dot-product decoding, and then the activity of the decoding population is scaled or normalized by some negative feedback through inhibitory interneurons, then the effect of such correlated firing in the sending population is reduced, for the decoding effectively measures the relative firing of the different neurons in the population to be decoded. This is equivalent to measuring the angle between the current vector formed by the population of neurons firing, and a previously learned vector, stored in synaptic weights. Thus, with for example this biologically plausible decoding, it is not clear whether the correlation Zohary *et al.* describe would place a severe limit on the ability of the brain to utilize the information available in a population of neurons.

The main conclusion from this section is that the information available from a set or ensemble of temporal cortex visual neurons increases approximately linearly as more neurons are added to the sample. This is powerful evidence that distributed encoding is used by the brain; and the code can be read just by knowing the firing rates in a short time of the population of neurons. The fact that the code can be read off from the firing rates, and by a principle as simple and neuron-like as dot product decoding, provides strong support for the general approach taken in this book to brain function. It is possible that more information would be available in the relative time of occurrence of the spikes, either within the spike train of a single neuron, or between the spike trains of different neurons, and it is to this that we now turn.

10.4.6 Temporal encoding in a spike train of a single neuron

It is possible that there is information contained in the relative time of firing of spikes within the spike train of a single neuron. For example, stimulus 1 might regularly elicit a whole burst of action potentials in the period 100–150 ms after the stimulus is shown. Stimulus 2 might regularly elicit a whole burst of action potentials between say 150–200 ms, and 350–400 ms. If one took into account when the action potentials occurred, then one might have more information from the spike train than if one took only the mean number of action potentials over the same 100–500 ms period. This possibility was investigated in a pioneering set of investigations by Optican and Richmond (1987) (see also Richmond and Optican, 1987) for inferior temporal cortex neurons. They used a set of black and white (Walsh) patterns as the stimuli. The way they assessed temporal encoding was by setting up the spikes to each stimulus as a time series with, for example, 64 bins each 6 ms long. With the response to each stimulus on each trial as such a time series, they calculated the first few principal components of the variance of spike trains across trials. The first principal component of the variance was in fact similar to the average response to all stimuli, and the second and higher components had waveforms that reflected more detailed differences in the response to the different stimuli. They found that several principal components were needed to capture most of the variance in their response set. They went on to calculate the information about which stimulus was shown that was available from the first principal component (which was similar to that in the firing rate), and if the second, third, etc. principal components were added. Their results appeared to indicate that considerably more information was available if the second principal component was added (50% more), and more if the third was added (37% more). They interpreted this as showing that temporal encoding in the spike train was used.

In subsequent papers, the same authors (Richmond and Optican, 1990; Optican *et al.*, 1991; Eskandar *et al.*, 1992) realized the need for a refinement of the analysis procedure, and independently also Tovee, Rolls, Treves and Bellis (1993) reinvestigated this issue, using different faces as stimuli while recording from temporal cortical face-selective neurons. They showed that a correction needed to be made in the information analysis routine to correct for the limited number of trials (see Appendix A2). When the correction was incorporated, Tovee *et al.* (1993) found that rather little additional information was available if the second (19% more) and third (8% more) principal components were used in addition to the first. Moreover, they showed that if information was available in the second principal component in their data, it did not necessarily reflect interesting evidence about which stimulus had been seen, but frequently reflected latency differences between the different stimuli. (For example, some of their stimuli were presented parafoveally, and the longer onset latency that this produced was often reflected in the second principal component.)

The conclusion at this stage was that little additional evidence about which stimulus had been presented was available if temporal aspects of the spike train were taken into account. To pursue the point further, Rolls and colleagues next reasoned that if information was encoded in the time of arrival of spikes over a 400 ms period, much less information would be available if a short temporal epoch was taken. There would simply not be time within a short temporal epoch for the hypothesized characteristic pattern of firing to each stimulus to be evident. Moreover, it might well matter when the short temporal epoch was taken. Tovee *et al.* (1993) and Tovee and Rolls (1995) therefore analysed the information available in short temporal epochs (for example 100, 50, and 20 ms) of the spike train. They found that considerable information was present even with short epochs. For example, in periods of 20 ms, 30% of the information present in 400 ms using temporal encoding with the first three principal components was available. Moreover, the exact time when the epoch was taken was not crucial, with the main effect being that rather more information was available if information was measured near the start of the spike train, when the firing rate of the neuron tended to be highest (see Fig. 10.19). The conclusion was that much information was available when temporal encoding could not be used easily, that is in very short time epochs of 20 or 50 ms. (It is also useful to note from Figs 10.19 and 10.12 the typical time course of the responses of many temporal cortex visual neurons in the awake behaving primate. Although the firing rate and availability of information is highest in the first 50–100 ms of the neuronal response, the firing is overall well sustained in the 500 ms stimulus presentation period. Cortical neurons in the temporal lobe visual system, in the taste cortex, and in the olfactory cortex, do not in general have rapidly adapting neuronal responses to sensory stimuli. This may be important for associative learning: the outputs of these sensory systems can be maintained for sufficiently long while the stimuli are present for synaptic modification to occur. Although rapid synaptic adaptation within a spike train is seen in some experiments in brain slices (Markram and Tsodyks, 1996; Abbott *et al.*, 1997), it is not a very marked effect in at least some brain systems *in vivo*, when they operate in normal physiological conditions with normal levels of acetylcholine, etc.)

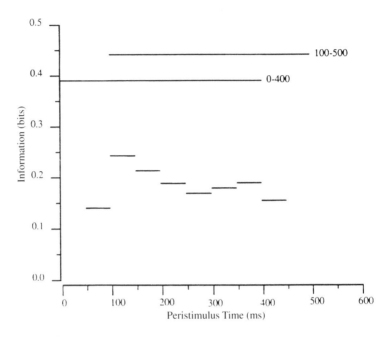

Fig. 10.19 The average information *I*(*S,R*) available in short temporal epochs of the spike trains of single neurons about which face had been shown. (From Tovee and Rolls, 1995.)

To pursue this issue even further, Rolls, Tovee, Purcell *et al.* (1994) and Rolls and Tovee (1994) limited the period for which visual cortical neurons could respond by using backward masking. In this paradigm, a short (16 ms) presentation of the test stimulus (a face) was followed after a delay of 0, 20, 40, 60, etc. ms by a masking stimulus (which was a high contrast set of letters). They showed that the mask did actually interrupt the neuronal response, and that at the shortest interval between the stimulus and the mask (a delay of 0 ms, or a 'stimulus onset asynchrony' of 20 ms), the neurons in the temporal cortical areas fired for approximately 30 ms (see Fig. 10.20). Under these conditions, the subjects could identify which of five faces had been shown much better than chance. Thus when the possibility of temporal encoding within a spike train was minimized by allowing temporal cortex visual neurons to fire for only approximately 30 ms, visual identification was still possible (though perception was certainly not perfect).

Thus most of the evidence indicates that temporal encoding within the spike train of a single neuron does not add much to the information that is present in the firing rate of the neuron. Most of the information is available from the firing rate, even when short temporal

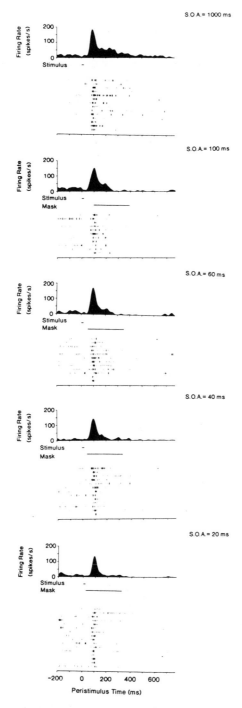

Fig. 10.20 Firing of a temporal cortex cell to a 20 ms presentation of a face stimulus when the face was followed with different Stimulus Onset Asynchronies (SOAs) by a masking visual stimulus. At an SOA of 20 ms, when the mask immediately followed the face, the neuron fired for only approximately 30 ms, yet identification of the face at this SOA by human observers was possible (Rolls and Tovee, 1994; Rolls, Tovee, Purcell et al., 1994.)

epochs are taken. Thus a neuron in the next cortical area would obtain considerable information within 20–50 ms of measuring the firing rate of a single neuron. Moreover, if it took a short sample of the firing rate of many neurons in the preceding area, then very much information is made available, as shown in the preceding section.

10.4.7 Temporal synchronization of the responses of different cortical neurons

Von der Malsburg (1973; see 1990 and von der Malsburg and Schneider, 1986) has proposed that information might be encoded by the relative time of firing of different neurons. In particular, he suggested this to help with the feature binding problem, as discussed in Section 8.7. Temporal synchronization of one subset of neurons, and separate temporal synchronization of a different subset, might allow the activity of different subsets of neurons to be kept apart. The decoding process might be performed by neurons with a strong sensitivity to temporal co-occurrence of inputs (which is a property of neurons). Singer, Engel, Konig and colleagues (see Engel *et al.*, 1991, 1992; Gray *et al.*, 1989, 1992) have obtained some evidence that when features must be bound, synchronization of neuronal populations can occur. At first the evidence suggested that oscillations might occur in the visual system during object recognition. However, these oscillations were most evident in the primary visual cortex of the anaesthetized cat, and were best evoked by moving stimuli. Indeed, the frequency of the oscillations was a function of the velocity of the moving stimulus. In the auditory cortex, the action potentials of neurons can become synchronized (deCharms and Merzenich, 1996). Tovee and Rolls (1992) found no evidence, however, for oscillations in neurons or groups of neurons in the temporal cortical visual areas of awake behaving macaques performing a visual fixation task while being shown sets of stationary face and non-face stimuli. Thus at least for form perception of static stimuli in primates, oscillations may not be necessary. Oscillations are likely to occur in any circuit with positive feedback. One possibility is that in evolution the feedback systems for maintaining cortical stability have improved, and that in line with this oscillations are less evident in the awake behaving monkey than in the anaesthetized cat. Oscillations may also be less likely to occur in systems processing static stimuli, such as what an object is that is presented for at least 0.5 s, than in systems concerned with processing dynamically changing stimuli, such as the motion of visual stimuli.

Engel *et al.* (1991, 1992) do, however, present impressive evidence that under some circumstances, in some animals, synchronization between different subsets of neurons may occur. It remains an open issue whether this is an important part of the code in some parts of the visual system (see further Ferster and Spruston, 1995; Singer, 1995). It may perhaps be used under conditions where ambiguous stimuli such as a Necker cube or with figure-ground ambiguity are presented, and there is plenty of time (perhaps hundreds of ms) to code, and decode, the different possibilities by different synchronization of different subsets of neurons. The evidence described above and in Chapter 8 suggests though that normally each cortical area can operate fast, in 20–50 ms, and that much of the information is available from the firing rates of ensembles of neurons, without taking into account the relative time of firing. The extent to which synchronization of the action potentials of different subsets of cortical neurons is required for visual object recognition is an open issue.

The neuronal synchronization scheme does has some disadvantages and incongruities. One disadvantage of the scheme is that it may not be sufficiently fast in operation (both to become set up, and to allow appropriate decoding) for the whole process to be sufficiently complete in 20–30 ms of cortical processing time per cortical area that the neurophysiological results above show is sufficient for object recognition. Another argument that the scheme may not at least be used in many parts of the visual system is that the information about which object is present can be read off with an enormous capacity from the firing rates of a small number (for example 10–50) of neurons, as shown above. (The information capacity calculations of neuronal populations performed by Rolls, Treves and Tovee (1997) and Abbott, Rolls and Tovee (1996) used *only* measurements of firing rates.) Third, the scheme is computationally very powerful. In a two-layer network, von der Malsburg has shown that it could provide the necessary feature linking to perform object recognition with relatively few neurons, because they can be reused again and again, linked differently for different objects. In contrast, the primate uses a considerable part of its brain, perhaps 50% in monkeys, for visual processing, with therefore what must be in the order of 10^9 neurons and 10^{13} synapses involved, so that the solution adopted by the real visual system may be one which relies on many neurons with simpler processing than arbitrary syntax implemented by synchronous firing of separate assemblies suggests. Instead, many neurons are used at each stage of, for example, the visual system, and the reason suggested for this is that syntax or binding is represented by low-order combinations of what is represented in previous layers of the cortical architecture (see Chapter 8). The low-order combinations (i.e. combinations of a small number of inputs) are suggested to occur because otherwise there would be a massive combinatorial explosion of the number of neurons required. This is why it is suggested there are so many neurons devoted to vision, which would not be required if a more powerful system utilizing temporal binding were actually implemented (Singer, 1995). It will be fascinating to see how research on these different approaches to processing in the primate visual system develops. For the development of both approaches, the use of well-defined neuronal network models will be very important.

Abeles *et al.* (1990, 1993; see also Abeles, 1991) have some evidence that occasionally (perhaps every few seconds or trials) groups of neurons emit action potentials with precise (1 or more ms) timing relationship to each other. They have argued that such synchrony could reflect a binding mechanism for different cell assemblies. They have termed these events Synfire chains. It is interesting that most of the evidence for such precise timing is from recordings in the frontal cortex, from areas in which the principles of operation might not coincide with those underlying the functioning of more sensory parts of the neocortex.

10.4.8 Conclusions on cortical encoding

The exponential rise in the number of stimuli that can be decoded when the firing rates of different numbers of neurons are analysed shows that encoding of information using firing rates is a very powerful coding scheme used by the cerebral cortex. Quantitatively, it is likely to be far more important than temporal encoding, in terms of the number of stimuli that can be encoded. Moreover, the information available from an ensemble of cortical neurons when only the firing rates are read, that is with no temporal encoding within or between neurons, is made available very rapidly (see Fig. 10.18). Further, the neuronal responses in most brain areas of behaving monkeys show sustained firing rate differences to different stimuli (see for example Fig. 10.12), so that it may not usually be necessary to invoke temporal encoding for the information about the stimulus to be present. Temporal encoding may, however, be used as well as rate coding, for example to help solve the binding problem (although there are alternatives to solve the binding problem) (see Chapter 8).

10.5 Functions of different cortical areas

The operation of a number of different cortical areas has been considered in detail in this book. The temporal cortical visual areas, and to some extent earlier visual processing, have been considered in Chapter 8 (see further Maunsell, 1995). Mechanisms for different types of short-term memory in the frontal and temporal lobes have been described above in Section 10.3 (see further Ungerleider, 1995, for neuroimaging studies in humans). Olfactory and taste processing, and the operation of the orbitofrontal cortex, have been considered in Chapter 7. The functions of some other cortical regions are introduced next, to provide the reader who is not familiar with these areas a view of how different cortical processing systems may be related to each other, and to introduce some of the computational problems in which their neuronal networks are involved.

10.5.1 Motor and premotor cortices, and somatosensory cortical areas

The primary somatosensory cortical areas (S–I) are located in areas 3a and 3b, which have projections to areas 1 and 2 (see Figs 1.8 and 1.13). Each of these areas has a body map. Some specialization of these areas is present, in that the dominant input to area 3a is from muscle stretch receptors, to area 3b from cutaneous touch, to area 1 from rapidly adapting cutaneous touch receptors, and to area 2 from deep pressure receptors (see Kandel and Jessell, 1991). A hierarchical principle of operation appears to be implemented, in that after processing in 3a and 3b, where receptive fields are small, in areas 1 and 2 the receptive fields are larger, and may respond to more complex stimuli such as edges. Further, S–I projects to area 5, and then to area 7b, which are both parts of the parietal cortex (see Fig. 1.13). In areas 5 and 7b, neurons respond to combinations of the inputs represented in areas 1, 2, and 3. For example neurons in 5 and 7b may respond to a combination of joint positions (not just to a single joint position as in areas 1, 2, and 3), and may thus encode the position of the limb in space with respect to the body (Iwamura, 1993). There are also neurons in areas 5 and 7b that respond to complex action-related inputs which occur for example during reaching with the arm or the manipulation of small objects with the hand. Outputs from this system project both to the striatum (Graziano and Gross, 1993) and to the motor and premotor areas (see Passingham, 1993). In addition to this pathway, which may be called a 'dorsal somatosensory pathway', there are outputs from S–I to the insula, which in turn projects to the amygdala and orbitofrontal cortex (Mesulam and Mufson, 1982; Kaas, 1993; Morecraft et al., 1992), which may be a 'ventral somatosensory pathway' concerned with the identification of felt objects, and in its latter stages, with the affective components of touch (Rolls, Francis et al., 1997; see Figs 1.8–1.13).

The primary motor cortex, area 4, receives projections from the lateral premotor cortex, area 6. This area receives inputs from the parietal areas 5 and 7b (see Figs 1.8 and 1.13), and may be involved in learned movements initiated by external signals (see Passingham, 1993). The representation of information in this area may be in a low-dimensional space, in that population vectors of neuronal activity can be easily interpreted as indicating the direction in which an arm movement may be made (Georgopoulos, 1995). In particular, neurons related to arm movement in this region fire in proportion to the cosine of the angle of their preferred direction of arm movement with the actual arm movement made. The result is that in this low-dimensional space of arm movements, the actual arm movement made is closely related to that predicted from the firing rate of a population of motor cortex (area 4) neurons and the preferred direction of firing of each neuron (Georgopoulos, 1995). A similar approach can be applied to the arm movement direction-related firing of populations of neurons in the premotor cortex (area 6) and areas 5 and 7b of the parietal cortex (see Georgopoulos, 1995). The supplementary motor area (also called the medial premotor area, which is the medial part of area 6), has strong inputs from the basal ganglia, and may be involved in self-initiated and self-timed movements, when there are no external cues for the initiation of the movement (Passingham, 1993). Premotor area 8, in the anterior part of the arcuate sulcus of the macaque, is the premotor area involved in the selection of eye movements. It receives from parietal areas 7a and LIP. It is also called the frontal eye field. In addition to this lateral part of area 8, there is also a 'supplementary eye field', more dorsally and medially (see Passingham, 1993).

10.5.2 Parietal cortex

In addition to the somatosensory projections to parietal areas 5 and 7b, there is a set of inputs from visual area 1 (V1) and V2 to area MT and thus to a whole set of visual cortical areas such as MST, LIP, etc. (see Figs 1.8, 1.11 and 1.12) concerned with processing visual motion, and where stimuli are in egocentric space. In one example, Movshon and colleagues (see Heeger, Simonelli and Movshon, 1996) have shown that whereas local motion is signalled by neurons in V1, global motion, for example of a plaid pattern, is reflected in the responses of neurons in area MT. Such a computation is implemented by a hierarchically organized convergence onto neurons in MT. In another example, Andersen *et al.* (see Andersen, 1995) have described how area 7a may help to implement a coordinate transform from retinal to head-based coordinates, using eye position information. Their hypothesis is that neurons that respond to combinations of retinal location and eye position (the angle of the eye in the head) would effectively encode a particular position relative to the head (Andersen, 1995). The neurons are described as having gain fields, in which eye position acts as a multiplier on the response of a neuron to a visual stimulus at a given place on the retina. In a similar way, head position can act to produce head gain fields in which the responses of area 7a and LIP neurons to a visual stimulus on the retina may depend on head (and in some cases also on eye) position. The nature of the encoding here is that the responses of such neurons can reflect the position of a visual stimulus in space relative to the head and even to the body, though different neurons are activated for different combinations of retinal angle and eye/head position. Thus this type of encoding is useful as an intermediate stage in a coordinate transformation computation. It would be relatively simple in principle to map the outputs of all the different neurons which map a given position relative to the head to some motor control signal. Vestibular signals may also modulate the responses of parietal neurons in a similar way to eye position, so that there is the possibility that the responses of some parietal neurons reflect the position of visual stimuli in world coordinates (see Andersen, 1995). Such an allocentric encoding of space has been described for the hippocampus and parahippocampal gyrus (see Chapter 6).

In area LIP (the lateral intraparietal area, see Figs 1.11 and 1.12), neuronal responses are related to eye movements, particularly to saccades. The majority of neurons here respond prior to saccades (whereas in area 7a, the responses of neurons related to saccades are more likely to be postsaccadic, see Andersen, 1995). In area LIP, neurons are found which are tuned to the distance of visual stimuli, and here the vergence angle of the eyes was found to modulate the magnitude of visually evoked responses (Gnadt, 1992). Also in area LIP, neurons are active in a memory delay period for a saccade, and appear to code for the plan of action, in that they respond even when a visual stimulus does not fall on their response field in a double saccade task (see Andersen, 1995). In area MST (medial superior temporal, see Figs 1.11 and 1.12), neuronal activity related to smooth pursuit eye movements is found.

The parietal system can be seen as a system for processing information about space, gradually performing the coordinate transforms required to proceed from the receptors (on for example the retina) to representations of positions in egocentric space (Hyvarinen, 1981; Andersen, 1995), and eventually, by projections to the premotor areas, to coordinate systems appropriate for producing motor responses implemented by muscle contractions.

10.5.3 Prefrontal cortex

The macaque dorsolateral prefrontal cortex, area 46, within and adjacent to the principal sulcus, is involved in delayed spatial responses. Its function as a working memory for spatial responses has been discussed in Section 10.3. It receives inputs from the parietal cortex, areas 7a and 7b, which are part of the 'where' system (see Figs 1.8 and 1.13 and Ungerleider, 1995). Such inputs are appropriate for a system involved in a short-term memory for visual and body responses. Area 46 sends outputs to the premotor cortex, area 6. It is not clear whether area 9, just dorsal to area 46, has a different function from area 46 (Passingham, 1993). Monkeys with damage to areas 46 and 9 have difficulty in remembering the order in which items were presented (Petrides, 1991).

The more ventral prefrontal cortex, in the inferior convexity (area 12) receives inputs from the temporal lobe visual cortical areas, TE and TEm (Barbas, 1988; Seltzer and Pandya, 1989), and is thus connected to the 'what' visual system (Ungerleider, 1995) (see Fig. 1.9). This ventral prefrontal area contains visual neurons which have responses associated with taste reward (see Chapter 7; Rolls, Critchley, Mason and Wakeman, 1996); and neurons with activity in the delay period of delay match to sample tasks (see Fuster, 1989; Wilson et al., 1993). This inferior prefrontal convexity cortex, and a third main part of the prefrontal cortex, the orbitofrontal cortex, are described in Chapter 7.

Reviews of the function of different parts of the dorsal prefrontal cortex are contained in Passingham (1993) and the special issue of the Philosophical Transactions of the Royal Society (1996, Vol. B 351, Number 1346); and of the orbitofrontal cortex by Rolls (1996b) and in Chapter 7.

10.5.4 Temporal lobe: the organization of representations in higher parts of the cortical visual system, and the representation of semantic knowledge

Processing in parts of the visual system from the primary visual cortex to the inferior temporal visual cortex is described in Chapter 8 (see Figs 1.8 and 1.9). Farah (1990, 1996; Farah, Meyer and McMullen, 1996) has reviewed evidence from the perceptual agnosias that provides indications on how information is represented in the higher parts of the visual system and its interface to semantic knowledge systems in humans. The evidence comes from patients with what is often classified as associative agnosia following brain damage to these regions. Associative agnosia is manifest as a deficit in knowing the attributes such as the name or function of objects, in the absence of low-level perceptual deficits. Such a patient might be able to copy a drawing of an object, providing evidence of reasonable low-level perception, yet be unable to name the object, or its function, or to use it correctly. It is contrasted with apperceptive agnosia, in which there are deficits in matching objects or copying drawings (see McCarthy and Warrington, 1990; Farah, 1990; Lissauer, 1890).

The associative agnosias can be category specific. For example, some patients have deficits in naming (or retrieving information about) non-living objects, and others with living things (Warrington and Shallice, 1984; Farah, Meyer and McMullen, 1996; Tippett, Glosser and Farah, 1996). This double dissociation, with damage to different areas of the brain differentially affecting different functions, is strong evidence for separable processing (although not necessarily totally independent, see Farah, 1994). The basis for the separation of these deficits may be the different inputs that are especially important in the categorization. For non-living objects, the function of the object, that is the use to which it is put, is frequently important, and therefore connections from representations of objects to systems concerned with function and motor use may be important. In comparison, the categories formed for living things, for example animals, reflect less function, and more the visual attributes and identifiers of each type of animal. Thus, connections to visual representations may be especially important. It is thus suggested that these category-specific impairments reflect in fact different modality-specific requirements for the support or inputs to the two types of semantic category. Given that the different types of support will be separately represented in the brain, and that maps of units concerned with similar functions tend to be formed due to short-range connections between the cells (see Section 4.6), the systems involved with the different categories of living and non-living things will tend to be topologically at least partly separate. This would provide the basis for brain lesions in slightly different parts of the temporal lobe to affect living versus non-living categorizations separately.

Another type of associative agnosia is prosopagnosia, a difficulty in recognizing people by the sight of their face. The deficit can occur without a major impairment in recognizing non-face objects. The deficit is not just because there is general damage to the temporal cortical visual areas and the slightly greater difficulty of face compared to object recognition, for the impairment in face recognition is much greater than in object recognition, and there is some evidence for a double dissociation of function, with other patients more impaired at object than face recognition (Farah, 1996; Farah *et al.*, in preparation). The implication of this is that there is at least partially separate processing of face and object information in the brain.

This is consistent with the evidence that in the temporal cortical visual areas of macaques, the proportion of visually responsive neurons responding to faces is higher in the cortex in the superior temporal sulcus than in the TE cortex on the lateral surface of the temporal lobe (Baylis, Rolls and Leonard, 1987; Rolls, Booth and Treves, 1996; see Chapter 8), even though there is certainly not complete segregation at the neuronal level. In addition, the face-responsive neurons are themselves divided into two groups, one in the inferior lip of the superior temporal sulcus (areas TEa and TEm) especially involved in encoding face identity, and another in the depths of the superior temporal sulcus more concerned with encoding face expression, and face movement/gesture (Hasselmo, Rolls and Baylis, 1989; Rolls, 1992b).

The separability of the neural systems involved in face and object processing raises the interesting question of whether the processing of these types of visual stimuli involves different types of computation in neural networks. Might the neuronal network architecture described for visual recognition in Chapter 8 be suitable for only objects or faces? The suggestion made here is that qualitatively the same type of neuronal network processing may be involved in face and object processing, but that the way the stimuli are represented is different, arising from the differences in the properties of face and object visual stimuli. Consider the fact that different objects are composed of different combinations of different parts. The parts can occur independently of each other, frequently combined in different ways to make different objects. This would lead in an architecture such as VisNet (Chapter 8) to the separate representation of the parts of objects (perhaps at an intermediate stage of processing), and (later) to representations of different objects as neurons responding to different combinations of the parts. In this sense, the representation of objects would be compositional. Consider now the representation of faces. The same types of part are usually present, for example in most views eyes, nose, mouth, and hair. What distinguishes different faces is not the types of part, but instead the relative spacing, size, etc. of the different parts. (Also, certainly, the texture and colour of the parts can differ.) Thus face recognition may be more holistic, relying on the metrics of the parts, such as their size and spacing, rather than on which parts are present. Moreover, the parts are not usually seen in different spatial relations to each other. The mouth, for example, is normally seen non-inverted in a non-inverted head. We are not as humans very used to seeing upside-down (inverted) faces, and may not set up holistic analysers for these, especially with the mouth non-inverted in an inverted head. So we do not notice the oddness until the image is presented non-inverted (with an inverted mouth, as in the demonstration with M. Thatcher's face). This type of evidence suggests that the representations set up to analyse faces, because of the fact that face parts are not normally seen separately, and are seen in standard relations to each other, are holistic. This type of representation would be built by a network such as VisNet if it were presented stimuli with the physical properties of faces, in which what differentiated between the stimuli was not which parts were present, but instead the metric of the relations between the same types of part. Thus the same architecture might lead to different types of representation for faces and objects. Topology of the systems, with face-processing neurons close to each other, and neurons representing parts of objects and objects, would be built in spatially separate areas because of the self-organizing map principle described in Chapter 4. Such self-organization into somewhat spatially separate neuronal populations would also be stimulated to some extent by the different support, that is input from other processing subsystems, involved in face and non-face processing. For example, face expression is represented in an area where moving visual stimuli also influence some of the neurons, and a reason for this may be that movement is often an important part of face expression decoding and use (Hasselmo, Rolls and Baylis, 1989; Baylis, Rolls and Leonard, 1987).

Another possible example of this principle of the stimulus properties themselves leading to spatially separable representations is the separate representation of letters and digits, which may attain a separate representation in the cerebral cortex because of temporal similarity, with digits frequently occurring close together in time, and letters frequently appearing close together in time (Polk and Farah, 1995).

10.6 Principles of cortical computation

The principles of cortical computation that are implicit in the approach taken here include the following.

First, we would like if possible to see cortical computation performed with a local learning rule, where the factors that determine the alterations in synaptic weight to implement learning are present in the presynaptic firing rate and the postsynaptic firing rate or activation. Algorithms such as error backpropagation which do not use local learning rules are implausible for the brain (see Chapter 5).

Second, cortical computation does not involve a principle such as compressing the representation into as few neurons as possible in 'hidden layers', in order to facilitate generalization as occurs in backpropagation of error networks. Instead, there is no clear compression of neuronal numbers from area to area of cortical processing, but instead the numbers are relatively constant from area to area. This suggests that the principle of operation is different to that used in backpropagation networks.

Third, the principles of operation of competitive networks described in Chapter 4 may underlie at least partly the capability of the cortex to perform categorization (see also Chapter 8). Another way in which categorization could be implemented in cortical networks is by utilizing the lateral connections between cortical pyramidal cells to implement inhibition between the cortical pyramidal cells which increases in strength if the cortical neurons have correlated activity (Foldiak, 1991). This would be an anti-Hebbian rule. It would enable sparse distributed representations with low redundancy to be learned. However, this particular scheme is implausible for the cerebral neocortex, because the connections between cortical pyramidal cells are excitatory. An equivalent scheme would utilize excitatory connections between nearby cortical pyramidal cells, and would then require that the synaptic weights between strongly coactive cortical pyramidal cells would decrease in strength, again an antiHebbian rule. This is also implausible in the cortex, at least given current understanding of cortical synaptic modification rules (see Singer, 1995).

Fourth, a short temporal memory (e.g. lasting for 0.5 s) may help cortical networks to form invariant representations, as described in Chapter 8. In VisNet, the trace learning rule was implemented on the feedforward inputs to the neurons in the next layer, by having a trace in the postsynaptic neurons. This trace could also be implemented by short-term memory effects implemented in cortico-cortical attractor network connections. The relevant connections could be those within a layer between nearby pyramidal cells (Parga and Rolls, 1997); or in the forward and backprojections between adjacent areas in the cortical processing hierarchy. This would enable pairwise association of, for example, different views of an object, with the relevant synapses being those in the forward inputs to the next cortical layer.

Fifth, attractor networks, to help with completion, generalization, and short-term (working) memory, and the short-term memory required for a trace rule may be implemented by Hebbian learning involving the excitatory connections between nearby pyramidal cells within a cortical area, or involving the forward and backprojection connections between pyramidal cells in connected cortical areas. These attractor nets would implement local attractors within a small part of a cortical area. A single attractor network for the whole cortex is not plausible, because the cortical connections are strong between nearby cortical cells, and effectively very dilute on average across the cerebral cortex; and because a single attractor network would have an impossibly low total number of attractors or different memories for the whole cerebral cortex, as the total number of memories would be determined mainly by the number of modifiable connections onto each cortical neuron (O'Kane and Treves, 1992). The numbers of neurons in the local attractor networks may need to be quite large, in the order of 100,000, because attractor networks with diluted connectivity, sparse representations, and low firing rates may only operate well if there are large numbers of neurons in the attractor network (cf. Amit, 1995; Simmen, Treves and Rolls, 1996).

Sixth, much cortical processing may take place utilizing firing rate encoding across populations of neurons, without requiring temporal encoding within the spike train of individual neurons or between the spike trains of different neurons.

Seventh, we would like much cortical computation to proceed without an explicit tutor or target for each cortical neuron to provide it with a training signal as happens in single-layer perceptrons, and in multilayer perceptrons in the output stages. The reason for this is that an explicit tutor or target for each cortical pyramidal cell does not appear to be present. The implication of this is that much cortical computation might be unsupervised, without an explicit tutor (see Chapter 4). The challenge is perhaps to discover how powerful the learning may be even when there is no explicit tutor for each neuron in the output layer. Powerful learning algorithms which are unsupervised or use local learning rules are starting to appear (Hinton *et al.*, 1995; Bell and Sejnowski, 1995; O'Reilly, 1996), but they do not yet seem to be completely neurophysiologically plausible.

Eighth, convergence onto single cortical neurons from several parts of the preceding cortical area, or from different cortical areas, allows information to be brought together that can be useful in many ways. For example, in Chapter 8 it was shown how convergence from area to area in the ventral visual system can help to produce neurons with large receptive fields. In another example, in Chapter 7, it was shown how after separate processing of stimuli in different modalities to the object level, processing systems for the sight of objects, for taste, and for smell, then converge in the orbitofrontal cortex. This convergence then allows correspondences of inputs across modalities at the object level to be detected, using associative learning. An important aspect of the way in which convergence operates is that sufficient inputs which are likely to be coactive must be brought together onto single neurons, which typically may have in the order of 10 000 inputs. Given that cortical neurons do not have vast numbers of input connections, and that a reasonable number probably need to be coactive to enable a cortical neuron to learn, considerable architectural convergence is needed in the cortex. The convergence onto single neurons is required so that local learning rules can operate effectively. The formation of topographic maps may be part of a mechanism which helps to keep coactive neurons in the same vicinity, so that relevant information using considerable numbers of connections can be exchanged between them easily. Conversely, complementary divergence from cortical neurons is needed, in order to provide the required convergence in further cortical processing stages. The complement of topographic maps which put similar neurons close to each other in the cortex is the segregation of neurons with different types of function, leading to cortical 'modularity'.

A challenge for the future, given that we are now starting to understand how brain computation could occur in at least simple brain processing systems, is to discover how complex representations could be built, and computations performed, using large numbers of neurons, and relatively simple, local, learning rules without an explicit tutor for each neuron. At the same time, it is important to continue to investigate whether more complex learning rules might be implemented in real neuronal systems, in case the range of algorithms considered neurophysiologically plausible can be extended. It will also be important to consider further how syntax, necessary for linguistic operations, and perhaps also important in the implementation of consciousness in neural networks (see Rolls, 1997a, 1997b), could be implemented in neuronal networks in the brain. It appears that we may have the tools now, as described in this book, for understanding how the processing implemented in a monkey's brain to solve perception, memory, emotion, and motivation, and the initiation of action, might be implemented; but that the principles of the processes that implement language in the human brain are not yet clear. An introduction to connectionist approaches to language is provided by McLeod, Plunkett and Rolls (1998).

A1 Introduction to linear algebra for neural networks

In this appendix we review some simple elements of linear algebra relevant to understanding neural networks. This will provide a useful basis for a quantitative understanding of how neural networks operate. A more extensive review is provided by Jordan (1986).

A1.1 Vectors

A vector is an ordered set of numbers. An example of a vector is the set of numbers

$$\begin{bmatrix} 7 \\ 4 \end{bmatrix}.$$

If we denote the jth element of this vector as w_j, then $w_1 = 7$, and $w_2 = 4$. We can denote the whole vector by $\mathbf{w}$. This notation is very economical. If the vector has 10 000 elements, then we can still refer to it in mathematical operations as $\mathbf{w}$. $\mathbf{w}$ might refer to the vector of 10 000 synaptic weights on the dendrites of a neuron. Another example of a vector is the set of firing rates of the axons that make synapses onto a dendrite, as shown in Fig. 1.2. The rate r' of each axon forming the input the vector can be indexed by j, and is denoted by r'_j. The vector would be denoted by $\mathbf{r}'$. (We use the prime after the r simply because we have used r' to refer elsewhere in this book to the firing of an input axon, and the algebra we introduce next is appropriate for understanding the operations performed on vectors made up by sets of axons.)

Certain mathematical operations can be performed with vectors. We start with the operation which is fundamental to simple models of neural networks, the inner product or dot product of two vectors.

A1.1.1 The inner or dot product of two vectors

Recall the operation of computing the activation h of a neuron from the firing rate on its input axons multiplied by the corresponding synaptic weight:

$$h = \Sigma_j r'_j w_j \qquad (A1.1)$$

where Σ_j indicates that the sum is over the C input axons indexed by j. Denoting the firing rate vector as $\mathbf{r}'$ and the synaptic weight vector as $\mathbf{w}$, we can write

$$h = \mathbf{r}' \cdot \mathbf{w} \qquad (A1.2)$$

If the weight vector is

$$\mathbf{w} = \begin{bmatrix} 9 \\ 5 \\ 2 \end{bmatrix}$$

and the firing rate input vector is

$$\mathbf{r'} = \begin{bmatrix} 3 \\ 6 \\ 7 \end{bmatrix}$$

then we can write

$$\mathbf{r'} \cdot \mathbf{w} = (3 \cdot 9) + (6 \cdot 5) + (7 \cdot 2)$$
$$= 71.$$

Thus in the inner or dot product, we multiply the corresponding terms, and then sum the result. As this is the simple mathematical operation that is used to compute the activation h in the most simplified abstraction of a neuron (see Chapter 1), we see that it is indeed the fundamental operation underlying many types of neural networks. We will shortly see that some of the properties of neuronal networks can be understood in terms of the properties of the dot product. We next review a number of basic aspects of vectors and inner products between vectors.

There is a simple geometrical interpretation of vectors, at least in low-dimensional spaces. If we define, for example, x and y axes at right angles to each other in a two-dimensional space, then any two-component vector can be thought of as having a *direction* and *length* in that space which can be defined by the values of the two elements of the vector. If the first element is taken to correspond to x and the second to y, then the x axis lies in the direction [1,0] in the space, and the y axis in the direction [0,1], as shown in Fig. A1.1. The line to point [1,1] in the space then lies at 45° to both axes, as shown in Fig. A1.1.

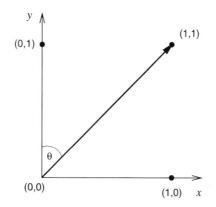

Fig. A1.1 Illustration of a vector in a two-dimensional space. The basis for the space is made up of the x axis in the [1,0] direction, and the y axis in the [0,1] direction. (The first element of each vector is then the x value, and the second the y value. The values of x and y for different points, marked by a dot, in the space are shown. The origins of the axes are at point 0,0.) The [1,1] vector projects in the [1,1] (or 45°) *direction* to the point 1,1, with *length* 1.414.

A1.1.2 The length of a vector

Consider taking the inner product of a vector $\mathbf{w} = \begin{bmatrix} 4 \\ 3 \end{bmatrix}$

with itself. Then

$$\|\mathbf{w}\| = \sqrt{\mathbf{w} \cdot \mathbf{w}} = \sqrt{4^2 + 3^2} = \sqrt{25} = 5$$

This is the length of the vector. We can represent this operation in the two-dimensional graph shown in Fig. A1.1. In this case, the coordinates where vector $\mathbf{w}$ ends in the space are [1,1]. The length of the vector (from [0,0]) to [1,1] is obtained by Pythagoras' theorem. Pythagoras' theorem states that the length of the vector $\mathbf{w}$ is equal to the square root of the sum of the squares of the two sides. Thus we define the length of the vector $\mathbf{w}$ as

$$\|\mathbf{w}\| = (\mathbf{w} \cdot \mathbf{w})^{1/2} \tag{A1.3}$$

In the [1,1] case, this value is $\sqrt{2} = 1.414$.

A1.1.3 Normalizing the length of a vector

We can scale a vector in such a way that its length is equal to 1 by dividing it by its length. If we form the dot product of two normalized vectors, its maximum value will be 1, and its minimum value −1.

A1.1.4 The angle between two vectors: the normalized dot product

The angle between two vectors $\mathbf{r}'$ and $\mathbf{w}$ is defined in terms of the inner product as follows:

$$\cos \theta = \frac{\mathbf{r}' \cdot \mathbf{w}}{\|\mathbf{r}'\| \, \|\mathbf{w}\|} \tag{A1.4}$$

For example, the angle between two vectors $\mathbf{r}' = \begin{bmatrix} 0 \\ 1 \end{bmatrix}$ and $\mathbf{w} = \begin{bmatrix} 1 \\ 1 \end{bmatrix}$

where the length of vector $\mathbf{r}'$ is $(0 \cdot 0 + 1 \cdot 1)^{1/2} = 1$ and of vector $\mathbf{w}$ is $(1 \cdot 1 + 1 \cdot 1)^{1/2} = 2$, is

$$\cos \theta = \frac{(0 \cdot 1) + (1 \cdot 1)}{1 \cdot \sqrt{2}} = 0.707 \tag{A1.5}$$

Thus $$\theta = \cos^{-1}(0.707) = 45°.$$

We can give a simple geometrical interpretation of this as shown in Fig. A1.1. However, Eq. A1.4 is much easier to use in a high-dimensional space!

The dot product reflects the similarity between two vectors. Once the length of the vectors is fixed, the higher their dot product, the more similar are the two vectors. By normalizing the dot product, that is by dividing by the lengths of each vector as shown in Eq. A1.4, we obtain a value which varies from -1 to $+1$. This *normalized dot product* is then just the cosine of the angle between the vectors, and is a very useful measure of the similarity between any two vectors, because it always lies in the range -1 to $+1$. It is closely related to the (Pearson product-moment) correlation between any two vectors, as we see if we write the equation in terms of its components

$$\cos \theta = \frac{\Sigma_j \, r'_j \, w_j}{(\Sigma_j \, r'^2_j)^{1/2} \, (\Sigma_j \, w_j^2)^{1/2}} \qquad (A1.6)$$

which is just the formula for the correlation coefficient between two sets of numbers with zero mean (or with the mean value removed by subtracting the mean of the components of each vector from each component of that vector).

Now consider two vectors which have a dot product of zero, that is where $\cos \theta = 0$ or the angle between the vectors is 90°. Such vectors are described as orthogonal (literally at right angles) to each other. If our two orthogonal vectors were r' and w, then the activation of the neuron, measured by the dot product of these two vectors, would be zero. If our two orthogonal vectors each had a mean of zero, their correlation would also be zero: the two vectors can then be described as unrelated or independent.

If instead the two vectors had zero angle between them, that is if $\cos \theta = 1$, then the dot product would be maximal (given the vectors' lengths), the normalized dot product would be 1, and the two vectors would be described as identical to each other apart from their length. Note that in this case their correlation would also be 1, even if the two vectors did not have zero mean components.

For intermediate similarities of the two vectors, the degree of similarity would be expressed by the relative magnitude of the dot product, or by the normalized dot product of the two vectors, which is just the cosine of the angle between them. These measures are closely related to the correlation between two vectors.

Thus we can think of the simple operation performed by neurons as measuring the similarity between their current input vector and their weight vector. Their activation, h, is this dot product. It is because of this simple operation that neurons can generalize to similar inputs; can still produce useful outputs if some of their inputs or synaptic weights are damaged or missing, that is they can show graceful degradation or fault tolerance; and can be thought of as learning to point their weight vectors towards input patterns, which is very useful in enabling neurons to categorize their inputs in competitive networks.

A1.1.5 The outer product of two vectors

Let us take a row vector having as components the firing rates of a set of output neurons in a pattern associator or competitive network, which we might denote as r, with components r_i and the index i running from 1 to the number N of output neurons. r is then a shorthand for writing down each component, e.g. [7,2,5,2, ...], to indicate that the firing rate of neuron 1 is

7, etc. To avoid confusion, we continue in the following to denote the firing rate of *input* neurons as r'_j, with a prime. Now recall how the synaptic weights are formed in a pattern associator using a Hebb rule as follows:

$$\delta w_{ij} = k\, r_i r'_j$$

where δw_{ij} is the change of the synaptic weight w_{ij} which results from the simultaneous (or conjunctive) presence of presynaptic firing r'_j and postsynaptic firing (or activation) r_i, and k is a learning rate constant which specifies how much the synapses alter on any one pairing. In a more compact vector notation, this expression would be

$$\delta \mathbf{w}_i = k\, r_i \mathbf{r'}$$

where the firing rates on the axons form a column vector with the values, for example, as follows:

$$\mathbf{r'} = \begin{bmatrix} 2 \\ 0 \\ 3 \end{bmatrix}$$

$$.....$$

The weights are then updated by a change proportional (the k factor) to the following matrix

$$\begin{array}{c c c c c}
 & [\ 7 & 2 & 5 & ..... \ ] \\
[\ 2\ ] & 14 & 4 & 10 & ..... \\
[\ 0\ ] & 0 & 0 & 0 & ..... \\
[\ 3\ ] & 21 & 6 & 15 & ..... \\
..... & ..... & ..... & ..... & .....
\end{array}$$

This multiplication of the two vectors is called the outer, or tensor, product, and forms a matrix, in this case of (alterations to) synaptic weights. In the most compact notation, among those used in linear algebra, if $\mathbf{W}$ refers to the matrix of synaptic weights, we can write

$$\delta \mathbf{W} = k\, \mathbf{r'} \otimes \mathbf{r}^{\mathrm{T}} \qquad\qquad (\text{A}1.7)$$

where the T superscript is useful because, since the normal definition of a vector we have assumed defines column vectors, if we wish to indicate instead that we are using a row vector, we should indicate the transpose operation. (The row vector [7,2,5] can be defined as the transpose of the column vector

$$\begin{bmatrix} 7 \\ 2 \\ 5 \end{bmatrix}.)$$

Thus we see that the operation of altering synaptic weights in a network can be thought of as forming a matrix of weight changes, which can then be used to alter the existing matrix of synaptic weights.

A1.2 Linear and non-linear systems

The operations with which we have been concerned in this appendix so far are linear operations. We should note that if two matrices operate linearly, we can form their product by matrix multiplication, and then replace the two matrices with the single matrix that is their product. We can thus effectively replace two synaptic matrices in a linear multilayer neural network with one synaptic matrix, the product of the two matrices. For this reason, multilayer neural networks if linear cannot achieve more than can be achieved in a single-layer linear network. It is only in non-linear networks that more can be achieved, in terms of mapping input vectors through the synaptic weight matrices, to produce particular mappings to output vectors. Much of the power of many networks in the brain comes from the fact that they are multilayer non-linear networks (in that the computing elements in each network, the neurons, have non-linear properties such as thresholds, and saturation with high levels of input). Because the matrix by matrix multiplication operations of linear algebra cannot be applied directly to the operation of neural networks in the brain, we turn instead back to other aspects of linear algebra, which can help us to understand which classes of pattern can be successfully learned by different types of neural network.

A1.2.1 Linear combinations of vectors, linear independence, and linear separability

We can multiply a vector by a scalar (a single value, e.g. 2) thus:

$$2 \begin{bmatrix} 4 \\ 1 \\ 3 \end{bmatrix} = \begin{bmatrix} 8 \\ 2 \\ 6 \end{bmatrix}.$$

We can add two vectors thus:

$$\begin{bmatrix} 4 \\ 1 \\ 3 \end{bmatrix} + \begin{bmatrix} 2 \\ 7 \\ 2 \end{bmatrix} = \begin{bmatrix} 6 \\ 8 \\ 5 \end{bmatrix}.$$

The sum of the two vectors is an example of a linear combination of two vectors, which is in general a weighted sum of several vectors, component by component. Thus, the linear combination of vectors v_1, v_2, . . . to form a vector v_S is expressed by the sum

$$v_S = c_1 v_1 + c_2 v_2 + . . . \tag{A1.8}$$

where c_1 and c_2 are scalars.

By adding vectors in this way, we can produce any vector in the space spanned by a set of vectors as a linear combination of vectors in the set. If in a set of n vectors at least one can be written as a linear combination of the others, then the vectors are described as linearly dependent. If in a set of n vectors none can be written as a linear combination of the others, then the vectors are described as linearly independent. A linearly independent set of vectors has the properties that any vector in the space spanned by the set can be written in only one way as a linear combination of the set, and the space has dimension $d = n$. In contrast, a vector in a space spanned by a linearly dependent set can be written in an infinite number of equivalent ways, and the dimension d of the space is less than n.

Consider a set of linearly dependent vectors and the d-dimensional space they span. Two subsets of this set are described as linearly separable if the vectors of one subset (that is, their endpoints) can be separated from those of the other by a hyperplane, that is a subspace of dimension $d-1$. Subsets formed from a set of linearly independent vectors are always linearly separable. For example, the four vectors

$$\begin{bmatrix} 0 \\ 0 \end{bmatrix}, \begin{bmatrix} 0 \\ 1 \end{bmatrix}, \begin{bmatrix} 1 \\ 0 \end{bmatrix}, \begin{bmatrix} 1 \\ 1 \end{bmatrix}$$

are linearly dependent, because the fourth can be formed by a linear combination of the second and third (and also because the first, being the null vector, can be formed by multiplying any other vector by zero—a specific linear combination). In fact, $n = 4$ and $d = 2$. If we split this set into subset A including the first and fourth vector, and subset B including the second and third, the two subsets are not linearly separable, because there is no way to draw a line (which is the subspace of dimension $d-1 = 1$) to separate the two subsets A and B. We have encountered this set of vectors in Chapter 4, and this is the geometrical interpretation of why a one-layer, one-output unit network cannot separate these patterns. Such a network (a simple perceptron) is equivalent to its (single) weight vector, and in turn the weight vector defines a set of parallel $d-1$ dimensional hyperplanes. (Here $d = 2$, so a hyperplane is simply a line, any line perpendicular to the weight vector.) No line can be found which separates the first and fourth vector from the second and third, whatever the weight vector the line is perpendicular to, and hence no perceptron exists which performs the required classification (see Fig. 5.4). To separate such patterns, a multilayer network with non-linear neurons is needed (see Chapter 5).

Any set of linearly independent vectors comprise the basis of the space they span, and they are called basis vectors. All possible vectors in the space spanned by these vectors can be formed as linear combinations of these vectors. If the vectors of the basis are in addition mutually orthogonal, the basis is an orthogonal basis, and it is, further, an orthonormal basis if the vectors are chosen to be of unit length. Given any space of vectors with a preassigned meaning to each of their components (for example the space of patterns of activation, in which each component is the activation of a particular unit) the most natural, *canonical* choice for a basis is the set of vectors in which each vector has one component, in turn, with value 1, and all the others with value 0. For example, in the $d = 2$ space considered earlier the natural choice is to take as basis vectors

$$\begin{bmatrix} 1 \\ 0 \end{bmatrix}$$

and

$$\begin{bmatrix} 0 \\ 1 \end{bmatrix},$$

from which all vectors in the space can be created. This can be seen from Fig. A1.1. (A vector in the [−1, −1] direction would have the opposite direction of the vector shown in Fig. A1.1.)

If we had three vectors that were all in the same plane in a three-dimensional (x,y,z) space, then the space they spanned would be less than three-dimensional. For example, the three vectors

$$
\begin{matrix}
[\ 1\ ] & [\ 0\ ] & [-1] \\
[\ 0\ ] & [\ 1\ ] & [-1] \\
[\ 0\ ], & [\ 0\ ], & [\ 0\ ]
\end{matrix}
$$

all lie in the same z plane, and span only a two-dimensional space. (All points in the space could be shown in the plane of the paper in Fig. A1.1.)

A1.2.2 Application to understanding simple neural networks

The operation of simple one-layer networks can be understood in terms of these concepts. A network with a single binary unit can implement a classification between two subspaces of a space of possible input patterns provided that the p actual patterns given as examples of the correct classification are linearly separable. The binary output unit is equivalent to a hyperplane (the hyperplane orthogonal to its synaptic weight vector) that divides the input space in two. The input space is obviously of dimension d, if d is the number of input axons. A one-layer network with a number n of binary output units is equivalent to n hyperplanes, that could potentially divide the input space into as many as 2^n regions, each corresponding to input patterns leading to a different output. However the number p of *arbitrary* examples of the correct classification (each example consisting of an input pattern and its required correct output) that the network may be able to implement is well below 2^n, and in fact depends on d not on n. This is because for p too large it will be impossible to position the n weight vectors such that all examples of input vectors for which the first output unit is required to be 'on' fall on one side of the hyperplane associated with the first weight vector, all those for which it is required to be 'off' fall on the other side, and simultaneously the same holds with respect to the second output unit (a different dichotomy), the third, and so on. The limit on p, which can be thought of also as the number of independent associations implemented by the network, when this is viewed as a heteroassociator (i.e. pattern associator) with binary outputs, can be calculated with the Gardner method and depends on the statistics of the patterns. For input patterns that are also binary, random and with equal probability for each of the two states on every unit, the limit is $p_c = 2d$ (for a fully connected system, otherwise the limit is twice the number of connections per output unit; see further Chapter 2 and Appendix A3).

These concepts also help one to understand further the limitation of linear systems, and the power of non-linear systems. Consider the dot product operation by which the neuronal activation h is computed:

$$ h = \Sigma_j r'_j w_j. $$

If the output firing is just a linear function of the activation, any input pattern will produce a non-zero output unless it happens to be exactly orthogonal to the weight vector. For positive-only firing rates and synaptic weights, being orthogonal means taking non-zero values only on non-corresponding components. Since with distributed representations the non-zero components of different input firing vectors will in general be overlapping (i.e. some corresponding components in both firing rate vectors will be on, that is the vectors will overlap), this will result effectively in interference between any two different patterns that for example have to be associated to different outputs. Thus a basic limitation of linear networks is that they can perform pattern association perfectly only if the input patterns r' are orthogonal; and for positive-only patterns that represent actual firing rates only if the different firing rate vectors are non-overlapping. Further, linear networks cannot of course perform any classification, just because they act linearly. (Classification implies producing output states that are clearly defined as being in one class, and not in other classes.) For example, in a linear network, if a pattern is presented which is intermediate between two patterns v_1 and v_2, such as $c_1 v_1 + c_2 v_2$, then the output pattern will be a linear combination of the outputs produced by v_1 and v_2 (e.g. $c_1 o_1 + c_2 o_2$), rather than being classified into o_1 or o_2. In contrast, with non-linear neurons, the patterns need not be orthogonal, only linearly separable, for a one-layer network to be able to correctly classify the patterns (provided that a sufficiently powerful learning rule is used, see Chapter 5).

A2 Information theory

Information theory provides the means for quantifying *how much* neurons communicate to other neurons, and thus provides a quantitative approach to fundamental questions about information processing in the brain. If asking what in neural activity carries information, one must compare the amounts of information carried by different codes, that is different descriptions of the same activity, to provide the answer. If asking about the speed of information transmission, one must define and measure information rates from neuronal responses. If asking to what extent the information provided by different cells is redundant or instead independent, again one must measure amounts of information in order to provide a quantitative study.

This appendix briefly introduces the fundamental elements of information theory in the first section. A more complete treatment can be found in many books on the subject (e.g. Abramson, 1963; Hamming, 1990; Cover and Thomas, 1991), among which the recent one by Rieke, Warland, de Ruyter van Steveninck and Bialek (1996) is specifically about information transmitted by neuronal firing. The second section discusses the extraction of information measures from neuronal activity, in particular in experiments with mammals, in which the central issue is how to obtain accurate measures in conditions of limited sampling. The third section summarizes some of the results obtained so far. The material in the second and third section has not yet been treated in any book, and as concerns the third section, it is hoped that it will be considerably updated in future editions of this volume, to include novel findings. The essential terminology is summarized in a Glossary at the end of the appendix.

A2.1 Basic notions and their use in the analysis of formal models

Although information theory was a surprisingly late starter as a mathematical discipline, having being developed and formalized in 1948 by C. Shannon, the intuitive notion of information is immediate to us. It is also very easy to understand why, in order to quantify this intuitive notion, of how much we know about something, we use logarithms, and why the resulting quantity is always defined in relative rather than absolute terms. An introduction to information theory is provided next, with a more formal summary given in the third subsection.

A2.1.1 The information conveyed by definite statements.

Suppose somebody, who did not know, is told that Reading is a town west of London. How much information is he given? Well, that depends. He may have known it was a town in

England, but not whether it was east or west of London; in which case the new information amounts to the fact that of *two* a priori (i.e. initial) possibilities (E or W), one holds (W). It is also possible to interpret the statement in the more precise sense, that Reading is west of London, rather than east, north or south, i.e. one out of *four* possibilities; or else, west rather that north-west, north, etc. Clearly, the larger the number k of a priori possibilities, the more one is actually told, and a measure of information must take this into account. Moreover, we would like independent pieces of information to just add together. For example, our fellow may also be told that Cambridge is, out of l possible directions, north of London. Provided nothing was known on the mutual location of Reading and Cambridge, there were overall $k \times l$ a priori possibilities, only one of which remains a posteriori (after receiving the information). We then define the amount I of information gained, such that

$$I(k) = \log_2 k \tag{A2.1}$$

because the log function has the basic property that

$$I(kl) = \log_2 kl = \log_2 k + \log_2 l = I(k) + I(l), \tag{A2.2}$$

i.e. the information about Cambridge adds up to that about Reading. We choose to take logarithms in base 2 as a mere convention, so that the answer to a yes/no question provides one unit, or bit, of information. Here it is just for the sake of clarity that we used different symbols for the number of possible directions with respect to which Reading and Cambridge are localized; if both locations are specified for example in terms of E, SE, S, SW, W, NW, N, NE, then obviously $k = l = 8$, $I(k) = I(l) = 3$ bits, and $I(kl) = 6$ bits. An important point to note is that the *resolution* with which the direction is specified determines the amount of information provided, and that in this example, as in many situations arising when analysing neuronal codings, the resolution could be made progressively finer, with a corresponding increase in information proportional to the log of the number of possibilities.

A2.1.2 The information conveyed by probabilistic statements.

The situation becomes slightly less trivial, and closer to what happens among neurons, if information is conveyed in less certain terms. Suppose for example that our friend is told, instead, that Reading has odds of 9 to 1 to be west, rather than east, of London (considering now just two a priori possibilities). He is certainly given some information, albeit less than in the previous case. We might put it this way: out of 18 equiprobable a priori possibilities (9 west + 9 east), 8 (east) are eliminated, and 10 remain, yielding

$$I = \log_2 18/10 = \log_2 9/5 \tag{A2.3}$$

as the amount of information given. It is simpler to write this in terms of probabilities

$$I = \log_2 P^{\text{posterior}}(W)/P^{\text{prior}}(W) \tag{A2.4}$$
$$= \log_2 (9/10)/(1/2) = \log_2 9/5.$$

This is of course equivalent to saying that the amount of information given by an uncertain statement is equal to the amount given by the absolute statement

$$I = -\log_2 P^{prior}(W)$$

minus the amount of uncertainty remaining after the statement, $I = -\log_2 P^{posterior}(W)$.
  A successive clarification that Reading is indeed west of London carries

$$I' = \log_2 (1)/(9/10) \tag{A2.5}$$

bits of information, because 9 out of 10 are now the a priori odds, while a posteriori there is certainty, $P^{posterior}(W) = 1$. In total we would seem to have

$$I^{TOTAL} = I + I' = \log_2 9/5 + \log_2 10/9 = 1 \text{ bit} \tag{A2.6}$$

as if the whole information had been provided at one time. This is strange, given that the two pieces of information are clearly *not* independent, and only independent information should be additive. In fact, we have cheated a little. Before the clarification, there was still one residual possibility (out of 10) that the answer was 'east', and this must be taken into account by writing

$$I = P^{posterior}(W) \log_2 P^{posterior}(W)/P^{prior}(W) + P^{posterior}(E) \log_2 P^{posterior}(E)/P^{prior}(E) \tag{A2.7}$$

as the information contained in the first message. This little detour should serve to emphasize two aspects which it is easy to forget when reasoning intuitively about information, and that in this example cancel each other. In general, when uncertainty remains, that is there is more than one possible a posteriori state, one has to average information values for each state with the corresponding a posteriori probability measure. In the specific example, the sum $I + I'$ totals slightly *more* than 1 bit, and the amount in excess is precisely the information 'wasted' by providing *correlated* messages.

A2.1.3 Information sources, information channels, and information measures

In summary, the expression quantifying the information provided by a definite statement that event s, which had an a priori probability $P(s)$, has occurred is

$$I(s) = \log_2 (1/P(s)) = -\log_2 P(s), \tag{A2.8}$$

whereas if the statement is probabilistic, that is several a posteriori probabilities remain non-zero, the correct expression involves summing over all possibilities with the corresponding probabilities:

$$I = \Sigma_s [P^{posterior}(s) \log_2 P^{posterior}(s)/P^{prior}(s)]. \tag{A2.9}$$

When considering a discrete set of mutually exclusive events, it is convenient to use the metaphor of a set of *symbols* comprising an *alphabet S*. The occurrence of each event is then referred to as the emission of the corresponding symbol by an information *source*. The *entropy* of the source, H, is the average amount of information per source symbol, where the average is taken across the alphabet, with the corresponding probabilities

$$H(S) = -\Sigma_{s \in S} P(s) \log_2 P(s). \tag{A2.10}$$

An information *channel* receives symbols s from an alphabet S and emits symbols s' from alphabet S'. If the *joint* probability of the channel receiving s and emitting s' is given by the product

$$P(s,s') = P(s)P(s') \tag{A2.11}$$

for any pair s,s', then the input and output symbols are *independent* of each other, and the channel transmits zero information. Instead of joint probabilities, this can be expressed with conditional probabilities: the conditional probability of s' given s is written $P(s'|s)$, and if the two variables are independent, it is just equal to the unconditional probability $P(s')$. In general, and in particular if the channel does transmit information, the variables are not independent, and one can express their joint probability in two ways in terms of conditional probabilities

$$P(s,s') = P(s'|s)P(s) = P(s|s')P(s'), \tag{A2.12}$$

from which it is clear that

$$P(s'|s) = P(s|s')P(s')/P(s), \tag{A2.13}$$

which is called Bayes' theorem (although when expressed as here in terms of probabilities it is strictly speaking an identity rather than a theorem). The information transmitted by the channel conditional to its having emitted symbol s' (or specific transinformation, $I(s')$) is given by Eq. A2.9, once the unconditional probability $P(s)$ is inserted as the prior, and the conditional probability $P(s|s')$ as the posterior:

$$I(s') = \Sigma_s P(s|s') \log_2 P(s|s')/P(s). \tag{A2.14}$$

Symmetrically, one can define the transinformation conditional to the channel having received symbol s

$$I(s) = \Sigma_{s'} P(s'|s) \log_2 P(s'|s)/P(s'). \tag{A2.15}$$

Finally, the average transinformation, or *mutual* information, can be expressed in fully symmetrical form

$$I = \Sigma_s \ P(s) \ \Sigma_{s'} \ P(s'\,|\,s) \ \log_2 P(s'\,|\,s)/P(s')$$
$$= \Sigma_{s,s'} \ P(s,s') \ \log_2 P(s,s')/[P(s)P(s')]. \qquad (A2.16)$$

The mutual information can also be expressed as the entropy of the source using alphabet S minus the *equivocation* of S with respect to the new alphabet S' used by the channel, written

$$I = H(S) - H(S\,|\,S') \equiv H(S) - \Sigma_{s'} \ P(s') \ H(S\,|\,s'). \qquad (A2.17)$$

A channel is characterized, once the alphabets are given, by the set of conditional probabilities for the output symbols, $P(s'\,|\,s)$, whereas the unconditional probabilities of the input symbols $P(s)$ depend of course on the source from which the channel receives. Then, the *capacity* of the channel can be defined as the maximal mutual information across all possible sets of input probabilities $P(s)$. Thus, the information transmitted by a channel can range from zero to the lower of two independent upper bounds: the entropy of the source, and the capacity of the channel.

A2.1.4 The information carried by a neuronal response and its averages

Considering the processing of information in the brain, we are often interested in the amount of information the response r of a neuron, or of a population of neurons, carries about an event happening in the outside world, for example a stimulus s shown to the animal. Once the inputs and outputs are conceived of as sets of symbols from two alphabets, the neuron(s) may be regarded as an information channel. We may denote with $P(s)$ the a priori probability that the particular stimulus s out of a given set was shown, while the conditional probability $P(s\,|\,r)$ is the a posteriori probability, that is updated by the knowledge of the response r. The response-specific transinformation

$$I(r) = \Sigma_s \ P(s\,|\,r) \ \log_2 P(s\,|\,r)/P(s) \qquad (A2.18)$$

takes the extreme values of $I(r) = -\log_2 P(s(r))$ if r unequivocally determines $s(r)$ (that is, $P(s\,|\,r)$ equals 1 for that one stimulus and 0 for all others); and $I(r) = \Sigma_s P(s) \log_2 P(s)/P(s) = 0$ if there is no relation between s and r, that is they are independent, so that the response tells us nothing new about the stimulus.

This is the information conveyed by each particular response. One is usually interested in further averaging this quantity over all possible responses r,

$$<I> = \Sigma_r \ P(r) \ [\ \Sigma_s \ P(s\,|\,r) \ \log_2 P(s\,|\,r)/P(s) \ ]. \qquad (A2.19)$$

The angular brackets $< \, >$ are used here to emphasize the averaging operation, in this case over responses. Denoting with $P(s,r)$ the joint probability of the pair of events s and r, and using Bayes' theorem, this reduces to the symmetric form (Eq. A2.16) for the mutual information

$$<I> = \Sigma_{s,r} \ P(s,r) \ \log_2 P(s,r)/P(s)P(r) \qquad (A2.20)$$

which emphasizes that responses tell us about stimuli just as much as stimuli tell us about responses. This is, of course, a general feature, independent of the two variables being in this instance stimuli and neuronal responses. In fact, what is of interest, beside the mutual information of Eqs A2.19–A2.20, is often the information specifically conveyed about each stimulus,

$$I(s) \; = \; \Sigma_r \; P(r \,|\, s) \; \log_2 \; P(r \,|\, s)/P(r) \qquad\qquad (A2.21)$$

which is a direct quantification of the variability in the responses elicited by that stimulus, compared to the overall variability. Since $P(r)$ is the probability distribution of responses averaged across stimuli, it is again evident that the stimulus-specific information measure of Eq. A2.21 depends not only on the stimulus s, but also on all other stimuli used. Likewise, the mutual information measure, despite being of an average nature, is dependent on what set of stimuli has been used in the average. This emphasizes again the relative nature of all information measures. More specifically, it underscores the relevance of using, while measuring the information conveyed by a given neuronal population, stimuli that are either representative of real-life stimulus statistics, or of particular interest for the properties of the population being examined.

A numerical example

To make these notions clearer, we can consider a specific example in which the response of a cell to the presentation of, say, one of four odours (A, B, C, D) is recorded for 10 ms, during which the cell emits either 0, 1, or 2 spikes, but no more. Imagine that the cell tends to respond more vigorously to smell B, less to C, even less to A, and never to D, as described by the following table of conditional probabilities $P(r \mid s)$:

	$r=0$	$r=1$	$r=2$
$s=A$	0.6	0.4	0.0
$s=B$	0.0	0.2	0.8
$s=C$	0.4	0.5	0.1
$s=D$	1.0	0.0	0.0

then, if different odours are presented with equal probability, the table of joint probabilities $P(s,r)$ will be

	$r=0$	$r=1$	$r=2$
$s=A$	0.15	0.1	0.0
$s=B$	0.0	0.05	0.2
$s=C$	0.1	0.125	0.025
$s=D$	0.25	0.0	0.0

From these two tables one can compute various information measures by directly applying the definitions above. Since odours are presented with equal probability, $P(s) = 1/4$, the entropy of the stimulus set, which corresponds to the maximum amount of information any transmission channel, no matter how efficient, could convey on the identity of the odours, is $H_s = -\Sigma_s[P(s)\log_2 P(s)] = -4[(1/4)\log_2(1/4)] = \log_2 4 = 2$ bits. There is a more stringent upper bound on the mutual information that this cell's responses convey on the odours, however, and this second bound is the channel capacity T of the cell. Calculating this quantity involves maximizing the mutual information across prior odour probabilities, and it is a bit complicated to do, in general. In our particular case the maximum information is obtained when only odours B and D are presented, each with probability 0.5. The resulting capacity is $T = 1$ bit. We can easily calculate, in general, the entropy of the responses. This is not an upper bound characterizing the source, like the entropy of the stimuli, nor an upper bound characterizing the channel, like the capacity, but simply a bound on the mutual information for this specific combination of source (with its related odour probabilities) and channel (with its conditional probabilities). Since only three response levels are possible within the short recording window, and they occur with uneven probability, their entropy is considerably lower than H_s, at $H_r = -\Sigma_r P(r)\log_2 P(r) = -P(0)\log_2 P(0) - P(1)\log_2 P(1) - P(2)\log_2 P(2) = -0.5\log_2 0.5 - 0.275\log_2 0.275 - 0.225\log_2 0.225 = 1.496$ bits. The actual average information I that the responses transmit about the stimuli, which is a measure of the correlation in the variability of stimuli and responses, does not exceed the absolute variability of either stimuli (as quantified by the first bound) or responses (as quantified by the last bound), nor the capacity of the channel. An explicit calculation using the joint probabilities of the second table into expression A2.20 yields $I = 0.733$ bits. This is of course only the average value, averaged both across stimuli and across responses.

The information conveyed by a particular response can be larger. For example, when the cell emits two spikes it indicates with a relatively large probability odour B, and this is reflected in the fact that it then transmits, according to expression A2.18, $I(r=2) = 1.497$ bits, more than double the average value.

Similarly, the amount of information conveyed about each individual odour varies with the odour, depending on the extent to which it tends to elicit a differential response. Thus, expression A2.21 yields that only $I(s=C) = 0.185$ bits are conveyed on average about odour C, which tends to elicit responses with similar statistics to the average statistics across odours, and therefore not easily interpretable. On the other hand, exactly 1 bit of information is conveyed about odour D, since this odour never elicits any response, and when the cell emits no spike there is a probability of 1/2 that the stimulus was odour D.

A2.1.5 The information conveyed by continuous variables

A general feature, relevant also to the case of neuronal information, is that if, among a *continuum* of a priori possibilities, only one, or a discrete number, remains a posteriori, the information is strictly infinite. This would be the case if one were told, for example, that Reading is exactly 10' west, 1' north of London. The a priori probability of precisely this set of coordinates among the continuum of possible ones is zero, and then the information diverges to infinity. The problem is only theoretical, because in fact, with continuous distributions, there is always one or several factors that limit the resolution in the a posteriori knowledge, rendering the information finite. Moreover, when considering the mutual information in the conjoint probability of occurrence of two sets, e.g. stimuli and responses, it suffices that at least one of the sets is discrete to make matters easy, that is, finite. Nevertheless, the identification and appropriate consideration of these resolution-limiting factors in practical cases may require careful analysis.

Example: the information retrieved from an autoassociative memory

One example is the evaluation of the information that can be retrieved from an autoassociative memory. Such a memory stores a number of firing patterns, each one of which can be considered, as in Chapter 3, as a vector $\mathbf{r}^\mu$ with components the firing rates $\{r_i^\mu\}$, where the subscript i indexes the cell (and the superscript μ indexes the pattern). In retrieving pattern μ, the network in fact produces a distinct firing pattern, denoted for example simply as $\mathbf{r}$. The quality of retrieval, or the similarity between $\mathbf{r}^\mu$ and $\mathbf{r}$, can be measured by the average mutual information

$$\langle I(\mathbf{r}^\mu,\mathbf{r}) \rangle = \Sigma_{\mathbf{r}',\mathbf{r}}\, P(\mathbf{r}^\mu,\mathbf{r})\, \log_2 P(\mathbf{r}^\mu,\mathbf{r})/P(\mathbf{r}^\mu)P(\mathbf{r}) \approx$$
$$\Sigma_i \Sigma_{r',r}\, P(r_i^\mu,r_i)\, \log_2 P(r_i^\mu,r_i)/P(r_i^\mu)P(r_i). \qquad (A2.22)$$

In this formula the 'approximately equal' sign $\approx$ marks a simplification which is not necessarily a reasonable approximation. If the simplification is valid, it means that in order to extract an information measure, one need not compare whole vectors (the entire firing patterns) with each other, and may instead compare the firing rates of individual cells at storage and retrieval, and sum the resulting single-cell information values. The validity of the simplification is a matter which will be discussed later and which has to be verified, in the end, experimentally, but for the purposes of the present discussion we can focus on the single-cell terms. If either r_i or r_i^μ has a continuous distribution of values, as it will if it represents not the number of spikes emitted in a fixed window, but more generally the firing rate of neuron i computed by convolving the firing train with a smoothing kernel, then one has to deal with probability densities, which we denote as $p(r)dr$, rather than the usual probabilities $P(r)$. Substituting $p(r)dr$ for $P(r)$ and $p(r^\mu,r)drdr^\mu$ for $P(r^\mu,r)$, one can write for each single-cell contribution (omitting the cell index i)

$$\langle I(r^\mu,r) \rangle_i = \int dr^\mu dr\, p(r^\mu,r)\, \log_2 p(r^\mu,r)/p(r^\mu)p(r) \qquad (A2.23)$$

and we see that the differentials $dr^\mu dr$ cancel out between numerator and denominator inside the logarithm, rendering the quantity well defined and finite. If, however, r^μ were to *exactly* determine r, one would have

$$p(r^\mu,r)\, dr^\mu dr = p(r^\mu)\delta(r-r(r^\mu))dr^\mu dr = p(r^\mu)dr^\mu \qquad (A2.24)$$

and, by losing one differential on the way, the mutual information would become infinite. It is therefore important to consider what prevents r^μ from fully determining r in the case at hand — in other words, to consider the sources of noise in the system. In an autoassociative memory storing an extensive number of patterns (see Appendix A4), one source of noise always present is the interference effect due to the concurrent storage of all other patterns. Even neglecting other sources of noise, this produces a finite resolution width ρ, which allows one to write some expression of the type $p(r \mid r^\mu)\, dr = \exp{-(r-r(r^\mu))^2/2\rho^2}\, dr$ which ensures that the information is finite as long as the resolution ρ is larger than zero.

One further point that should be noted, in connection with estimating the information retrievable from an autoassociative memory, is that the mutual information between the current distribution of firing rates and that of the stored pattern does not coincide with the information *gain* provided by the memory device. Even when firing rates, or spike counts, are all that matters in terms of information carriers, as in the networks considered in this book, one more term should be taken into account in evaluating the information gain. This term, to be subtracted, is the information contained in the external input that elicits the retrieval. This may vary a lot from the retrieval of one particular memory to the next, but of course an efficient memory device is one that is able, when needed, to retrieve much more information than it requires to be present in the inputs, that is, a device that produces a large information gain.

Finally, one should appreciate the conceptual difference between the information a firing pattern carries about another one (that is, about the pattern stored), as considered above, and two different notions: (a) the information produced by the network in selecting the correct memory pattern and (b) the information a firing pattern carries about something in the outside world. Quantity (a), the information intrinsic to selecting the memory pattern, is ill defined when analysing a real system, but is a well-defined and particularly simple notion when considering a formal model. If p patterns are stored with equal strength, and the selection is errorless, this amounts to $\log_2 p$ bits of information, a quantity often, but not always, small compared to the information in the pattern itself. Quantity (b), the information conveyed about some outside correlate, is not defined when considering a formal model that does not include an explicit account of what the firing of each cell represents, but is well defined and measurable from the recorded activity of real cells. It is the quantity considered in the numerical example with the four odours, and it can be generalized to the information carried by the activity of several cells in a network, and specialized to the case that the network operates as an associative memory. One may note, in this case, that the capacity to retrieve memories with high fidelity, or high information content, is only useful to the extent that the representation to be retrieved carries that amount of information about something relevant — or, in other words, that it is pointless to store and retrieve with great care largely meaningless messages. This type of argument has been used to discuss the role of the mossy fibres in the operation of the CA3 network in the hippocampus (Treves and Rolls, 1992; and Chapter 6).

A2.2 Estimating the information carried by neuronal responses

A2.2.1 The limited sampling problem

We now discuss more in detail the application of these general notions to the information transmitted by neurons. Suppose, to be concrete, that an animal has been presented with stimuli drawn from a discrete set $\mathbb{S}$, and that the responses of a set of C cells have been recorded following the presentation of each stimulus. We may choose any quantity or set of quantities to characterize the responses, for example let us assume that we consider the firing rate of each cell, r_i, calculated by convolving the spike response with an appropriate smoothing kernel. The response space is then C times the continuous set of all positive real numbers, $(\mathbf{R}/2)^C$. We want to evaluate the average information carried by such responses about which stimulus was shown. In principle, it is straightforward to apply the above formulas, e.g. in the form

$$<I(s,\mathbf{r})> \; = \; \Sigma_s \, P(s) \int \Pi_i \, dr_i \; p(\mathbf{r}\,|\,s) \, \log_2 p(\mathbf{r}\,|\,s)/p(\mathbf{r}) \qquad (A2.25)$$

where it is important to note that $p(\mathbf{r})$ and $p(\mathbf{r}\,|\,s)$ are now probability densities defined over the high-dimensional vector space of multi-cell responses. The product sign Π signifies that this whole vector space has to be integrated over, along all its dimensions. $p(\mathbf{r})$ can be calculated as $\Sigma_s \, p(\mathbf{r}\,|\,s)\, P(s)$, and therefore, in principle, all one has to do is to estimate, from the data, the conditional probability densities $p(\mathbf{r}\,|\,s)$ — the distributions of responses following each stimulus. In practice, however, in contrast to what happens with formal models, in which there is usually no problem in calculating the exact probability densities, real data come in limited amounts, and thus sample only sparsely the vast response space. This limits the accuracy with which from the experimental *frequency* of each possible response we can estimate its *probability*, in turn seriously impairing our ability to estimate $<I>$ correctly. We refer to this as the limited sampling problem. This is a purely technical problem that arises, typically when recording from mammals, because of external constraints on the duration or number of repetitions of a given set of stimulus conditions. With computer simulation experiments, but also with recordings from, for example, insects, sufficient data can usually be obtained that straightforward estimates of information are accurate enough (Strong, Koberle, de Ruyter van Steveninck and Bialek, 1996; Golomb, Kleinfeld, Reid, Shapley and Shraiman, 1994). The problem is, however, so serious in connection with recordings from monkeys and rats, that it is worthwhile to discuss it, in order to appreciate the scope and limits of applying information theory to neuronal processing.

In particular, if the responses are continuous quantities, the probability of observing exactly the same response twice is infinitesimal. In the absence of further manipulation, this would imply that each stimulus generates its own set of unique responses, therefore any response that has actually occurred could be associated unequivocally with one stimulus, and the mutual information would always equal the entropy of the stimulus set. This absurdity shows that in order to estimate probability densities from experimental frequencies one has to resort to some *regularizing* manipulation, such as smoothing the point-like response values by convolution with suitable kernels, or binning them into a finite number of discrete bins.

A numerical example

The problem is illustrated in Fig. A2.1. In the figure, we have simulated a simple but typical situation, in which we try to estimate how much information, about which of a set of two stimuli was shown, can be extracted from the firing rate of a single cell. We have assumed that the true underlying probability density of the response to each stimulus is Gaussian, with

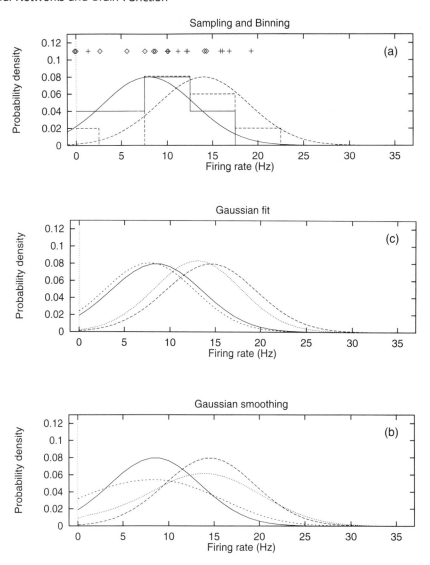

Fig. A2.1 The probability distributions used in the example in the text. (a) The two original Gaussian distributions (solid curve for the one centred at 8 Hz, long-dashed curve for the one centred at 14 Hz), the 10 points which provided the sampling of each (◇ and +, respectively), and the distributions obtained by binning the samples into 5 Hz bins. (b) The two original distributions compared with those obtained by smoothing the samples with Gaussian kernels (the short-dash curve applies to the lower distribution and the dotted curve to the higher one). (c) The original distributions compared with those obtained after a Gaussian fit (with the same curves as in (b)).

width $\sigma = 5$ Hz and mean value $r_1 = 8$ Hz and $r_2 = 14$ Hz respectively; and that we have sampled each Gaussian 10 times (i.e. 10 trials with each stimulus). How much information is present in the response, on average? If we had the underlying probability density available, we could easily calculate $<I>$, which with our choice of parameters would turn out to be 0.221

bits. Our problem, however, is how to estimate the underlying probability distributions from the 10 trials available for each stimulus. Several strategies are possible. One is to discretize the response space into bins, and estimate the probability density as the histogram of the fraction of trials falling into each bin. Choosing bins of width $\Delta r = 1$ Hz would yield $<I> = 0.900$ bits; the bins are so narrow that almost every response is in a different bin, and then the estimated information is close to its maximum value of 1 bit. Choosing bins of larger width, $\Delta r = 5$ Hz, one would end up with $<I> = 0.324$; a value closer to the true one but still overestimating it by half, even though the bin width now matches the standard deviation of each underlying distribution. Alternatively, one may try to 'smooth' the data by convolving each response with a Gaussian kernel of width the standard deviation measured for each stimulus. In our particular case the measured standard deviations are very close to the true value of 5 Hz, but one would get $<I> = 0.100$, i.e. one would underestimate by more than half the true information value, by oversmoothing. Yet another possibility is to make a bold assumption as to what the general shape of the underlying densities should be, for example a Gaussian. Even in our case, in which the underlying density was indeed chosen to be Gaussian, this procedure yields a value, $<I> = 0.187$, off by 16% (and the error would be much larger, of course, if the underlying density had been of very different shape).

The effects of limited sampling

The crux of the problem is that, whatever procedure one adopts, limited sampling tends to produce distortions in the estimated probability densities. The resulting mutual information estimates are intrinsically biased. The bias, or average error of the estimate, is upward if the raw data have not been regularized much, and is downward if the regularization procedure chosen has been heavier. The bias can be, if the available trials are few, much larger that the true information values themselves. This is intuitive, as fluctuations due to the finite number of trials available would tend, on average, to either produce or emphasize differences among the distributions corresponding to different stimuli, differences that are preserved if the regularization is 'light', and that are interpreted in the calculation as carrying genuine information. This is particularly evident if one considers artificial data produced by using the 'same' distribution for both stimuli, as illustrated in Fig. A2.2. Here, the 'true' information should be zero, yet all of our procedures — many more are possible, but the same effect would emerge — yield finite amounts of information. The actual values obtained with the four methods used above are, in order, $<I> = 0.462, 0.115, 0.008$, and 0.007 bits. Clearly, 'heavier' forms of regularization produce less of such artefactual information; however, if the regularization is heavy, even the underlying meaningful differences between distributions can be suppressed, and information is underestimated. Choosing the right amount of regularization, or the best regularizing procedure, is not possible a priori. Hertz et al. (1992) have proposed the interesting procedure of using an artificial neural network to regularize the raw responses. The network can be trained on part of the data using backpropagation, and then used on the remaining part to produce what is in effect a clever data-driven regularization of the responses. This procedure is, however, rather computer intensive and not very safe, as shown by some self-evident inconsistency in the results (Heller et al., 1995). Obviously, the best way to deal with the limited sampling problem is to try and use as many trials as possible.

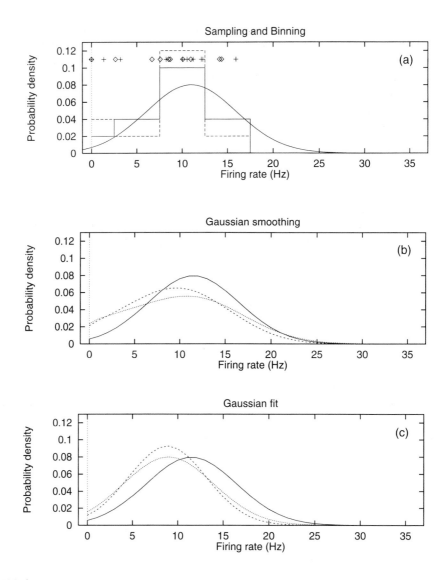

Fig. A2.2 The same manipulations applied in Fig. A2.1 are applied here to samples derived from the *same* distribution, a Gaussian centred at 11 Hz. Notation as in Fig. A2.1. The information values obtained from the various manipulations are reported in the text.

The improvement is slow, however, and generating as many trials as would be required for a reasonably unbiased estimate is often, in practice, impossible.

A2.2.2 Correction procedures for limited sampling

The above observation, that data drawn from a single distribution, when artificially paired, at random, to different stimulus labels, results in 'spurious' amounts of apparent information, suggests a simple way of checking the reliability of estimates produced from real data (Optican et al., 1991). One can disregard the true stimulus associated with each response, and generate a randomly reshuffled pairing of stimuli and responses, which should therefore, being not linked by any underlying relationship, carry no mutual information about each other. Calculating, with some procedure of choice, the spurious information obtained in this way, and comparing with the information value estimated with the same procedure for the real pairing, one can get a feeling for how far the procedure goes into eliminating the apparent information due to limited sampling. Although this spurious information, I_s, is only indicative of the amount of bias affecting the original estimate, a simple heuristic trick (called 'bootstrap'[1]) is to subtract the spurious from the original value, to obtain a somewhat 'corrected' estimate. The following table shows to what extent this trick — or a similar correction based on subtracting the 'square' of the spurious fraction of information (Optican et al., 1991) — works on our artificial data of Fig. A2.1, when averaged over 1000 different random instantiations of the two sets of ten responses. The correct value, again, is $<I> = 0.221$ bits.

Method	Binning $\Delta r = 1$ Hz	Binning $\Delta r = 5$ Hz	Gaussian smoothing	Gaussian fit
I (raw)	0.900	0.324	0.100	0.187
$I - I_s$	0.421	0.203	0.091	0.179
$I - (I_s)^2/I$	0.645	0.279	0.099	0.187

A different correction procedure (called 'jack-knife') is based on the assumption that the bias is proportional to $1/N$, where N is the number of responses (data points) used in the estimation. One computes, beside the original estimate $<I_N>$, N auxiliary estimates $<I_{N-1}>_k$, by taking out from the data set response k, where k runs across the data set from 1 to N. The corrected estimate

$$<I> = N<I_N> - (1/N) \Sigma_k (N-1)<I_{N-1}>_k \qquad (A2.26)$$

is free from bias (to leading order in $1/N$), if the proportionality factor is more or less the same in the original and auxiliary estimates. This procedure is very time-consuming, and it suffers from the same imprecision of any algorithm that tries to determine a quantity as the result of the subtraction of two large and nearly equal terms; in this case the terms have been made large on purpose, by multiplying them by N and $N-1$.

A more fundamental approach (Miller, 1955) is to derive an analytical expression for the bias (or, more precisely, for its leading terms in an expansion in $1/N$, the inverse of the sample size). This allows the estimation of the bias from the data itself, and its subsequent subtraction, as discussed in Treves and Panzeri (1995) and Panzeri and Treves (1996). Such

---

[1] In technical usage bootstrap procedures utilize random pairings of responses with stimuli with replacement, while shuffling procedures utilize random pairings of responses with stimuli without replacement.

a procedure produces satisfactory results, thereby lowering the size of the sample required for a given accuracy in the estimate by about an order of magnitude (Golomb *et al.*, 1997). However, it does not, in itself, make possible measures of the information contained in very complex responses with few trials. As a rule of thumb, the number of trials per stimulus required for a reasonable estimate of information, once the subtractive correction is applied, is of the order of the effectively independent (and utilized) bins in which the response space can be partitioned (Panzeri and Treves, 1996).

A2.2.3 Decoding procedures for multiple cell responses

The bias of information measures grows with the dimensionality of the response space, and for all practical purposes the limit on the number of dimensions that can lead to reasonably accurate direct measures, even when applying a correction procedure, is quite low, two to three. This implies, in particular, that it is not possible to apply Eq. A2.25 to extract the information content in the responses of several cells (more than two to three) recorded simultaneously. One way to address the problem is then to apply some strong form of regularization to the multiple cell responses. Smoothing has already been mentioned as a form of regularization which can be tuned from very soft to very strong, and which preserves the structure of the response space. Binning is another form, which changes the nature of the responses from continuous to discrete, but otherwise preserves their general structure, and which can also be tuned from soft to strong. Other forms of regularization involve much more radical transformations, or changes of variables. Of particular interest for information estimates is a change of variables that transforms the response space into the stimulus set, by applying an algorithm that derives a predicted stimulus from the response vector, i.e. the firing rates of all the cells, on each trial. Applying such an algorithm is called *decoding*. Of course, the predicted stimulus is not necessarily the same as the actual one. Therefore the term decoding should not be taken to imply that the algorithm works successfully, each time identifying the actual stimulus. The predicted stimulus is simply a function of the response, as determined by the algorithm considered. Just as with any regularizing transform, it is possible to compute the mutual information between actual and predicted stimuli, instead of the original one between stimuli and responses. Since information about (real) stimuli can only be lost and not be created by the transform, the information measured in this way is bound to be lower in value than the real information in the responses. If the decoding algorithm is efficient, it manages to preserve nearly all the information contained in the raw responses, while if it is poor, it loses a large portion of it. If the responses themselves provided *all* the information about stimuli, *and* the decoding is optimal, then predicted stimuli coincide with the actual stimuli, and the information extracted equals the entropy of the stimulus set.

The procedure of extracting information values after applying a decoding algorithm is schematized below

$$s \Longrightarrow r \longrightarrow s'$$
$$\llcorner\!-I\,(s,r)\!-\!\lrcorner$$
$$\llcorner\!\rule{1cm}{0pt}\!-I\,(s,s')\!\rule{1cm}{0pt}\!\lrcorner$$

where the double line indicates the transformation from stimuli to responses operated by the nervous system, while the single line indicates the further transformation operated by the decoding procedure.

A slightly more complex variant of this procedure is a decoding step that extracts from the response on each trial not a single predicted stimulus, but rather probabilities that each of the possible stimuli was the actual one. The joint probabilities of actual and posited stimuli can be averaged across trials, and information computed from the resulting probability matrix ($S \times S$). Computing information in this way takes into account the relative uncertainty in assigning a predicted stimulus to each trial, an uncertainty that is instead not considered by the previous procedure based solely on the identification of the maximally likely stimulus (Treves, 1997). *Maximum likelihood* information values I_{ml} tend therefore to be higher than *probability* information values I_p, although in very specific situations the reverse could also be true.

The same correction procedures for limited sampling can be applied to information values computed after a decoding step. Values obtained from maximum likelihood decoding, I_{ml}, suffer from limited sampling more than those obtained from probability decoding, I_p, since each trial contributes a whole 'brick' of weight $1/N$ (N being the total number of trials), whereas with probabilities each brick is shared among several slots of the ($S \times S$) probability matrix. The neural network procedure devised by Hertz *et al.* (1992) and co-workers can in fact be thought of as a decoding procedure based on probabilities, which deals with limited sampling not by applying a correction but rather by strongly regularizing the original responses. When decoding is used, the rule of thumb becomes that the minimal number of trials per stimulus required for accurate information measures is roughly equal to the size of the stimulus set, if the subtractive correction is applied (Panzeri and Treves, 1996).

Decoding algorithms

Any transformation from the response space to the stimulus set could be used in decoding, but of particular interest are the transformations that either approach optimality, so as to minimize information loss and hence the effect of decoding, or else are implementable by mechanisms that could conceivably be operating in the real system, so as to extract information values that *could* be extracted by the system itself.

The optimal transformation is in theory well-defined: one should estimate from the data the conditional probabilities $P(r \mid s)$, and use Bayes' rule to convert them into the conditional probabilities $P(s' \mid r)$. Having these for any value of r, one could use them to estimate I_p, and, after selecting for each particular real response the stimulus with the highest conditional probability, to estimate I_{ml}. To avoid biasing the estimation of conditional probabilities, the responses used in estimating $P(r \mid s)$ should not include the particular response for which $P(s' \mid r)$ is going to be derived (jack-knife cross validation). In practice, however, the estimation of $P(r \mid s)$ in usable form involves the fitting of some simple function to the responses. This need for fitting, together with the approximations implied in the estimation of the various quantities, prevents us from defining the really optimal decoding, and leaves us with various algorithms, depending essentially on the fitting function used, which are hopefully close to optimal in some conditions. We have experimented extensively with two such algorithms, that both approximate Bayesian decoding (Rolls, Treves and Tovee, 1997). Both these algorithms fit the response vectors produced over several trials by the cells being recorded to a product of conditional probabilities for the response of each cell given the stimulus. In one case the single cell conditional probability is assumed to be Gaussian (truncated at zero), in the other it is assumed to be Poisson (with an additional weight at zero). Details of these algorithms are given by Rolls, Treves and Tovee (1997).

Biologically plausible decoding algorithms are those that limit the algebraic operations used to types that could be easily implemented by neurones, e.g. dot product summations, thresholding and other single-cell nonlinearities, competition and contrast enhancement among the outputs of nearby cells. There is then no need for ever fitting functions or other sophisticated approximations, but of course the degree of arbitrariness in selecting a particular algorithm remains substantial, and a comparison among different choices based on which yields the higher information values may favour one choice in a given situation and another choice with a different data set.

To summarize, the key idea in decoding, in our context of estimating information values, is that it allows substitution of a possibly very high-dimensional response space (which is difficult to sample and regularize) with a reduced object much easier to handle, that is with a discrete set equivalent to the stimulus set. The mutual information between the new set and the stimulus set is then easier to estimate even with limited data, and if the assumptions about population coding, underlying the particular decoding algorithm used, are justified, the value obtained approximates the original target, the mutual information between stimuli and responses. For each response recorded, one can use all the responses except for that one to generate estimates of the average response vectors (the response for each unit in the population) to each stimulus. Then one considers how well the selected response vector matches the average response vectors, and uses the degree of matching to estimate, for all stimuli, the probability that they were the actual stimuli. The form of the matching embodies the general notions about population encoding, for example the 'degree of matching' might be simply the dot product between the current vector and the average vector ($\mathbf{r}^{av}$), suitably normalized over all average vectors to generate probabilities

$$P(s' \mid \mathbf{r}(s)) = \mathbf{r}(s) \cdot \mathbf{r}^{av}(s')/(\Sigma_{s''} \mathbf{r}(s) \cdot \mathbf{r}^{av}(s'')) \tag{A2.27}$$

where s'' is a dummy variable. One ends up, then, with a table of conjoint probabilities $P(s,s')$, and another table obtained by selecting for each trial the most likely (or predicted) stimulus s^p, $P(s,s^p)$. Both s' and s^p stand for all possible stimuli, and hence belong to the same set S. These can be used to estimate mutual information values based on probability decoding (I_p) and on maximum likelihood decoding (I_{ml})

$$<I_p> = \Sigma_{s,s'} P(s,s') \log_2 P(s,s')/P(s)P(s') \tag{A2.28}$$

and

$$<I_{ml}> = \Sigma_{s,s^p} P(s,s^p) \log_2 P(s,s^p)/P(s)P(s^p) \tag{A2.29}$$

A2.3 Main results obtained from applying information-theoretic analyses

Although information theory provides the natural mathematical framework for analysing the performance of neuronal systems, its applications in neuroscience have been for many years rather sparse and episodic (e.g. MacKay and McCulloch, 1952; Eckhorn and Pöpel, 1974, 1975; Eckhorn *et al.*, 1976). One reason for this lukewarm interest in information theory applications has certainly been the great effort that was apparently required, due essentially to the limited sampling problem, in order to obtain reliable results. Another reason has been the hesitation towards analysing as a single complex 'black-box' large neuronal systems all the way from some external, easily controllable inputs, down to neuronal outputs in some central cortical area of interest, for example including all visual stations from the periphery to the end of the ventral visual stream in the temporal lobe. In fact, two important bodies of work, that have greatly helped revive interest in applications of the theory in recent years, both sidestep these two problems. The problem with analyzing a huge black-box is avoided by considering systems at the sensory periphery; the limited sampling problem is avoided either by working with insects, in which sampling can be extensive (Bialek *et al.*, 1991; de Ruyter van Steveninck and Laughlin, 1996; Rieke *et al.*, 1996) or by utilizing a formal model instead of real data (Atick and Redlich, 1990; Atick, 1992). Both approaches have provided insightful quantitative analyses that are in the process of being extended to more central mammalian systems (see e.g. Atick *et al.*, 1996).

A2.3.1 Temporal codes versus rate codes (at the single unit level)

In the third of a series of papers which analyse the response of single units in the primate inferior temporal cortex to a set of static visual stimuli, Optican and Richmond (1987) have applied information theory in a particularly direct and useful way. To ascertain the relevance of stimulus-locked temporal modulations in the firing of those units, they have compared the amount of information about the stimuli that could be extracted from just the firing rate, computed over a relatively long interval of 384 ms, with the amount of information that could be extracted from a more complete description of the firing, that included temporal modulation. To derive this latter description (the temporal *code*) they have applied principal component analysis (PCA) to the temporal response vectors recorded for each unit on each trial. A temporal response vector was defined as a vector with as components the firing rates in each of 64 successive 6 ms time bins. The (64 × 64) covariance matrix was calculated across all trials of a particular unit, and diagonalized. The first few eigenvectors of the matrix, those with the largest eigenvalues, are the principal components of the response, and the weights of each response vector on these four to five components can be used as a reduced description of the response, which still preserves, unlike the single value giving the mean firing rate along the entire interval, the main features of the temporal modulation within the interval. Thus a four to five-dimensional temporal code could be contrasted with a one-dimensional rate code, and the comparison made quantitative by measuring the respective values for the mutual information with the stimuli. Although the initial claim, that the temporal code carried nearly three times as much information as the rate code, was later found to be an artefact of limited sampling, and more recent analyses tend to minimize the additional information in the temporal description (Optican *et al.*, 1991; Eskandar *et al.*, 1992; Tovee *et al.*, 1993; Heller *et al.*, 1995), this type of application has immediately appeared straightforward and important, and it has led to many developments. By concentrating on the code expressed in the output rather than on the characterization of the neuronal channel itself, this approach is not affected much by the potential complexities of the preceding black box. Limited sampling, on the other hand, *is* a problem, particularly because it affects much more codes with a larger number of components, for example the four to five components of the PCA temporal description, than the one-dimensional firing rate code. This is made evident in the Heller *et al.* (1995) paper, in which the comparison is extended to several more detailed temporal descriptions, including a binary vector description in which the presence or not of a spike in each 1 ms bin of the response constitutes a component of a 320-dimensional vector. Obviously, this binary vector must contain at least all the information present in the reduced descriptions, whereas in the results of Heller *et al.* (1995), despite the use of the sophisticated neural network procedure to control limited sampling biases, the binary vector appears to be the code that carries the least information of all. In practice, with the data samples available in the experiments that have been done, and even when using the most recent procedures to control limited sampling, reliable comparison can be made only with up to two- to three-dimensional codes.

Overall, the main result of these analyses applied to the responses to static stimuli in the temporal visual cortex of primates is that not much more information (perhaps only up to 10% more) can be extracted from temporal codes than from the firing rate measured over a judiciously chosen interval (Tovee *et al.*, 1993; Heller *et al.*, 1995). In earlier visual areas this additional fraction of information can be larger, due especially to the increased relevance, earlier on, of precisely locked *transient* responses (Kjaer *et al.*, 1994; Golomb *et al.*, 1994; Heller *et al.*, 1995). This is because if the responses to some stimuli are more transient and to others more sustained, this will result in more information in the temporal modulation of the response. A similar effect arises from differences in the mean response *latency* to different stimuli (Tovee *et al.*, 1993). However, the relevance of more substantial temporal codes for static visual stimuli remains to be demonstrated. For non-static visual stimuli and for other cortical systems, similar analyses have largely yet to be carried out, although clearly one expects to find much more prominent temporal effects e.g. in the auditory system (Nelken *et al.*, 1994; deCharms and Merzenich, 1996).

A2.3.2 The speed of information transfer

It is intuitive that if short periods of firing of single cells are considered, there is less time for temporal modulation effects. The information conveyed about stimuli by the firing rate and that conveyed by more detailed temporal codes become similar in value. When the firing periods analysed become shorter than roughly the mean interspike interval, even the statistics of firing rate values on individual trials cease to be relevant, and the information content of the firing depends solely on the mean firing rates across all trials with each stimulus. This is expressed mathematically by considering the amount of information provided as a function of the length t of the time window over which firing is analysed, and taking the limit for $t \to 0$ (Skaggs *et al.*, 1993; Panzeri *et al.*, 1996b). To first order in t, only two responses can occur in a short window of length t: either the emission of an action potential, with probability tr_s, where r_s is the mean firing rate calculated over many trials using the same window and stimulus; or no action potential, with probability $1-tr_s$. Inserting these conditional probabilities into Eq. A2.21, taking the limit and dividing by t, one obtains for the derivative of the stimulus-specific transinformation

$$dI(s)/dt = r_s \log_2(r_s/<r>) + (<r>-r_s)/\ln 2, \quad (A2.30)$$

where $<r>$ is the grand mean rate across stimuli. This formula thus gives the rate, in bits/s, at which information about a stimulus begins to accumulate when the firing of a cell is recorded. Such information rate depends only on the mean firing rate to that stimulus and on the grand mean rate across stimuli. As a function of r_s, it follows the U-shaped curve in Fig. A2.3. The curve is universal, in the sense that it applies irrespective of the detailed firing statistics of the cell, and it expresses the fact that the emission or not of a spike in a short window conveys information inasmuch as the mean response to a given stimulus is above or below the overall mean rate. No information is conveyed about those stimuli the mean response to which is the same as the overall mean. In practice, although the curve describes only the universal behaviour of the initial slope of the specific information as a function of time, it approximates well the full specific information computed even over rather long periods (Rolls, Critchley and Treves, 1996; Rolls, Treves, Tovee and Panzeri, 1997).

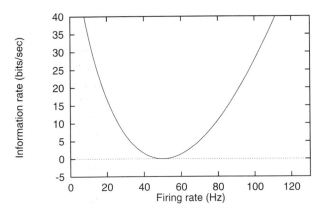

Fig. A2.3 Time derivative of the stimulus specific information as a function of firing rate, for a cell firing at a grand mean rate of 50 Hz. For different grand mean rates, the graph would be simply rescaled.

Averaging Eq. A2.30 across stimuli one obtains the time derivative of the mutual information. Further dividing by the overall mean rate yields the adimensional quantity

$$\chi = \Sigma_s P(s)(r_s/<r>) \log_2(r_s/<r>) \quad (A2.31)$$

which measures, in bits, the mutual information per spike provided by the cell (Bialek et al., 1991; Skaggs et al., 1993). One can prove that this quantity can range from 0 to $\log_2 (1/a)$

$$0 < \chi < \log_2(1/a), \quad (A2.32)$$

where a is the sparseness defined in Chapter 1. For mean rates r_s distributed in a nearly binary fashion, χ is close to its upper limit $\log_2(1/a)$, whereas for mean rates that are nearly uniform, or at least unimodally distributed, χ is relatively close to zero (Panzeri et al., 1996b). In practice, whenever a large number of more or less 'ecological' stimuli are considered, mean rates are not distributed in arbitrary ways, but rather tend to follow stereotyped distributions (Panzeri et al., 1996a), and as a consequence χ and a (or, equivalently, its logarithm) tend to covary rather than to be independent variables (Skaggs and McNaughton, 1992). Therefore, measuring sparseness is in practice nearly equivalent to measuring information per spike, and the rate of rise in mutual information, $\chi <r>$, is largely determined by the sparseness a and the overall mean firing rate $<r>$. Another quantity measuring which is equivalent to measuring the information per spike is the breadth of tuning B (Smith and Travers, 1979; see Section 10.4, where the breadth of tuning index is defined), at least when n stimuli are presented with equal frequency. It is easy to show that

$$\chi = (1-B) \log_2 n \qquad (A2.33)$$

so that extreme selectivity corresponds to $B=0$ and $\chi = \log_2 n$, whereas absence of any tuning corresponds to $B=1$ and $\chi =0$. Equations A2.32 and A2.33 can be turned into an inequality for the breadth of tuning as a function of sparseness, or vice versa, e.g.

$$a < n^{B-1}. \qquad (A2.34)$$

The important point to note about the single-cell information rate $\chi <r>$ is that, to the extent that different cells express non-redundant codes, as discussed below, the instantaneous *information flow* across a population of C cells can be taken to be simply $C\chi <r>$, and this quantity can easily be measured directly without major limited sampling biases, or else inferred indirectly through measurements of the sparseness a. Values for the information rate $\chi <r>$ that have been published range from 2–3 bits/s for rat hippocampal cells (Skaggs et al., 1993) to 10–30 bits/s for primate temporal cortex visual cells (Rolls, Treves, Tovee and Panzeri, 1997), and could be compared with analogous measurements in the sensory systems of frogs and crickets, in the 100–300 bits/sec range (Rieke et al., 1993).

If the first time-derivative of the mutual information measures information flow, successive derivatives characterize, at the single-cell level, different firing modes. This is because whereas the first derivative is universal and depends only on the mean firing rates to each stimulus, the next derivatives depend also on the variability of the firing rate around its mean value, across trials, and take different forms in different firing regimes. Thus they can serve as a measure of discrimination among firing regimes with limited variability, for which, for example, the second derivative is large and positive, and firing regimes with large variability, for which the second derivative is large and negative. Poisson firing, in which in every short period of time there is a fixed probability of emitting a spike irrespective of previous firing, is an example of large variability, and the second derivative of the mutual information can be calculated to be

$$d^2I/dt^2 = [\ln a + (1-a)]<r>^2/(a \ln 2), \qquad (A2.35)$$

where a is the sparseness defined in Chapter 1. This quantity is always negative. Strictly periodic firing is an example of zero variability, and in fact the second time-derivative of the mutual information becomes infinitely large in this case (although actual information values measured in a short time interval remain of course finite even for exactly periodic firing, because there is still some variability, ± 1, in the number of spikes recorded in the interval). Measures of mutual information from short intervals of firing of temporal cortex visual cells have revealed a degree of variability intermediate between that of periodic and of Poisson regimes (Rolls, Treves, Tovee and Panzeri, 1997). Similar measures can also be used to contrast the effect of the graded nature of neuronal responses, once they are analysed over a finite period of time, with the information content that would characterize neuronal activity if it reduced to a binary variable (Panzeri et al., 1996b). A binary variable with the same degree of variability would convey information at the same instantaneous rate (the first derivative being universal), but in for example 20–30% reduced amounts when analysed over times of the order of the interspike interval or longer.

A2.3.3 Redundancy versus independence across cells

The rate at which a single cell provides information translates into an instantaneous information flow across a population (with a simple multiplication by the number of cells) only to the extent that different cells provide different (independent) information. To verify whether this condition holds, one cannot extend to multiple cells the simplified formula for the first time-derivative, because it is made simple precisely by the assumption of independence between spikes, and one cannot even measure directly the full information provided by multiple (more than two to three) cells, because of the limited sampling problem discussed above. Therefore one has to analyse the degree of independence (or conversely of redundancy) either directly among pairs — at most triplets — of cells, or indirectly by using decoding procedures to transform population responses. Obviously, the results of the analysis will vary a great deal with the particular neural system considered and the particular set of stimuli, or in general of neuronal correlates, used. For many systems, before undertaking to quantify the analysis in terms of information measures, it takes only a simple qualitative description of the responses to realize that there is a lot of redundancy and very little diversity in the responses. For example, if one selects pain-responsive cells in the somatosensory system and uses painful electrical stimulation of different intensities, most of the recorded cells are likely to convey pretty much the same information, signalling the intensity of the stimulation with the intensity of their single-cell response. Therefore, an analysis of redundancy makes sense only for a neuronal system which functions to represent, and enable discriminations between, a large variety of stimuli, and only when using a set of stimuli representative, in some sense, of that large variety.

Redundancy can be defined with reference to a multiple channel of capacity $T(C)$ which can be decomposed into C separate channels of capacities T_i, $i = 1, \ldots, C$:

$$R = 1 - T(C) / \Sigma_i T_i \qquad (A2.36)$$

so that when the C channels are multiplexed with maximal efficiency, $T(C) = \Sigma_i T_i$ and $R = 0$. What is measured more easily, in practice, is the redundancy defined with reference to a specific source (the set of stimuli with their probabilities). Then in terms of mutual information

$$R' = 1 - I(C) \,/\, \Sigma_i I_i. \qquad (A2.37)$$

Gawne and Richmond (1993) have measured the redundancy R' among pairs of nearby primate inferior temporal cortex visual neurons, in their response to a set of 32 Walsh patterns. They have found values with a mean $<R'> = 0.1$ (and a mean single-cell transinformation of 0.23 bits). Since to discriminate 32 different patterns takes 5 bits of information, in principle one would need at least 22 cells providing each 0.23 bits of strictly orthogonal information to represent the full entropy of the stimulus set. Gawne and Richmond have reasoned, however, that, because of the overlap, y, in the information they provided, more cells would be needed than if the redundancy had been zero. They have constructed a simple model based on the notion that the overlap, y, in the information provided by *any* two cells in the population always corresponds to the average redundancy measured for *nearby* pairs. A redundancy $R' = 0.1$ corresponds to an overlap $y = 0.2$ in the information provided by the two neurons, since, counting the overlapping information only once, two cells would yield 1.8 times the amount transmitted by one cell alone. If a fraction $1-y = 0.8$ of the information provided by a cell is novel with respect to that provided by another cell, a fraction $(1-y)^2$ of the information provided by a third cell will be novel with respect to what was known from the first pair, and so on, yielding an estimate of $I(C) = I(1) \, \Sigma_{i=0}^{C-1}(1-y)^i$ for the total information conveyed by C cells. However such a sum saturates, in the limit of an infinite number of cells, at the level $I(\infty) = I(1)/y$, implying in their case that even with very many cells, no more than $0.23/0.2 = 1.15$ bits could be read off their activity, or less than a quarter of what was available as entropy in the stimulus set! Gawne and Richmond have concluded, therefore, that the average overlap among non-nearby cells must be considerably lower than that measured for cells close to each other.

The model above is simple and attractive, but experimental verification of the actual scaling of redundancy with number of cells entails collecting the responses of several cells interspersed in a population of interest. Gochin *et al.* (1994) have recorded from up to 58 cells in the primate temporal visual cortex, using sets of two to five visual stimuli, and have applied decoding procedures to measure the information content in the population response. The recordings were not simultaneous, but comparison with simultaneous recordings from a reduced number of cells indicated that the effect of recording the individual responses on separate trials was minor. The results were expressed in terms of the *novelty* N in the information provided by C cells, which being defined as the ratio of such information to C times the average single-cell information, can be expressed as

$$N = 1 - R' \qquad (A2.38)$$

and is thus the complement of the redundancy. An analysis of two different data sets, which included three information measures per data set, indicated a behaviour $N(C) \approx 1/\sqrt{C}$, reminiscent of the improvement in the overall noise-to-signal ratio characterizing C independent processes contributing to the *same* signal. The analysis neglected however to consider limited sampling effects, and more seriously it neglected to consider saturation effects due to the information content approaching its ceiling, given by the entropy of the stimulus set. Since this ceiling was quite low, for 5 stimuli at $\log_2 5 = 2.32$ bits, relative to the mutual information values measured from the population (an average of 0.26 bits, or 1/9 of the ceiling, was provided by single cells), it is conceivable that the novelty would have taken much larger values if larger stimulus sets had been used.

A simple formula describing the approach to the ceiling, and thus the saturation of information values as they come close to the entropy of the stimulus set, can be derived from a natural extension of the Gawne and Richmond (1993) model. In this extension, the information provided by single cells, measured as a fraction of the ceiling, is taken to *coincide* with the average overlap among pairs of randomly selected, not necessarily nearby, cells from the population. The actual value measured by Gawne and Richmond would have been, again, $1/22 = 0.045$, below the overlap among nearby cells, $y = 0.2$. The assumption that y, measured across any pair of cells, would have been as low as the fraction of information provided by single cells is equivalent to conceiving of single cells as 'covering' a random portion y of information space, and thus of randomly selected pairs of cells as overlapping in a fraction $(y)^2$ of that space, and so on, as postulated by the Gawne and Richmond (1993) model, for higher numbers of cells. The approach to the ceiling is then described by the formula

$$I(C) \approx H\{1-\exp[C \ln(1-y)]\} \qquad (A2.39)$$

that is, a simple exponential saturation to the ceiling. This simple law indeed describes remarkably well the trend in the data analysed by Rolls, Treves and Tovee (1997). Although the model has no reason to be exact, and therefore its agreement with the data should not be expected to be accurate, the crucial point it embodies is that deviations from a purely *linear* increase in information with the number of cells analysed are due *solely* to the ceiling effect. Aside from the ceiling, due to the sampling of an information space of finite entropy, the information contents of different cells' responses are independent of each other. Thus, in the model, the observed redundancy (or indeed the overlap) is purely a consequence of the finite size of the stimulus set. If the population were probed with larger and larger sets of stimuli, or more precisely with sets of increasing entropy, and the amount of information conveyed by single cells were to remain approximately the same, then the fraction of space 'covered' by each cell, again y, would get smaller and smaller, tending to eliminate redundancy for very large stimulus entropies (and a fixed number of cells). The actual data were obtained with limited numbers of stimuli, and therefore cannot probe directly the conditions in which redundancy might reduce to zero. The data are consistent, however, with the hypothesis embodied in the simple model, as shown also by the near exponential approach to *lower* ceilings found for information values calculated with reduced subsets of the original set of stimuli (Rolls, Treves and Tovee, 1997).

The picture emerging from this set of analyses, all performed towards the end of the ventral visual stream of the monkey, is that the representation of at least some classes of objects in those areas is achieved with minimal redundancy by cells that are allocated each to analyse a different aspect of the visual stimulus. This minimal redundancy is what would be expected of a self-organizing system in which different cells acquired their response selectivities through a random process, with or without local competition among nearby cells (see Chapter 4). At the same time, such low redundancy could also very well result in a system that is organized under some strong teaching input, so that the emerging picture is *compatible* with a simple random process, but by no means represents evidence in favour of its occurrence. What appears to be more solid evidence is that towards the end of one part of the visual system redundancy may be effectively minimized, a finding consistent with the general idea that one of the functions of the early visual system is indeed that of progressively minimizing redundancy in the representation of visual stimuli (Attneave, 1954; Barlow, 1961).

A2.3.4 The metric structure of neuronal representations

Further analyses can be made on the results obtained by extracting information measures from population responses, using decoding algorithms. One such analysis is that of the *metric content* of the representation of a set of stimuli, and is based on the comparison of information and percentage correct measures. The percentage correct, or fraction of correct decodings, is immediately extracted from decoding procedures by collating trials in which the decoded stimulus coincided with the actual stimulus presented. Mutual information measures, as noted above, take into account further aspects of the representation provided by the population of cells than percent correct measures, because they depend on the distribution of wrong decodings among all the other stimuli in the set (the mutual information I_p based on the distribution of probabilities also takes into account the likelihood with which stimuli are decoded, but we focus here on maximum likelihood mutual information measures I_{ml}, see above). For a given value of percentage correct f_{corr}, the mutual information, which can be written

$$I = H(S) + f_{corr} \log_2 f_{corr} + <\Sigma_{s \neq s'} P(s \,|\, s') \log_2 P(s \,|\, s')>_{s'}, \quad (A2.40)$$

can range from a minimum to a maximum value. The minimum value I_{min} is attained when wrong decodings are distributed equally among all stimuli except the correct one, thus providing maximum entropy to the distribution of wrong decodings, and in this sense interpreting all stimuli as being equally similar or dissimilar from each other. The maximum transinformation value I_{max}, in contrast, is attained when all wrong decodings are concentrated, with minimal entropy, on a subset of stimuli which thus comprise a neighbourhood of the correct stimulus, while all remaining stimuli are then more distant, or dissimilar, from the correct one. The position of the actual value found for the mutual information within this range can be parametrized by the metric content

$$\lambda_m = (I - I_{min})/(I_{max} - I_{min}) \quad (A2.41)$$

which measures the extent to which relations of similarity or dissimilarity, averaged across the stimulus set, are relevant in generating the distribution of wrong decodings found in the analysis. λ_m thus ranges from 0 to 1, and represents a global measure of the entropy in decoding, which can be extracted whatever the value found for f_{corr}. Of particular interest are the values of metric content found for the representations of the same set of stimuli by different cortical areas, as they indicate, beyond the accuracy with which stimuli are represented (measured by f_{corr}), the extent to which the representation has a tight or instead a loose metric structure. For example, a comparison of the representations of spatial views by hippocampal and parahippocampal cells indicates more metric content for the latter, consistent with a more semantic, possibly less episodic encoding of space 'out there' by the neocortical cells in comparison with their hippocampal counterparts (Treves, Panzeri, *et al.*, 1996).

Further descriptors of the detailed structure of the representations embodied in neuronal activity can be derived from the analysis of the decoding afforded by population responses (Treves, 1997).

Information theory terms — a short glossary

1. The **amount of information**, or **surprise**, in the occurrence of an event (or symbol) s_i of probability $P(s_i)$ is

$$I(s_i) = \log 1/P(s_i) = -\log P(s_i)$$

(the measure is in bits if logs to the base 2 are used). This is also the amount of **uncertainty** removed by the occurrence of the event.

2. The average amount of information per source symbol over the whole alphabet (S) of symbols is the **entropy**,

$$H(S) = \Sigma P(s_i) \log P(s_i)$$

(or *a priori* entropy).

3. The probability of the *pair* of symbols s_i and s_j is denoted $P(s_i, s_j)$, and is $P(s_i) P(s_j)$ only when the two symbols are **independent**.

4. Bayes theorem (given the output s', what was the input s?) states that

$$P(s \mid s') = \frac{P(s' \mid s) \ P(s)}{P(s')}$$

where $P(s'|s)$ is the **forward** conditional probability (given the input s, what will be the output s'?), and $P(s|s')$ is the **backward** conditional probability (given the output s', what was the input s?).

5. Mutual information. Prior to reception of s', the probability of the input symbol s was $P(s)$. This is the a priori probability of s. After reception of s', the probability that the input symbol was s becomes $P(s|s')$, the conditional probability that s was sent given that s' was received. This is the a posteriori probability of s. The difference between the a priori and a posteriori uncertainties measures the gain of information due to the reception of s'. Once averaged across the values of both symbols s and s', this is the **mutual information**, or **transinformation**

$$I(S,S') = \Sigma_{s,s'} \ P(s,s') \ \{ \log [1/P(s)] - \log [1/P(s|s')] \} = \Sigma_{s,s'} \ P(s,s') \log [P(s|s')/P(s)]$$

Alternatively,

$$I(S,S') = H(S) - H(S|S').$$

$H(S|S')$ is sometimes called the **equivocation** (of S with respect to S').

A3 Pattern associators

Chapters 2 and 3 provide introductory accounts of the operation of pattern associators and of autoassociative memories; networks comprising a few units, often just binary ones, are used to demonstrate the basic notions with minimal complication. In later chapters, instead, these notions are argued to be relevant to understanding how networks of actual neurons in parts of the brain operate. It is obvious that real networks are much more complex objects than are captured by the models used as illustrative examples. Here we address some of the issues arising when considering the operation of large networks implemented in the brain.

Beyond providing an intuitive introduction to the real systems, a second important reason for striving to simplify the complexities of neurobiology is that formal, mathematical models of sufficient simplicity are amenable to mathematical analysis, which is much more powerful, particularly in extracting quantitative relationships, than either intuition or computer simulation (which also requires a degree of simplification, anyway). Providing all the necessary tools for formal analyses of neural networks is largely outside the scope of this book, and many of these tools can be found, for example, in the excellent books by Amit (1989), and by Hertz, Krogh and Palmer (1991), and in the original and review literature. Nevertheless, we shall sketch some of the lines of the mathematical approaches and refer back to them in discussing issues of realism and simplification, because some appreciation of the analytical methods is important for an understanding of the domain of validity of the results we quote.

A3.1 General issues in the modelling of real neuronal networks

A3.1.1 Small nets and large nets

In Chapter 2 we used as an example of an associative memory a network with 6 input axons and 4 output cells. It is easy to simulate larger nets on a computer, and to check that the same mechanisms, associative storage and retrieval, with generalization, fault tolerance, etc., can operate just as successfully. In the brain, we may be considering local populations comprising tens of thousands of neurons, each receiving thousands of synaptic inputs just from the sensory stream carrying the conditioned stimulus. Formal mathematical analyses can be worked out in principle for nets of any size, but they are always much simpler and more straightforward (and this is a point which may not be known to experimentalists) in the limit when the size approaches infinity. In practice, infinite means very large, so that finite size effects can be neglected and the central limit theorem applied to probabilities, whenever needed. Usually, what makes life simpler by being very large is not just the number of cells in the population, but also the number of inputs per cell. This is because the precise identity and behaviour of the individual cells feeding into any given neuron become unimportant, and what remain important are only the distributions characterizing, for example, input firing rates and synaptic efficacies; often, not even the full distributions but only their means and maybe a few extra moments. Therefore, while for those conducting very detailed conductance-based simulations, or even more for those developing artificial networks in hardware, the problem tends to be that of reaching sizes comparable to those of brain circuits, and the number of units or of inputs per unit simulated is a score of success for the enterprise, for the theoretician the problem with size, if there is any, is to check that actual networks in the brain are large enough to operate in the same way as the infinitely large formal models—which is usually concluded to be the case.

A3.1.2 What is the output of a neuron?

Neurons may communicate, or more in general interact with each other in a variety of ways. The only form of communication considered in this book is the emission of action potentials carried by the axon and resulting in release of neurotransmitter at synaptic terminals. This is clearly the most important way in which central neurons affect one another, but one should note that there are alternatives that appear to be important in certain systems (see for example Shepherd, 1988), such as dendro-dendritic synapses (described in the olfactory bulb by Rall and Shepherd, 1968), gap junctions in various parts of the developing nervous system, or ephaptic interactions, that is the (minor) electrical couplings produced by sheer proximity, for example among the tightly packed cell bodies of pyramidal neurons in the hippocampus. The role that these alternative forms of interaction (which, to be noted, are still *local*) may play in operations implemented in the corresponding networks remains to a great extent to be elucidated.

One should also note that an experimenter can access a larger set of parameters describing the 'state' of the neuron than is accessible to the postsynaptic neurons that receive inputs from it. Thus, one can measure membrane potential with intracellular electrodes in the soma, calcium concentration, various indicators of metabolic activity, etc., all variables that may be correlated to some degree with the rate of emission of action potentials. Action potentials are what triggers the release of neurotransmitter that is felt, via conductance changes, by receiving neurons, that is they are the true output, and since they are to a good approximation self-similar, only their times of occurrence are relevant. Therefore, the fuller description of the output of cell i that we consider here is the list of times $\{t^k\}_i$ for the emission of each of the action potentials, or spikes, which are indexed by k.

Reduced descriptions of the output of a cell are often sufficient, or thought to be sufficient. For example, if one looks at a given cell for 20 ms, records its action potentials with 1 ms resolution, and the cell never fires at more than 500 Hz, the full description of the output would be a vector of 20 binary elements (each signifying whether there is a spike or not in the corresponding 1 ms bin), which could in principle have as many as $2^{20} \approx 1\,000\,000$ different configurations, or values (most of which will never be accessed in practice, particularly if the cell never fires more than 10 spikes in the prescribed 20 ms window). If the precise emission time of each spike is unimportant to postsynaptic cells, a reduced description of the cell's output is given by just the number of spikes emitted in the time window, that is by specifying one of the 11 values from 0 to 10, with a nearly 10^5 reduction in the number of possible outputs.

Whether a reduction in the output space to be considered is justified is a question that can be addressed experimentally case by case, by using the information theoretic measures introduced in Appendix A2. One may ask what proportion of the information conveyed by the full output is still conveyed by a particular reduced description (Optican and Richmond, 1987), and whether any extra information that is lost would in any case be usable, or decodable, by receiving neurons (Rolls, Treves and Tovee, 1997). In higher sensory cortices, at least in parts of the visual, olfactory, and taste systems, it appears (Rolls, Critchley and Treves, 1996; Tovee and Rolls, 1995; Tovee, Rolls, Treves and Bellis, 1993) that the simple firing rate, or equivalently the number of spikes recorded in a time window of a few tens of ms, is indeed a reduced description of neuronal output that preserves nearly all the information present in the spike emission times (at the level of single cells). This finding does not of course preclude the possibility that more complex descriptions, involving the detailed time course of neuronal firing, may be necessary to describe the output of single cells in other, for example more peripheral, systems; nor does it preclude the possibility that populations of cells convey some information in ways dependent on the precise relative timing of their firing, and that could not be revealed by reporting only the firing rate of individual cells, as suggested by findings by Abeles and collaborators (Abeles *et al.*, 1993) in frontal cortex and by Singer and collaborators (Gray *et al.*, 1989) in primary visual cortex.

Turning to the question of what is decodable by receiving neurons, that is of course dependent on the type of operation performed by the receiving neurons. If this operation is simple pattern association, then the firing rates of individual input cells measured over a few tens of ms is all the output cells are sensitive to, because the fundamental operation in a pattern associator is just the dot product between the incoming axonal pattern of activity and the synaptic weight vector of each receiving cell. This implies that each input cell contributes a term to a weighted sum (and not a factor that interacts in more complex ways with other inputs) and precise emission times are unimportant, as thousands of inputs are effectively integrated over, in space along the dendritic tree, and in time (allowing for leakage) between one action potential of the output cell and the next.

The firing rate of each cell i, denoted r_i, is therefore what will be considered as a suitable description of its output, for pattern associators. Since these networks, moreover, are feedforward, and there are no loops whereby the effect of the firing of a given cell is felt back by the cell itself, that is there is no recurrent *dynamics* associated with the operation of the network, it does not really matter whether the inputs are considered to persist only briefly or for a prolonged period. Neglecting time-dependent effects such as adaptation in the firing, the outputs are simply a function of the inputs, which does not involve the time dimension. Pattern associators, like other types of feedforward nets, can thus be analysed, to a first approximation, without having to describe the detailed time course of a cell's response to inputs that vary in time: the inputs can be considered to be carried by steady firing rates, and likewise the outputs.

A3.1.3 Input summation and the transfer function

Specifying the way inputs are summed and transduced into outputs means in this case assigning functions that transform a set of input rates $\{r'_j\}$ into one rate for each output cell i, that is functions $r_i(\{r'_j\})$. This is to be a very compressed description of what in real neurons is a complex cascade of events, typically comprising the opening of synaptic conductances by neurotransmitter released by presynaptic spikes, the flow of synaptic currents into the cell, their transduction through the dendrites into the soma possibly in conjunction with active processes or non-linear interactions in the dendrites themselves, and the initiation of sodium spikes at the axon hillock. A simple summary of the above which is useful in basic formal models is as follows. Input rates from N presynaptic cells are summed with coefficients w representing the efficacy of the corresponding synapses into an activation variable h for each of N output cells:

$$h_i = \Sigma_{j=1,C} w_{ij} r'_j = w_{i1} r'_1 + \ldots + w_{ij} r'_j + \ldots + w_{iC} r'_C. \tag{A3.1}$$

One should note that more complex forms of input integration have been suggested as alternative useful models in the connectionist literature (cf. the Sigma–Pi units in Rumelhart and McClelland, 1986), and that for some of them it might even be argued that they capture what is observed in certain special neuronal systems.

The activation is then converted into the firing rate of the output cell via a transfer function that includes at least a non-linearity corresponding to the threshold for firing, and an approximately linear range above threshold. A simple one is

$$r_i = g_i(h_i - \theta_i) \; \Theta(h_i - \theta_i) \tag{A3.2}$$

where $\Theta(x)$ is the step function which equals 1 for $x > 0$ and 0 for $x < 0$, and g is the gain of the transfer function. Since both in the input and output one deals with steady rates, the activation is generally thought to represent the current flowing into the soma (Treves, 1990), and hence θ is a threshold for this current to elicit a non-zero steady response. If more transient responses were considered, it would be appropriate to consider, instead, the membrane potential and its corresponding threshold (Koch et al., 1995). This is because the triggering of a single action potential is due to the establishment, at the local membrane level, of a positive feedback process involving the opening of voltage-dependent sodium channels together with the depolarization of the membrane. Hence, what is critical, to enter the positive feedback regime, is reaching a critical level of depolarization, that is reaching a membrane potential threshold. What we call (in a somewhat idealized fashion) *steady* state, instead entails a constant current flowing into the soma and a constant rate of emission of action potentials, while the membrane potential particularly at the soma is *oscillating* with each emission. Hence, the threshold for firing is at steady state a threshold for the current. This point is actually quite important in view of the fact that, as explained in Appendix A5, currents are attenuated much less, in reaching the soma, than voltages. Therefore, the effectiveness of inputs far out on the apical dendrites and close to the soma of a pyramidal cell may be much more comparable, at steady state, than might be concluded from experiments in which afferent inputs are transiently activated in order to measure their strengths. In addition to being on average more similar relative to each other, both far and near inputs on a dendrite are stronger in absolute terms, when they are measured in terms of steady-state currents. This should, however, not be taken to imply that very few typical synaptic inputs are sufficient to fire a pyramidal cell; the number of required inputs may still be quite large, in the order of several tens. Precise estimates have been produced with studies *in vitro*, but in order to understand what the conditions are *in vivo*, one must take into account typical sizes of Excitatory Post-Synaptic Currents (EPSCs) elicited in the soma by synaptic inputs to different parts of a dendrite, and the effects of inhibitory inputs to the cell (see for example Tsodyks and Sejnowski, 1995).

For an isolated cell, the gain g_i (in Hz/mA) and the threshold θ_i (in mA) could be made to reproduce the corresponding parameters measured from the response to current injection, for example in a slice. Here, however, gain and threshold are supposed to account not only for the properties of the cell itself, but also for the effects of non-specific inputs not explicitly included in the sum of Eq. A3.1, which only extends over principal cells. That is, the networks we consider are usually networks of principal cells (both in the input and in the output), and local interneurons (for example those mediating feedforward and feedback inhibition) are not individually represented in the model, but only in terms of a gross description of their effect on the transduction performed by principal cells. Divisive inhibition may be represented as a modulation of the gain of pyramidal cells, and subtractive inhibition as a term adding to their threshold. Both may be made a function of the mean activity on the input fibres, to represent feedforward control of overall level of activation; and of the mean activity of output cells, to represent feedback effects. Likewise, the general, non-specific effects of neuromodulatory afferents may be represented as modifications in the parameters of the transfer function.

The threshold linear transfer function of Eq. A3.2 is the simplest one that reproduces the two basic features of a threshold and a graded range above threshold. Many other choices are of course possible, although they may be less easy to treat analytically in formal models. The two most widely used forms of transfer function are in fact even simpler, but at the price of renouncing representing one of those two features: the binary transfer function captures only the threshold but not the graded response above it, and the linear transfer function only the graded response and not the threshold. The so-called logistic (or sigmoid) function (Fig. 1.3) is derived as a 'smoothing' of the sharp threshold of a binary unit, and is also widely used especially in connectionist models. It can have the undesirable feature that the resulting firing rates tend to cluster around both zero and the maximal rate, producing an almost binary distribution of values that is most unlike the typical firing statistics found in real neurons.

A3.1.4 Synaptic efficacies and their modification with learning

The effect of the firing of one spike on the activation of a postsynaptic cell i is weighted by a single coefficient w_{ij}, parametrizing the efficacy of the synapse between the firing cell j and the receiving one. Even in more detailed models in which synaptic inputs are represented in terms of conductances and their respective equilibrium potentials, the size of the open conductance can still be parametrized by a weight coefficient. Modifiable synapses, in the models, are those in which these coefficients are taken to change in the course of time, depending for example on the pre- and postsynaptic cell activity and their relation in time. A mathematical representation of such modifiability is usually called a learning rule.

The modifiability expressed in formal models by learning rules is, at a general level, implemented in the real nervous system by various forms of synaptic plasticity. Among them, the one with the most interesting properties is the phenomenon, or group of phenomena, called long-term potentiation, or LTP (and its counterpart long-term depression, LTD), first discovered in the hippocampus by Bliss and Lomo (1973). LTP has long been a focus of attention because it appears to fulfil the desiderata for a learning mechanism formulated on a conceptual basis by the psychologist Donald Hebb (1949): essentially, its sustained nature and the fact that its induction depends in a conjunctive way on both activity in the presynaptic fibre and activation (depolarization) of the postsynaptic membrane. One crucial component that senses this conjunction, or AND function, is the so-called NMDA receptor. This receptor opens when glutamate released from an activated presynaptic terminal binds at the appropriate site, but to let ions pass through it also requires that magnesium ions be expelled by sufficient depolarization of the postsynaptic membrane. (At normal potentials magnesium ions block the channel by entering it from the outside of the cell and acting effectively as corks.) The entry of calcium ions through the unblocked channel appears to release a complex cascade of events that result in the LTP of the synapse, possibly expressed as a combination of increased average release of neurotransmitter by each incoming spike, and an increased postsynaptic effect of each quantum of neurotransmitter binding at the ordinary (AMPA) glutamate receptors (co-expressed on the postsynaptic membrane with the NMDA ones). The precise nature and relevance of the steps in the cascade, and in general the mechanisms underlying different subforms of LTP, the mechanisms involved in LTD, the differences between the plasticity evident during development and that underlying learning in the adult, are all very much topics of current attention, still at one of the frontiers of neurobiological research. New important phenomena may still be discovered if the appropriate experimental paradigms are employed, and overall, it is fair to say that present knowledge about synaptic plasticity does not tightly constrain theoretical ideas about its occurrence and its roles. Although the discoveries that have been made have provided powerful inspiration for the further development and refinement of theoretical notions, the latter may potentially exert an even more profound stimulus for the design and execution of the crucial experiments.

The main features of any learning rule are (a) which factors, local or not, the modification depends on, (b) what precise form it takes and (c) its time course of induction and expression. On point (a), experimental evidence does provide some indications, which are however much more vague on points (b) and (c).

A learning rule is called local if the factors determining the modification refer only to the pre- and postsynaptic cells; most commonly the firing rates of the two cells are taken to be these factors. It should be noted that whereas the firing rate of the presynaptic cell may be directly related to the size of the modification, in that each spike may have the potential for contributing to synaptic change, it is unclear whether the postsynaptic site may be sensitive to individual spikes emitted by the postsynaptic cell (and retrogradely transmitted up to the synapse as sharp variations in membrane potential), or rather more to a time averaged and possibly partly local value of the membrane potential, such as that which, in the operation of NMDA receptors, is thought to be responsible for relieving the magnesium block. It is noted in Appendix A5 that voltage transmission from the cell body or proximal part of the dendrite towards the apical part of the dendrite is not severely attenuated (Carnevale and Johnston, 1982). An implication of this is that much of the postsynaptic term in a learning rule can be felt throughout all the dendrites, so that depolarization of the cell soma associated with fast firing *in vivo* would lead to sufficient depolarization of the dendrite to relieve the block of NMDA receptors. In any case, the term local implies that the relevant factors are available at the synapse, but it does not rule out the necessity, under a strict analysis of the molecular mechanisms of the change, for short-distance messengers. Thus, to the extent that the change is expressed on the presynaptic site, there is a need for a retrograde messenger that conveys presynaptically the signal that the proper factors for induction are present postsynaptically. Nevertheless, local rules do not need long-distance messengers that bring in signals from third parties, such as the error signals backpropagating across cell populations required in the non-local and rather biologically implausible learning rule employed in backpropagation networks.

In the standard learning rules we consider, modifications are expressed as sums of individual synaptic changes Δw, each supposed to occur at a different moment in time. Δw is a product of a function of the presynaptic firing rate, and of a function of the postsynaptic rate at that moment in time. The two functions may be different from each other, but in one common form of learning rule they are in fact equal: they are just the rate itself minus its average

$$\Delta w_{ij} = \gamma(r_i - <r_i>)(r_j - <r_j>) \qquad (A3.3)$$

so that, in this simple model, any fluctuations from the average values, when occurring both pre- and postsynaptically, are sufficient to elicit synaptic changes. The plasticity γ is a parameter quantifying the average amount of change. The change Δw is then independent of the value of the synaptic efficacy at the time it occurs and of previous changes (the overall modification is just a sum of independent changes). The modification is to be conceived as being from a baseline value large enough to keep the synaptic efficacy, at any moment in time, a positive quantity, since it ultimately represents the size of a conductance (in fact, of an excitatory one). The important aspects of the precise form of Eq. A3.3 are the linear superposition of successive changes, and the fact that these are both positive and negative, and on average cancel out, thus keeping the synaptic value fluctuating around its baseline. Minor modifications that are broadly equivalent to Eq. A3.3 are expressions that maintain these two aspects. For example, the postsynaptic factor may be a different function of the postsynaptic rate, even a positive definite function (the presynaptic factor is sufficient to ensure the average cancellation of different contributions), and it may also be a function not of the postsynaptic rate but of a variable correlated with it, such as a time-averaged value of the postsynaptic membrane potential. This time average may even be slightly shifted with respect to the presynaptic factor, to represent what happens to the postsynaptic cell over the few tens of ms that follow the presynaptic spike. Learning rules, instead, that violate either the linear superposition or the average cancellation aspects represent major departures from (A3.3), and may lead to different functional performance at the network level. This is evident already in the later part of this appendix, where a signal-to-noise analysis reveals the importance of the average cancellation in the presynaptic factor, leading to Eq. A3.15. Thus some form of synaptic plasticity analogous to what is referred to as heterosynaptic LTD (see Chapter 2) is found to be computationally crucial in providing a balance for increases in synaptic efficacies, modelled after LTP. Learning rules that do not fall into this general class are discussed in this book under separate headings.

The third feature of the learning rule, its time course, is expressed in the standard rule we consider by taking each change to be induced over a short time (of less than 1 s, during which the synaptic efficacy is still effectively the one before the change), and then persisting until forgetting, if a forgetting mechanism is included, occurs. The simplest forgetting mechanism is just an exponential decay of the synaptic value back to its baseline, which may be exponential in time or in the number of changes incurred (Nadal *et al.*, 1986). This form of forgetting does not require keeping track of each individual change and preserves linear superposition. In calculating the storage capacity of pattern associators and of autoassociators, the inclusion or exclusion of simple exponential decay does not change significantly the calculation, and only results, as can be easily shown, in a different prefactor (one 2.7 times the other) for the maximum number of associations that can be stored. Therefore a forgetting mechanism as simple as exponential decay is normally omitted, and one has just to remember that its inclusion would reduce the critical capacity obtained by roughly 0.37. Another form of forgetting, which is potentially interesting in terms of biological plausibility, is implemented by setting limits to the range allowed for each synaptic efficacy or weight (Parisi, 1986). As a particular synapse hits the upper or lower limit on its range, it is taken to be unable to further modify in the direction that would take it beyond the limit. Only after modifications in the opposite direction have taken it away from the limit does the synapse regain its full plasticity. A forgetting rule of this sort requires a slightly more complicated formal analysis (since it violates linear superposition of different memories), but it effectively results in a progressive, exponential degradation of older memories similar to that produced by straight exponential decay of synaptic weights. A combined forgetting rule that may be particularly attractive in the context of modelling synapses between pyramidal cells (the ones we usually consider throughout this book) is implemented by setting a lower limit, that is, zero, on the excitatory weight (ultimately requiring that the associated conductance be a non-negative quantity!), and allowing exponential decay of the value of the weight with time (all the way down to zero, not just to the baseline as above). Again, this type of combined forgetting rule places demands on the analytical techniques that have to be used, but leads to functionally similar effects.

A3.1.5 The statistics of memories

As synaptic modification reflects, with the simplest learning rule, the firing rates of different cells at the same moment in time, the information stored in associative memories is in the form of distributions, or patterns, of firing rates. Such rates are of course the result of processing in the networks upstream of the one considered, and in that sense they do not comprise completely arbitrary sets of numbers. In fact, we have argued that interesting constraints arise, in autoassociators, from the way patterns of firing rates emerge in the learning of new memories (Treves and Rolls, 1992). Specifically, the autoassociator posited to be implemented in the CA3 region of the hippocampus may require a special preprocessing device, the dentate gyrus, that ensures the decorrelation of patterns of firing that must be stored in CA3 (see Chapter 6). Nevertheless, when considering solely the *retrieval* of information from associative memories, and not its *encoding*, one usually avoids the need to analyse how patterns of firing arise, and takes them as given, with certain statistics. In pattern associators, each firing pattern is actually a pair, with one pattern for the afferent axons and another for the output cells. We shall consider that p patterns have been stored at a given time, and label them $\mu = 1, \ldots, p$. Each pattern is assumed to have been generated independently of others. This is an approximation, whose validity can be checked in specific cases. Further, the firing of each cell (we use the index i for output cells and j for input ones) in each pattern is also assumed to be independent of that of other cells. This is also an approximation, which is both very handy for the analysis of mathematical models and appears to be an excellent one for some firing rates recorded in higher sensory and memory areas. Probably the agreement is due to the high number of (weak) inputs each real neuron receives, which effectively removes substantial correlations among different cells in those cases. In any case, the probability of a given firing pattern is expressed mathematically, given our independence assumption, as

$$P\left(\{r_j^{\mu}\},\{r_i^{\mu}\}\right) = \Pi_{\mu}\Pi_j \ P(r_j^{\mu})\,\Pi_i \ P(r_i^{\mu}). \tag{A3.4}$$

For simplicity, the probability distribution for the rates of each cell in each pattern is usually taken to be the same, hence denoted simply as $P(r)$.

By considering different forms for $P(r)$ one can to some extent explore the effect of different firing statistics on the efficient coding of memories in pattern associators (the full range of possibilities being restricted by the assumptions of independence and homogeneity among cells, as explained above). As any probability distribution, $P(r)$ is constrained to integrate to 1, and since it is a distribution of firing rates, it takes values above zero only for $r \geq 0$, therefore $\int_0^{\infty} dr\, P(r) = 1$. Different values for the first moment of the distribution, that is for the mean firing rate, are equivalent to simple rescalings, or, when accompanied by similar rescalings for thresholds and synaptic efficacies, are equivalent to using different units than s^{-1}. They are not therefore expected to affect storage capacity or accuracy, and in fact, when threshold-linear model units are considered, these limits on performance (which presuppose optimal values for thresholds and gain) are independent of the mean rate in $P(r)$. Other aspects of the probability distribution are instead important, and it turns out that the most important one is related to the second moment of the distribution. It is convenient to define the *sparseness* (see Chapter 1) as

$$a = \left(\int_0^{\infty} dr\, r\, P(r)\right)^2 / \int_0^{\infty} dr\, r^2\, P(r) = <r>^2 / <r^2>. \tag{A3.5}$$

The sparseness ranges from 0 to 1 and it parametrizes the extent to which the distribution is biased towards zero: $a \approx 0$ characterizes distributions with a large probability concentrated close to zero and a thin tail extending to high rates, and $a \approx 1$ distributions with most of the weight away from zero.

Defined in this way, in terms of a theoretical probability distribution of firing rates assumed to be common to all cells in a given population, the sparseness can be interpreted in two complementary ways: either as measuring roughly the proportion of very active cells at any particular moment in time, or the proportion of times that a particular cell is very active (in both cases, by saying *very active* we wish to remind the reader that a actually measures the ratio of the average rate squared to the square rate averaged, and not simply the fraction of active cells or active times, which is just what the definition reduces to for binary units). Usually, when talking about sparse coding, or, conversely, about distributed coding, one refers to an activity pattern occurring at a moment in time, and in which few or many cells fire, or fire strongly; and when instead referring to the distribution of rates of one cell at different times or, for example, in response to different stimuli, one sometimes uses the terms fine tuning, or broad tuning.

Measuring the sparseness of real neuronal activity

In analysing formal network models, especially when something like Eq. A3.4 is taken to hold, the two interpretations are essentially equivalent. In the observation of real firing statistics, on the other hand, the interpretation that considers at a moment in time the firing across cells has been applied less, partly because the simultaneous recording of many cells is a relatively recent technique, partly because it is non-trivial to identify and 'count' cells that fire rarely or never, and last because the firing statistics of even similar-looking cells is expected a priori to be somewhat different due to differences in their exact physiological characteristics. What one does is thus normally to record the distribution of firing rates, say collected as the number of spikes in a fixed time window, of a particular cell across different time windows. This procedure can itself be carried out in at least two main ways. One is to record from the cell as the animal interacts with a large variety of external correlates (for example, in the case of visual stimulation, sees a movie), and to collect a large number K of time windows, for example indexed by k. The sparseness of the distribution is then

$$a = [\Sigma_k r_k / K]^2 / [\Sigma_k r^2{}_k / K]. \tag{A3.6}$$

This particular way of measuring sparseness is affected by the length of the time window considered, in the sense that if the window is much shorter than the typical interspike interval, most of the times it will contain no spikes, and as a result the distribution will appear artificially sparse; progressively increasing the window the measured sparseness parameter, a, increases, typically saturating when the window is long enough that the above artefact does not occur. The value obtained is also obviously affected by the nature and variety of the external correlates used.

   A second way to measure sparseness is to use a fixed set of S well-defined correlates, each one occurring for a number of repetitions (for example, in the case of visual stimulation, these could be a number of static visual stimuli presented each for a number of trials, with the number of spikes emitted in a given peristimulus interval quantifying the response). In this case the sparseness can be measured from the mean response rate r_s to each stimulus s, (averaged across trials with the same stimulus), as

$$a = [\Sigma_s r_s / S]^2 / [\Sigma_s r^2{}_s / S]. \tag{A3.7}$$

This measure of sparseness is less affected by using a short time window, because the average across trials implies that non-zero mean responses can be produced even to stimuli for which most trials carry no spikes in the window; on the other hand, this measure is affected in the opposite direction by the number of correlates in the set, in the sense that the minimal value of the sparseness parameter measured in this way is just $a = 1/S$, and therefore one will find an artificially high value if using a small set.

Both artefacts can be easily controlled by measuring the effect of the size of the time window and of the size of the set of correlates. In both cases, the sparseness value obtained will nevertheless remain, obviously, a measure *relative* to those particular correlates.

Specific models of firing probability distributions

One particular type of probability distribution, which arises inevitably when considering binary model units, is a binary one, with the whole weight of the distribution concentrated on just two values, one of which is usually set to zero (a non-firing unit) and the other to one (the maximal rate in arbitrary units). The sparseness a is then just the fraction of the weight at $r = 1$; in the usual mathematical notation

$$P(r) = (1-a)\,\delta(r) + a\,\delta(r-1) \tag{A3.8}$$

where the δ-functions indicate concentrated unitary weight on the value that makes their argument zero. Binary distributions (with one of the output values set at zero firing) maximize the information that can be extracted from each spike (see Appendix A2 and Panzeri *et al.*, 1996b), among distributions with a fixed sparseness. Since the information per spike is closely related to the breadth of tuning, it is easy to see that binary distributions also minimize the breadth of tuning (again, among distributions with a fixed sparseness). In this quantitative sense, binary distributions provide what amounts to a rather precise code; on the other hand, they do not allow exploiting the graded or nearly continuous range of firing rate that a real neuron has at its disposal in coding its output message.

Other types of probability distributions that we have considered, still parametrized by their sparseness, include *ternary* ones such as

$$P(r) = (1-4a/3)\,\delta(r) + a\,\delta(r-1/2) + a/3\,\delta(r-3/2) \tag{A3.9}$$

and continuous ones such as the exponential distribution

$$P(r) = (1-2a)\,\delta(r) + 4a\exp(-2\,r). \tag{A3.10}$$

In both these distributions the first moment is conventionally set so that, as for the binary example, $<r> = a$. The ternary distribution is a conveniently simple form that allows analysis of the effects of departing from the more usual binary assumption, while the exponential form is one which is not far from what is observed in at least some parts of the brain. These three types of distribution are represented in Fig. A3.1. Other types of distribution can be considered, made to model in detail experimentally observed statistics.

The distributions observed for example in the primate temporal visual cortex are always unimodal, with the mode at or close to the spontaneous firing rate and sometimes at zero, and a tail extending to higher rates, which is often close to exponential (Baddeley *et al.*, 1998). It may be possible to understand this typical form of the firing distribution in terms of a random, normally distributed input activation, with additional normally distributed fast noise, resulting in an asymmetric rate distribution after passing through a threshold-linear input–output transform (Panzeri *et al.*, 1996a).

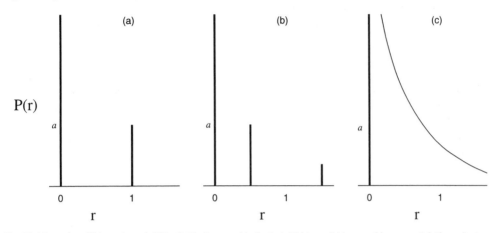

Fig. A3.1 Examples of firing rate probability distributions used in the text: (a) binary; (b) ternary; (c) exponential. The ordinate shows probability, and the abscissa the firing rate. (From Treves and Rolls, 1991, Fig. 3.)

A3.2 Quantitative analyses of performance

Having made a series of simplifying assumptions, and postponing judgement on whether each such assumption is appropriate or not when the results are applied to specific networks of real neurons, one may now evaluate the performance of pattern associators, instantiated as precisely defined mathematical models. Two aspects of their performance are to be examined: how many associations between input and output firing patterns they can store (so that each input pattern can retrieve its output one), and how accurate the retrieval of the output pattern is (which of course depends on the accuracy of the input). A third aspect, that of the time over which retrieval occurs, is not really relevant for pattern associators, because the operation of the feedforward network adds no dynamics of its own to that characterizing its constituent units.

Storage capacity and retrieval accuracy (the asymmetry in the traditional terminology hiding the fact that both aspects imply both the storage and retrieval of associations) are two faces of the same coin. This is because if more memories are stored one has poorer retrieval, that is an output with a weaker resemblance to the stored output. With nets whose dynamics leads to a steady state, such as autoassociators, the progressive deterioration is abruptly interrupted when, above a critical storage load, the network fails to evolve to a retrieval state. This is because the iterative dynamics of autoassociators converges towards one of its attractors, which can be pictured as the states lying at the bottom of valleys—their basins of attraction—with the network rolling down the slopes as if pulled by gravity. Not all of the attractors correspond to retrieval of one firing pattern, and increasing the storage load leads indeed to more retrieval attractors, but with progressively shallower valleys, until at a critical load they cease to be attractors at all; the network then ends up in states with no strong correlation to any of the stored patterns. With pattern associators that whole picture is meaningless, if anything because the starting state, that is the input firing pattern, belongs to a different panorama, as it were, from that of the output state, and there is no rolling down because the whole dynamical trajectory reduces to a single step. The absence of a critical load can still be understood intuitively, however, by realizing that by making one step from a state with some correlation to an input pattern, the network will still be within some correlation, stronger or weaker, from the corresponding output pattern. No matter how many associations are stored, which result in interfering influences on the direction of that one step, a single step will always maintain some initial correlation, if altered in magnitude. Therefore the storage capacity of a pattern associator can only be defined as that storage load which preserves, *up to a given degree*, the correlations present in the input; and since the degree of preservation (or enhancement) depends on the input correlations themselves, one usually defines capacity in terms of the constraint that the network should at least preserve *full* input correlations. In other words, this amounts to using a complete and noiseless input pattern and checking what is the maximal load such that the network still produces a noiseless output. (When using continuous output units, a strictly noiseless output is non-physical, and obviously one should set a finite tolerance level.)

A3.2.1 Signal-to-noise ratios

A signal-to-noise analysis is a simple approach to analyse how retrieval accuracy depends on storage load, and on all the other characteristic parameters of the network. A pattern associator is considered to have stored $p + 1$ associations with the learning rule

$$w_{ij} = w^0 + \gamma \Sigma_\mu F(r_i^\mu) G(r_j^\mu) \tag{A3.11}$$

which is a somewhat more general expression than Eq. A3.3 above, with F and G arbitrary pre- and postsynaptic factors. The averages and variances of these factors over the distribution of firing rates on the input (for G) and output (for F) are denoted as

$$<F> = m_F \qquad\qquad <F^2> - <F>^2 = \sigma_F^2$$
$$<G> = m_G \qquad\qquad <G^2> - <G>^2 = \sigma_G^2. \tag{A3.12}$$

The network is taken to receive in the input one of the stored input patterns, say $\mu = 1$. What each output cell receives from the C input cells that feed into it is, due to the linear superposition in the learning rule, Eq. A3.11, a signal S coming from the $\mu = 1$ term in the sum, plus noise N coming from the other p stored associations. The mean value of these noise terms, plus the w^0 baseline terms, can be subtracted out by appropriate thresholds. Their variance, instead, can be evaluated by using the assumption of lack of correlations among cells in the same pattern and among patterns, yielding

$$\sigma_N^2 = p\gamma^2 C<r>^2 \{C\sigma_F^2 m_G^2 + p(1/a-1)m_F^2 m_G^2 + (m_F^2 + \sigma_F^2)\sigma_G^2/a + (1/a-1)\sigma_F^2 m_G^2\}. \tag{A3.13}$$

The variance of the noise has to be compared with the mean square amplitude of the signal, which can be taken to be the square difference of the signal received by a cell that fires at the conventional unitary rate in the pattern being retrieved, and that of a cell that does not fire. This is

$$S^2 = \gamma^2 C^2 [F(1) - F(0)]^2 < G(r) r >^2. \tag{A3.14}$$

The argument now is that, for any given level of required retrieval accuracy, the signal must be at least of the same order of magnitude as the noise. If p is a large number, this can only happen if the first two terms in the mean square noise vanish: in that case the noise is of order $(pC)^{\frac{1}{2}}$ and the signal of order C, and they can be comparable up to p of order C. Clearly this requires that the mean of the presynaptic G factor be negligible: $m_G \approx 0$. If this mean is really close to zero, only the third term in the noise is substantial, and the signal-to-noise ratio is of order $(C/p)^{\frac{1}{2}}$. To be more precise, one may specify for the presynaptic factor the form $G(r) = r - <r>$, to find

$$S/N \approx [C(1-a)/p]^{\frac{1}{2}} [F(1) - F(0)] / [<F(r)^2>]^{\frac{1}{2}} \tag{A3.15}$$

which indicates the potentially beneficial effect of sparseness in the output pattern: if it is sparse, and the postsynaptic factor is such that

$$[<F(r)^2>]^{\frac{1}{2}} \ll [F(1) - F(0)] \tag{A3.16}$$

then the number of stored associations p can be much larger than the number of inputs per unit C while preserving the signal-to-noise ratio of order one. While Eq. A3.15 is valid for any form of the postsynaptic factor F and of the distribution of output firing rates, it is helpful to assume specific forms for these in order to gain an intuitive understanding of that equation. Taking output units to be binary, with a distribution of output rates of sparseness a, as in Eq.(A3.8), and the postsynaptic factor to be simply proportional to the output rate, $F(r) = r$, yields

$$S/N \approx [C(1-a)/p\,a]^{\frac{1}{2}} \tag{A3.17}$$

which is of order unity (implying that the signal is strong enough to effect retrieval despite the noise) for

$$p \approx C(1-a)/a. \tag{A3.18}$$

The above analysis illustrates the origin of the enhancement in storage capacity (as measured by the number of associations stored) produced by sparseness in the output distribution. The exact same effect holds for an autoassociator, in which, however, capacity is measured by the number of memory patterns stored, and the sparseness that matters is just the sparseness of the patterns, since there is no distinction between an input pattern and an output pattern being associated with it. More careful analyses have been worked out for binary units, especially in the limiting case of very sparse coding, both for autoassociators with a covariance learning rule (Tsodyks and Feigel'man, 1988; Buhmann, Divko and Schulten, 1989; Evans, 1989) and for pattern associators or autoassociators with an optimal synaptic matrix, following the approach mentioned below (Gardner, 1988). These more precise calculations enable one to extract in more detail the exact dependence of the maximum value allowed for p, on the firing sparseness, in the form

$$p \approx C / [a \ln(1 / a)] \tag{A3.19}$$

which is valid in the limit $a \to 0$. Eq. A3.19 has been found to be valid also for autoassociators with graded response units (Treves, 1990; Treves and Rolls, 1991), as described in Appendix A4. For pattern associators, the lack of a well-defined notion of storage capacity, independent of retrieval quality and of cue quality, makes the approaches based on binary units not applicable to more realistic systems, as discussed earlier in this section, but the semiquantitative signal-to-noise analysis shows that the important determinants of the capacity are the same.

A3.2.2 Information-theoretic measures of performance

The basic signal-to-noise analysis sketched above may be elaborated in specific cases and made more accurate, even without having to resort to binary units. To obtain more precise quantitative results, one needs to look not at the *input* to the output cells, but at the *output* of the output cells, and consider a definite measure of the performance of the system, in particular a measure of the correlation between the output firing at retrieval and the firing in the associated output pattern. The correct measures to use are information measures, as discussed in Appendix A2, and in particular for this correlation the right measure is the mutual information between the two patterns, or, expressed at the single cell level, the average information any firing rate $r_R{}^\mu$ which the output cell sustains during retrieval of association μ conveys on the firing rate r^μ during the storage of that association

$$I = \int dr \int dr_R \, P(r, r_R) \ln_2[P(r, r_R) / P(r) P(r_R)]. \tag{A3.20}$$

In addition, one should also consider the information provided by the cue, which is to be subtracted from the above to quantify the actual 'yield' (per cell) of the memory system. Moreover, some information is also associated with the selection of the pattern being retrieved. This last quantity is at most (when the *correct* pattern is always selected) of order $\log(p)$ and therefore usually much smaller than that reflecting the full correlation in the retrieved pattern, or that in the cue, which are both quantities expected to be proportional to the number of cells involved. However, there are cases (Frolov and Murav'ev, 1993) when binary units are used with very sparse coding, in which all such quantities could be of similar magnitude; and, more importantly, it is also true that what can be measured in neurophysiological experiments is, in practice, the information associated with the selection of the output pattern, not the one expressed by Eq. A3.20 (since the firing rates recorded are typically only those, as it were, at retrieval, r_R).

In general, one is interested in the value of I, in the equation above, averaged over a large number of variables whose specific values are irrelevant: for example, the firing rates of both output and input cells in all the other stored associations. One is therefore faced with the computational problem of calculating the average of a logarithm, which is not trivial. This is where mathematical techniques imported from theoretical physics become particularly useful, as a body of methods (including the so-called replica method) have been developed in statistical physics specifically to calculate averages of logarithms. These methods will be briefly sketched in Appendix A4, on autoassociators, but the motivation for resorting to them in the quantitative analysis of formal models is essentially identical in the case of pattern associators. An example of the application of the replica method to the calculation of the information retrieved by a pattern associator is given by Treves (1995). In that case, the pattern associator is in fact a slightly more complex system, designed as a model of the CA3 to CA1 Schaffer collateral connections, and cells are modelled as threshold-linear units (this is one of the few calculations carried out with formal models whose units are neither binary nor purely linear). The results, which in broad terms are of course consistent with a simple signal-to-noise analysis, detail the dependence of the information gain on the various parameters describing the system, in particular its degree of plasticity, as discussed in Chapter 6.

A3.2.3 Special analyses applicable to networks of binary units

Networks of binary units, which of necessity support the storage and retrieval of binary firing patterns, have been most intensely studied both with simulations and analytically, generating an abundance of results, some of which have proved hard to generalize to more realistic nets with continuously graded units. One of the peculiarities of memories of binary units operating with binary patterns is that one can define what one means by precise, exact retrieval of a memorized pattern: when all units are in the correct 0 or 1 state, bit by bit. This greatly helps intuition, and we have used this fact in the illustratory examples of Chapter 2, but it may also generate misconceptions, such as thinking that 'correct retrieval', a notion which makes sense in the technological context of digital transmission of information, is a meaningful characterization of the operation of realistic, brain-like networks.

The prototypical pattern associator, or Associative Net, originally introduced by Willshaw, Buneman and Longuet-Higgins (1969), and often referred to as the Willshaw net; and the slightly different version used at about the same time in the extensive analyses by the late David Marr (1969, 1970, 1971), are both rather extreme in their reliance on binary variables, in that not only the processing units, but also the synaptic weights are taken to be binary. This is in contrast with the models considered by Little (1974) and Hopfield (1982), and analysed mathematically by Amit, Gutfreund and Sompolinsky (1985, 1987), in which the units are taken to be binary, but the synaptic weights, which are determined by a Hebbian covariance rule, are in principle real-valued. The learning rule employed in the Willshaw net may be considered a clipped version of a Hebbian, but non-LTD-balanced, learning rule. A synapse in the Willshaw associator can be in one of two states only, off or on, and it is on if there is at least one input–output association in which both pre- and postsynaptic units are on, and it is off otherwise. If a is, again, the average proportion of units on in both the input and output patterns, the number of associations that can be stored and retrieved correctly is

$$p \approx 1/a^2 \tag{A3.21}$$

which can be understood intuitively by noting that the probability that a particular synapse be off is, from simple combinatorics,

$$P(w_{ij} = 0) = (1 - a^2)^p \approx \exp(-p\,a^2) \tag{A3.22}$$

and that this probability must be of order unity for the network to be in a regime far from saturation of its potential for synaptic modification. A detailed analysis of the Willshaw net (interpreted as an autoassociator) has been carried out by Golomb, Rubin and Sompolinsky (1990) using the full formal apparatus of statistical physics. A particularly insightful analysis is that of Nadal and Toulouse (1990), where they compare a binary pattern associator operating with a purely incremental, Hebbian rule and one operating with a clipped version of this rule, that is a Willshaw net. The interesting message to take home from the analysis is that, in the regime of very sparse patterns in which both nets operate best, in terms of being able to retrieve many associations, the clipping has a very minor effect, resulting in a slightly inferior capacity. In terms of information, if the pattern associator with graded synapses is shown to utilize up to $1/(2 \ln 2) \simeq 0.721$ bits per synapse, the Willshaw net is shown to utilize $\ln 2 \simeq 0.693$ bits per synapse, with a reduction of only 4% (Nadal 1991; Nadal and Toulouse 1990). This can be converted into a statement about the precision with which a biological mechanism such as LTP would have to set synaptic weights in associative learning, that is, it would not have to be very precise. In general, when a learning rule that includes hetero-synaptic LTD is used, it remains true that not much resolution is required in the synaptic weights to achieve near optimal storage capacity (cf. Sompolinsky, 1987). Having very few bits available produces only a moderate decrease in the number of associations that can be stored or in the amount of information stored on each synapse. In the particular case of the Willshaw type of learning rule, however, it should be noted that the near equivalence between binary-valued and continuously-valued synapses only holds in the very extreme sparse coding limit. Moreover, in the Willshaw type of network a price for being allowed to use weights with only 1 bit resolution is that the thresholds have to be set very precisely for the network to operate successfully.

The way in which the binary nature of the units explicitly enters the calculation of how many associations can be stored in a binary pattern associator is typically in finding the best value of the threshold that can separate the input activations that should correspond to a quiescent output from those that should correspond to a firing output. (Differences in specifications arise in whether this threshold is allowed to vary between different units, and in the course of learning, or rather it is held fixed, and/or equal across output units.) In a Willshaw net if the input is exactly one of the stored patterns, the activation of the output units that should be 'on' is exactly the same (a distribution of width zero), because all the synapses from active input units are themselves in the 'on' state, and the number of active inputs is, at least in the original Willshaw net, fixed exactly. This allows setting the threshold immediately below this constant activation value, and the only potential for interference is the extreme tail of the distribution of activation values among output units that should be 'off' in the pattern. Only if at least one of these output units happens to have all the synaptic weights from all active inputs on, will an error be generated. Thus one factor behind the surprisingly good capacity achieved by the Willshaw net in the very sparse coding regime is its exploiting the small amount of probability left at the very tail of a binomial distribution (the one giving the probability of being 'on' across all synapses of the same output unit). This condition allowing relatively large capacity would seem not to carry over very naturally to more realistic situations in which *either* synaptic values or firing rates are taken to have not fully binary distributions, but rather continuous distributions with large widths around each peak. In fact, realistic distributions of values for *both* synaptic efficacies and firing rates are more likely to be unimodal, and not to show any more than one peak). A quantitative analysis of the Willshaw net, adapted to include finite resolution widths around two dominant modes of synaptic values, has never been carried out, but in any case, binary or nearly binary-valued synaptic weights would be expected to function reasonably only in the limit of very sparse coding. This can be understood intuitively by noting again that with more distributed coding, any particular synaptic value would have a fair chance of being modified with the learning of each new association, and it would quickly become saturated. Even if synaptic values are allowed to modify downwards, the availability of only two synaptic levels implies that few successive associations are sufficient to 'wash out' any information previously stored on a particular synapse, unless the coding is very sparse.

The Gardner approach

The approach developed by the late Elizabeth Gardner (1987, 1988) represented an entirely novel way to analyse the storage capacity of pattern associators. Instead of considering a model network defined in terms of a given learning rule, she suggested considering the 'space' of *all* possible networks, that is of all possible synaptic matrices, independently of their having been produced, or not, by a learning mechanism, and selecting out among all those the ones that satisfied certain constraints. For a pattern associator with binary units, these constraints can be specified by requiring that for each association the net is taken to have stored, the activation resulting from the presentation of the correct input pattern be above threshold for those output units that should be 'on', and below threshold for those output units that should be 'off'. The average number of synaptic matrices that satisfy these requirements, for each of a number p of stored associations, can be literally counted using statistical physics techniques, and the maximum p corresponds to when this average number of 'viable' nets reduces to one. Such an approach, which is too technical to be described more fully here (but an excellent account is given in the book by Hertz *et al.* (1991)) has been generalized in an endless variety of ways, generating a large body of literature. Ultimately, however, the approach can only be used with binary units, and this limits its usefulness in understanding real networks in the brain.

Nevertheless, the Gardner calculations, although strictly valid only for binary units, are often indicative of results that apply to more general models as well. An example is the increase in storage capacity found with sparse coding, Eq. A3.19 above, which is a result first obtained by Gardner (1987, 1988) without reference to any specific learning rule, later found to hold for fully connected autoassociators with Hebbian covariance learning (Tsodyks and Feigel'man, 1988; Buhmann, Divko and Schulten, 1989), then to apply also to autoassociators with sparse connectivity (Evans, 1989), and subsequently generalized (modulo the exact proportionality factor) to non-binary units and non-binary patterns of activity (e.g.

Treves and Rolls, 1991). A point to be noted in relation to the specific dependence of p on a, expressed by Eq. A3.19, is that the original Gardner (1988) calculation was based on steady-state equations that would apply equally to a pattern associator and to an autoassociator, and hence did not distinguish explicitly between the sparseness of the input pattern and that of the output pattern. In the Hertz *et al.* (1991) account of the same calculation, a different notation is used for input and output patterns, but unfortunately, when sparseness is introduced it mistakenly appears that it refers to the input, whereas by repeating the calculation one easily finds that it is the sparseness of the output pattern that matters, and that enters Eq. A3.19. One can also be convinced that this should be the case by considering a single binary output unit (the exact calculation is in fact carried out independently for each output unit) and what the learning of different associations entails for the connection weights to this unit. If the output statistics is very sparse, the unit has to learn to respond (that is to be in the 'on' state) to very few input patterns, which can be effected simply by setting a large threshold and the weights appropriate for a superposition of a few AND-like functions among the input lines (the functions would be AND-like and not AND proper in the sense that they would yield positive output for a specific conjunction of 'on' and 'off', not just 'on', inputs).

The crucial point is that just superimposing the weights appropriate to each AND-like operation only works if the functions to be superimposed are few, and they are few if the output statistics is sparse. The full calculation is necessary to express this precisely in terms of the $[a \ln(1/a)]^{-1}$ factor.

The Gardner approach has been fruitful in leading to many results on networks with binary output units (see Hertz, Krogh and Palmer, 1991).

A4 Autoassociators

The formal analysis of the operation of autoassociative memories is more developed than for other types of network, which are either simpler (and mathematically not as interesting) to describe quantitatively, or, at the opposite end, tend to be mathematically intractable, at least with sufficient generality. Autoassociative memories, instead, combine conceptual simplicity with a requirement for some sophisticated mathematics, and have therefore provided material for a rather large amount of formal, analytical work. At a deeper level, some intriguing analogies with notions arising in statistical physics, such as the reduction of the dynamical behaviour, over long times, to a description of attractor states, have appealed to physicists, who have been able to apply to the study of model networks analytical techniques developed in their field.

A4.1 Autoassociation versus short-term memory: the role of feedback

As described in Chapter 3, the prototypical autoassociative memory network is based on recurrent collaterals, and thus its neurons can sustainedly activate each other even after the termination of any external input. The network can thus serve as a short-term memory, where the attribute of short-term implies, more than a definite time-scale, precisely the fact that it is implemented as the sustained activation of one specific neuronal firing pattern. This is in addition to the network functioning as a long-term memory (for many patterns simultaneously), implemented via specific modifications of a matrix of synaptic efficacies.

There are, however, several different possible architectures for an autoassociative memory, and not all, or even most of them, can operate as short-term memories. What is required, to produce sustained activation, is obviously a substantial degree of internal feedback, and one element that differentiates between autoassociative networks is therefore the relative amount of feedback.

We have suggested that many, of all possible autoassociative architectures, can be synthetically described as points on a two-dimensional continuum (Treves and Rolls, 1991; Fig. A4.1). In the lower part of Fig. A4.1, one dimension (vertical) corresponds to the degree to which there exists a preferred direction for information flow within the network, from none at the top to fully directional at the bottom. The second dimension (horizontal in the lower part of Fig. A4.1) corresponds to the degree to which synaptic connections exist, among all possible pairs of cells in the network. This is referred to as the degree of 'dilution' and ranges from the full connectivity of the prototypical architecture (M1) to the extremely sparse one (M3). One may realize that the existence of a preferred direction (in the connectivity, without regard to the values of the synaptic weights) in itself implies that not quite all the possible connections are present; and also that in the limit of extreme dilution, the extent to which a preferred direction is imposed a priori becomes less relevant, since the dilution already limits the possible paths for information flow. Hence, the continuum is depicted in the lower part of Fig. A4.1 as a triangle. The vertices of the triangle are three limiting (extreme) cases, which have been studied extensively in the literature, and which we refer to as the prototypical architecture (M1), the fully connected but layered architecture (M2), and the one with extremely sparse connectivity (M3). These are, again, conceptual extremes and any biologically relevant architecture is likely to fall somewhere in the middle. At the latter two vertices, and on the line joining them, there is no feedback, that is they are purely feedforward networks. Some feedback is present as soon as there are loops whereby a cell can indirectly affect itself, that is anywhere in the inside of the triangle (in the lower part of Fig. A4.1). Substantial feedback, sufficient for sustained activation, is only present towards the top left corner.

Two other effects are caused by feedback, beyond its potential for subserving short-term memory, and in addition a third effect is related in part to feedback. The first effect is that it makes a substantially more complex analysis required, in order to understand the operation of the network quantitatively. Any degree of feedback is enough to produce this complication. The second effect is that interference effects among concurrently stored memories are more serious, increasing with more feedback but decreasing as the coding becomes sparser. The third effect, related only in part to feedback, is that the learning phase is much simpler to implement when all units in the network receive external inputs setting, at least in broad terms, a firing level for them to encode in the memory. Recapitulating, feedback

(1) makes short-term memory possible, but only if it is dominant;
(2) helps make learning easy, with the appropriate afferent connectivity;
(3) makes interference, or cross-talk among memories, more serious;
(4) requires more sophisticated formal analyses.

Effect (1) is self-evident. Effect (2) is intuitive, but has not been properly quantified in general. Effect (3) has been quantified, and will be illustrated at the end of this appendix. Effect (4) is addressed here in the following.

A4.2 Feedback requires the use of self-consistent statistics

Analysing the retrieval operation performed by a feedforward memory model amounts to nothing more than elementary statistics. The system is fully described by equations of the type

$$r_i = f(h_i) \tag{A4.1}$$

and

$$h_i = \Sigma_j r'_j w_{ij} \tag{A4.2}$$

and to find the output firing rates one must simply start from the input rates, work through the intermediate units (if any), and get to the output by successive multiplications, summations, and

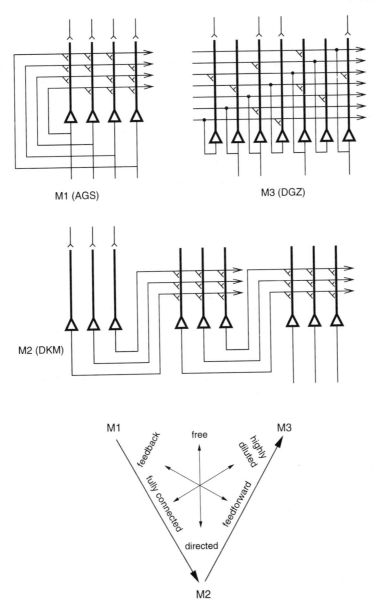

Fig. A4.1 Schematic representation of three limiting architectures used in the analysis. In the three diagrams M1, M2 and M3, external inputs come to the networks from above, and outputs leave from below. Lower diagram: the thin arrows illustrate the way the schemes M1–M3 relate to usual qualitative descriptions, while the thick arrows are in the direction of increasing capacity (see text). The acronyms refer to the formalisms used in analysing the corresponding binary units models as follows: AGS, Amit, Gutfreund and Sompolinsky, 1987; DKM, Domany, Kinzel and Meir, 1989; DGZ, Derrida, Gardner and Zippelius, 1987. (After Treves and Rolls, 1991, Fig. 2.)

applications of the single-unit transform, or activation function, $f(h)$. The need for statistics is just in the fact that usually one is not interested in the output given an input exactly specified in each incoming firing rate (as was the case with the illustratory examples of Chapter 2), but rather in the likely output given some more generic specification of the input. For example, the input may be specified only in its correlation or overlap with one or several input firing patterns previously learned by the net. These correlations or overlaps act as 'macroscopic' parameters, while 'microscopic' details (individual input firing rates) are left free to vary, provided they are consistent with the specified parameters. Doing statistics means, then, averaging over those unspecified details. Also on the output side, global parameters, such as the overlap of the output rates with a previously learned pattern, are usually more interesting than individual rates, and to find them one needs only to carry out the corresponding sum.

When feedback loops are present, some of the units whose rates are given as the left side of Eq. A4.1 are the same as some whose rates appear on the right side of Eq. A4.2. At steady state, their rates are also the same. (We have mentioned in Chapter 3 why the operation of autoassociators is best considered at steady state, which, if the system has functioned properly, is just a recall state.) Therefore, the same variables appear on both sides of a system of equations, and as a consequence the system has to be *solved*, not just read out from right to left as before. One has to find a solution, that is a set of rates that are mutually consistent with each other (and with the true inputs, those which are never on the left side). Again, one normally wants answers for a generic specification of macroscopic input parameters, and also in terms of macroscopic output parameters. An example of a macroscopic input parameter that one may want to specify is the correlation of the cue, implemented by the firing r' of those axons that are true input lines, and not recurrent ones, with a specific stored pattern. We may denote such input parameters symbolically as $IP(r')$, since they depend on the input rates r', and the specific values they are required to take as IP. An example, instead, of a macroscopic output parameter, which characterizes the steady recall state reached by the network, is the final correlation with that same stored pattern that the cue was correlated with. We may denote such output parameters symbolically as $OP(r)$, since they are functions of the output firing rates r, and the values they end up taking as OP.

The way, in the formal analysis, that one obtains the solutions is quite brutal: one sums up (integrates) all the possible values of each firing rate (times all the units in the system), and also over all values of the global output parameters, and one just throws away any set of rates that does not satisfy the following types of constraints:

(1) that rates are mutually consistent, that is Eqs. A4.1 and A4.2 are satisfied, denoted symbolically as $r = F(r)$;

(2) that the input parameters take the required values, that is $IP(r') = IP$;
and also one throws away inconsistent values for output parameters, that is enforces

(3) that output parameters take the values corresponding to solutions, $OP = OP(r)$.

The constraints are enforced by inserting in the overall integration δ-functions:

$$\int \dots \int \prod dOP \dots dr \; \delta \left(IP(r') - IP \right) \delta \left(r - F(r) \right) \delta \left(OP - OP(r) \right) \qquad (A4.3)$$

where the δ-functions are by definition zero when their argument differs from zero, but yield one when integrated over values that allow the argument to be zero. This is the basic step in performing self-consistent statistics; the remaining steps are largely complicated technical-ities. For example, usually the next step is to write the δ-functions, in turn, as integrals over some new variables, for example

$$\delta(r - F(r)) = \int d\lambda \exp i\lambda(r - F(r)) \qquad (A4.4)$$

using a standard procedure of mathematical physics. Another usual ingredient of the analysis is the use of saddle-point approximations. Whereas all values of microscopic variables have the potential for appearing in a solution, and therefore have to be duly integrated over, generally values of the macroscopic ones, that correspond to solutions, are all tightly concentrated around one or more points (the saddle-points in the complex plane) which give the dominating contribution to the integrals over op. Those integrals are then not carried out, in practice, but estimated with the value of the integrand at the saddle point(s), using again a standard technique. Details of these procedures can be learned from the Amit (1989) and Hertz, Krogh and Palmer (1991) books, or, even better, by following the calculations in the original articles, such as the seminal one by Amit, Gutfreund and Sompolinsky (1987). A collection of important papers on the physics of disordered systems, which includes some on neural networks and is complete with explicatory text by the authors, is the book by Mezard, Parisi and Virasoro (1987).

A4.3 Averaging over fixed parameters and the replica trick

Apart from the need to check the stability of the solutions found, the above would more or less describe the methods required, were it not for the need to average over microscopic parameters, whose values are fixed but unspecified. In autoassociative memories such parameters are the input firing rates r', if they are sustained during the retrieval opera-tion, but also the values of the synaptic weights w. The latter are in turn taken to be determined by the firing rates produced in the cells during the learning of each memory pattern. One may give some statistics of these learned firing patterns; for example, if they are binary, the probability that each unit is 'on' in a pattern. Much more than that, it is in general not possible to specify. Even if it were possible, one would be in a situation in which the formal analysis would yield answers that are not valid in general, for a class of networks described by certain overall parameters, but are only valid for a specific network, which has gone through its own specific history of learning, and is therefore loaded with its own 'baggage' of microscopic parameters. A new calculation would have to be performed for each individual network, much as, in the minuscule-size examples of Chapters 2 and 3, one has to calculate the properties for each example separately, in order to yield numbers.

Fortunately, what saves the day for analytical calculations, and allows them to be carried out only once (or rather, once for each *type* of network, instead of once for each *individual* network) is that some of the quantitative properties of systems characterized by a large number of unspecified parameters are known to *self-average*. This term means that the values attained by those quantities, if the system is sufficiently large, coincide with their average over many equal systems, that differ only in the unspecified details. Therefore, the answer produced by a calculation, which includes an average over those details, is not only an average answer for the type of systems (as such it would be of limited use), but it is *the* answer for each system, provided each system is large enough (in what is referred in the physicists' jargon as the thermodynamic limit). Not all quantities self-average, however; in general, those that are expected to self-average are extensive quantities, that is quantities whose magnitude is proportional to the size of the system.

A4.3.1 Free energy and mutual information

Some models of autoassociative memory networks (those of the class introduced by Hopfield (1982), and which elicited the interest of many statistical physicists) are designed to operate, at retrieval, in close analogy with physical systems that reach thermal equilibrium. In particular, their synaptic connections are chosen to be always reciprocal, and with equal weights, which results in the possibility to define an energy function that is minimal when the system has reached equilibrium. To complete the correspondence, such model networks can be considered to operate under the influence of a temperature, T, which represents a special way to account for some noise in the input–output transform operated by each unit. In such networks, extensive quantities are the value E taken by the energy function, and the so called free-energy, F, which is defined for each equilibrium state

$$F = <E> - TS. \tag{A4.5}$$

In this formula, the fact that noise is present (T not zero) implies that several (for low noise very similar) *configurations* of the system concur in the equilibrium *state* (a configuration is the set of values taken by all the variables in the system, in the case of the network by the firing rates); therefore, the energy function needs to be averaged over those configurations, and to obtain the free-energy one has to further subtract the temperature times the entropy S of the state, which is roughly the logarithm of the number of configurations that contribute to it. We refer for example to the Amit (1989) book for a thorough discussion of the analogy between certain network models and physical systems. The only point to be made here is that the retrieval operation in such models can be described using thermodynamic notions, and that the free-energy is the important quantity to consider, not only because thermodynamic theory shows that many other interesting quantities can be obtained by taking derivatives of the free-energy, but also because it is an extensive quantity (like, in fact, the energy and the entropy) that, as such, is expected to self-average.

Another quantity that is extensive, and therefore should self-average, is mutual information, for example in an autoassociative network the mutual information between the current firing pattern and some previously stored pattern that the net is retrieving. Mutual information is also related to entropy, in fact it can be written simply as an entropy difference. Some type of mutual information can usually be defined for any associative network model, independently of whether the analogy with physical systems holds, and moreover, mutual information is, more directly than free-energy, a measure of the performance of the network as an information processing system.

In any case, what makes both free-energy (when definable) and mutual information extensive quantities also makes them depend logarithmically on probabilities. For mutual information this can be immediately traced to its being an entropy difference. To see that the same applies to the free-energy, which is an energy minus an entropy, one has to remember that the probabilities of different configurations in an equilibrium state are exponential in their energies, and therefore the energy, and the free-energy, are logarithmic in the probabilities.

A4.3.2 Averaging the logarithms

The bottom line is that in systems with 'microscopic' unspecified parameters one wants to calculate self-averaging quantities, and that these are generally logarithmic functions of probabilities. Now, averaging over probabilities is usually something that can be done directly, whereas averaging over logarithms of probabilities cannot be done directly. One uses therefore the replica trick, which amounts to writing the log as the limit of a power

$$\log P = \lim_{n \to 0} [P^n - 1]/n \qquad (A4.6)$$

where the power of the probability implies that one must fictitiously consider n identical copies ('replicas') of the system, calculate the required probabilities, average them over microscopic details, and, at the end, take the limit in which the number of replicas tends to zero! The replica trick was originally introduced in physics as an analytical expedient, but it yields answers that appear to be correct, and may in some cases be confirmed with independent analytical methods (and with computer simulations). Many technicalities are involved in the use of the replica trick, which would take us beyond the scope of this book (see Mezard *et al.*, 1987; Amit, 1989; Hertz *et al.*, 1991).

A last remark on the analogy between models of associative memory networks and certain physical systems is the special role, among the latter, historically played by spin-glasses. A spin-glass is a system of magnetic moments that interact with each other with coupling energies that are positive or negative as a complicated function of the exact location of each pair of spins. As a result, for all practical purposes each coupling can be taken to be random (but fixed in value) and the system is said to be disordered. As such it must be described in generic terms, that is in terms of macroscopic parameters and probability distributions for microscopic ones, leaving out what we called unspecified details (the value of each and every coupling). Spin-glasses, because of these characteristics, resisted a theoretical understanding based on ordinary notions and techniques of thermodynamics, and stimulated the refinement of new concepts (and techniques) that, following the Hopfield (1982) paper, have been found applicable to associative networks models as well. The *mathematics* of spin-glasses has thus proven very useful for developing formal analyses of autoassociative nets. Their *physics*, however, that is the peculiar properties of equilibrium spin-glass states at low temperature, which initially had seemed to be relevant to neuronal nets too (Amit, Gutfreund and Sompolinsky, 1987; Crisanti, Amit and Gutfreund, 1986), has turned out to be less relevant once elements of realism such as graded response are introduced in the models previously strictly adherent to the physical analogy (Treves, 1991a,b). An example is what is referred to as the peculiar way in which spin-glasses *freeze* into one of several disordered states, unrelated by any symmetry, due to the *frustration* in their couplings (Toulouse, 1977). This was thought to apply also to autoassociative nets, that would, whenever failing to reach a recall state, get stuck, as it were, into a meaningless spin-glass state, in which each unit would carry on firing at its own steady-state firing rate, but the overall firing pattern would not match any stored pattern. This indeed occurs, but only in autoassociators that are fairly similar to the original Hopfield model, with binary units which have both excitatory and inhibitory influences on other units. In autoassociators made up of graded-response excitatory units, typically the network, whenever it fails to retrieve a stored pattern, falls into a *unique* uniform state, in which no firing pattern is discernible (Treves, 1990, 1991a,b).

A4.4 Results from the analysis: storage capacity and information content

Several quantitative relationships can be extracted from the analysis of any particular model of autoassociator. Of these, three represent fundamental properties that characterize the performance of the network, and turn out to be relatively independent of the exact model of autoassociator analysed. One of these three basic properties is the time required for the operation of the network. Its quantitative analysis requires, however, a discussion of the appropriate level of realism at which to model individual processing units, and is the subject of Appendix A5. The other basic properties are the number of memories that can be stored in the network, and the amount of information each of them contains.

We show here graphs that describe the results of our analyses (Treves and Rolls, 1991) based on the use of threshold linear units as models of single neurons, and in which a number of other minor specifications of the models have been made with biological realism in mind. These specifications concern, for example, the way the thresholds are set, uniform across units, or the way the learning of different patterns produces purely superimposed modifications on a common baseline value for synaptic weights. Very similar results, qualitatively identical, were obtained with earlier analyses based on binary units (Amit, Gutfreund and Sompolinsky, 1987; Derrida, Gardner and Zippelius, 1987; Tsodyks and Feigel'man, 1988, Buhmann, Divko and Schulten, 1989; Evans, 1989). Minor differences in the specifications do produce minor differences in the results. For example, optimizing the threshold independently for each unit results in some improvements over what can be achieved with uniform thresholds, equal for all the units (Perez-Vicente and Amit, 1989). Even relatively major differences, such as constraining synaptic efficacies to take only two values, as in the Willshaw model of a pattern associator (Willshaw *et al.*, 1969), yield much smaller differences in performance than might have been expected, once the appropriate comparisons are made (Nadal and Toulouse, 1990; Nadal, 1991). The deep reason for this relative insensitivity to exact specifications appears to be that in all the models the amount of information each synapse can store and contribute at retrieval is nearly the same, a fraction of a bit, and largely independent of the range of values the synaptic efficacy can take, continuous or discrete or binary. This aspect will be considered again in the following.

A4.4.1 The number of memories that can be stored

For partly historical reasons, this has traditionally been the main aim of physicists' analyses, the target that could not be reached with connectionist approaches based on the use of examples (McClelland and Rumelhart, 1986) and even not quite with quantitative approaches based on less suitable mathematics (Weisbuch and Fogelman-Souliè, 1985; McEliece, Posner, Rodemich and Venkatesh, 1987). It is referred to as the storage capacity calculation, although the information content of the memories (see below) is also an element to be taken into account in discussing the capacity of the network—sometimes referred to as the information capacity.

The number of memories that can be simultaneously in storage, and that can be individually retrieved from the autoassociative net, is usually calculated in the situation, discussed in Appendix A3, in which all memories are discrete representations, independent of each other, and each unit codes for independent information, that is its firing rate is statistically independent from that of other units, in every aspect except for the correlation induced by the memory firing patterns themselves. These conditions simplify the calculation, and also fill the need, in the absence of more structured a priori hypotheses, for a complete definition of the model. Exceptions that have been considered include the situations in which memories are organized into classes (Dotsenko, 1985; Parga and Virasoro, 1986), or in which they are correlated spatially or temporally (Monasson, 1992; Tarkowski and Loewenstein 1993) or, but then one deals with something different from an autoassociator, are organized into temporal sequences (see Chapter 3). In addition, (a) the firing of each unit across memory patterns is taken to follow the same statistics, that is it is described by the same probabilities or probability distributions for different firing levels; and (b) each memory pattern is taken to be encoded with the same mean strength. These are just convenient assumptions, which could be lifted without effort and substituted with (a) several different probability distributions for different cells, or even with a probability distribution of probability distributions; and (b) a distribution of coding strengths or, as one does to model the effects of forgetting, with coding strengths that decay in time.

As for pattern associators, if one demands absolute or nearly absolute fidelity in the retrieval of a previously stored memory pattern, the number of patterns that can be retrieved is a function of the fidelity required. Absolute fidelity is in any case an undefined concept in the case of continuously graded firing rates, and even for, for example, binary rates it is defined only with certain qualifications. Contrary to pattern associators, however, in autoassociators there exists a critical loading level, that is a critical number of memories in storage, beyond which no memory can be retrieved at all, in the sense that the retrieval operation, instead of completing the information present in the cue, erases it down to zero. Thus one can define a storage capacity in terms of this critical load. Of course it remains to be analysed how well completion operates, if it does, when the network is loaded close to the critical level. This second aspect will be discussed in the following subsection.

Within these conditions, the most important factor that determines the storage capacity is the sparseness a of the distribution of firing rates in each memory pattern, as defined in Appendix A3. Recall that for binary firing patterns, in which one of the two levels corresponds to no firing, the sparseness is just the fraction of the distribution at the other level, that is the firing one; whereas in general, if $< \ldots >$ denotes averages over the distribution of rates, the sparseness was defined as

$$a = <r>^2 / <r^2>$$ \hfill (A4.7)

that is simply in terms of the first two moments of the distribution. Sparser encodings lead to higher capacity. The reason can be approximately understood by considering again the signal-to-noise analysis sketched in Appendix A3 for pattern associators, which can be reproduced here for autoassociators: In conditions of efficient storage, that is when the learning rule is such as to make the mean of the presynaptic G factor vanish, the (mean square) noise is proportional to a power of the sparseness higher by one than the (mean square) signal (the precise powers depend on the normalization). This can be seen by substituting in Eqs. A3.11 and A3.12 for example $F(r) = r$ and $G(r) = r - <r>$. Apart from constant factors, the mean square signal S^2 is then proportional to $C^2 <r^2>^2 (1-a)^2$, while the mean square noise N^2 is proportional to $pC <r^2>^3 (1-a) \sim apS^2/C$. Therefore, to maintain a constant signal-to-noise ratio the number of memories p has to be of the order of the number of inputs per cell C divided by the sparseness a. The full analysis yields

$$p_{max} \sim C/[a \ln(1/a)] \tag{A4.8}$$

in the limit in which $a \ll 1$ (Treves and Rolls, 1991).

The analysis based on the signal-to-noise ratio can be refined and used for architectures with no feedback, in particular the M2 and M3 models of Fig. A4.1. When feedback is present, a self-consistent analysis is required, as discussed above. This could be theoretically obtained for any architecture that includes feedback (see the analysis developed by Fukai and Shiino, 1992), but in practice it has been done extensively only in cases in which one can use the thermodynamics formalism, that is when intrinsic connections are always reciprocal and of equal strength (the so-called *symmetric* networks). Using a common notation one can see, however, that the thermodynamics formalism also reduces to similar equations yielding the storage capacity, with differences in some terms that express the effect of reverberating noise through feedback loops (Treves and Rolls, 1991). Formula A4.8 holds for all the architectures in the two-dimensional continuum introduced above, in the sparse coding limit. In this limit the architectures do not differ in the number of patterns they can store; whereas differences, which can be large, appear if the coding is not sparse. The reason is that the reverberation of the noise is negligible for very sparse coding, while it increases interference and therefore lowers the critical value of p/C for non-sparse coding. When $a \sim 1$, architecture M1 has a substantially lower capacity than M2, and this in turn is lower than M3, with intermediate cases falling in the middle (Amit, Gutfreund and Sompolinsky, 1987; Derrida, Gardner and Zippelius, 1987; Domany *et al.*, 1989; Treves and Rolls, 1991). A plot of the critical values of p/C versus a for the extreme cases M1 and M3 is given in Fig. A4.2. The common nearly inverse proportionality for $a \sim 0$ is evident, along with the marked differences for larger values of a. It should also be noted that for very distributed codes, not only the exact type of connectivity, but also the exact type of pattern statistics strongly influences the storage capacity. As an example of 'realistic' statistics, we report in the graph the result obtained for an exponential distribution of firing rates, as shown in Fig. A3.1. However, probability distributions with $a \sim 0.7$, a single peak around the spontaneous firing rate, and a long nearly exponential tail, which are common in the primate temporal cortex (Rolls and Tovee, 1995a; Rolls, Treves, Tovee and Panzeri, 1997) cannot quite be modelled by the purely exponential form of Eq. A3.10 (the one used in these graphs), which is peaked at strictly zero rate, and which has a maximal possible $a = 0.5$. Also, the exact way in which the threshold is set would be important in this situation. One can easily insert any specification of the statistics, of the threshold setting, etc., in the formulas determining the capacity (see Treves and Rolls, 1991), but the main point here is that for non-sparse codes only a relatively limited number of memories can be stored in an autoassociator. Therefore, an exact calculation of the capacity for a non-sparse coding situation is likely to be not very relevant, since one is considering a network unlikely to be optimized in terms of the number of patterns it could store.

It should be noted that, while the calculations above were carried out using the independent pattern model, essentially the same results are expected to apply to models in which the representations stored are not independent firing patterns, but are instead organized in multiple spatial maps, or *charts* (Samsonovich and McNaughton, 1996). Work now in progress (Battaglia and Treves, unpublished) confirms that the maximum number of such charts that can be stored remains proportional to C and inversely proportional to the appropriate parameter defining the sparseness of each chart.

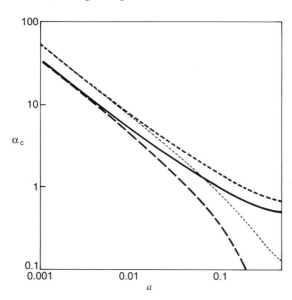

Fig. A4.2 The maximum number of patterns per input connection $\alpha_c \equiv p_c/C$ that can be stored in an autoassociative memory as a function of the sparseness a. This is shown separately for a fully connected feedback net (long-dashed line for binary patterns, and dotted line for exponential patterns); and for a net with extremely diluted connectivity (full line for binary patterns, and dashed line for exponential patterns). (After Treves and Rolls, 1991, Fig. 5a.)

A4.4.2 Information content of each memory, and information capacity

The notion of the information in a representation has been introduced in Appendix A2, and discussed again in Appendix A3. In this appendix on autoassociators, in discussing the information in a memory pattern, we assume, without further justification, that different cells code independent aspects of the representation, and the information they provide just adds up. Thus the total information is proportional to the number of cells present, $I = Ni$, and what is considered is the proportionality constant i. For a binary unit, the information it contributes on average to each representation is just a function of the entropy of the states the unit can take, $i = a\log_2(1/a) + (1-a)\log_2[1/(1-a)]$. The information *retrieved* is not identical to this quantity, as it depends also on the imprecision in the retrieval. If, on average, retrieval results in a fraction ab of bits that should be 'on' and instead are 'off', and in a fraction $(1-a)c$ of bits that should be 'off' and instead are 'on', the information retrieved is (see for example Nadal and Toulouse, 1990)

$$i_r = a[b\log_2(b) + (1-b)\log_2(1-b)] + (1-a)[c\log_2(c) + (1-c)\log_2(1-c)]$$
$$-[ab + (1-a)(1-c)]\log_2[ab + (1-a)(1-c)]$$
$$-[a(1-b) + (1-a)c]\log_2[a(1-b) + (1-a)c] \qquad (A4.9)$$

which coincides with the information stored if $b = c = 0$. In the more realistic situation of continuously-valued units, the information stored in the representation cannot in general even be defined, as it is strictly infinite. The value obtained after subtracting the average information lost in retrieval is, however, finite, and it is just the mutual information between the distribution of stored rates and that of firing rates at retrieval

$$i_r = \int dr \int dr_R\, P(r,r_R) \ln_2[P(r,r_R)/P(r)P(r_R)]. \qquad (A4.10)$$

as in Eq. A3.20 of Appendix A3.

The amount of information retrieved for each memory representation depends on the storage load, but what turns out to be the case for networks with a substantial load is that this amount, per cell, when graded response units are considered, is again in the order of what would be provided by binary units, or approximately

$$i_r \approx a\log_2(1/a) \qquad (A4.11)$$

which is the trend plotted in Fig. A4.3 as a function of a. The reason lies in the fact that the *total* information retrievable from the net is relatively insensitive to the type of units considered, and amounts to 0.2–0.3 bits per synapse. The total information retrievable, without considering how much information is provided with the retrieval cue, is just what can be retrieved of each pattern times the number of patterns stored simultaneously,

$$I = pNi_r \qquad (A4.12)$$

and its maximal value per synapse (that is, divided by CN) is an indication of the efficiency, in information terms, of the network as a storage device. One can write $i_r = (I/CN)/(p/C)$, and use Eq. A4.8 together with the approximately constant value for I_m (the maximum value of I/CN) to derive Eq. A4.11, apart from a numerical factor which turns out to be of order unity.

Again, the important point is that from all quantitative analyses performed so far on autoassociators that function on the same basic lines we have been discussing, the total

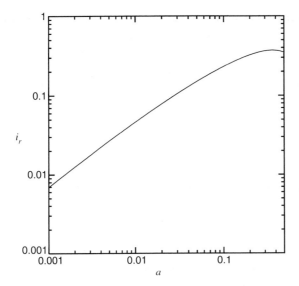

Fig. A4.3 The approximate amount of information that can be recovered from the firing of each neuron (bits/cell), i_r, as a function of the sparseness a of the firing, see Eq. A4.11.

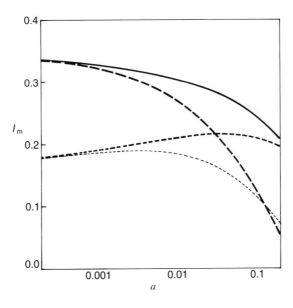

Fig. A4.4 The maximum amount of retrievable information per neuron (per modifiable synapse onto each neuron) I_m that can be retrieved from an autoassociative memory as a function of the sparseness a. This is shown separately for a fully connected feedback net (long-dashed line for binary patterns, and dotted line for exponential patterns); and for a net with extremely diluted connectivity (full line for binary patterns, and dashed line for exponential patterns). (After Treves and Rolls, 1991, Fig. 5b.)

information per synapse has always resulted to be, at its maximum, a similar fraction (0.2–0.3) of a bit. It should be noted, though, that it is only for binary-valued *synapses* that a bit is a self-evident upper limit; apart from binary synapses, which are of very remote interest in the context of the brain, it is only for binary-valued *units* (again, a rather artificial situation) that an upper bound (of 2 bits per synapse) can in some sense be derived using the Gardner (1987) approach. In this situation, therefore, simple binary models that are not in direct correspondence with biological networks have been used to demonstrate with mathematical rigour quantitative properties that have been later, and with much less rigour, found to apply also to more realistic models.

Generally, across models, as the storage load p increases, I increases initially proportionally to p, but eventually saturates and then goes down, as more load results in more interference, less precision at retrieval, and a greater fraction of the stored information i lost in the retrieved information i_r. The maximum for I is generally attained when the number of representations stored is of the order, but not very close to, the critical load value corresponding to the storage capacity. This is true independently of the sparseness.

In summary, therefore, the information retrieved, even when representations are encoded in terms of continuous variables, is roughly the same as in the case of binary variables. This is illustrated in Fig. A4.4, in which I_m is shown as a function of the sparseness a. It is evident how the effects of sparse coding on the critical value p_c for p, Fig. A4.2 and Eq. A4.8, and the effects of sparse coding on i_r, described by Fig. A4.3 and Eq. 4.11, nearly cancel out in the what is essentially their product, the total information encoded by the net (Fig. A4.4; Treves and Rolls, 1991).

A5 Recurrent dynamics

The methods described in the previous Appendices were all of use for the quantitative analysis of properties dependent only on the steady state activation of networks of neuron-like units. Those may be referred to as 'static' properties, in the sense that they do not involve the time dimension. In order to address 'dynamical' questions, the time dimension has to be reintroduced into the formal models used, and the adequacy of the models themselves has to be reconsidered in view of the specific properties to be discussed.

Consider for example a real network whose operation in memory has been described by an autoassociative formal model which acquires, with learning, a given attractor structure. How does the net approach, in real time during a retrieval operation, one of those attractors? How long does it take? How does the amount of information that can be read off the network's activity evolve with time? Also, which of the potential steady states is indeed a stable state that can be reached asymptotically by the net? How is the stability of different states modulated by external agents? These are examples of dynamical properties, which to be studied require the use of models endowed with some *dynamics*.

A5.1 From discrete to continuous time

Already at the level of simple models in which each unit is described by an input-output relation, one may introduce equally simple 'dynamical' rules, in order both to fully specify the model, and to simulate it on computers. These rules are generally formulated in terms of 'updatings': time is considered to be discrete, a succession of time *steps*, and at each time step the output of one or more of the units is set, or updated, to the value corresponding to its input variable. The input variable may reflect the outputs of other units in the net as updated at the previous time step or, if delays are considered, the outputs as they were at a prescribed number of time steps in the past. If all units in the net are updated together, the dynamics is referred to as *parallel*; if instead only one unit is updated at each time step, the dynamics is *sequential* (one main difference between the Hopfield (1982) model of an autoassociator and a similar model considered earlier by Little (1974) is that the latter was based on parallel rather than sequential dynamics). Many intermediate possibilities obviously exist, involving the updating of groups of units at a time. The order in which sequential updatings are performed may for instance be chosen at random at the beginning and then left the same in successive *cycles* across all units in the net; or it may be chosen anew at each cycle; yet a third alternative is to select at each time step a unit, at random, with the possibility that a particular unit may be selected several times before some of the other ones are ever updated. The updating may also be made probabilistic, with the output being set to its new value only with a certain probability, and otherwise remaining at the current value.

Variants of these dynamical rules have been used for decades in the analysis and computer simulation of physical systems in statistical mechanics (and field theory). They can reproduce in simple but effective ways the stochastic nature of transitions among discrete quantum states, and they have been subsequently considered appropriate also in the simulation of neural network models in which units have outputs that take discrete values, implying that a change from one value to another can only occur in a sudden jump. To some extent different rules are equivalent, in that they lead, in the evolution of the activity of the net along successive steps and cycles, to the same set of possible steady states. For example, it is easy to realize that when no delays are introduced, states that are stable under parallel updating are also stable under sequential updating. The reverse is not necessarily true, but on the other hand states that are stable when updating one unit at a time are stable irrespective of the updating order. Therefore, static properties, which can be deduced from an analysis of stable states, are to some extent *robust* against differences in the details of the dynamics assigned to the model (this is a reason for using these dynamical rules in the study of the thermodynamics of physical systems). Such rules, however, bear no relation to the actual dynamical processes by which the activity of real neurons evolves in time, and are therefore inadequate for the discussion of dynamical issues in neural networks.

A first step towards realism in the dynamics is the substitution of discrete time with continuous time. This somewhat parallels the substitution of the discrete output variables of the most rudimentary models with continuous variables representing firing rates. Although continuous output variables may evolve also in discrete time, and as far as static properties are concerned differences are minimal, with the move from discrete to continuous outputs the main raison d'être for a dynamics in terms of sudden updatings ceases to exist, since continuous variables can change continuously in continuous time. A paradox arises immediately, however, if a continuous time dynamics is assigned to firing rates. The paradox is that firing rates, although in principle continuous if computed with a generic time-kernel, tend to vary in jumps as new spikes—essentially discrete events—come to be included in the kernel. To avoid this paradox a continuous time dynamics can be assigned, instead, to instantaneous continuous variables like membrane potentials. Hopfield (1984), among others, has introduced a model of an autoassociator in which the output variables represent membrane potentials and evolve continuously in time, and has suggested that under certain conditions the stable states attainable by such a network are essentially the same as for a network of binary units evolving in discrete time. If neurons in the central nervous system communicated with each other via the transmission of graded membrane potentials, as they do in some peripheral systems, this model could be an excellent starting point. The fact that, centrally, transmission is primarily via the emission of discrete spikes makes a model based on membrane·potentials as output variables inadequate to correctly represent spiking dynamics.

A5.2 Continuous dynamics with discontinuities

In principle, a solution would be to keep the membrane potential as the basic dynamical variable, evolving in continuous time, and to use as the output variable the spike emission times, as determined by the rapid variation in membrane potential corresponding with each spike. A point-neuron-like processing unit in which the membrane potential V is capable of undergoing spikes is the one described by equations of the Hodgkin–Huxley type

$$C\frac{dV}{dt} = g_0(V_{rest} - V) + g_{Na}mh^3(V_{Na} - V) + g_K n^4(V_K - V) + I$$

$$\tau_m \frac{dm}{dt} = m_\infty(V) - m$$

$$\tau_h \frac{dh}{dt} = h_\infty(V) - h \qquad\qquad (A5.1)$$

$$\tau_n \frac{dm}{dt} = n_\infty(V) - n$$

in which changes in the membrane potential, driven by the input current I, interact with the opening and closing of intrinsic conductances (here a sodium conductance, whose channels are gated by the 'particles' m and h and a potassium conductance, whose channels are gated by n; Hodgkin and Huxley, 1952). These equations provide an effective description, phenomenological but broadly based on physical principles, of the conductance changes underlying action potentials, and they are treated in any standard neurobiology text. From the point of view of formal models of neural networks, this level of description is too complicated to be the basis for an analytical understanding of the operation of network, and it must be simplified. The most widely used simplification is the so-called *integrate-and-fire* model (see for example MacGregor, 1987), which is legitimized by the observation that (sodium) action potentials are typically brief and self-similar events. If, in particular, the only relevant variable associated with the spike is its time of emission (at the soma, or axon hillock), which essentially coincides with the time the potential V reaches a certain threshold level V_{thr}, then the conductance changes underlying the rest of the spike can be omitted from the description, and substituted with the *ad hoc* prescription that (i) a spike is emitted, with its effect on receiving units and on the unit itself, and (ii) after a brief time corresponding to the duration of the spike plus a refractory period, the membrane potential is reset and resumes its integration of the input current I. After a spike the membrane potential is taken to be reset to a value V_{ahp} (for after hyperpolarization). This type of simplified dynamics of the membrane potentials is thus in continuous time with added discontinuities: continuous in between spikes, with discontinuities occurring at different times for each unit in a population, every time a unit emits a spike.

Although the essence of the simplification is in omitting the description of the evolution in time of intrinsic conductances, in order to model the phenomenon of adaptation in the firing rate, prominent especially with pyramidal cells, it is necessary to include at least an intrinsic (potassium-like) conductance (Brown, Gähwiler, Griffith and Halliwell, 1990). This can be done in a rudimentary way with minimal complication, by specifying that this conductance, which if open tends to shunt the membrane and thus to prevent firing, opens by a fixed amount with the potential excursion associated with each spike, and then relaxes exponentially to its closed state. In this manner sustained firing driven by a constant input current occurs at lower rates after the first few spikes, in a way similar, if the relevant parameters are set appropriately, to the behaviour observed *in vitro* of many pyramidal cells (for example, Lanthorn, Storm and Andersen, 1984; Mason and Larkman, 1990).

The equations for the dynamics of each unit, which replace the input–output transduction function, are then

$$C\frac{\mathrm{d}V(t)}{\mathrm{d}t} = g_0\left(V_{\mathrm{rest}}-V(t)\right)+g_K(t)\left(V_K-V(t)\right)+I(t)$$

$$\frac{\mathrm{d}g_K(t)}{\mathrm{d}t} = -\frac{g_K(t)}{\tau_K}+\sum_k \Delta g_K \delta\left(t-t_k\right)$$

(A5.2)

supplemented by the prescription that when at time $t = t_{k+1}$ the potential reaches the level V_{thr}, a spike is emitted, and hence included also in the sum of the second equation, and the potential resumes its evolution according to the first equation from the reset level V_{ahp}. The resulting behaviour is exemplified in Fig. A5.1a, while Fig. A5.1b shows the input–output transform (current to frequency transduction) operated by an integrate-and-fire unit of this type. One should compare with the transduction operated by real cells, as exemplified for example in Fig. 1.3e.

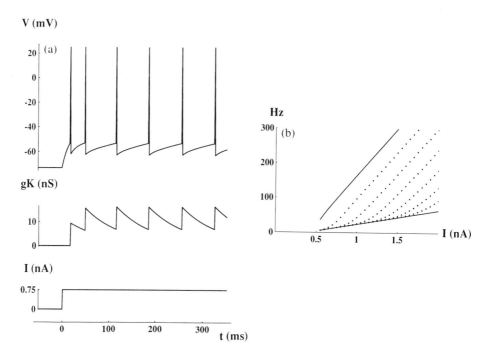

Fig. A5.1 (a) Model *in vitro* behaviour of an integrate-and-fire unit: the membrane potential and adaptation-producing potassium conductance in response to a step of injected current. The spikes are added to the graph by hand, as they do not emerge from the simplified voltage equation. (From Fig. 1 of Treves, 1993). (b) Current-to-frequency transduction in a pyramidal cell modelled as an integrate-and-fire unit. The top solid curve is the firing frequency in the absence of adaptation, $\Delta g_K = 0$. The dotted curves are the instantaneous frequencies computed as the inverse of the ith interspike interval (top to bottom, $i = 1, \ldots$,6). The bottom solid curve is the adapted firing curve ($i \to \infty$). With or without adaptation, the input–output transform is close to threshold-linear. (From Treves, 1993, Fig. 2.)

A5.3 Conductance dynamics for the input current

Besides the input–output transduction, the neuron-like units of the static formulations are characterized by the weighed summation of inputs, which models the quasi-linear (Rall and Segev, 1987) summation of currents of synaptic origin flowing into the soma. This description can be made dynamical by writing each current, at the synapse, as the product of a conductance, which varies in time, with the difference between the membrane potential and the equilibrium potential characteristic of each synapse

$$I(t) = \sum_\alpha I_\alpha(t) = \sum_\alpha g_\alpha(t)(V_\alpha - V(t)) \tag{A5.3}$$

The opening of each synaptic conductance is driven by the arrival of spikes at the presynaptic terminal, and its closing can often be described as a simple exponential process. A simplified equation for the dynamics of $g_\alpha(t)$ is then

$$\frac{dg_\alpha(t)}{dt} = -\frac{g_\alpha(t)}{\tau_\alpha} + \Delta g_\alpha \sum_l \delta(t - \Delta t - t_l) \tag{A5.4}$$

According to the above equation, each conductance opens instantaneously by a fixed amount Δg a time Δt after the emission of the presynaptic spike at t_l. Δt summarizes delays (axonal, synaptic, and dendritic), and each opening superimposes linearly, without saturating, on previous openings. An alternative equation which incorporates also a finite synaptic opening time, and which is fairly accurate at least for certain classes of synapses, is the one that leads to the so-called α-function for the fraction of open channels

$$\tau_\alpha \frac{d^2 g_\alpha(t)}{dt^2} + 2 \frac{dg_\alpha(t)}{dt} = -\frac{g_\alpha(t)}{\tau_\alpha} + \Delta g_\alpha \sum_l \delta(t - \Delta t - t_l) \tag{A5.5}$$

The main feature of this second equation is that it yields a finite opening time without introducing a second time parameter alongside τ_α. More complex descriptions of synaptic dynamics may be more appropriate especially, for example, when the interaction between transmitter receptor and channel is mediated by second messengers, but they introduce the complication of additional parameters, and often additional dynamical variables as well (such as calcium concentration).

It should be noted that conductance dynamics is not always included in integrate-and-fire models: sometimes it is substituted with *current* dynamics, which essentially amounts to neglecting non-linearities due to the appearance of the membrane potential in the driving force for synaptic action (see for example, Amit and Tsodyks, 1991; Gerstner, 1995); and sometimes it is simplified altogether by assuming that the membrane potential undergoes small sudden jumps when it receives instantaneous pulses of synaptic current (see the review in Gerstner, 1995). The latter simplification is quite drastic and changes the character of the dynamics markedly; whereas the former can be a reasonable simplification in some circumstances, but it produces serious distortions in the description of inhibitory GABA$_A$ currents, which, having an equilibrium (Cl$^-$) synaptic potential close to the operating range of the membrane potential, are quite sensitive to the instantaneous value of the membrane potential itself.

A5.4 Main inaccuracies at this level of description

The above discussion is about how to incorporate into neural network models, which are to be studied in their behaviour in time, the three most basic elements of (i) continuous time, (ii) output signal as spikes, and (iii) synaptic action via conductance changes. Many additional features of real neuronal dynamics are still missing at this still rather abstract level, the most obvious of which stem from the *point-like* character of the units in this description, that is the fact that the finite spatial extent of real neurons is not reproduced.

In talking about the membrane potential, we have been referring to it as if it were a quantity associated with the neuron as a whole, whereas in fact the potential difference across the membrane is a function not only of time but also, of course, of the point of the membrane where it is measured. The inaccuracy is not so prominent as long as, in Eqs. A5.2, the membrane potential can be intended as measured at the soma, which to a good approximation may be isopotential. It may become more serious when, in Eq. A5.3, the same variable is used as a measure of the membrane potential at a particular synapse. This synapse may be far away on a dendrite at a considerable electrotonic distance from the soma, and it is well known (Rall and Rinzel, 1973) that, for example, EPSPs measured in a spine out on the dendritic tree may be attenuated hundreds of times before they reach the soma. The details of potential and current propagation within a neuron can be understood and quantified, once the geometry is known, in the framework of Rall's theory (Rall 1959; 1962; and see other papers collected in Segev, Rinzel and Shepherd, 1994), at least as far as passive properties are concerned (in the absence of voltage-activated intrinsic conductances). One important simple principle to keep in mind (Carnevale and Johnston, 1982) is that for a given point on the dendrite of the neuron, the attenuation in the potential as it reaches the soma is the same as the attenuation in the current on the reverse path, from the soma to the point on the dendrite (and they are both typically large). By contrast, the quantities that are of primary interest for the simplified dynamical formulation are the attenuation coefficient in the current from the dendrite to the soma, which affects Eq. A5.3, and the attenuation coefficient in the potential from soma to dendrite, which also affects Eq. A5.3. These are also equal but much weaker than in the reverse directions (although some attenuation does occur, for example in the current as it is leaked away along the path). Therefore, the approximation implied in Eq. A5.3, in which we have been rather cavalier as to where the variables are supposed to be measured, may be less crude than it would appear at first sight. Moreover, three factors further contribute to the adequacy of the approximation: (a) a constant attenuation in the current from the dendrite to the soma may of course be reabsorbed into a scaling factor for the corresponding conductance; (b) for excitatory synapses, many of which are often electrotonically distant, the driving force is typically much larger than the typical excursion of $V(t)$ below threshold, therefore it does not matter very much whether the potential is measured locally, as formally required, or at the soma, as done in practice when applying Eq. A5.3; (c) for many inhibitory synapses, which have an equilibrium potential close to or within the operational subthreshold range, the electrotonic distance from the soma is negligible, and therefore the local potential can be taken to be the one at the soma. These factors notwithstanding, the validity of the point-like approximation, with respect to where the I and V variables are meant to be measured, deserves careful attention, also in view of the frequency dependence of the attenuation coefficients (see further Rall and Segev, 1987; Tsai, Carnevale, Claiborne and Brown, 1994). The reader will find useful further information on the spatio-temporal effects of dendritic processes in Koch and Segev (1989) and in Segev, Rinzel and Shepherd (1994).

A further limit on the validity of the point-like approximation is with respect to *when* the variables are meant to be measured, in that the finite time of propagation of electric signals across finite distances in the spatial extent of a real neuron has to be somehow captured even in a formal unit which has been shrunk to a single point. Considering for example the propagation of a synaptic current from a spine on a dendrite to the soma, the latter will receive not only an attenuated but also a *delayed* version of the original signal, and like the attenuation coefficient the delay will also be a function of frequency. The wave function describing a postsynaptic current, which is not an eigenfunction of the propagation (that is, is not kept similar to itself by the process), will be spread both in space and time as it travels to the soma. An important result obtained recently (Agmon-Snir and Segev, 1993) is that the average time of any wavefunction propagating passively through a dendrite is delayed by an amount that depends solely on the geometry of the dendrite and not on the shape of the wavefunction. This applies to the average time, and does not apply for example to the peak time, or to any time variable characterizing the rapidity of onset or decay of the signal (current or potential; Major, Evans and Jack, 1994). This result implies that given any real neuron, with its geometry and the location of its synaptic connections, it is possible to define for each synapse a characteristic time delay due to dendritic propagation, which will be the delay with which the mean time of any passively propagating signal will travel to the soma.

In the reduced dynamical formulation considered in this appendix there is one delay parameter available for any synaptic input (hence one for any pre- and postsynaptic pair of connected units), that is Δt, or Δt_{ij} if adding indices for the two units. This parameter has to reflect axonal delays from the soma to the axon terminal, which due to the properties of active axon potential propagation are fairly well defined, as well as synaptic delays originating from the different processes implied in synaptic transmission (vescicle release, neurotransmitter diffusion and binding). At the same time, it can incorporate in an effective way also dendritic delays as defined by the propagation of the mean time of a postsynaptic current. This description remains an approximation, if anything because the delay is applied to the firing driving the synaptic conductance, in Eq. A5.4 or A5.5, and not to the current itself, which is not treated as an independent dynamical variable. Nevertheless, it is a description that allows one to take into account delays of different origin within a simple analysable formalism. For a slightly different and elegant formalism formulated in terms of time kernels, see the recent work of Gerstner (1995).

In practice, the refinement of more accurate descriptions of delays and their dependence on the spatial structure of real neurons is made less useful also by the fact that one wants to study models that incorporate only the generic, statistical properties of a population of neurons, and not the infinite details which remain in any case unknown. The important statistical properties of the delays may well reduce to their mean value and their non-uniformity, and in a computer simulation, for example, these may well be captured by generating random values for Δt_{ij} for example from a Gaussian distribution with a given mean and standard deviation. In considering local populations, axonal and synaptic delays are within 1–2 ms, and the delays that are most relevant to time scales in the range of tens up to hundreds of ms are dendritic delays, which, however, may still be considered in the order of 10 ms or less

(Agmon-Snir and Segev, 1993). Overall more important than the mean value could be the non-uniformity of delays, which can help prevent, in the models, *artefactual synchronization* in the firing (as typical of discretized time models) and, on the other hand, ensure that if synchronized firing sets in it is a robust phenomenon.

A5.5 The mean-field treatment

Units whose potential and conductances follow the equations above can be assembled together in a model network of any composition and architecture. It is convenient to imagine that units are grouped into classes, such that the parameters quantifying the electrophysiological properties of the units are uniform, or nearly uniform, within each class, while the parameters assigned to synaptic connections are uniform or nearly uniform for all connections from a given presynaptic class to another given postsynaptic class. The parameters that have to be set in a model at this level of description are quite a few, as listed in Tables A5.1 and A5.2.

In the limit in which the parameters are constant within each class or pair of classes, a mean-field treatment can be applied to analyse a model network, by summing equations that describe the dynamics of individual units to obtain a more limited number of equations that describe the dynamical behaviour of groups of units (Frolov and Medvedev, 1986). The treatment is exact in the further limit in which very many units belong to each class, and is an approximation if each class includes just a few units. Suppose that N_C is the number of classes defined. Summing Eqs A5.2 across units of the same class results in N_C functional equations describing the evolution in time of the fraction of cells of a particular class that at a given instant have a given membrane potential. In other words, from a treatment in which the

Table A5.1 Cellular parameters (chosen according to the class of each unit)

V_{rest}	Resting potential
V_{thr}	Threshold potential
V_{ahp}	Reset potential
V_K	Potassium conductance equilibrium potential
C	Membrane capacitance
τ_K	Potassium conductance time constant
g_0	Leak conductance
Δg_K	Extra potassium conductance following a spike
Δg	Overall transmission delay

Table A5.2 Synaptic parameters (chosen according to the classes of pre- and postsynaptic units)

V_α	Synapse equilibrium potential
τ_α	Synaptic conductance time constant
Δg_α	Conductance opened by one presynaptic spike
Δt_α	Delay of the connection

evolution of the variables associated with each unit is followed separately one moves to a treatment based on density functions, in which the common behaviour of units of the same class is followed together, keeping track solely of the portion of units at any given value of the membrane potential. Summing Eq. A5.4 or A5.5 across connections with the same class of origin and destination results in $N_C \times N_C$ equations describing the dynamics of the overall summed conductance opened on the membrane of a cell of a particular class by all the cells of another given class. A more explicit derivation of mean-field equations is given by Treves (1993).

The system of mean-field equations can have many types of asymptotic solutions for long times, including chaotic, periodic, and stationary ones. The stationary solutions are stationary in the sense of the mean fields, but in fact correspond to the units of each class firing tonically at a certain rate. They are of particular interest as the dynamical equivalent of the steady states analysed by using non-dynamical model networks. In fact, since the neuronal current-to-frequency transfer function resulting from the dynamical equations is rather similar to a threshold linear function (see Fig. A5.1), and since each synaptic conductance is constant in time, the stationary solutions are essentially the same as the states described using model networks made up of threshold linear, non-dynamical units. Thus the dynamical formulation reduces to the simpler formulation in terms of steady-state rates when applied to asymptotic stationary solutions; but, among simple rate models, it is equivalent only to those that allow description of the continuous nature of neuronal output, and not to those, for example based on binary units, that do not reproduce this fundamental aspect. The *advantages* of the dynamical formulation are that (i) it enables one to describe the character and prevalence of other types of asymptotic solutions, and (ii) it enables one to understand how the network reaches, in time, the asymptotic behaviour.

A5.6 The approach to stationary solutions

To describe the full course of the evolution of a network from an arbitrary initial state to a stationary asymptotic state one has to resort to computer simulations, because handling the full set of dynamical equations is a formidable task, even with the simplification afforded by mean-field theory. The *approach* to the stationary state, however, can be described analytically by linearizing the mean-field equations. This refers to the dynamics close to the stationary state, which is characterized by transient modes that decay with an exponential time dependence, $\exp(\lambda t)$. The earlier dynamical evolution, it should be stressed, cannot be analysed in this way.

Linearizing the equations is a straightforward mathematical procedure, and the linearized system can be further manipulated (Treves, 1993) into a final equation that yields the possible values of λ, that is the time constants of the transients. In general, there is a very high number of different transient modes, and correspondingly of different values for the time constant λ. The equation which admits as solutions the different values for λ is for uniform delays of the form

$$\Delta = |\mathbf{Q}(\lambda) - e^{\lambda \Delta t} \mathbf{1}| = 0 \qquad (A5.6)$$

where Δ denotes the determinant of a matrix of rank N_C composed of the matrix $\mathbf{Q}(\lambda)$ and the identity matrix $\mathbf{1}$. The matrix $\mathbf{Q}(\lambda)$ results from the analytical procedure sketched above and detailed in Treves (1993), and incorporates the parameters of the dynamical model. Here its explicit form is given for the relatively simpler case in which no firing adaptation effects are included, that is the intrinsic potassium conductance is omitted from the single cell model, and in which synaptic conductances follow Eq. A5.5. It simplifies the analysis to express all characteristic potentials in terms of adimensional variables, such as

$$x(t) = \frac{V(t) - V_{\text{ahp}}}{V_{\text{thr}} - V_{\text{ahp}}} \tag{A5.7}$$

which measure the excursion of the membrane potential between spikes. This variable, for cells of a given class F, follows the equation

$$\dot{x}(t) = A_F(t) - x(t) B_F(t) \tag{A5.8}$$

where $A_F(t)$ and $B_F(t)$ represent the appropriate combinations of leak and synaptically driven conductances. At steady state each cell of class F fires at a constant frequency

$$r_F^0 = B_F^0 \left[\ln \frac{A_F^0}{A_F^0 - B_F^0} \right]^{-1} \tag{A5.9}$$

which results in a system of self-consistent equations, given that A_F^0 and B_F^0 depend in turn on the distribution of frequencies $\{r_F^0\}$. The matrix yielding the time constants of the transient modes turns out (Treves, 1993) to have the form

$$Q_F^G(\lambda) = \frac{r_F^0 \omega_F^G}{\tau_F^G (\lambda + 1/\tau_F^G)^2 (B_F^0 + \lambda)} \left\{ 1 + \lambda \frac{1 + x_F^G [e^{(\lambda + B_F^0)/r_F^0} - 1]}{[A_F^0 - B_F^0][e^{-\lambda/r_F^0} - 1]} \right\} \tag{A5.10}$$

where in the synaptic parameters x (adimensional equilibrium potentials), τ (time constants), and ω (which measure efficacy and have the dimensions of a frequency), the superscript denotes the class of the presynaptic cell and the subscript that of the postsynaptic one.

The collection of solutions (values for λ) of Eq. A5.6 can be referred to as the *spectrum* of time constants, that set the pace for the relaxation into steady states.

A5.7 Recurrent connections and associative memory

Although the dynamical equations can be used with any architecture, they are of particular interest in model networks that include or are even dominated by recurrent connections, since the dynamical behaviour is then non-trivial. With purely feedforward connections, instead, the dynamical behaviour is rather simple, in the same sense in which, as discussed in previous appendices, the steady states are simple to analyse: each output unit can be taken on its own, and just averaged with other units.

Among networks with recurrent connections, the specific type considered sets the appropriate way in which to group units into classes. Most of the relatively few analyses that use formal dynamical models similar to what has been described here dwell on the simplest case of recurrent nets with two classes of units, excitatory and inhibitory (the latter may occasionally be further subdivided into fast and slow units, mediating the analogue of $GABA_A$ and $GABA_B$ inhibition). Cellular parameters are then assigned to the two (or three) classes, and synaptic parameters occur in just four (or nine) potentially different values. In particular, connections among excitatory units are taken to be uniform, which deprives the net of any possible functional role. The interest of these analyses is not, then, in relating dynamics to function, but in studying dynamics *per se* on a relatively simple and controlled test model. For example, they can suggest under which conditions the network may tend to switch from asynchronous to synchronized firing (Abbott and VanVreswijk, 1993), or under which conditions the inhibition is sufficient to prevent an excitatory explosion of activity (Amit and Brunel, 1997; VanVreeswijk and Hasselmo, 1997).

An autoassociative memory network with feedback is a network that performs a useful function, that is to store memories with great efficiency, and that can at the same time be analysed in its dynamics within this formalism. The minimal requirement to carry out this analysis is to impose a matrix memory on the connections between excitatory units, and therefore a further division into (sub)classes, based solely on connection strengths, of the physiologically homogeneous group of excitatory units. In other words, to keep the model simple, all cellular parameters of the excitatory units can be kept identical (modelling for example an average pyramidal cell), and also the synaptic time constant, delays and equilibrium potential for excitatory–excitatory connections can be identical across pairs of units, whereas only the synaptic efficacies (parametrized by ω in our notation) can be modulated to reflect the memory structure. Since the memory matrix is determined by the firing rates of each unit in all the memory patterns, there are in principle as many classes in this further subdivision as units, and the mean-field treatment, which assumes instead very many units per class, becomes an approximation. On the other hand, the homogeneity of all parameters except synaptic efficacy across these classes with a single element makes the approximation reasonable. Further, the main features of the spectrum of time constants describing the approach to steady state can be shown (Treves, 1993) to be relatively independent of the exact memory structure, as described next. After that, computer simulations will be discussed that support the validity of the analytical treatment; in order to introduce the simulations, however, we shall have to mention a conflict which arises, in implementing the network, between memory function and dynamical stability, and the way in which such conflict can be avoided. The results of the simulations will be expressed in terms of information quantities that correspond to the quantities that are measured from recording the activity of real cells in the brain.

A5.8 The spectrum of time constants

The set of λ values obtained as solutions of Eq. A5.6 reflects the parameters that describe cells and synapses, and thus it explicitly links the single unit dynamics (the level at which the model attempts to capture relevant biophysical features) with network dynamics (where contact with neurophysiological recording experiments can be made).

The dependence of the spectrum on the underlying parameters is in general very complicated, but may be characterized as follows. It can be represented on the complex plane in which the abscissa is the imaginary component $\Im m(\lambda)$ and the ordinate the real component $\Re e(\lambda)$ of the λ values, both measured in hertz (Fig. A5.2). The inverse of the absolute value of the real part, if it is negative, gives the true time constant of the exponential decay of the corresponding mode, while the imaginary part, if different from zero (solutions then come always in conjugate pairs, since the equation is real) gives the frequency of the oscillations that accompany the decay of the mode. If there is at least one solution with positive real part, the steady state is of course unstable, and it is only marginally stable if there are solutions with zero real part. The spectrum presents, if parameters in a reasonable range are chosen for the dynamical model, a gross, a fine, and a hyperfine structure. The gross structure consists of those λ values that satisfy Eq. A5.6, whose real part is in the kilohertz range and beyond. These values correspond thus to fast time scales, and are determined by fast time parameters (for example 1 ms) such as the delay Δt. They are also very sensitive to global conditions like the detailed balance of excitation and inhibition, and in fact an instability associated with an imbalance of this sort may show up as one or more of the λ values of the gross structure acquiring a positive real part. There are two reasons, however, for not focusing one's attention on the gross structure of the spectrum. One is that, inasmuch as fast phenomena characterizing the real neural system have either been omitted altogether from the model (such as the Hodgkin–Huxley details of spike emission) or crudely simplified (such as unifying all delays into a unique and uniform delay parameter Δt), the model itself is not likely to be very informative about the fast dynamics of the real system. Second, in the presence of a stable stationary solution and of transients (those associated with the fine and hyperfine structures) lasting longer than, say, 5 ms, the faster transients are not very significant anyway.

The fine and hyperfine structures both consist of orderly series of λ values satisfying Eq. A5.6, whose real parts are similar (within a series), while the imaginary parts take discrete values ranging all the way to infinity, with approximate periodicity $\Delta m(\lambda) = 2\pi r$, where r is

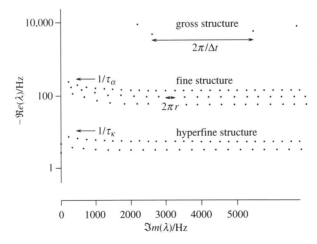

Fig. A5.2 Schematic representation of the spectrum of time constants characterizing collective dynamics, representing the dependence of the families of transient modes on the main parameters of the model. (From Treves, Rolls and Simmen, 1997, Fig. 2.)

the firing rate of a particular group of cells. The (negative) real parts are complicated functions of all the parameters, but in broad terms they are determined by (and similar in value to) the inverse of the conductance time constants (Treves, 1993). The fine and hyperfine structures differ in the magnitude of their real parts, that is in their time constants, in that relatively fast inactivating conductances (for example those associated with excitatory transmission via AMPA receptors) produce the fine structure, while slower conductances (for example the intrinsic potassium conductances underlying firing rate adaptation) produce the hyperfine structure. The fine structure, which therefore corresponds to time scales intermediate between those of the gross and hyperfine structures, is perhaps the most interesting, as it covers the range (10–20 ms) in which the formal model best represents dynamical features of real neuronal networks.

In principle, having assigned a definite starting configuration for the model network, that is the distribution of membrane potentials of the units and the degree of synaptic activation, at a given time, one should be able to follow in detail the dynamical evolution, thereby quantifying the relative importance of different time scales. In practice, however, the explicit characterization of the modes, with their spectrum of time constants, is only exact asymptotically close to a stationary state, and thus applies only to starting configurations infinitesimally close to the stationary state itself. For arbitrary initial conditions, behaviour increasingly similar to that described by the modes will progressively set in, provided the network approaches a stationary state of the type considered here, but it will be impossible to predict the relative weight of the different modes.

The main conclusions of this analysis are that (i) the way the model approaches stable firing configurations is governed by a variety of time scales of widely different orders of magnitude; and (ii) an important subset of those time scales is determined by the inactivation times of synaptic conductances. These conclusions are expected to be valid also for the dynamics of local cortical networks with recurrent connections, whether they are comparable to auto-associators or to other types of networks. With excitatory synapses between pyramidal cells the time course of inactivation, even taking into account some additional spread due to the finite electrical size of the postsynaptic cell (part of this effect, as discussed above, can be included in the connection delay), may be as short as 5–10 ms. Therefore a substantial component in the *settling down* of the distribution of firing activity into a stable distribution, under the influence of interactions mediated by lateral connections, may be very rapid, that is it may occur in little more than 5–10 ms.

5.9 The analysis does not support intuitions based on simpler models

The conclusions reached by this type of analysis are very partial, due to the limitations of using mean-field theory, and more importantly of having to be limited to the linearized equations. Nevertheless, they yield a very different insight, from what could have been said of the process by relying on intuitive considerations or even discrete-time models. Intuitive considerations suggest that (i) the stabilization of a firing configuration, when it is mediated by recurrent connections, is a genuine feedback effect; (ii) for real feedback to take place, the minimal requirement is that a firing cell contributes to the firing of another cell, then receives back (directly or polysynaptically) a spike from that cell and fires again, using up in the order of two interspike intervals (one each at the current firing rate of each of the two cells); (iii) the process is iterated a few times in order to lead to a steady activity distribution, taking up in the order of several hundreds of ms.

This type of reasoning, which neglects both the spread in the latencies of activation of the cells in the local population, and more importantly their ongoing spontaneous activity, is a sequel of three incorrect steps. In fact (i) recurrent connections may contribute much of their effect well before real feedback occurs, in fact even before many of the cells have had the time to fire a single action potential in response to a new stimulus. For example, a spike emitted by cell A may immediately influence the firing of a cell B that happened to be close to threshold, while more time may be needed for the firing of B to have a feedback effect on the emission of a subsequent spike by A. (ii) The influence, reciprocated or not, of a given cell onto another is not important *per se*, but only as part of a collective behaviour of the whole population, which is determined by the concerted action of the thousands of different inputs to each cell. (iii) Reasoning in terms of time steps, or iteration cycles, is meaningless anyway, as the dynamics of the real system is in continuous time and, unless a special synchronization mechanism sets in, is asynchronous.

We have seen how a formal model based on more careful considerations leads to different conclusions. These may be appreciated intuitively, by noting that within a large population of cells, firing at low but non-zero rate, as occurs in cerebral cortex (Abeles, Vaadia and Bergman, 1990; Rolls, 1984a, 1992b; Rolls and Tovee, 1995a), incoming afferent inputs will find some cells that are already close to threshold. The slight modification in the firing times of those cells (brought forward or postponed depending on whether each cell is predomi-nantly excited or inhibited by the afferents) will be rapidly transmitted through local recurrent connections, and (for example in autoassociative networks) contribute to initiate a self-amplifying process of adjusting the firing pattern that would have been determined by the afferent inputs alone. The process will proceed at a pace set by the time course of synaptic action, which in practice, in a situation in which delays are small, means set by the time constants of synaptic conductances. The fact that the speed of this *collateral effect* (Marr, 1971) is essentially independent of prevailing firing rates is somewhat analogous to, and no more puzzling than, the propagation speed of mechanical waves being distinct from the speed of motion of the individual particles carrying the wave.

The rapid settling down into steady firing by a local population of neurons may therefore well occur even with the contribution of recurrent connections, even though the use of such connections does not necessarily imply, for any given cell, the propagation of spikes around feedback loops. The sequential activation of most feedback loops will require certainly more that 10–20 ms at typical cortical firing rates, but the population response may already within that time be substantially dependent on the activation of recurrent connections. The ability of recurrent connections to contribute usefully to autoassociative retrieval, or in general to information processing, is due to the fact that part of their effects occurs within a time window determined primarily by the time constants of the synaptic conductances. Slower time scales, partly intrinsic and partly also associated with recurrent connections, such as those underlying adaptation processes and those involving slow inhibition, might produce slower adjustments in the pattern of firing activity. Their influences for example in autoassociative retrieval are likely to be very minor, in that the information relative to the memory is thought to be stored solely in the matrix of excitatory–excitatory connections, which, to the extent that AMPA transmission dominates over NMDA–receptor transmission, operate very fast.

A5.10 Dynamical stability versus memory capacity

A single solution λ with a positive real part is enough to destabilize a steady state, and therefore the stability analysis has to take into account the whole spectrum when assessing the states of a model network, and *a fortiori* when studying the dynamical behaviour of real networks of neurons. Since more additional modes are present in the spectrum of the model the more detailed is the description implemented by the dynamical equations used, reduced models as mentioned in the beginning of this appendix are not very useful to discuss dynamical stability. Nevertheless, there is one important source of instability which is present also in reduced models, and can thus be approximately but conveniently analysed in a simplified framework. This is the instability that corresponds to the excitation self-reinforcing itself beyond inhibitory control into an epileptic explosion, and to avoid it requires a low effective coupling among excitatory cells. On the other hand, in an autoassociative memory a minimum effective coupling is needed for the retrieval operation to be successful. In a matrix memory that accumulates via linear superposition of contributions from memorized patterns, and where purely subtractive inhibition is used to control the spread of excitatory activity, the two requirements on the coupling can be satisfied together only if very few patterns are stored concurrently. This would represent a major handicap for the operation of autoassociative memories, invalidating the result that each synapse could be used to store a finite amount (a fraction of a bit, as discussed in Appendix A4) of information. It is worthwhile to understand this conflict between dynamical stability and memory capacity in more detail, so as to appreciate the requirements on the type of inhibition which avoids the conflict.

Consider an extremely reduced description in which the only two dynamical variables are the mean activation, that is the open fraction, of excitatory (g_E) and inhibitory (g_I) conductances. These are driven by the firing of excitatory and inhibitory neurons, as in Eq. A5.4. Instead of describing the firing process through the integrate-and-fire model, the mean excitatory and inhibitory firing rates are given by the equations valid at steady state, which are further simplified by taking the mean-field perspective. The opening of the conductances g_E and g_I is then driven by rates given by equations similar to Eq. A5.9. Moreover, since a linearization around steady state is going to be performed anyway in order to check its stability, these equations can be approximated by their close linear analogue, the corresponding threshold-linear equations for the mean frequencies. The reduced dynamical system reads

$$\dot{z}_E \equiv \frac{d(g_E/g_{0_E})}{dt} = -\frac{(g_E/g_{0_E})}{\tau_E} + B_E \left[\ln \frac{A_E}{A_E - B_E} \right]^{-1} = -\frac{z_E}{\tau_E} + [A_E - B_E/2]^+$$

$$\dot{z}_E \equiv \frac{d(g_I/g_{0_I})}{dt} = -\frac{(g_I/g_{0_I})}{\tau_I} + B_I \left[\ln \frac{A_I}{A_I - B_I} \right]^{-1} = -\frac{z_I}{\tau_I} + [A_I - B_I/2]^+$$

(A5.11)

where the notation of Treves (1993) has been used, with

$$
\begin{aligned}
A_E - B_E/2 &= r_E^{\text{ffwd}} + \omega_E^E(x_E^E - 1/2)z_E + \omega_E^I(x_E^I - 1/2)z_I \\
A_I - B_I/2 &= r_I^{\text{ffwd}} + \omega_I^E(x_I^E - 1/2)z_E + \omega_I^I(x_I^I - 1/2)z_I
\end{aligned}
\tag{A5.12}
$$

and r^{ffwd} are the firing rates that would have been produced by feedforward activation alone. This is a dynamical system in just the two variables z_E and z_I, and the stability of a fixed point is ensured by having the two eigenvalues of the stability matrix (essentially, the λ values of the transient modes) both with negative real part. This condition is equivalent to enforcing that

$$
\begin{aligned}
(J_E^E - 1)/\tau_E &< (J_I^I + 1)/\tau_I \\
(J_E^E - 1)(J_I^I + 1) &< J_E^I J_I^E
\end{aligned}
\tag{A5.13}
$$

where the overall couplings J are positive quantities defined as

$$
\begin{aligned}
J_F^E &= \tau_F \omega_F^E (x_F^E - 1/2) \\
J_F^I &= \tau_F \omega_F^I (1/2 - x_F^E).
\end{aligned}
\tag{A5.14}
$$

Equations A5.13 provide a simple mathematical expression of the intuitive condition that excitatory–excitatory couplings must be controlled by inhibition for the activity not to explode while reverberating over the recurrent collateral connections. Overall, then, the conditions constrain the strength of connections among excitatory cells to be beneath a certain value. In practice, it is possible to find an ample and robust parameter region in which stability holds only if J_E^E is of order 1 (see further Battaglia and Treves, 1997).

In the simple models of autoassociators discussed in Appendix A4, the connections among principal cells were due to the linear superpositions of modification terms on a baseline value, which had to be large enough that the sum was in any case positive (being ultimately equivalent to the size of a conductance)

$$
w_{ij} = w^0 + \frac{1}{Ca^2} \sum_{\mu=1}^{p} (r_i^\mu - a)(r_j^\mu - a).
\tag{A5.15}
$$

The equivalent of the coupling J_E^E is obtained in these models by multiplying the baseline value w^0 by the number C of excitatory connections each unit receives and by the gain γ of the transfer function

$$
J_E^E \approx \gamma C w^0.
\tag{A5.16}
$$

The modification terms average to zero and need not be included. On the other hand, w^0 must be at least equal to the opposite of the minimum value taken by the modification terms, to ensure that the overall conductance is positive. Over p patterns this implies

$$
w^0 \geqslant (p/C)(1/a - 1).
\tag{A5.17}
$$

At the same time, the capacity analysis mentioned in Appendix A4 indicates that for retrieval to succeed close to the capacity limit, the gain γ of the transfer function must be at or above the value, set by the sparseness, $a/(1-a)$ (Treves and Rolls, 1991). Putting pieces together one finds

$$J_E^E \geqslant p \qquad (A5.18)$$

which is clearly incompatible with the stability condition that J_E^E be of order 1, unless very few patterns are stored!

How can this conflict be avoided? The important point is to note that in considering Eqs. A5.12 the point-like approximation was used, in particular in tacitly assuming that inhibitory inputs only provide an additive (or, more correctly, subtractive) term to the activation produced by excitatory inputs. A more realistic description of inhibitory inputs, which takes into account their relative locations on the dendritic tree, would have allowed also for a different type of effect of inhibition, one that acts *multiplicatively* on excitatory activation (see for example Abbott, 1991). Multiplicative inhibition arises, in part, from the small potential difference driving GABA$_A$ currents (due to the Cl$^-$ equilibrium potential being close to the average membrane potential) and, in part, from the relative position of many inhibitory inputs on the dendrites close to the cell body. Hence, it cannot be fully modelled in a mathematical formulation of single-neuron dynamics that neglects dendritic geometry, but it can be incorporated by hand by taking ω_E^E to be an *effective* conductance, which is modulated by inhibitory activity, or in other words by considering J_E^E to be a function of z_I. The stability equations take the same form (Eq. A5.13), as they are obtained after linearization, but now the parameters, in particular J_E^E, are not constant but rather have to be calculated at the fixed point. It is easy to realize that for purely multiplicative inhibition and for no external drive on excitatory cells, $r_E^{\text{ffwd}} = 0$, the fixed point of Eqs. A5.11–A5.12 occurs for $J_E^E = 1$, and therefore stability is automatically ensured (Battaglia and Treves, 1996). For realistic forms of inhibition, including a mixture of subtractive and multiplicative effects, and for arbitrary values of the firing level due to the external drive, J_E^E will tend to be above 1 and no simple formula guarantees dynamical stability, but the more multiplicative the inhibition the more likely one is to be able to retrieve from a large number of patterns while keeping the network stable. In a finite surround of the ideal case of purely multiplicative inhibition, the constraint on the number of patterns in storage will indeed be that produced by the capacity analysis, and not the constraint due to the instability discussed above (Battaglia and Treves, 1997).

A5.11 Information dynamics

From the preceding section it is clear that simulations designed to study the temporal course of activity in a recurrent associative memory must include predominantly multiplicative inhibition, otherwise it will be impossible to find parameters allowing for stable retrieval states if more than a handful of patterns are stored (Simmen, Rolls and Treves, 1996a; Treves, Rolls and Simmen, 1997). This implies that such simulations will not match exactly the details of the analytical treatment described earlier in the appendix, but still the main result on the dependence of the characteristic time to reach a retrieval state on the underlying parameters is expected to hold. This is what is illustrated by the simulations described briefly in the following. Fuller details can be found in Battaglia and Treves (1997).

A network of 800 excitatory and 200 inhibitory cells was simulated in its retrieval of one of several memory patterns stored on the synaptic weights representing excitatory-to-excitatory recurrent connections. The memory patterns were assigned at random, drawing the value of each unit in each of the patterns from a binary distribution with sparseness 0.1, that is a probability of 1 in 10 for the unit to be active in the pattern. No baseline excitatory weight was included, but the modifiable weights were instead constrained to remain positive (by clipping at zero modifications that would make a weight negative) and a simple exponential decay of the weight with successive modifications was also applied, to prevent runaway synaptic modifications within a rudimentary model of forgetting. Both excitation on inhibitory units and inhibition were mediated by non-modifiable uniform weights, with values chosen so as to satisfy stability conditions of the type of Eqs. A5.13. Both inhibitory and excitatory units were of the general integrate-and-fire type, but excitatory units had in addition an extended dendritic cable, and they received excitatory inputs only at the more distal end of the cable, and inhibitory inputs spread along the cable. In this way, inhibitory inputs reached the soma of excitatory cells with variable delays, and in any case earlier than synchronous excitatory inputs, and at the same time they could shunt the excitatory inputs, resulting in a largely multiplicative form of inhibition (Abbott, 1991). The uniform connectivity was not complete, but rather each type of unit could contact units of the other type with probability 0.5, and the same was true for inhibitory-to-inhibitory connections. After 100 (simulated) ms of activity evoked by external inputs uncorrelated with any of the patterns, a cue was provided which consisted of the external input becoming correlated with one of the patterns, at various level of correlation, for 300 ms. After that, external inputs were removed, but when retrieval operated successfully the activity of the units remained strongly correlated with the memory pattern, or even reached a higher level of correlation if a rather corrupted cue had been used, so that, if during the 300 ms the network had stabilized into a state rather distant from the memory pattern, it got much closer to it once the cue was removed. All correlations were quantified using information measures (see Appendix A2), both in terms of mutual information between the firing rate pattern across units and the one memory pattern being retrieved, or in terms of mutual information between the firing rate of one unit and the set of patterns, or, finally, in terms of mutual information between the decoded firing rates of a subpopulation of 10 excitatory cells and the set of memory patterns. The same algorithms were used to extract information measures as were used for example by Rolls, Treves, Tovee and Panzeri (1997) with real data. The firing rates were measured over sliding windows of 30 ms, after checking that shorter windows produced noisier measures. The effect of using a relatively long window, 30 ms, for measuring rates is an apparent *linear* early rise in information values with time. Nevertheless, in the real system the activity of these cells is 'read' by other cells receiving inputs from them, and that in turn have their own membrane capacitance characteristic time for integrating input activity, a time broadly in the order of 30 ms; using such a time window for integrating firing rates does not therefore artificially slow down the read-out process.

The time course of different information measures did not depend significantly on the firing rates prevailing at the retrieval state, nor on the *RC* membrane time 'constants' of the units (note that the resistance R across the membrane was not really a constant, as it varied with the number of input conductances open at any moment in time). Figure A5.3 shows that the rise in information after providing the cue followed a roughly exponential approach to its steady-state value, which remained until, with the removal of the cue, the steady state switched to a new value. The time constant of the approach to the first steady state was a linear function, as shown in the Figure, of the time constant for excitatory conductances, as predicted by the analysis reported above (the proportionality factor in the Figure is 2.5, or a collective time constant 2.5 times longer than the synaptic time constant). The approach to

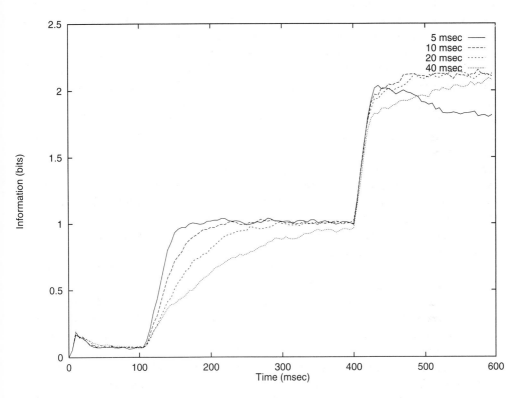

Fig. A5.3 Time course of the transinformation about which memory pattern had been selected, as decoded from the firing rates of 10 randomly selected excitatory units. Excitatory conductances closed exponentially with time constants of 5, 10, 20 and 40 ms (curves from top to bottom) A cue of correlation 0.2 with the memory pattern was presented from 100 to 400 ms, uncorrelated external inputs with the same mean strength and sparseness as the cue were applied at earlier times, and no external inputs were applied at later times.

the second steady-state value was more rapid, and the early apparent linear rise prevented the detection of a consistent exponential mode. Therefore, it appears that the cue leads to the basin of attraction of the correct retrieval state by activating transient modes, whose time constant is set by that of excitatory conductances; once the network is in the correct basin, its subsequent reaching the 'very bottom' of the basin after the removal of the cue is not accompanied by any prominent transient mode (see further Battaglia and Treves, 1997). Overall, these simulations confirm that recurrent networks, in which excitation is mediated mainly by fast (AMPA) channels, can reach asynchronous steady firing states very rapidly, over a few tens of milliseconds, and the rapid approach to steady state is reflected in the relatively rapid rise of information quantities that measure the speed of the operation in functional terms. An analysis based on integrate-and-fire model units thus indicates that recurrent dynamics can be so fast as to be practically indistinguishable from purely feedforward dynamics, in contradiction with what simple intuitive arguments would suggest. This makes it impossible to draw conclusions on the underlying circuitry on the basis of the experimentally observed speed with which selective responses arise, as attempted by Thorpe and Imbert (1989).

References

Abbott, L.F (1991) Realistic synaptic inputs for model neural networks. *Network* **2**: 245–258.

Abbott, L.F. and van Vreeswijk, C. (1993) Asynchronous states in networks of pulse-coupled oscillators. *Physical Review E* **48**: 1483–1490.

Abbott, L.F., Rolls, E.T. and Tovee, M.J. (1996) Representational capacity of face coding in monkeys. *Cerebral Cortex* **6**: 498–505.

Abbott, L.F., Varela, J.A., Sen, K. and Nelson, S.B. (1997) Synaptic depression and cortical gain control. *Science* **275**: 220–224.

Abeles, M. (1991) *Corticonics: Neural Circuits of the Cerebral Cortex*. Cambridge: Cambridge University Press.

Abeles, M., Vaadia, E. and Bergman, H. (1990) Firing patterns of single units in the prefrontal cortex and neural network models. *Network* **1**: 13–25.

Abeles, M., Bergman, H., Margalit, E. and Vaadia, E. (1993) Spatiotemporal firing patterns in the frontal cortex of behaving monkeys. *Journal of Neurophysiology* **70**: 1629–1638.

Abramson, N. (1963) *Information Theory and Coding*. McGraw-Hill: New York.

Ackley, D.H., Hinton, G.E. and Sejnowski, T.J. (1985) A learning algorithm for Boltzmann machines. *Cognitive Science* **9**: 147–169.

Aggleton, J.P. (Ed.) (1992) *The Amygdala*. Wiley-Liss: New York.

Aggleton, J.P. (1993) The contribution of the amygdala to normal and abnormal emotional states. *Trends in Neurosciences* **16**: 328–333.

Aggleton, J.P. and Passingham, R.E. (1981) Syndrome produced by lesions of the amygdala in monkeys (*Macaca mulatta*). *Journal of Comparative Physiology* **95**: 961–977.

Agmon-Snir, H. and Segev, I. (1993) Signal delay and input synchronization in passive dendritic structures. *Journal of Neurophysiology* **70**: 2066–2085.

Aigner, T.G., Mitchell, S.J., Aggleton, J.P., DeLong, M.R., Struble, R.G., Price, D.L., Wenk, G.L., Pettigrew, K.D. and Mishkin, M. (1991) Transient impairment of recognition memory following ibotenic-acid lesions of the basal forebrain in macaques. *Experimental Brain Research* **86**: 18–26.

Albus, J.S. (1971) A theory of cerebellar function. *Mathematical Biosciences* **10**: 25–61.

Alexander, G.E., Crutcher, M.D. and DeLong, M.R. (1990) Basal ganglia thalamo-cortical circuits: parallel substrates for motor, oculomotor, 'prefrontal' and 'limbic' functions. *Progress in Brain Research* **85**: 119–146.

Amaral, D.G. (1986) Amygdalohippocampal and amygdalocortical projections in the primate brain. Pp. 3–18 in *Excitatory Amino Acids and Epilepsy*, eds. R.Schwarcz and Y.Ben-Ari. Plenum Press: New York.

Amaral, D.G. (1987) Memory: Anatomical organization of candidate brain regions. Pp. 211–294 in *Higher Functions of the Brain. Handbook of Physiology, Part I*, eds. F.Plum and V.Mountcastle. American Physiological Society: Washington, DC.

Amaral, D.G. (1993) Emerging principles of intrinsic hippocampal organization. *Current Opinion in Neurobiology* **3**: 225–229.

Amaral, D.G. and Price, J.L. (1984) Amygdalo-cortical projections in the monkey (*Macaca fascicularis*). *Journal of Comparative Neurology* **230**: 465–496.

Amaral, D.G. and Witter, M.P. (1989) The three-dimensional organization of the hippocampal formation: a review of anatomical data. *Neuroscience* **31**: 571–591.

Amaral, D.G., Ishizuka, N. and Claiborne, B. (1990) Neurons, numbers and the hippocampal network. *Progress in Brain Research* **83**: 1–11.

Amaral, D.G., Price, J.L., Pitkanen, A. and Carmichael, S.T. (1992) Anatomical organization of the primate amygdaloid complex. Ch. 1, pp. 1–66 in *The Amygdala*, ed. J.P.Aggleton. Wiley-Liss: New York.

Amari, S. (1982) Competitive and cooperative aspects in dynamics of neural excitation and self-organization. Ch. 1, pp. 1–28 in *Competition and Cooperation in Neural Nets*, eds. S.Amari and M.A.Arbib. Springer: Berlin.

Amari, S., Yoshida, K., and Kanatani, K.-I. (1977) A mathematical foundation for statistical neurodynamics. *SIAM Journal of Applied Mathematics* **33**: 95–126.

Amit, D.J. (1989) *Modelling Brain Function*. Cambridge University Press: New York.

Amit, D.J. (1995) The Hebbian paradigm reintegrated: local reverberations as internal representations. *Behavioral and Brain Sciences* **18**: 617–657.

Amit, D.J., Gutfreund, H. and Sompolinsky, H. (1985) Spin-glass models of neural networks. *Physical Review, A* **32**: 1007–1018.

Amit, D.J., Gutfreund, H. and Sompolinsky, H. (1987) Statistical mechanics of neural networks near saturation. *Annals of Physics (New York)* **173**: 30–67.

Amit, D.J. and Tsodyks, M.V. (1991) Quantitative study of attractor neural network retrieving at low spike rates. I. Substrate – spikes, rates and neuronal gain. *Network* **2**: 259–273.

Amit, D.J. and Brunel, N. (1997) Global spontaneous activity and local structured (learned) delay period activity in cortex. *Cerebral Cortex* **7**: in press.

Anderson, J.A. and Rosenfeld, E. (eds.) (1988) *Neurocomputing: Foundations of Research*. MIT Press: Cambridge, Mass.

Anderson, M.E. (1978) Discharge patterns of basal ganglia neurons during active maintenance of postural stability and adjustment to chair tilt. *Brain Research* **143**: 325–338.

Andersen, P., Dingledine, R., Gjerstad, L., Langmoen, I.A. and Laursen, A.M. (1980) Two different responses of hippocampal pyramidal cells to application of gamma-amino butyric acid. *Journal of Physiology* **307**: 279–296.

Andersen, R.A. (1995) Coordinate transformations and motor planning in the posterior parietal cortex. Ch. 33, pp. 519–532 in *The Cognitive Neurosciences*, ed. M.S.Gazzaniga. MIT Press: Cambridge, Mass.

Angeli, S.J., Murray, E. and Mishkin, M. (1993) Hippocampectomized monkeys can remember one place but not two. *Neuropsychologia* **31**: 1021–1030.

Arbib, M.A. (1964) *Brains, Machines, and Mathematics*. McGraw-Hill. (2nd edn. 1987, Springer: New York)

Artola, A. and Singer, W. (1993) Long-term depression: related mechanisms in cerebellum, neocortex and hippocampus. Ch. 7, pp. 129–146 in *Synaptic Plasticity: Molecular, Cellular and Functional Aspects*, eds. M.Baudry, R.F.Thompson and J.L.Davis. MIT Press: Cambridge, Mass.

Atick, J.J. (1992) Could information theory provide an ecological theory of sensory processing? *Network* **3**: 213–251.

Atick, J.J. and Redlich, A.N. (1990) Towards a theory of early visual processing. *Neural Computation* **2**: 308–320.

Atick, J.J. Griffin, P.A. and Relich, A.N. (1996) The vocabulary of shape: principal shapes for probing perception and neural response. *Network* **7**: 1–5.

Attneave, F. (1954) Some informational aspects of visual perception. *Psychological Review* **61**: 183–193.

Baddeley, R.J. (1995) Topographic map formation as statistical inference. Pp. 86–96 in *Neural Computation and Psychology*, ed. L.Smith. Springer: London.

Baddeley, R.J., Abbott, L.F., Booth, M.J.A., Sengpiel, F., Freeman, T., Wakeman, E.A., and Rolls, E.T. (1998) Responses of neurons in primary and inferior visual cortices to natural scenes. *Proceedings of the Royal Society, B*, in press.

Ballard, D.H. (1990) Animate vision uses object-centred reference frames. Pp. 229–236 in *Advanced Neural Computers*, ed. R.Eckmiller. North-Holland: Amsterdam.

Ballard, D.H. (1993) Subsymbolic modelling of hand eye co-ordination. Ch. 3, pp. 71–102 in *The Simulation of Human Intelligence*, ed. D.E.Broadbent. Blackwell: Oxford.

Barbas, H. (1988) Anatomic organization of basoventral and mediodorsal visual recipient prefrontal regions in the rhesus monkey. *Journal of Comparative Neurology* **276**: 313–342.

Barlow, H.B. (1961) Possible principles underlying the transformation of sensory messages. In *Sensory Communication*, ed. W.Rosenblith. MIT Press: Cambridge, Mass.

Barlow, H.B. (1972) Single units and sensation: A neuron doctrine for perceptual psychology. *Perception* **1**: 371–394.

Barlow, H.B. (1985) Cerebral cortex as model builder. Pp. 37–46 in *Models of the Visual Cortex*, eds. D.Rose and V.G.Dobson. Wiley: Chichester.

Barlow, H.B. (1989) Unsupervised Learning. *Neural Computation* **1**: 295–311.

Barlow, H. (1995) The neuron doctrine in perception. Ch 26, pp. 415–435 in *The Cognitive Neurosciences*, ed. M.S.Gazzaniga. MIT Press: Cambridge, Mass.

Barlow, H.B., Kaushal, T.P. and Mitchison, G.J. (1989) Finding minimum entropy codes. *Neural Computation* **1**: 412–423.

Barnes, C.A., McNaughton, B.L., Mizumori, S.J. and Lim, L.H. (1990) Comparison of spatial and temporal characteristics of neuronal activity in sequential stages of hippocampal processing. *Progress in Brain Research* **83**: 287–300.

Barnes, C.A., Treves, A., Rao, G. and Shen, J. (1994) Electrophysiological markers of cognitive aging: region specificity and computational consequences. *Seminars in the Neurosciences* **6**: 359–367.

Barto, A.G. (1985) *Learning by Statistical Cooperation of Self-Interested Neuron-Like Computing Elements* (COINS Tech. Rep. 85 11). University of Massachusetts, Department of Computer and Information Science: Amherst.

Barto, A.G. (1995) Adaptive critics and the basal ganglia. Ch. 11, pp. 215–232 in *Models of Information Processing in the Basal Ganglia*, eds. J.C.Houk, J.L.Davis and D.G.Beiser. MIT Press: Cambridge, Mass.

Battaglia, F.P. and Treves, A. (1996) Information dynamics in associative memories with spiking neurons. *Society for Neuroscience Abstracts* **22**: 445.4.

Battaglia, F.P. and Treves, A. (1997) Stable and rapid recurrent processing in realistic autoassociative memories. In preparation.

Baylis, G.C., Rolls, E.T. and Leonard, C.M. (1985) Selectivity between faces in the responses of a population of neurons in the cortex in the superior temporal sulcus of the monkey. *Brain Research* **342**: 91–102.

Baylis, G.C. and Rolls, E.T. (1987) Responses of neurons in the inferior temporal cortex in short term and serial recognition memory tasks. *Experimental Brain Research* **65**: 614–622.

Baylis, G.C., Rolls, E.T. and Leonard, C.M. (1987) Functional subdivisions of temporal lobe neocortex. *Journal of Neuroscience* **7**: 330–342.

Baylis, L.L. and Gaffan, D. (1991) Amygdalectomy and ventromedial prefrontal ablation produce similar deficits in food choice and in simple object discrimination learning for an unseen reward. *Experimental Brain Research* **86**: 617–622.

Baylis, L.L., Rolls, E.T. and Baylis, G.C. (1994) Afferent connections of the orbitofrontal cortex taste area of the primate. *Neuroscience* **64**: 801–812.

Bear, M.F. and Singer, W. (1986) Modulation of visual cortical plasticity by acetylcholine and noradrenaline. *Nature* **320**: 172–176.

Becker, S. and Hinton, G.E. (1992) Self-organizing neural network that discovers surfaces in random-dot stereograms. *Nature* **355**: 161–163.

Bell, A.J. and Sejnowski, T.J. (1995) An information-maximation approach to blind separation and blind deconvolution. *Neural Computation* **7**: 1129–1159.

Bennett, A. (1990) Large competitive networks. *Network* **1**: 449–462.

Berger, T.W., Yeckel, M.F. and Thiels, E. (1996) Network determinants of hippocampal synaptic plasticity. In *Long-Term Potentiation*, Volume 3, eds. M.Baudry and J.L.Davis. Cambridge, Mass: MIT Press.

Bialek, W., Rieke, F., de Ruyter van Steveninck, R.R. and Warland, D. (1991) Reading a neural code. *Science* **252**: 1854–1857.

Bienenstock, E.L., Cooper, L.N. and Munro, P.W. (1982) Theory for the development of neuron selectivity: orientation specificity and binocular interaction in visual cortex. *Journal of Neuroscience* **2**: 32–48.

Bishop, C.M. (1995) *Neural Networks for Pattern Recognition*. Clarendon: Oxford.

Blaney, P.H. (1986) Affect and memory: a review. *Psychological Bulletin* **99**: 229–246.

Bliss, T.V.P. and Lomo, T. (1973) Long-lasting potentiation of synaptic transmission in the dentate area of the anaesthetized rabbit following stimulation of the perforant path. *Journal of Physiology* **232**: 331–356.

Bliss, T.V.P. and Collingridge, G.L. (1993) A synaptic model of memory: long-term potentiation in the hippocampus. *Nature* **361**: 31–39.

Block, H.D. (1962) The perceptron: a model for brain functioning. *Reviews of Modern Physics* **34**: 123–135.

Bloomfield, S. (1974) Arithmetical operations performed by nerve cells. *Brain Research* **69**: 115–124.

Borsini, F. and Rolls, E.T. (1984) Role of noradrenaline and serotonin in the basolateral region of the amygdala in food preferences and learned taste aversions in the rat. *Physiology and Behavior* **33**: 37–43.

Boussaoud, D., Desimone, R. and Ungerleider, L.G. (1991) Visual topography of area TEO in the macaque. *Journal of Comparative Neurology* **306**: 554–575.

Braitenberg, V. and Schuz, A. (1991) *Anatomy of the Cortex*. Springer-Verlag: Berlin.

Bridle, J.S. (1990) Probabilistic interpretation of feedforward classification network outputs, with relationships to statistical pattern recognition. Pp. 227–236 in *Neurocomputing: Algorithms, Architectures and Applications*, eds. F.Fogelman Soulie and J.Herault. Springer-Verlag: New York.

Brodmann, K. (1925) *Vergleichende Localisationslehre der Grosshirnrinde*, 2nd Edition. Barth: Leipzig.

Brooks, L.R. (1978) Nonanalytic concept formation and memory for instances. In *Cognition and Categorization*, eds. E.Rosch and B.B.Lloyd. Erlbaum: Hillsdale, NJ.

Brown, D.A., Gähwiler, B.H., Griffith, W.H. and Halliwell, J.V. (1990) Membrane currents in hippocampal neurons. *Progress in Brain Research* **83**: 141–160.

Brown, T.H. and Zador, A. (1990) The hippocampus. Pp. 346–388 in *The Synaptic Organization of the Brain*, ed. G.Shepherd. Oxford University Press: New York.

Brown, T.H., Kairiss, E.W. and Keenan, C.L. (1990) Hebbian synapses: biophysical mechanisms and algorithms. *Annual Review of Neuroscience* **13**: 475–511.

Buerger, A.A., Gross, C.G. and Rocha-Miranda, C.E. (1974) Effects of ventral putamen lesions on discrimination learning by monkeys. *Journal of Comparative and Physiological Psychology* **86**: 440–446.

Buhl, E.H., Halasy, K. and Somogyi, P. (1994) Diverse sources of hippocampal unitary inhibitory postsynaptic potentials and the number of synaptic release sites. *Nature* **368**: 823–828.

Buhmann, J., Divko, R. and Schulten, K. (1989) Associative memory with high information content. *Physical Review A* **39**: 2689–2692.

Buhmann, J., Lades, M. and von der Malsburg, C. (1990) Size and distortion invariant object recognition by hierarchical graph matching. Pp. 411–416 in *International Joint Conference on Neural Networks*. IEEE: New York.

Burgess, N., Recce, M. and O'Keefe, J. (1994) A model of hippocampal function. *Neural Networks* **7**: 1065–1081.

Burton, M.J., Rolls, E.T. and Mora, F. (1976) Visual responses of hypothalamic neurones. *Brain Research* **107**: 215–216.

Butter, C.M. (1969) Perseveration in extinction and in discrimination reversal tasks following selective prefrontal ablations in Macaca mulatta. *Physiology and Behavior* **4**: 163–171.

Butter, C.M., McDonald, J.A. and Snyder, D.R. (1969) Orality, preference behavior, and reinforcement value of non-food objects in monkeys with orbital frontal lesions. *Science* **164**: 1306–1307.

Butter, C.M., Snyder, D.R. and McDonald, J.A. (1970) Effects of orbitofrontal lesions on aversive and aggressive behaviors in rhesus monkeys. *Journal of Comparative and Physiological Psychology* **72**: 132–144.

Butter, C.M. and Snyder, D.R. (1972) Alterations in aversive and aggressive behaviors following orbitofrontal lesions in rhesus monkeys. *Acta Neurobiologica Experimentalis* **32**: 525–565.

Caan, W., Perrett, D.I. and Rolls, E.T. (1984) Responses of striatal neurons in the behaving monkey. 2. Visual processing in the caudal neostriatum. *Brain Research* **290**: 53–65.

Cador, M., Robbins, T.W. and Everitt, B.J. (1989) Involvement of the amygdala in stimulus-reward associations: interaction with the ventral striatum. *Neuroscience* **30**: 77–86.

Cahusac, P.M.B., Miyashita, Y. and Rolls, E.T. (1989) Responses of hippocampal formation neurons in the monkey related to delayed spatial response and object-place memory tasks. *Behavioural Brain Research* **33**: 229–240.

Cahusac, P.M.B., Rolls, E.T. and Marriott, F.H.C. (1991) Potentiation of neuronal responses to natural visual input paired with postsynaptic activation in the hippocampus of the awake monkey. *Neuroscience Letters* **124**: 39–43.

Cahusac, P.M.B., Rolls, E.T., Miyashita, Y. and Niki, H. (1993) Modification of the responses of hippocampal neurons in the monkey during the learning of a conditional spatial response task. *Hippocampus* **3**: 29–42.

Calabresi, P., Maj, R., Pisani, A., Mercuri, N.B. and Bernardi,G (1992) Long-term synaptic depression in the striatum: physiological and pharmacological characterization. *Journal of Neuroscience* **12**: 4224–4233.

Calvert, G.A., Bullmore, E.T., Brammer, M.J., Campbell, R., Williams, S.C.R., McGuire,

P.K., Woodruff, P.W.R., Iversen, S.D. and David, A.S. (1997) Activation of auditory cortex during silent lip-reading. *Science* **276**: 593–596.

Campbell, N.C., Ekerot, C.-F., Hesslow, G. and Oscarsson, O. (1983) Dendritic plateau potentials evoked in Purkinje cells by parallel fibre volleys. *Journal of Physiology* **340**: 209–223.

Carmichael, S.T., Clugnet, M.-C. and Price, J.L. (1994) Central olfactory connections in the macaque monkey. *Journal of Comparative Neurology* **346**: 403–434.

Carnevale, N.T. and Johnston, D. (1982) Electrophysiological characterization of remote chemical synapses. *Journal of Neurophysiology* **47**: 606–621.

Christie, B.R. (1996) Long-term depression in the hippocampus. *Hippocampus* **6**: 1–2.

Churchland, P.S. and Sejnowski, T.J. (1992) *The Computational Brain*. MIT Press: Cambridge, Mass.

Collingridge, G.L. and Bliss, T.V.P. (1987) NMDA receptors: their role in long-term potentiation. *Trends in Neurosciences* **10**: 288–293.

Collingridge, G.L. and Singer, W. (1990) Excitatory amino acid receptors and synaptic plasticity. *Trends in Pharmacological Sciences* **11**: 290–296.

Cortes, C., Jaeckel, L.D., Solla, S.A., Vapnik, V. and Denker, J.S. (1996) Learning curves: asymptotic values and rates of convergence. *Neural Information Processing Systems* **6**: 327–334.

Cover, T.M. (1965) Geometrical and statistical properties of systems of linear inequalities with applications in pattern recognition. *IEEE Transactions on Electronic Computers* **14**: 326–334.

Cover, T.M. and Thomas, J.A. (1991) *Elements of Information Theory*. Wiley: New York.

Cowey, A. (1979) Cortical maps and visual perception. *Quarterly Journal of Experimental Psychology* **31**: 1–17.

Crisanti, A., Amit, D.J. and Gutfreund, H. (1986) Saturation level of the Hopfield model for neural network. *Europhysics Letters* **2**: 337–345.

Critchley, H.D. and Rolls, E.T. (1996a) Hunger and satiety modify the responses of olfactory and visual neurons in the primate orbitofrontal cortex. *Journal of Neurophysiology* **75**: 1673–1686.

Critchley, H.D. and Rolls, E.T. (1996b) Olfactory neuronal responses in the primate orbitofrontal cortex: analysis in an olfactory discrimination task. *Journal of Neurophysiology* **75**: 1659–1672.

Crutcher, M.D. and DeLong, M.R. (1984a) Single cell studies of the primate putamen. I. Functional organization. *Experimental Brain Research* **53**: 233–243.

Crutcher, M.D. and DeLong, M.R. (1984b) Single cell studies of the primate putamen. II. Relations to direction of movements and pattern of muscular activity. *Experimental Brain Research* **53**: 244–258.

Davis, M. (1992) The role of the amygdala in conditioned fear. Ch. 9, pp. 255–305 in *The Amygdala*, ed. J.P.Aggleton. Wiley-Liss: New York.

Davis, M., Rainnie, D. and Cassell, M. (1994) Neurotransmission in the rat amygdala related to fear and anxiety. *Trends in Neurosciences* **17**: 208–214.

Dawkins, R. (1989) *The Selfish Gene* 2nd edn. Oxford University Press: Oxford.

Dayan, P. and Hinton, G.E. (1996) Varieties of Helmholtz machine. *Neural Networks* **9**: 1385–1403.

deCharms, R.C. and Merzenich, M.M. (1996) Primary cortical representation of sounds by the coordination of action-potential timing. *Nature* **381**: 610–613.

DeLong, M.R., Crutcher, M.D. and Georgopoulos, A.P. (1983) Relations between movement and single cell discharge in the substantia nigra of the behaving monkey. *Journal of Neuroscience* **3**: 1599–1606.

DeLong, M.R., Georgopoulos, A.P., Crutcher, M.D., Mitchell, S.J., Richardson, R.T. and Alexander, G.E. (1984) Functional organization of the basal ganglia: contributions of single-cell recording studies. Pp. 64–78 in *Functions of the Basal Ganglia*, Ciba Foundation Symposium 107. Pitman: London.

Demer, J.L. and Robinson, D.A. (1982) Effects of reversible lesions and stimulation of olivocerebellar system on vestibuloocular reflex plasticity. *Journal of Neurophysiology* **47**: 1084–1107.

Derrick, B.E. and Martinez, J.L. (1996) Associative, bidirectional modifications at the hippocampal mossy fibre CA3 synapse. *Nature* **381**: 429–434.

Derrida, B., Gardner, E. and Zippelius, A. (1987) An exactly soluble asymmetric neural network model. *Europhysics Letters* **4**: 167–174.

de Ruyter van Steveninck, R.R. and Laughlin, S.B. (1996) The rates of information transfer at graded-potential synapses. *Nature* **379**: 642–645.

De Sieno, D. (1988) Adding a conscience to competitive learning. Vol. 1, pp. 117–124 in *IEEE International Conference on Neural Networks* (San Diego 1988). IEEE: New York.

Diamond, M.E., Huang, W. and Ebner, F.F. (1994) Laminar comparison of somatosensory cortical plasticity. *Science* **265**: 1885–1888.

Divac, I. (1975) Magnocellular nuclei of the basal forebrain project to neocortex, brain stem, and olfactory bulb. Review of some functional correlates. *Brain Research* **93**: 385–398.

Divac, I., Rosvold, H.E. and Szwarcbart, M.K. (1967) Behavioral effects of selective ablation of the caudate nucleus. *Journal of Comparative and Physiological Psychology* **63**: 184–190.

Divac, I. and Oberg, R.G.E. (1979) Current conceptions of neostriatal functions. Pp. 215–230 in *The Neostriatum*, eds. I.Divac and R.G.E.Oberg. Pergamon: New York.

Dolan, R.J., Fink, G.R., Rolls, E.T., Booth, M., Frackowiak, R.S.J. and Friston, K.J. (1997) How the brain learns to see objects and faces in an impoverished context. *Nature*, in press.

Domany, E., Kinzel, W. and Meir, R. (1989) Layered neural networks. *Journal of Physics A* **22**: 2081–2102.

Dotsenko, V. (1985) Ordered spin glass: a hierarchical memory machine. *Journal of Physics C* **18**: L1017–L1022.

Douglas, R.J. and Martin, K.A.C. (1990) Neocortex. Ch. 12, pp. 389–438 in *The Synaptic Organization of the Brain*, 3rd edn., ed. G.M.Shepherd. Oxford University Press: Oxford.

Douglas, R.J., Mahowald, M.A. and Martin, K.A.C. (1996) Microarchitecture of cortical columns. In *Brain Theory: Biological Basis and Computational Theory of Vision*, eds. A.Aertsen and V.Braitenberg. Elsevier: Amsterdam.

Dunn, L.T. and Everitt, B.J. (1988) Double dissociations of the effects of amygdala and insular cortex lesions on conditioned taste aversion, passive avoidance, and neophobia in the rat using the excitotoxin ibotenic acid. *Behavioral Neuroscience* **102**: 3–23.

Dunnett, S.B. and Iversen, S.D. (1981) Learning impairments following selective kainic acid-induced lesions within the neostriatum of rats. *Behavioural Brain Research* **2**: 189–209.

Dunnett, S.B. and Iversen, S.D. (1982a) Sensorimotor impairments following localised kainic acid and 6-hydroxydopamine lesions of the neostriatum. *Brain Research* **248**: 121–127.

Dunnett, S.B. and Iversen, S.D. (1982b) Neurotoxic lesions of ventrolateral but not anteromedial neostriatum impair differential reinforcement of low rates (DRL) performance. *Behavioural Brain Research* **6**: 213–226.

Durbin, R. and Mitchison, G. (1990) A dimension reduction framework for understanding cortical maps. *Nature* **343**: 644–647.

Eccles, J.C. (1984) The cerebral neocortex: a theory of its operation. Ch. 1, pp. 1–36 in *Cerebral Cortex*, Vol. 2, *Functional Properties of Cortical Cells*, eds. E.G.Jones and A.Peters. Plenum: New York.

Eccles, J.C., Ito, M. and Szentágothai, J. (1967) *The Cerebellum as a Neuronal Machine*. Springer-Verlag: New York.

Eckhorn, R. and Pöpel, B. (1974) Rigorous and extended application of information theory to the afferent visual system of the cat. I. Basic concepts. *Kybernetik* **16**: 191–200.

Eckhorn, R. and Pöpel, B. (1975) Rigorous and extended application of information theory to the afferent visual system of the cat. II. Experimental results. *Kybernetik* **17**: 7–17.

Eckhorn, R., Grüsser, O.-J., Kröller, J., Pellnitz, K. and Pöpel, B. (1976) Efficiency of different neural codes: information transfer calculations for three different neuronal systems. *Biological Cybernetics* **22**: 49–60.

Eichenbaum, H., Otto, T. and Cohen, N.J. (1992) The hippocampus what does it do? *Behavioural and Neural Biology* **57**: 2–36.

Ekerot, C.-F. (1985) Climbing fibre actions of Purkinje cells—plateau potentials and long-lasting depression of parallel fibre responses. Pp. 268–274 in *Cerebellar Functions*, eds. J.R.Bloedel, J.Dichgans and W.Precht. Springer-Verlag: New York.

Ekerot, C.-F. and Kano, M. (1983) Climbing fibre induced depression of Purkinje cell responses to parallel fibre stimulation. *Proceedings of the International Union of Physiological Sciences Sydney, Vol. 15*, 393: 470.03.

Ekerot, C.-F. and Oscarsson, O. (1981) Prolonged depolarization elicited in Purkinje cell dendrites by climbing fibre impulses in the cat. *Journal of Physiology* **318**: 207–221.

Engel, A.K., Konig, P. and Singer, W. (1991) Direct physiological evidence for scene segmentation by temporal encoding. *Proceedings of the National Academy of Sciences of the USA* **88**: 9136–9140.

Engel, A.K., Konig, P., Kreiter, A.K., Schillen, T.B. and Singer, W. (1992) Temporal coding in the visual system: new vistas on integration in the nervous system. *Trends in Neurosciences* **15**: 218–226.

Eskandar, E.N., Richmond, B.J. and Optican, L.M. (1992) Role of inferior temporal neurons in visual memory. I. Temporal encoding of information about visual images, recalled images, and behavioural context. *Journal of Neurophysiology* **68**: 1277–1295.

Evans, M.R. (1989) Random dilution in a neural network for biased patterns. *Journal of Physics A* **22**: 2103–2118.

Evarts, E.V. and Wise, S.P. (1984) Basal ganglia outputs and motor control. Pp. 83–96 in *Functions of the Basal Ganglia*. Ciba Foundation Symposium 107. Pitman: London.

Everitt, B.J. and Robbins, T.W. (1992) Amygdala-ventral striatal interactions and reward-related processes. Pp. 401–430 in *The Amygdala*, ed. J.P.Aggleton. Wiley: Chichester.

Fabre, M., Rolls, E.T., Ashton, J.P. and Williams, G. (1983) Activity of neurons in the ventral tegmental region of the behaving monkey. *Behavioural Brain Research* **9**: 213–235.

Fahy, F.L., Riches, I.P. and Brown, M.W. (1993) Neuronal activity related to visual recognition memory and the encoding of recency and familiarity information in the primate anterior and medial inferior temporal and rhinal cortex. *Experimental Brain Research* **96**: 457–492.

Farah, M.J. (1990) *Visual Agnosia*. MIT Press: Cambridge, Mass.

Farah, M.J. (1994) Neuropsychological inference with an interactive brain: a critique of the 'locality' assumption. *Behavioral and Brain Sciences* **17**: 43–104.

Farah, M.J. (1996) Is face recognition special? Evidence from neuropsychology. *Behavioural Brain Research* **76**: 181–189.

Farah, M.J., O'Reilly, R.C. and Vecera, S.P. (1993) Dissociated overt and covert recognition as an emergent property of a lesioned neural network. *Psychological Review* **100**: 571–588.

Farah, M.J., Meyer, M.M. and McMullen, P.A. (1996) The living/nonliving dissociation is not an artifact: giving an a priori implausible hypothesis a strong test. *Cognitive Neuropsychology* **13**: 137–154.

Fazeli, M.S. and Collingridge, G.L. (ed.) (1996). *Cortical Plasticity: LTP and LTD*. Bios: Oxford.

Feigenbaum, J.D. and Rolls, E.T. (1991) Allocentric and egocentric spatial information processing in the hippocampal formation of the behaving primate. *Psychobiology* **19**: 21–40.

Ferster, D. and Spruston, N. (1995) Cracking the neuronal code. *Science* **270**: 756–757.

Foldiák, P. (1990) Forming sparse representations by local anti-Hebbian learning. *Biological Cybernetics* **64**: 165–170.

Foldiák, P. (1991) Learning invariance from transformation sequences. *Neural Computation* **3**: 193–199.

Foster, T.C., Castro, C.A. and McNaughton, B.L. (1989). Spatial selectivity of rat hippo-campal neurons: dependence on preparedness for movement. *Science* **244**: 1580–1582.

Frolov, A.A. and Medvedev, A.V. (1986) Substantiation of the 'point approximation' for describing the total electrical activity of the brain with use of a simulation model. *Biophysics* **31**: 332–337.

Frolov, A.A. and Murav'ev, I.P. (1993) Informational characteristics of neural networks capable of associative learning based on Hebbian plasticity. *Network* **4**: 495–536.

Fukai, T. and Shiino, M. (1992) Study of self-inhibited analogue neural networks using the self-consistent signal-to-noise analysis. *Journal of Physics A* **25**: 4799–4811.

Fukushima, K. (1975) Cognitron: a self-organizing neural network. *Biological Cybernetics* **20**: 121–136.

Fukushima, K. (1980) Neocognitron: a self-organizing neural network model for a mechan-ism of pattern recognition unaffected by shift in position. *Biological Cybernetics* **36**: 193–202.

Fukushima, K. (1989) Analysis of the process of visual pattern recognition by the neocogni-tron. *Neural Networks* **2**: 413–420.

Fukushima, K. (1991) Neural networks for visual pattern recognition. *IEEE Transactions E* **74**: 179–190.

Funahashi, S., Bruce, C.J. and Goldman-Rakic, P.S. (1989) Mnemonic coding of visual space in monkey dorsolateral prefrontal cortex. *Journal of Neurophysiology* **61**: 331–349.

Fuster, J.M. (1973) Unit activity in prefrontal cortex during delayed-response performance: neuronal correlates of transient memory. *Journal of Neurophysiology* **36**: 61–78.

Fuster, J.M. (1989) *The Prefrontal Cortex*, 2nd edn. Raven Press: New York.

Fuster, J.M. and Jervey, J.P. (1982) Neuronal firing in the inferotemporal cortex of the monkey in a visual memory task. *Journal of Neuroscience* **2**: 361–375.

Gaffan, D. (1992) The role of the hippocampo-fornix-mammillary system in episodic memory. Pp. 336–346 in *Neuropsychology of Memory*, 2nd edn., eds. L.R.Squire and N.Butters. Guilford: New York.

Gaffan, D. (1993) Additive effects of forgetting and fornix transection in the temporal gradient of retrograde amnesia. *Neuropsychologia* **31**: 1055–1066.

Gaffan, D. (1994) Scene-specific memory for objects: a model of episodic memory impair-ment in monkeys with fornix transection. *Journal of Cognitive Neuroscience* **6**: 305–320.

Gaffan, D., Saunders, R.C., Gaffan, E.A., Harrison, S., Shields, C. and Owen, M.J. (1984) Effects of fornix transection upon associative memory in monkeys: role of the hippo-campus in learned action. *Quarterly Journal of Experimental Psycholology* **26B**: 173–221.

Gaffan, D. and Saunders, R.C. (1985). Running recognition of configural stimuli by fornix transected monkeys. *Quarterly Journal of Experimental Psychology* **37B**: 61–71.

Gaffan, D. and Harrison, S. (1987) Amygdalectomy and disconnection in visual learning for auditory secondary reinforcement by monkeys. *Journal of Neuroscience* **7**: 2285–2292.

Gaffan, E.A., Gaffan, D. and Harrison, S. (1988) Disconnection of the amygdala from visual association cortex impairs visual reward-association learning in monkeys. *Journal of Neuroscience* **8**: 3144–3150.

Gaffan, D., Gaffan, E.A. and Harrison, S. (1989) Visual visual associative learning and reward-association learning: the role of the amygdala. *Journal of Neuroscience* **9**: 558–564.

Gaffan, D. and Harrison, S. (1989a) A comparison of the effects of fornix section and sulcus principalis ablation upon spatial learning by monkeys. *Behavioural Brain Research* **31**: 207–220.

Gaffan, D. and Harrison, S. (1989b) Place memory and scene memory: effects of fornix transection in the monkey. *Experimental Brain Research* **74**: 202–212.

Gaffan, D. and Murray, E.A. (1992) Monkeys (*Macaca fascicularis*) with rhinal cortex ablations succeed in object discrimination learning despite 24-hr intertrial intervals and fail at matching to sample despite double sample presentations. *Behavioral Neuroscience* **106**: 30–38.

Gardner, E. (1987) Maximum storage capacity in neural networks. *Europhysics Letters* **4**: 481–485.

Gardner, E. (1988) The space of interactions in neural network models. *Journal of Physics A* **21**: 257–270.

Gardner-Medwin, A.R. (1976) The recall of events through the learning of associations between their parts. *Proceedings of the Royal Society of London, Series B* **194**: 375–402.

Gawne, T.J. and Richmond, B.J. (1993) How independent are the messages carried by adjacent inferior temporal cortical neurons? *Journal of Neuroscience* **13**: 2758–2771.

Georgopoulos, A.P. (1995) Motor cortex and cognitive processing. Ch. 32, pp. 507–517 in *The Cognitive Neurosciences*, ed. M.S.Gazzaniga. MIT Press: Cambridge, Mass.

Gerstner, W. (1995) Time structure of the activity in neural network models. *Physical Review E* **51**: 738–758.

Ghelarducci, P., Ito, M. and Yagi, N. (1975) Impulse discharges from flocculus Purkinje cells of alert rabbits during visual stimulation combined with horizontal head rotation. *Brain Research* **87**: 66–72.

Gilbert, P.F.C. and Thach, W.T. (1977) Purkinje cell activity during motor learning. *Brain Research* **128**: 309–328.

Gnadt, J.W. (1992) Area LIP: three-dimensional space and visual to oculomotor transformation. *Behavioral and Brain Sciences* **15**: 745–746.

Gochin, P.M., Colombo, M., Dorfman, G.A., Gerstein, G.L. and Gross, C.G. (1994) Neural ensemble encoding in inferior temporal cortex. *Journal of Neurophysiology* **71**: 2325–2337.

Goldman, P.S. and Nauta, W.J.H. (1977) An intricately patterned prefronto-caudate projection in the rhesus monkey. *Journal of Comparative Neurology* **171**: 369–386.

Goldman-Rakic, P.S. (1987) Circuitry of primate prefrontal cortex and regulation of behavior by representational memory. Pp. 373–417 in *Handbook of Physiology, Section*

1, *The Nervous System, Vol. V, Higher Functions of the Brain, Part* 1. American Physiological Society: Bethesda, Md.

Goldman-Rakic, P.S. (1996) The prefrontal landscape: implications of functional architecture for understanding human mentation and the central executive. *Philosophical Transactions of the Royal Society of London, Series B* **351**: 1445–1453.

Golomb, D., Rubin, N. and Sompolinsky, H. (1990) Willshaw model: associative memory with sparse coding and low firing rates. *Physical Review A* **41**: 1843–1854.

Golomb, D., Kleinfeld, D., Reid, R.C., Shapley, R.M. and Shraiman, B. (1994) On temporal codes and the spatiotemporal response of neurons in the lateral geniculate nucleus. *Journal of Neurophysiology* **72**: 2990–3003.

Golomb, D., Hertz, J.A., Panzeri, S., Treves, A. and Richmond, B.J. (1997) How well can we estimate the information carried in neuronal responses from limited samples? *Neural Computation* **9**: 649–665.

Gonshor, A. and Melvill-Jones, G. (1976) Extreme vestibulo-ocular adaptation induced by prolonged optical reversal of vision. *Journal of Physiology* **256**: 381–414.

Gray, C.M., Konig, P., Engel, A.K. and Singer, W. (1989) Oscillatory responses in cat visual cortex exhibit inter-columnar synchronization which reflects global stimulus properties. *Nature* **338**: 334–337.

Gray, C.M., Engel, A.K. and Singer, W. (1992) Synchronization of oscillatory neuronal responses in cat striate cortex: temporal properties. *Visual Neuroscience* **8**: 337–347.

Gray, J.A. (1975) *Elements of a Two-Process Theory of Learning*. Academic Press: London.

Gray, J.A. (1987) *The Psychology of Fear and Stress*, 2nd edn. Cambridge University Press: Cambridge.

Graybiel, A.M. and Kimura, M. (1995) Adaptive neural networks in the basal ganglia. Ch. 5, pp. 103–116 in *Models of Information Processing in the Basal Ganglia*, eds. J.C.Houk, J.L.Davis and D.G.Beiser. MIT Press: Cambridge, Mass.

Graziano, M.S. and Gross, C.G. (1993) A bimodal map of space: somatosensory receptive fields in the macaque putamen with corresponding visual receptive fields. *Experimental Brain Research* **97**: 96–109.

Green, A.R. and Costain, D.W. (1981) *Pharmacology and Biochemistry of Psychiatric Disorders*. Wiley: Chichester.

Gross, C.G., Desimone, R., Albright, T.D. and Schwartz, E.L. (1985) Inferior temporal cortex and pattern recognition. *Experimental Brain Research, Supplement* **11**: 179–201.

Grossberg, S. (1976a) Adaptive pattern classification and universal recoding: I. Parallel development and coding of neural feature detectors. *Biological Cybernetics* **23**: 121–134.

Grossberg, S. (1976b) Adaptive pattern classification and universal recoding: II. Feedback, expectation, olfaction, illusions. *Biological Cybernetics* **23**: 187–202.

Grossberg, S. (1988) Nonlinear neural networks: principles, mechanisms, and architectures. *Neural Networks* **1**: 17–61.

Groves, P.M. (1983) A theory of the functional organization of the neostriatum and the neostriatal control of voluntary movement. *Brain Research Reviews* **5**: 109–132.

Groves, P.M., Garcia-Munoz, M., Linder, J.C., Manley, M.S., Martone, M.E. and Young, S.J. (1995) Elements of the intrinsic organization and information processing in the neostriatum. Ch. 4, pp. 51–96 in *Models of Information Processing in the Basal Ganglia*, eds. J.C.Houk, J.L.Davis and D.G.Beiser. MIT Press: Cambridge, Mass.

Grusser, O.-J. and Landis, T. (1991) *Visual Agnosias*, Ch. 21. MacMillan, London.

Habib, M. and Sirigu, A. (1987) Pure topographical disorientation: a definition and anatomical basis. *Cortex* **23**: 73–85.

Haddad, G.M., Demer, J.L. and Robinson, D.A. (1980) The effect of lesions of the dorsal cap of the inferior olive and the vestibulo-ocular and optokinetic systems of the cat. *Brain Research* **185**: 265–275.

Hamming, R.W. (1990) *Coding and Information Theory*. 2nd Edition. Prentice-Hall: Englewood Cliffs, New Jersey.

Hardiman, M.J., Ramnani, N. and Yeo, C.H. (1996) Reversible inactivations of the cerebellum with muscimol prevent the acquisition and extinction of conditioned nictitating membrane responses in the rabbit. *Experimental Brain Research* **110**: 235–247.

Hasselmo, M.E., Rolls, E.T. and Baylis, G.C. (1989) The role of expression and identity in the face-selective responses of neurons in the temporal visual cortex of the monkey. *Behavioural Brain Research* **32**: 203–218.

Hasselmo, M.E., Rolls, E.T., Baylis, G.C. and Nalwa, V. (1989) Object-centered encoding by face-selective neurons in the cortex in the superior temporal sulcus of the monkey. *Experimental Brain Research* **75**: 417–429.

Hasselmo, M.E. and Bower, J.M. (1993) Acetylcholine and memory. *Trends in Neurosciences* **16**: 218–222.

Hasselmo, M.E., Schnell, E. and Barkai, E. (1995) Learning and recall at excitatory recurrent synapses and cholinergic modulation in hippocampal region CA3. *Journal of Neuroscience* **15**: 5249–5262.

Hebb, D.O. (1949) *The Organization of Behavior*. Wiley: New York.

Heeger, D.J., Simonelli, E.P. and Movshon, D.J. (1996) Computational models of cortical visual processing. *Proceedings of the National Academy of Sciences of the USA* **93**: 623–627.

Heller, J., Hertz, J.A., Kjaer, T.W. and Richmond, B.J. (1995) Information flow and temporal coding in primate pattern vision. *Journal of Computational Neuroscience* **2**: 175–193.

Hertz, J.A., Krogh, A. and Palmer, R.G. (1991) *Introduction to the Theory of Neural Computation*. Addison-Wesley: Wokingham, UK.

Hertz, J.A., Kjaer, T.W., Eskander, E.N. and Richmond, B.J. (1992) Measuring natural neural processing with artificial neural networks. *International Journal of Neural Systems* **3** (Suppl.): 91–103.

Hessler, N.A., Shirke, A.M. and Malinow, R. (1993) The probability of transmitter release at a mammalian central synapse. *Nature* **366**: 569–572.

Higuchi, S.-I. and Miyashita, Y. (1996) Formation of mnemonic neuronal responses to visual paired associates in inferotemporal cortex is impaired by perirhinal and entorhinal lesions. *Proceedings of the National Academy of Sciences of the USA* **93**: 739–743.

Hinton, G.E. (1981) A parallel computation that assigns canonical object based frames of reference. In *Proceedings of the* **9**th *International Joint Conference on Artificial Intelligence*. Reviewed in Rumelhart and McClelland (1986), Vol. 1, Ch. 4.

Hinton, G.E. (1989) Deterministic Boltzmann learning performs steepest descent in weight-space. *Neural Computation* **1**: 143–150.

Hinton, G.E. and Anderson, J.A. (ed.) (1981) *Parallel Models of Associative Memory*. Erlbaum: Hillsdale, NJ.

Hinton, G.E. and Sejnowski, T.J. (1986) Learning and relearning in Boltzmann machines. Ch. 7, pp. 282–317 in *Parallel Distributed Processing*, Vol. 1, ed. D.Rumelhart and J.L.McClelland. MIT Press: Cambridge, Mass.

Hinton, G.E., Dayan, P., Frey, B.J. and Neal, R.M. (1995) The 'wake-sleep' algorithm for unsupervised neural networks. *Science* **268**: 1158–1161.

Hodgkin, A.L. and Huxley, A.F. (1952) A quantitative description of membrane current and its application to conduction and excitation in nerve. *Journal of Physiology* **117**: 500–544.

Hoebel, B.G., Rada, P., Mark, G.P., Parada, M., Puig de Parada, M., Pothos, E. and Hernandez, L. (1996) Hypothalamic control of accumbens dopamine: a system for feeding reinforcement. Vol. 5, pp. 263–280 in *Molecular and Genetic Aspects of Obesity*, ed. G.Bray and D.Ryan. Louisiana State University Press.

Hopfield, J.J. (1982) Neural networks and physical systems with emergent collective computational abilities. *Proceedings of the National Academy of Sciences of the U.S.A.* **79**: 2554–2558.

Hopfield, J.J. (1984) Neurons with graded response have collective computational properties like those of two-state neurons. *Proceedings of the National Academy of Sciences of the USA* **81**: 3088–3092.

Hornak, J., Rolls, E.T. and Wade, D. (1996) Face and voice expression identification in patients with emotional and behavioural changes following ventral frontal lobe damage. *Neuropsychologia* **34**: 247–261.

Hornykiewicz, O. (1973) Dopamine in the basal ganglia: its role and therapeutic implications. *British Medical Bulletin* **29**: 172–178.

Houk, J.C., Adams, J.L. and Barto, A.C. (1995) A model of how the basal ganglia generates and uses neural signals that predict reinforcement. Ch. 13, pp. 249–270 in *Models of Information Processing in the Basal Ganglia*, eds. J.C.Houk, J.L.Davies and D.G.Beiser. MIT Press: Cambridge, Mass.

Huang, Y.Y., Kandel, E.R., Varshavsky, L., Brandon, E.P., Qi, M., Idzerda, R.L., McKnight, G.S. and Bourtchouladze, R. (1995) A genetic test of the effects of mutations in PKA on mossy fiber LTP and its relation to spatial and contextual learning. *Cell* **83**: 1211–22.

Hubel, D.H. and Wiesel, T.N. (1962) Receptive fields, binocular interaction, and functional architecture in the cat's visual cortex. *Journal of Physiology* **160**: 106–154.

Hummel, J.E. and Biederman, I. (1992) Dynamic binding in a neural network for shape recognition. *Psychological Review* **99**: 480–517.

Hyvarinen, J. (1981) Regional distribution of functions in parietal association area 7 of monkeys. *Brain Research* **206**: 287–303.

Insausti, R., Amaral, D.G. and Cowan, W.M. (1987) The entorhinal cortex of the monkey. II. Cortical afferents. *Journal of Comparative Neurology* **264**: 356–395.

Ishizuka, N., Weber, J. and Amaral, D.G. (1990) Organization of intrahippocampal projections originating from CA3 pyramidal cells in the rat. *Journal of Comparative Neurology* **295**: 580–623.

Ito, M. (1970) Neurophysiological aspects of the cerebellar motor control system. *International Journal of Neurology* **7**: 162–176.

Ito, M. (1972) Neural design of the cerebellar motor control system. *Brain Research* **40**: 81–84.

Ito, M. (1974) Central mechanisms of cerebellar motor system. Pp. 293–303 in *The Neurosciences, 3rd Study Program*, eds. F.O.Schmidt and F.G.Worden. MIT Press: Cambridge, Mass.

Ito, M. (1979) Neuroplasticity. Is the cerebellum really a computer ? *Trends in Neurosciences* **2**: 122–126.

Ito, M. (1982) Experimental verification of Marr–Albus' plasticity assumption for the cerebellum. *Acta Biologica Academiae Scientiarum Hungaricae* **33**: 189–199.

Ito, M. (1984) Neuronal network model. Ch. 10, pp. 115–130 in *The Cerebellum and Neural Control*. Raven Press: New York.

Ito, M. (1989) Long-term depression. *Annual Review of Neuroscience* **12**: 85–102.

Ito, M. (1993a) Synaptic plasticity in the cerebellar cortex and its role in motor learning. *Canadian Journal of Neurological Science, Suppl.* **3**: S70–74.

Ito, M. (1993b) Cerebellar mechamisms of long-term depression. Ch. 6, pp. 117–128 in *Synaptic Plasticity: Molecular, Cellular and Functional Aspects*, eds. M.Baudry, R.F.Thompson and J.L.Davis. MIT Press: Cambridge, Mass.

Ito, M. and Miyashita, Y. (1975) The effects of chronic destruction of the inferior olive upon visual modification of the horizontal vestibulo-ocular reflex of rabbits. *Proceedings of the Japan Academy* **51**: 716–720.

Ito, M., Jastreboff, P.J. and Miyashita, Y. (1982) Specific effects of unilateral lesions in the flocculus upon eye movements in albino rabbits. *Experimental Brain Research* **45**: 233–242.

Ito, M., Sakurai, M. and Tongroach, P. (1982) Climbing fibre induced depression of both mossy fibre responsiveness and glutamate sensitivity of cerebellar Purkinje cells. *Journal of Physiology* **324**: 113–134.

Iversen, S.D. (1979) Behaviour after neostriatal lesions in animals. Pp. 195–210 in *The Neostriatum*, ed. I.Divac and R.G.E.Oberg. Pergamon: Oxford.

Iversen, S.D. (1984) Behavioural effects of manipulation of basal ganglia neurotransmitters. Pp. 183–195 in *Functions of the Basal Ganglia*, Ciba Symposium 107. Pitman: London.

Iwamura, Y. (1993) Dynamic and hierarchical processing in the monkey somatosensory cortex. *Biomedical Research* **14**, *S4*: 107–111.

Jacoby, L.L. (1983a) Perceptual enhancement: persistent effects of an experience. *Journal of Experimental Psychology: Learning, Memory, and Cognition* **9**: 21–38.

Jacoby, L.L. (1983b) Remembering the data: analyzing interaction processes in reading. *Journal of Verbal Learning and Verbal Behavior* **22**: 485–508.

Jarrard, E.L. (1993) On the role of the hippocampus in learning and memory in the rat. *Behavioral and Neural Biology* **60**: 9–26.

Johnstone, S. and Rolls, E.T. (1990) Delay, discriminatory, and modality specific neurons in striatum and pallidum during short-term memory tasks. *Brain Research* **522**: 147–151.

Jones, B. and Mishkin, M. (1972) Limbic lesions and the problem of stimulus-reinforcement associations. *Experimental Neurology* **36**: 362–377.

Jones, E.G. (1981) Anatomy of cerebral cortex: columnar input-output organization. Pp. 199–235 in *The Organization of the Cerebral Cortex*, eds. F.O.Schmitt, F.G.Worden, G.Adelman and S.G.Dennis. MIT Press: Cambridge, Mass.

Jones, E.G. and Peters, A. (1984) (ed.) *Cerebral Cortex, Vol.* **2**, *Functional Properties of Cortical Cells*. Plenum: New York.

Jones, E.G. and Powell, T.P.S. (1970) An anatomical study of converging sensory pathways within the cerebral cortex of the monkey. *Brain* **93**: 793–820.

Jones, E.G., Coulter, J.D., Burton, H. and Porter, R. (1977) Cells of origin and terminal distribution of corticostriatal fibres arising in sensory motor cortex of monkeys. *Journal of Comparative Neurology* **181**: 53–80.

Jordan, M.I. (1986) An introduction to linear algebra in parallel distributed processing. Ch. 9, pp. 365–442 in *Parallel Distributed Processing*, Vol. 1, Foundations, eds. D.E.Rumelhart and J.L.McClelland. MIT Press: Cambridge, Mass.

Jung, M.W. and McNaughton, B.L. (1993) Spatial selectivity of unit activity in the hippocampal granular layer. *Hippocampus* **3**: 165–182.

Kaas, J.H. (1993) The functional organization of the somatosensory cortex in primates. *Anatomischer Anzeiger* **175**: 509–518.

Kammen, D.M. and Yuille, A.L. (1988) Spontaneous symmetry-breaking energy functions and the emergence of orientation selective cortical cells. *Biological Cybernetics* **59**: 23–31.

Kandel, E.R. (1991) Cellular mechanisms of learning. Ch. 65, pp. 1009–1031 in *Principles of Neural Science*, 3rd edn., eds. E.R.Kandel, J.H.Schwartz and T.H.Jessel. Elsevier: Amsterdam.

Kandel, E.R., Schwartz, J.H. and Jessel, T.H. (eds.) (1991) *Principles of Neural Science*, 3rd edn. Elsevier: Amsterdam.

Kandel, E.R. and Jessel, T.H. (1991) Touch. Ch. 26, pp. 367–384 in *Principles of Neural Science*, 3rd edn., eds. E.R.Kandel, J.H.Schwartz and T.H.Jessel. Elsevier: Amsterdam.

Kanter, I. and Sompolinsky, H. (1987) Associative recall of memories without errors. *Physical Review A* **35**: 380–392.

Kemp, J.M. and Powell, T.P.S. (1970) The cortico-striate projections in the monkey. *Brain* **93**: 525–546.

Kievit, J. and Kuypers, H.G.J.M. (1975) Subcortical afferents to the frontal lobe in the rhesus monkey studied by means of retrograde horseradish peroxidase transport. *Brain Research* **85**: 261–266.

Kirkwood, A., Dudek, S.M., Gold, J.T., Aizenman, C.D. and Bear, M.F. (1993) Common forms of synaptic plasticity in the hippocampus and neocortex *in vitro*. *Science* **260**: 1518–1521.

Kjaer, T.W., Hertz, J.A. and Richmond, B.J. (1994) Decoding cortical neuronal signals: networks models, information estimation and spatial tuning. *Journal of Computational Neuroscience* **1**: 109–139.

Kleinfeld, D. (1986) Sequential state generation by model neural networks. *Proceedings of the National Academy of Sciences of the USA* **83**: 9469–9473.

Kluver, H. and Bucy, P.C. (1939) Preliminary analysis of functions of the temporal lobes in monkeys. *Archives of Neurology and Psychiatry* **42**: 979–1000.

Koch, C. and Segev, I. (ed.) (1989) *Methods in Neuronal Modelling*. MIT Press: Cambridge, MA.

Koch, C., Bernander, O. and Douglas, R.J. (1995) Do neurons have a voltage or a current threshold for action potential initiation? *Journal of Computational Neuroscience* **2**: 63–82.

Koenderink, J.J. (1990) *Solid Shape*. MIT Press: Cambridge, Mass.

Koenderink, J.J. and Van Doorn, A.J. (1979) The internal representation of solid shape with respect to vision. *Biological Cybernetics* **32**: 211–217.

Kohonen, T. (1977) *Associative Memory: a System Theoretical Approach*. New York: Springer.

Kohonen, T. (1982) Clustering, taxonomy, and topological maps of patterns. Pp. 114–125 in *Proceedings of the Sixth International Conference on Pattern Recognition*, ed. M.Lang. IEEE Computer Society Press: Silver Spring, MD.

Kohonen, T. (1989) *Self-Organization and Associative Memory*, (3rd edn.). Springer-Verlag: Berlin (1984, 1st edn.; 1988, 2nd edn.)

Kohonen, T. (1995) *Self-Organizing Maps*. Springer-Verlag: Berlin.

Kolb, B. and Whishaw, I.Q. (1996) *Fundamentals of Human Neuropsychology*, 4th edn. Freeman: New York.

Kosslyn, S.M. (1994) *Image and Brain: the Resolution of the Imagery Debate*. MIT Press: Cambridge, Mass.

Krettek, J.E. and Price, J.L. (1974) A direct input from the amygdala to the thalamus and the cerebral cortex. *Brain Research* **67**: 169–174.

Krettek, J.E. and Price, J.L. (1977) The cortical projections of the mediodorsal nucleus and adjacent thalamic nuclei in the rat. *Journal of Comparative Neurology* **171**: 157–192.

Kubie, J.L. and Muller, R.U. (1991) Multiple representations in the hippocampus. *Hippocampus* **1**: 240–242.

Kunzle, H. (1975) Bilateral projections from precentral motor cortex to the putamen and other parts of the basal ganglia. *Brain Research* **88**: 195–209.

Kunzle, H. (1977) Projections from primary somatosensory cortex to basal ganglia and thalamus in the monkey. *Experimental Brain Research* **30**: 481–482.

Kunzle, H. (1978) An autoradiographic analysis of the efferent connections from premotor and adjacent prefrontal regions (areas 6 and 9) in *Macaca fascicularis*. *Brain Behavior and Evolution* **15**: 185–234.

Kunzle, H. and Akert, K. (1977) Efferent connections of area 8 (frontal eye field) in *Macaca fascicularis*. *Journal of Comparative Neurology* **173**: 147–164.

Lanthorn, T., Storn, J. and Andersen, P. (1984) Current-to-frequency transduction in CA1 hippocampal pyramidal cells: slow prepotentials dominate the primary range firing. *Experimental Brain Research,* **53**: 431–443.

Larkman, A.U. and Jack, J.J.B. (1995) Synaptic plasticity: hippocampal LTP. *Current Opinion in Neurobiology* **5**: 324–334.

LeDoux, J.E. (1994) Emotion, memory and the brain. *Scientific American* **220** *(June)*: 50–57.

Le Doux, J.E., Iwata, J., Cicchetti, J.P. and Reis, D.J. (1988) Different projections of the central amygdaloid nucleus mediate autonomic and behavioral correlates of conditioned fear. *Journal of Neuroscience* **8**: 2517–29.

Leonard, C.M., Rolls, E.T., Wilson, F.A.W, Baylis, G.C. (1985) Neurons in the amygdala of the monkey with responses selective for faces. *Behavioural Brain Research* **15**: 159–176.

Leonard, B.W. and McNaughton, B.L. (1990) Spatial representation in the rat: conceptual, behavioral and neurophysiological perspectives. In *Neurobiology of Comparative Cognition*, eds. R.P.Kesner and D.S.Olton. Lawrence Erlbaum Associates: Hillsdale, NJ.

Levitt, J.B., Lund, J.S. and Yoshioka, T. (1996) Anatomical substrates for early stages in cortical processing of visual information in the macaque monkey. *Behavioural Brain Research* **76**: 5–19.

Levy, W.B. (1985) Associative changes in the synapse: LTP in the hippocampus. Ch. 1, pp. 5–33 in *Synaptic Modification, Neuron Selectivity, and Nervous System Organization*, eds. W.B.Levy, J.A.Anderson and S.Lehmkuhle. Erlbaum: Hillsdale, NJ.

Levy, W.B. and Desmond, N.L. (1985) The rules of elemental synaptic plasticity. Ch. 6, pp. 105–121 in *Synaptic Modification, Neuron Selectivity, and Nervous System Organization*, eds. W.B.Levy, J.A.Anderson and S.Lehmkuhle. Erlbaum: Hillsdale, NJ.

Levy, W.B., Colbert, C.M. and Desmond, N.L. (1990) Elemental adaptive processes of neurons and synapses: a statistical/computational perspective. Ch. 5, pp. 187–235 in *Neuroscience and Connectionist Theory*, eds. M.Gluck and D.Rumelhart. Erlbaum: Hillsdale, NJ.

Levy, W.B., Colbert, C.M. and Desmond, N.L. (1995) Another network model bites the dust: entorhinal inputs are no more than weakly excitatory in the hippocampal CA1 region. *Hippocampus* **5**: 137–140.

Levy, W.B., Wu, X. and Baxter, R.A. (1995) Unification of hippocampal function via computational/encoding considerations. In Proceedings of the Third Workshop on Neural Networks: from Biology to High Energy Physics, *International Journal of Neural Systems* **6** *(Suppl.)*: 71–80.

Levy, W.B. and Baxter, R.A. (1996) Energy efficient neural codes. *Neural Computation* **8**: 531–543.

Linsker, E. (1986) From basic network principles to neural architecture. *Proceedings of the National Academy of Science of the USA* **83**: 7508–7512, 8390–8394, 8779–8783.

Linsker, E. (1988) Self-organization in a perceptual network. *Computer (March)*: 105–117.

Lissauer, H. (1890) Ein Fall von Seelenblindt nebst einem Beitrage zur Theorie derselben. *Archiv fur Psychiatrie und Nervenkrankheiten* **21**: 222–270.

Little, W.A. (1974) The existence of persistent states in the brain. *Mathematical Bioscience* **19**: 101–120.

Logothetis, N.K., Pauls, J., Bulthoff, H.H. and Poggio, T. (1994) View-dependent object recognition by monkeys. *Current Biology* **4**: 401–414.

Lund, J.S. (1984) Spiny stellate neurons. Ch. 7, pp. 255–308 in *Cerebral Cortex, Vol.* **1**, *Cellular Components of the Cerebral Cortex*, eds. A.Peters and E.G.Jones. Plenum: New York.

MacGregor, R.J. (1987) *Neural and Brain Modelling*. Academic Press: San Diego.

MacKay, D.J.C. and Miller, K.D. (1990) Analysis of Linsker's simulation of Hebbian rules. *Neural Computation* **2**: 173–187.

MacKay, D.M. and McCulloch, W.S. (1952) The limiting information capacity of a neuronal link. *Bulletin of Mathematical Biophysics* **14**: 127–135.

Mackintosh, N.J. (1983) *Conditioning and Associative Learning*. Oxford University Press: Oxford.

Major, G., Evans, J.D. and Jack, J.J.B. (1994) Solutions for transients in arbitrarily branching cables. I. Voltage recording with a somatic shunt. *Journal of Biophysics* **65**: 423–449.

Malach, R. and Graybiel, A.M. (1987) The somatic sensory corticostriatal projection: patchwork of somatic sensory zones in the extra-striasomal matrix. Pp. 11–16 in *Sensory Consideration for Basal Ganglia Functions*, eds. J.S.Schneider and T.I.Lidsky. Haber: New York.

Malsburg, C. von der (1973) Self-organization of orientation-sensitive columns in the striate cortex. *Kybernetik* **14**: 85–100.

Malsburg, C. von der (1990) A neural architecture for the representation of scenes. Ch. 19, pp. 356–372 in *Brain Organization and Memory: Cells, Systems and Circuits*, eds. J.L.McGaugh, N.M.Weinberger and G.Lynch. Oxford University Press: New York.

Malsburg, C. von der and Schneider, W. (1986) A neural cocktail-party processor. *Biological Cybernetics* **54**: 29–40.

Markram, H. and Siegel, M. (1992) The inositol 1,4,5 triphosphate pathway mediates cholinergic potentiation of rat hippocampal neuronal responses to NMDA. *Journal of Physiology* **447**: 513–533.

Markram, H. and Tsodyks, M. (1996) Redistribution of synaptic efficacy between neocortical pyramidal neurons. *Nature* **382**: 807–810.

Marr, D. (1969) A theory of cerebellar cortex. *Journal of Physiology* **202**: 437–470.

Marr, D. (1970) A theory for cerebral cortex. *Proceedings of The Royal Society of London, Series B* **176**: 161–234.

Marr, D. (1971) Simple memory: a theory for archicortex. *Philosophical Transactions of The Royal Society of London, Series B* **262**: 23–81.

Marr, D. (1982) *Vision*. Freeman: San Francisco.

Marshall, J.P., Richardson, J.S. and Teitelbaum, P. (1974) Nigrostriatal bundle damage and the lateral hypothalamic syndrome. *Journal of Comparative and Physiological Psychology* **87**: 808–830.

Martin, K.A.C. (1984) Neuronal circuits in cat striate cortex. Ch. 9, pp. 241–284 in *Cerebral Cortex, Vol 2, Functional Properties of Cortical Cells*, eds. E.G. Jones and A.Peters. Plenum: New York.

Mason, A. and Larkman, A. (1990) Correlations between morphology and electrophysiology of pyramidal neurones in slices of rat visual cortex. I. Electrophysiology. *Journal of Neuroscience* **10**: 1415–1428.

Maunsell, J.H.R. (1995) The brain's visual world: representation of visual targets in cerebral cortex. *Science* **270**: 764–769.

McCarthy, R.A. and Warrington, E.K. (1990) *Cognitive Neuropsychology*. Academic Press: London.

McClelland, J.L. and Rumelhart, D.E. (1986) A distributed model of human learning and memory. Ch. 17, pp. 170–215 in *Parallel Distributed Processing*, Vol. 2, eds. J.L.McClelland and D.E.Rumelhart. MIT Press: Cambridge, Mass.

McClelland, J.L. and Rumelhart, D.E. (1988) *Explorations in Parallel Distributed Processing*. MIT Press: Cambridge, Mass.

McClelland, J.L., McNaughton, B.L. and O'Reilly, R.C. (1995). Why there are complementary learning systems in the hippocampus and neocortex: insights from the successes and failures of connectionist models of learning and memory. *Psychological Review* **102**: 419–457.

McDonald, A.J. (1992) Cell types and intrinsic connections of the amygdala. Ch. 2, pp. 67–96 in *The Amygdala*, ed. J.P.Aggleton. Wiley-Liss: New York.

McEliece, R.J., Posner, E.C., Rodemich, E.R. and Venkatesh, S.S. (1987) The capacity of the Hopfield associative memory. *IEEE Trans. IT* **33**: 461.

McGinty, D. and Szymusiak, R. (1988) Neuronal unit activity patterns in behaving animals: brainstem and limbic system. *Annual Review of Psychology* **39**: 135–168.

McGurk, H. and MacDonald, J. (1976) Hearing lips and seeing voices. *Nature* **264**: 746–748.

McLeod, P., Plunkett, K. and Rolls, E.T. (1998) *Introduction to Connectionist Modelling of Cognitive Processes*. Oxford University Press: Oxford.

McNaughton, B.L., Barnes, C.A. and O'Keefe, J. (1983). The contributions of position, direction, and velocity to single unit activity in the hippocampus of freely-moving rats. *Experimental Brain Research* **52**: 41–49.

McNaughton, B.L., Barnes, C.A., Meltzer, J. and Sutherland, R.J. (1989) Hippocampal granule cells are necessary for normal spatial learning but not for spatially selective pyramidal cell discharge. *Experimental Brain Research* **76**: 485–496.

McNaughton, B.L. and Nadel, L. (1990) Hebb–Marr networks and the neurobiological representation of action in space. Pp. 1–64 in *Neuroscience and Connectionist Theory*, eds. M.A.Gluck and D.E.Rumelhart. Erlbaum: Hillsdale, NJ.

McNaughton, B.L., Chen, L.L. and Markus, E.J. (1991) 'Dead reckoning', landmark learning, and the sense of direction: a neurophysiological and computational hypothesis. *Journal of Cognitive Neuroscience* **3**: 190–202.

McNaughton, B.L., Knierim, J.J. and Wilson, M.A. (1995) Vector encoding and the vestibular foundations of spatial cognition: neurophysiological and computational mechanisms. Pp. 585–596 in *The Cognitive Neurosciences*, ed. M.S.Gazzaniga. MIT Press: Cambridge, Mass.

McNaughton, B.L., Barnes, C.A., Gerrard, J.L., Gothard, K., Jung, M.W., Knierim, J.J., Kudrimoti, H., Qin, Y., Skaggs, W.E., Suster, M. and Weaver, K.L. (1996) Deciphering the hippocampal polyglot: the hippocampus as a path integration system. *Journal of Experimental Biology* **199**: 173–185.

Medin, D.L. and Schaffer, M.M. (1978) Context theory of classification learning. *Psychological Review* **85**: 207–238.

Mesulam, M.-M. (1990) Human brain cholinergic pathways. *Progress in Brain Research* **84**: 231–241.

Mesulam, M.-M. and Mufson, E.J. (1982) Insula of the old world monkey. III. Efferent cortical output and comments on function. *Journal of Comparative Neurology* **212**: 38–52.

Mezard, M., Parisi, G. and Virasoro, M. (1987) *Spin Glass Theory and Beyond*. World Scientific: Singapore.

Middleton, F.A. and Strick, P.L. (1994) Anatomical evidence for cerebellar and basal ganglia involvement in higher cognitive function. *Science* **266**: 458–461.

Middleton, F.A. and Strick, P.L. (1996a) The temporal lobe is a target of output from the basal ganglia. *Proceedings of the National Academy of Sciences of the USA* **93**: 8683–8687.

Middleton, F.A. and Strick, P.L. (1996b) New concepts about the organization of the basal ganglia. In *Advances in Neurology: The Basal Ganglia and the Surgical Treatment for Parkinson's Disease*, ed. J.A.Obeso. Raven: New York.

Miles, R. (1988) Plasticity of recurrent excitatory synapses between CA3 hippocampal pyramidal cells. *Society for Neuroscience Abstracts* **14**: 19.

Millenson, J.R. (1967) *Principles of Behavioral Analysis*. MacMillan: New York.

Miller, E.K. and Desimone, R. (1994) Parallel neuronal mechanisms for short-term memory. *Science* **263**: 520–522.

Miller, G.A. (1955) Note on the bias of information estimates. *Information Theory in Psychology; Problems and Methods II-B*: 95–100.

Millhouse, O.E. (1986) The intercalated cells of the amygdala. *Journal of Comparative Neurology* **247**: 246–271.

Millhouse, O.E. and DeOlmos, J. (1983) Neuronal configuration in lateral and basolateral amygdala. *Neuroscience* **10**: 1269–1300.

Milner, B. and Petrides, M. (1984) Behavioral effects of frontal lobe lesions in man. *Trends in Neurosciences* **7**: 403–407.

Minsky, M.L. and Papert, S.A. (1969) *Perceptrons*. MIT Press: Cambridge, Mass. (expanded edition 1988).

Mishkin, M. (1978) Memory severely impaired by combined but not separate removal of amygdala and hippocampus. *Nature* **273**: 297–298.

Mishkin, M. (1982) A memory system in the monkey. *Philosophical Transactions of The Royal Society of London, Series B* **298**: 85–95.

Mishkin, M. and Aggleton, J. (1981) Multiple functional contributions of the amygdala in the monkey. Pp. 409–420 in *The Amygdaloid Complex*, ed. Y.Ben-Ari. Elsevier: Amsterdam.

Miyashita, Y. (1993) Inferior temporal cortex: where visual perception meets memory. *Annual Review of Neuroscience* **16**: 245–263.

Miyashita, Y. and Chang, H.S. (1988) Neuronal correlate of pictorial short-term memory in the primate temporal cortex. *Nature* **331**: 68–70.

Miyashita, Y., Rolls, E.T., Cahusac, P.M.B., Niki, H. and Feigenbaum, J.D. (1989) Activity of hippocampal neurons in the monkey related to a conditional spatial response task. *Journal of Neurophysiology* **61**: 669–678.

Monaghan, D.T. and Cotman, C.W. (1985) Distribution on N-methyl-D-aspartate-sensitive L-[3H]glutamate-binding sites in the rat brain. *Journal of Neuroscience* **5**: 2909–2919.

Monasson, R. (1992) Properties of neural networks storing spatially correlated patterns. *Journal of Physics A* **25**: 3701–3720.

Mora, F., Rolls, E.T. and Burton, M.J. (1976) Modulation during learning of the responses of neurones in the lateral hypothalamus to the sight of food. *Experimental Neurology* **53**: 508–519.

Mora, F., Mogenson, G.J. and Rolls, E.T. (1977) Activity of neurones in the region of the substantia nigra during feeding. *Brain Research* **133**: 267–276.

Morecraft, R.J., Geula, C. and Mesulam, M.M. (1992) Cytoarchitecture and neural afferents of the orbitofrontal cortex in the brain of the monkey. *Journal of Comparative Neurology* **323**: 341–358.

Morris, R.G.M. (1989) Does synaptic plasticity play a role in information storage in the vertebrate brain? Ch. 11, pp. 248–285 in *Parallel Distributed Processing: Implications for Psychology and Neurobiology*, ed. R.G.M.Morris. Oxford University Press: Oxford.

Morris, R.G.M. (1996) Spatial memory and the hippocampus: the need for psychological analyses to identify the information processing underlying spatial learning. Ch. 22, pp. 319–342 in *Perception, Memory and Emotion: Frontiers in Neuroscience*, eds. T.Ono, B.L.McNaughton, S.Molotchnikoff, E.T.Rolls and H.Nishijo. Elsevier: Amsterdam.

Mountcastle, V.B. (1984) Central nervous mechanisms in mechanoreceptive sensibility. Pp. 789–878 in *Handbook of Physiology, Section 1: The Nervous System, Vol. III, Sensory Processes*, Part 2, ed. I.Darian-Smith. American Physiological Society: Bethesda, MD.

Muir, J.L, Everitt, B.J. and Robbins, T.W. (1994) AMPA-induced excitotoxic lesions of the basal forebrain: a significant role for the cortical cholinergic system in attentional function. *Journal of Neuroscience* **14**: 2313–1326.

Mulkey, R.M. and Malenka, R.C. (1992) Mechanisms underlying induction of homosynaptic long-term depression in area CA1 of the hippocampus. *Neuron* **9**: 967–975.

Muller, R.U., Kubie, J.L., Bostock, E.M., Taube, J.S. and Quirk, G.J. (1991) Spatial firing correlates of neurons in the hippocampal formation of freely moving rats. Pp. 296–333 in *Brain and Space*, ed. J.Paillard. Oxford University Press: Oxford.

Murre, J.M.J. (1996) Tracelink: a model of amnesia and consolidation of memory. *Hippocampus* **6**: 675–684.

Nadal, J.-P. (1991) Associative memory: on the (puzzling) sparse coding limit. *Journal of Physics A* **24**: 1093–1102.

Nadal, J.-P., Toulouse, G., Changeux, J.-P. and Dehaene, S. (1986) Networks of formal neurons and memory palimpsests. *Europhysics Letters* **1**: 535–542.

Nadal, J.-P. and Toulouse, G. (1990) Information storage in sparsely coded memory nets. *Network* **1**: 61–74.

Nauta, W.J.H. (1964) Some efferent connections of the prefrontal cortex in the monkey. Pp. 397–407 in *The Frontal Granular Cortex and Behavior*, eds. J.M.Warren and K.Akert. McGraw Hill: New York.

Nauta, W.J.H. and Domesick, V.B. (1978) Crossroads of limbic and striatal circuitry: hypothalamo-nigral connections. Pp.75–93 in *Limbic Mechanisms*, ed. K.E.Livingston and O.Hornykiewicz. Plenum: New York.

Naya, Y., Sakai, K. and Miyashita, Y. (1996) Activity of primate inferotemporal neurons related to a sought target in pair-association task. *Proceedings of the National Academy of Sciences of the USA* **93**: 2664–2669.

Nelken, I., Prut, Y., Vaadia, E. and Abeles, M. (1994) Population responses to multi-frequency sounds in the cat auditory cortex: one- and two-parameter families of sounds. *Hearing Research* **72**: 206–222.

Nicoll, R.A. and Malenka, R.C. (1995) Contrasting properties of two forms of long-term potentiation in the hippocampus. *Nature* **377**: 115–118.

Nilsson, N.J. (1965) *Learning Machines*. McGraw-Hill: New York.

Nishijo, H., Ono, T. and Nishino, H. (1988) Single neuron responses in amygdala of alert monkey during complex sensory stimulation with affective significance. *Journal of Neuroscience* **8**: 3570–83.

O'Kane, D. and Treves, A. (1992) Why the simplest notion of neocortex as an autoassociative memory would not work. *Network* **3**: 379–384.

O'Keefe, J. (1979) A review of the hippocampal place cells. *Progress in Neurobiology* **13**: 419–439.

O'Keefe, J. (1983) Spatial memory within and without the hippocampal system. Pp. 375–403 in *Neurobiology of the Hippocampus*, ed. W.Seifert. Academic Press: London. -

O'Keefe, J. (1990) A computational theory of the cognitive map. *Progress in Brain Research* **83**: 301–312.

O'Keefe, J. (1991) The hippocampal cognitive map and navigational strategies. Ch. 16, pp. 273–295 in *Brain and Space*, ed. J.Paillard. Oxford University Press: Oxford.

O'Keefe, J. and Nadel, L. (1978). *The Hippocampus as a Cognitive Map*. Clarendon Press: Oxford.

O'Mara, S.M., Rolls, E.T., Berthoz, A. and Kesner, R.P. (1994) Neurons responding to whole-body motion in the primate hippocampus. *Journal of Neuroscience* **14**: 6511–6523.

O'Reilly, R.C. (1996) Biologically plausible error-driven learning using local activation differences: the generalized recirculation algorithm. *Neural Computation* **8**: 895–938.

Oberg, R.G.E and Divac,I (1979) 'Cognitive' functions of the neostriatum. Pp. 291–313 in *The Neostriatum*, eds. I.Divac and R.G.E.Oberg. Pergamon: New York.

Oja, E. (1982) A simplified neuron model as a principal component analyzer. *Journal of Mathematical Biology* **15**: 267–273.

Olshausen, B.A., Anderson, C. H. and Van Essen, D.C. (1993) A neurobiological model of visual attention and invariant pattern recognition based on dynamic routing of information. *Journal of Neuroscience* **13**: 4700–4719.

Olshausen, B.A., Anderson, C.H. and Van Essen, D.C. (1995) A multiscale dynamic routing circuit for forming size- and position-invariant object representations. *Journal of Computational Neuroscience* **2**: 45–62.

Olshausen, B.A. and Field, D.J. (1996) Emergence of simple-cell receptive field properties by learning a sparse code for natural images. *Nature* **381**: 607–609.

Ono, T., Nishino, H., Sasaki, K., Fukuda, M. and Muramoto, K. (1980) Role of the lateral hypothalamus and amygdala in feeding behavior. *Brain Research Bulletin (Suppl. 4)* **5**: 143–149.

Ono, T. and Nishijo, H. (1992) Neurophysiological basis of the Kluver–Bucy syndrome: responses of monkey amygdaloid neurons to biologically significant objects. Pp. 167–190 in *The Amygdala*, ed. J.P.Aggleton. Wiley: Chichester.

Ono, T., Tamura, R., Nishijo, H. and Nakamura, K. (1993) Neural mechanisms of recognition and memory in the limbic system. Ch. 19, pp. 330–355 in *Brain Mechanisms of Perception and Memory: From Neuron to Behavior*, eds. T.Ono, L.R.Squire, M.E.Raichle, D.I.Perrett and M.Fukuda. Oxford University Press: New York.

Optican, L.M. and Richmond, B.J. (1987) Temporal encoding of two-dimensional patterns by single units in primate inferior temporal cortex: III. Information theoretic analysis. *Journal of Neurophysiology* **57**: 162–178.

Optican, L.M., Gawne, T.J., Richmond, B.J. and Joseph, P.J. (1991) Unbiased measures of transmitted information and channel capacity from multivariate neuronal data. *Biological Cybernetics* **65**: 305–310.

Panzeri, S. and Treves, A. (1996) Analytical estimates of limited sampling biases in different information measures. *Network* **7**: 87–107.

Panzeri, S., Booth, M., Wakeman, E.A., Rolls, E.T. and Treves, A. (1996a) Do firing rate distributions reflect anything beyond just chance? *Society for Neuroscience Abstracts* **22**: 445.5.

Panzeri, S., Biella, G., Rolls, E.T., Skaggs, W.E. and Treves, A. (1996b) Speed, noise, information and the graded nature of neuronal responses. *Network* **7**: 365–370.

Parga, N. and Virasoro, M.A. (1986) The ultrametric organization of memories in a neural network. *Journal de Physique* **47**: 1857.

Parga, N. and Rolls, E.T. (1997) Transform invariant recognition by association in a recurrent network. *Neural Computation*.

Parisi, G. (1986) A memory which forgets. *Journal of Physics A* **19**: L617–619.

Parkinson, J.K., Murray, E.A. and Mishkin, M. (1988) A selective mnemonic role for the hippocampus in monkeys: memory for the location of objects. *Journal of Neuroscience* **8**: 4059–4167.

Passingham, R.E. (1993) *The Frontal Lobes and Voluntary Action*. Oxford University Press: Oxford.

Pennartz, C.M., Ameerun, R.F., Groenewegen, H.J. and Lopes da Silva, F.H. (1993) Synaptic plasticity in an *in vitro* slice preparation of the rat nucleus accumbens. *European Journal of Neuroscience* **5**: 107–117.

Percheron, G., Yelnik, J. and François, C. (1984a) A Golgi analysis of the primate globus pallidus. III. Spatial organization of the striato-pallidal complex. *Journal of Comparative Neurology* **227**: 214–227.

Percheron, G., Yelnik, J. and François, C. (1984b) The primate striato-pallido-nigral system: an integrative system for cortical information. Pp. 87–105 in *The Basal Ganglia: Structure and Function*, ed. J.S.McKenzie, R.E.Kemm and L.N.Wilcox. Plenum: New York.

Percheron, G., Yelnik, J., François, C., Fenelon, G. and Talbi, B. (1994) Informational neurology of the basal ganglia related system. *Revue Neurologique (Paris)* **150**: 614–626.

Perez-Vicente, C.J. and Amit, D.J. (1989) Optimized network for sparsely coded patterns. *Journal of Physics A* **22**: 559–569.

Perrett, D.I., Rolls, E.T. and Caan, W. (1982) Visual neurons responsive to faces in the monkey temporal cortex. *Experimental Brain Research* **47**: 329–342.

Perrett, D.I., Smith, P.A.J., Potter, D.D., Mistlin, A.J., Head, A.S., Milner, D. and Jeeves, M.A. (1985) Visual cells in temporal cortex sensitive to face view and gaze direction. *Proceedings of the Royal Society of London, Series B* **223**: 293–317.

Perrett, D.I., Mistlin, A.J. and Chitty, A.J. (1987) Visual neurons responsive to faces. *Trends in Neuroscience* **10**: 358–364.

Personnaz, L., Guyon, I. and Dreyfus, G. (1985) Information storage and retrieval in spin-glass-like neural networks. *Journal de Physique Lettres (Paris)* **46**: 359–365.

Peters, A. (1984a) Chandelier cells. Ch. 10, pp. 361–380 in *Cerebral Cortex, Vol. 1, Cellular Components of the Cerebral Cortex*, eds. A.Peters and E.G.Jones. Plenum: New York.

Peters, A. (1984b) Bipolar cells. Ch. 11, pp. 381–407 in *Cerebral Cortex, Vol. 1, Cellular Components of the Cerebral Cortex*, eds. A.Peters and E.G.Jones. Plenum: New York.

Peters, A. and Jones, E.G. (1984) (ed.) *Cerebral Cortex, Vol. 1, Cellular Components of the Cerebral Cortex*, Plenum: New York.

Peters, A. and Regidor, J. (1981) A reassessment of the forms of nonpyramidal neurons in area 17 of the cat visual cortex. *Journal of Comparative Neurology* **203**: 685–716.

Peters, A. and Saint Marie, R.L. (1984) Smooth and sparsely spinous nonpyramidal cells forming local axonal plexuses. Ch. 13, pp. 419–445 in *Cerebral Cortex, Vol. 1, Cellular Components of the Cerebral Cortex*, eds. A.Peters and E.G.Jones. Plenum: New York.

Peterson, C. and Anderson, J.R. (1987) A mean field theory learning algorithm for neural networks. *Complex Systems* **1**: 995–1015.

Petri, H.L. and Mishkin, M. (1994) Behaviorism, cognitivism, and the neuropsychology of memory. *American Scientist* **82**: 30–37.

Petrides, M. (1985) Deficits on conditional associative-learning tasks after frontal- and temporal-lobe lesions in man. *Neuropsychologia* **23**: 601–614.

Petrides, M. (1991) Functional specialization within the dorsolateral frontal cortex for serial order memory. *Proceedings of the Royal Society of London, Series B* **246**: 299–306.

Phillips, R.R., Malamut, B.L., Bachevalier, J. and Mishkin, M. (1988) Dissociation of the effects of inferior temporal and limbic lesions on object discrimination learning with 24-h intertrial intervals. *Behavioural Brain Research* **27**: 99–107.

Phillips, A.G. and Fibiger, H.C. (1990) Role of reward and enhancement of conditioned reward in persistence of responding for cocaine. *Behavioral Pharmacology* **1**: 269–282.

Phillips, W.A., Kay, J. and Smyth, D. (1995) The discovery of structure by multi-stream networks of local processors with contextual guidance. *Network* **6**: 225–246.

Poggio, T. and Edelman, S. (1990) A network that learns to recognize three-dimensional objects. *Nature* **343**: 263–266.

Polk, T.A. and Farah, M.J. (1995) Brain localization for arbitrary stimulus categories: a simple account based on Hebbian learning. *Proceedings of the National Academy of Sciences of the USA* **92**: 12370–12373.

Posner, M.I. and Keele, S.W. (1968) On the genesis of abstract ideas. *Journal of Experimental Psychology* **77**: 353–363.

Powell, T.P.S. (1981) Certain aspects of the intrinsic organisation of the cerebral cortex. Pp. 1–19 in *Brain Mechanisms and Perceptual Awareness*, eds. O.Pompeiano and C. Ajmone Marsan. Raven Press: New York.

Price, J.L., Carmichael, S.T., Carnes, K.M., Clugnet, M.-C. and Kuroda, M. (1991) Olfactory input to the prefrontal cortex. Pp. 101–120 in *Olfaction: a Model System for Computational Neuroscience*, eds. J.L.Davis and H.Eichenbaum. MIT Press: Cambridge, Mass.

Rall, W. (1959) Branching dendritic trees and motoneuron membrane resistivity. *Experimental Neurology* **1**: 491–527.

Rall, W. (1962) Theory of physiological properties of dendrites. *Annals of the New York Academy of Sciences* **96**: 1071–1092.

Rall, W. and Shepherd, G.M. (1968) Theoretical reconstruction of field potentials and dendrodendritic synaptic interactions in olfactory bulb. *Journal of Neurophysiology* **31**: 884–915.

Rall, W. and Rinzel, J. (1973) Branch input resistance and steady attenuation for input to one branch of a dendritic neuron model. *Biophysical Journal* **13**: 648–688.

Rall, W. and Segev, I. (1987) Functional possibilities for synapses on dendrites and dendritic spines. Pp. 605–636 in *Synaptic Function*, eds. G.M.Edelman, E.E.Gall and W.M.Cowan. Wiley: New York.

Raymond, J.L., Lisberger, S.G. and Mauk, M.D. (1996) The cerebellum: a neuronal learning machine? *Science* **272**: 1126–1131.

Richmond, B.J. and Optican, L. (1987) Temporal encoding of two-dimensional patterns by single units in primate inferior temporal cortex. II. Quantification of response waveform. *Journal of Neurophysiology* **57**: 147–161.

Richmond, B.J. and Optican, L. (1990) Temporal encoding of two dimensional patterns by single units in primate primary visual cortex. II. Information transmission. *Journal of Neurophysiology* **64**: 351–369.

Rieke, F., Warland, D. and Bialek, W. (1993) Coding efficiency and information rates in sensory neurons. *Europhysics Letters* **22**: 151–156.

Rieke, F., Warland, D., de Ruyter van Steveninck, R.R. and Bialek, W. (1996) *Spikes: Exploring the Neural Code*. MIT Press: Cambridge, MA.

Ripley, B.D. (1996) *Pattern Recognition and Neural Networks*. Cambridge University Press: Cambridge.

Robbins, T.W., Cador, M., Taylor, J.R. and Everitt, B.J. (1989) Limbic-striatal interactions in reward-related processes. *Neuroscience and Biobehavioral Reviews* **13**: 155–162.

Robinson, D.A. (1976) Adaptive gain control of vestibulo-ocular reflex by the cerebellum. *Journal of Neurophysiology* **39**: 954–969.

Roland, P.E. and Friberg, L. (1985) Localization of cortical areas activated by thinking. *Journal of Neurophysiology* **53**: 1219–1243.

Rolls, E.T. (1974) The neural basis of brain-stimulation reward. *Progress in Neurobiology* **3**: 71–160.

Rolls, E.T. (1975) *The Brain and Reward*. Pergamon Press: Oxford.

Rolls, E.T. (1976) The neurophysiological basis of brain-stimulation reward. Pp 65–87 in *Brain-Stimulation Reward*, eds. A.Wauquier and E.T.Rolls. North Holland: Amsterdam.

Rolls, E.T. (1979) Effects of electrical stimulation of the brain on behaviour. Pp. 151–169 in *Psychology Surveys* **2**. George Allen and Unwin: London.

Rolls, E.T. (1981a) Responses of amygdaloid neurons in the primate. Pp. 383–393 in *The Amygdaloid Complex*, ed. Y.Ben-Ari. Elsevier: Amsterdam.

Rolls, E.T. (1981b) Processing beyond the inferior temporal visual cortex related to feeding, learning, and striatal function. Ch. 16, pp. 241–269 in *Brain Mechanisms of Sensation*, eds. Y.Katsuki, R.Norgren and M.Sato. Wiley: New York.

Rolls, E.T. (1984a) Activity of neurons in different regions of the striatum of the monkey. Pp. 467–493 in *The Basal Ganglia: Structure and Function*, eds. J.S.McKenzie, R.E.Kemm and L.N.Wilcox. Plenum: New York.

Rolls, E.T. (1984b) Neurons in the cortex of the temporal lobe and in the amygdala of the monkey with responses selective for faces. *Human Neurobiology* **3**: 209–222.

Rolls, E.T. (1986) A theory of emotion, and its application to understanding the neural basis of emotion. Pp. 325–344 in *Emotions. Neural and Chemical Control*, ed. Y.Oomura. Japan Scientific Societies Press: Tokyo and Karger: Basel.

Rolls, E.T. (1987) Information representation, processing and storage in the brain: analysis at the single neuron level. Pp. 503–540 in *The Neural and Molecular Bases of Learning*, eds. J.-P.Changeux and M.Konishi. Wiley: Chichester.

Rolls, E.T. (1989a) Information processing in the taste system of primates. *Journal of Experimental Biology* **146**: 141–164.

Rolls, E.T. (1989b) Functions of neuronal networks in the hippocampus and neocortex in memory. Ch. 13, pp. 240–265 in *Neural Models of Plasticity: Experimental and Theoretical Approaches*, ed. J.H.Byrne and W.O.Berry. Academic Press: San Diego.

Rolls, E.T. (1989c) Parallel distributed processing in the brain: implications of the functional architecture of neuronal networks in the hippocampus. Ch. 12, pp. 286–308 in *Parallel Distributed Processing: Implications for Psychology and Neurobiology*, ed. R.G.M.Morris. Oxford University Press: Oxford.

Rolls, E.T. (1989d) Functions of neuronal networks in the hippocampus and cerebral cortex in memory. Pp. 15–33 in *Models of Brain Function*, ed. R.M.J.Cotterill. Cambridge University Press: Cambridge.

Rolls, E.T. (1989e) The representation and storage of information in neuronal networks in the primate cerebral cortex and hippocampus. Ch. 8, pp. 125–159 in *The Computing Neuron*, eds. R.Durbin, C.Miall and G.Mitchison. Addison-Wesley: Wokingham, UK.

Rolls, E.T. (1990a) Principles underlying the representation and storage of information in neuronal networks in the primate hippocampus and cerebral cortex. Ch. 4, pp. 73–90 in *An Introduction to Neural and Electronic Networks*, eds. S.F.Zornetzer, J.L.Davis and C.Lau. Academic Press: San Diego.

Rolls, E.T. (1990b) Theoretical and neurophysiological analysis of the functions of the primate hippocampus in memory. *Cold Spring Harbor Symposia in Quantitative Biology* **55**: 995–1006.

Rolls, E.T. (1990c) A theory of emotion, and its application to understanding the neural basis of emotion. *Cognition and Emotion* **4**: 161–190.

Rolls, E.T. (1990d) Functions of neuronal networks in the hippocampus and of back-projections in the cerebral cortex in memory. Ch. 9, pp. 184–210 in *Brain Organization and Memory: Cells, Systems and Circuits*, eds. J.L.McGaugh, N.M.Weinberger and G.Lynch. Oxford University Press: New York.

Rolls, E.T. (1990e) Functions of the primate hippocampus in spatial processing and memory. Ch. 12, pp. 339–362 in *Neurobiology of Comparative Cognition*, eds. D.S.Olton and R.P.Kesner. Lawrence Erlbaum: Hillsdale, N.J.

Rolls, E.T. (1991a) Functions of the primate hippocampus in spatial and non-spatial memory. *Hippocampus* **1**: 258–261.

Rolls, E.T. (1991b) Functions of the primate hippocampus in spatial processing and memory. Pp. 353–376 in *Brain and Space*, ed. J.Paillard. Oxford University Press: Oxford.

Rolls, E.T. (1991c) Neural organisation of higher visual functions. *Current Opinion in Neurobiology* **1**: 274–278.

Rolls, E.T. (1992a) Neurophysiology and functions of the primate amygdala. Ch. 5, pp. 143–165 in *The Amygdala*, ed. J.P.Aggleton. Wiley-Liss: New York.

Rolls, E.T. (1992b) Neurophysiological mechanisms underlying face processing within and beyond the temporal cortical visual areas. *Philosophical Transactions of the Royal Society* **335**: 11–21.

Rolls, E.T. (1992c) Networks in the brain. Ch. 4, pp. 103–120 in *The Simulation of Human Intelligence*, ed. D.E.Broadbent. Blackwell: Oxford.

Rolls, E.T. (1992d) The processing of face information in the primate temporal lobe. Ch. 3, pp. 41–68 in *Processing Images of Faces*, eds. V.Bruce and M.Burton. Ablex: Norwood, NJ.

Rolls, E.T. (1993) The neural control of feeding in primates. Ch. 9, pp. 137–169 in *Neurophysiology of Ingestion*, ed. D.A.Booth. Pergamon: Oxford.

Rolls, E.T. (1994a) Brain mechanisms for invariant visual recognition and learning. *Behavioural Processes* **33**: 113–138.

Rolls, E.T. (1994b) Neural processing related to feeding in primates. Ch. 2, pp. 11–53 in *Appetite: Neural and Behavioural Bases*, eds. C.R.Legg and D.A.Booth. Oxford University Press: Oxford.

Rolls, E.T. (1994c) Neurophysiological and neuronal network analysis of how the primate hippocampus functions in memory. Pp. 713–744 in *The Memory System of the Brain*, ed. J. Delacour. World Scientific: London.

Rolls, E.T. (1994d) Neurophysiology and cognitive functions of the striatum. *Revue Neurologique (Paris)* **150**: 648–660.

Rolls, E.T. (1995a) Central taste anatomy and neurophysiology. Ch. 24, pp. 549–573 in *Handbook of Olfaction and Gustation*, ed. R.L.Doty. Dekker: New York.

Rolls, E.T. (1995b) Learning mechanisms in the temporal lobe visual cortex. *Behavioural Brain Research* **66**: 177–185.

Rolls, E.T. (1995c) A model of the operation of the hippocampus and entorhinal cortex in memory. *International Journal of Neural Systems* **6**, *Suppl.*: 51–70.

Rolls, E.T. (1995d) A theory of emotion and consciousness, and its application to understanding the neural basis of emotion. Ch 72, pp. 1091–1106 in *The Cognitive Neurosciences*, ed. M.S.Gazzaniga. MIT Press: Cambridge, Mass.

Rolls, E.T. (1996a) Roles of long term potentiation and long term depression in neuronal network operations in the brain. Ch. 11, pp. 223–250 in *Cortical Plasticity: LTP and LTD*, eds. M.S.Fazeli and G.L.Collingridge. Bios: Oxford.

Rolls, E.T. (1996b) The orbitofrontal cortex. *Philosophical Transactions of the Royal Society of London, Series B* **351**: 1433–1444.

Rolls, E.T. (1996c) A theory of hippocampal function in memory. *Hippocampus* **6**: 601–620.

Rolls, E.T. (1996d) The representation of space in the primate hippocampus, and episodic memory. Ch. 25, pp. 375–400 in: *Perception, Memory and Emotion: Frontiers in Neuroscience*, eds. T.Ono, B.L.McNaughton, S.Molotchnikoff, E.T.Rolls and H.Nishijo. Elsevier: Amsterdam.

Rolls, E.T. (1996e) The representation of space in the primate hippocampus, and its relation to memory. Pp. 205–229 in: *Brain Processes and Memory*, eds. K.Ishikawa, J.L.McGaugh and H.Sakata. Elsevier: Amsterdam.

Rolls, E.T. (1997a) Brain mechanisms involved in perception and memory, and their relation to consciousness. Ch. 6, pp. 81–120 in: *Cognition, Computation, and Consciousness*, eds. M.Ito, Y.Miyashita and E.T.Rolls. Oxford University Press: Oxford.

Rolls, E.T. (1997b) Consciousness in Neural Networks? *Neural Networks*, in press.

Rolls, E.T. (1997c) Taste and olfactory processing in the brain. *Critical Reviews in Neurobiology*, in press.

Rolls, E.T. (1997d) A neurophysiological and computational approach to the functions of the temporal lobe cortical visual areas in invariant object recognition. Pp. 184–220, in *Computational and Psychophysical Mechanisms of Visual Coding*, eds. M.Jenkin and L.Harris. Cambridge University Press: Cambridge.

Rolls, E.T. and Rolls, B.J. (1973) Altered food preferences after lesions in the basolateral region of the amygdala in the rat. *Journal of Comparative and Physiological Psychology* **83**: 248–259.

Rolls, E.T., Burton, M.J. and Mora, F. (1976) Hypothalamic neuronal responses associated with the sight of food. *Brain Research* **111**: 53–66.

Rolls, E.T., Judge, S.J. and Sanghera, M. (1977) Activity of neurones in the inferotemporal cortex of the alert monkey. *Brain Research* **130**: 229–238.

Rolls, E.T., Sanghera, M.K. and Roper-Hall, A. (1979a) The latency of activation of neurons in the lateral hypothalamus and substantia innominata during feeding in the monkey. *Brain Research* **164**: 121–135.

Rolls, E.T., Thorpe, S.J., Maddison, S., Roper-Hall, A., Puerto, A. and Perrett, D. (1979b) Activity of neurones in the neostriatum and related structures in the alert animal. Pp. 163–182 in *The Neostriatum*, eds. I.Divac and R.G.E.Oberg. Pergamon Press: Oxford.

Rolls, E.T., Burton, M.J. and Mora, F. (1980) Neurophysiological analysis of brain-stimulation reward in the monkey. *Brain Research* **194**: 339–357.

Rolls, E.T., Perrett, D.I., Caan, A.W. and Wilson, F.A.W. (1982) Neuronal responses related to visual recognition. *Brain* **105**: 611–646.

Rolls, E.T., Thorpe, S.J. and Maddison, S.P. (1983) Responses of striatal neurons in the behaving monkey. 1. Head of the caudate nucleus. *Behavioural Brain Research* **7**: 179–210.

Rolls, E.T., Thorpe, S.J., Boytim, M., Szabo, I. and Perrett, D.I. (1984) Responses of striatal neurons in the behaving monkey. 3. Effects of iontophoretically applied dopamine on normal responsiveness. *Neuroscience* **12**: 1201–1212.

Rolls, E.T., Baylis, G.C. and Leonard, C.M. (1985) Role of low and high spatial frequencies in the face-selective responses of neurons in the cortex in the superior temporal sulcus. *Vision Research* **25**: 1021–1035.

Rolls, E.T. and Baylis, G.C. (1986) Size and contrast have only small effects on the responses to faces of neurons in the cortex of the superior temporal sulcus of the monkey. *Experimental Brain Research* **65**: 38–48.

Rolls, E.T. and Williams, G.V. (1987a) Sensory and movement-related neuronal activity in different regions of the striatum of the primate. Pp. 37–59 in *Sensory Considerations for Basal Ganglia Functions*, eds. J.S.Schneider and T.I.Lidsky. Haber: New York.

Rolls, E.T. and Williams, G.V. (1987b) Neuronal activity in the ventral striatum of the primate. Pp. 349–356 in *The Basal Ganglia II. Structure and Function — Current Concepts*, eds. M.B.Carpenter and A.Jayamaran. Plenum: New York.

Rolls, E.T., Baylis, G.C. and Hasselmo, M.E. (1987) The responses of neurons in the cortex in the superior temporal sulcus of the monkey to band-pass spatial frequency filtered faces. *Vision Research* **27**: 311–326.

Rolls, E.T., Scott, T.R., Sienkiewicz, Z.J. and Yaxley, S. (1988) The responsiveness of neurones in the frontal opercular gustatory cortex of the macaque monkey is independent of hunger. *Journal of Physiology* **397**: 1–12.

Rolls, E.T., Baylis, G.C., Hasselmo, M.E. and Nalwa, V. (1989) The effect of learning on the face-selective responses of neurons in the cortex in the superior temporal sulcus of the monkey. *Experimental Brain Research* **76**: 153–164.

Rolls, E.T., Miyashita, Y., Cahusac, P.M.B., Kesner, R.P., Niki, H., Feigenbaum, J. and Bach, L. (1989) Hippocampal neurons in the monkey with activity related to the place in which a stimulus is shown. *Journal of Neuroscience* **9**: 1835–1845.

Rolls, E.T., Sienkiewicz, Z.J. and Yaxley, S. (1989) Hunger modulates the responses to gustatory stimuli of single neurons in the caudolateral orbitofrontal cortex of the macaque monkey. *European Journal of Neuroscience* **1**: 53–60.

Rolls, E.T. and Treves, A. (1990) The relative advantages of sparse versus distributed encoding for associative neuronal networks in the brain. *Network* 1: 407–421.

Rolls, E.T., Yaxley, S. and Sienkiewicz, Z.J. (1990) Gustatory responses of single neurons in the orbitofrontal cortex of the macaque monkey. *Journal of Neurophysiology* 64: 1055–1066.

Rolls, E.T. and Johnstone, S. (1992) Neurophysiological analysis of striatal function. Ch. 3, pp. 61–97 in *Neuropsychological Disorders Associated with Subcortical Lesions*, eds. G.Vallar, S.F.Cappa and C.W.Wallesch. Oxford University Press: Oxford.

Rolls, E.T., Cahusac, P.M.B., Feigenbaum, J.D. and Miyashita, Y. (1993) Responses of single neurons in the hippocampus of the macaque related to recognition memory. *Experimental Brain Research* 93: 299–306.

Rolls, E.T. and O'Mara, S. (1993) Neurophysiological and theoretical analysis of how the hippocampus functions in memory. Ch. 17, pp. 276–300 in *Brain Mechanisms of Perception and Memory: From Neuron to Behavior*, eds. T.Ono, L.R.Squire, M.E.Raichle, D.I.Perrett and M.Fukuda. Oxford University Press: New York.

Rolls, E.T., Tovee, M.J. and Ramachandran, V.S. (1993) Visual learning reflected in the responses of neurons in the temporal visual cortex of the macaque. *Society for Neuroscience Abstracts* 19: 27.

Rolls, E.T. and Baylis, L.L. (1994) Gustatory, olfactory and visual convergence within the primate orbitofrontal cortex. *Journal of Neuroscience* 14: 5437–5452.

Rolls, E.T., Hornak, J., Wade, D. and McGrath, J. (1994) Emotion-related learning in patients with social and emotional changes associated with frontal lobe damage. *Journal of Neurology, Neurosurgery and Psychiatry* 57: 1518–1524.

Rolls, E.T. and Tovee, M.J. (1994) Processing speed in the cerebral cortex and the neurophysiology of visual masking. *Proceedings of the Royal Society of London, Series B* 257: 9–15.

Rolls, E.T., Tovee, M.J., Purcell, D.G., Stewart, A.L. and Azzopardi, P. (1994) The responses of neurons in the temporal cortex of primates, and face identification and detection. *Experimental Brain Research* 101: 474–484.

Rolls, E.T. and Tovee, M.J. (1995a) Sparseness of the neuronal representation of stimuli in the primate temporal visual cortex. *Journal of Neurophysiology* 73: 713–726.

Rolls, E.T. and Tovee, M.J. (1995b) The responses of single neurons in the temporal visual cortical areas of the macaque when more than one stimulus is present in the visual field. *Experimental Brain Research* 103: 409–420.

Rolls, E.T. and O'Mara, S.M. (1995) View-responsive neurons in the primate hippocampal complex. *Hippocampus* 5: 409–424.

Rolls, E.T., Robertson, R. and Georges-François, P. (1995) The representation of space in the primate hippocampus. *Society for Neuroscience Abstracts* 21: 586.10.

Rolls, E.T., Booth, M.C.A. and Treves, A. (1996) View-invariant representations of objects in the inferior temporal visual cortex. *Society for Neuroscience Abstracts* 22: 760.5.

Rolls, E.T., Critchley, H., Mason, R. and Wakeman, E.A. (1996) Orbitofrontal cortex neurons: role in olfactory and visual association learning. *Journal of Neurophysiology* **75**: 1970–1981.

Rolls, E.T., Critchley, H.D. and Treves, A. (1996) The representation of olfactory information in the primate orbitofrontal cortex. *Journal of Neurophysiology* **75**: 1982–1996.

Rolls, E.T., Critchley, H., Wakeman, E.A. and Mason, R. (1996) Responses of neurons in the primate taste cortex to the glutamate ion and to inosine 5'-monophosphate. *Physiology and Behavior* **59**: 991–1000.

Rolls, E.T., Treves, A. and Tovee, M.J. (1997) The representational capacity of the distributed encoding of information provided by populations of neurons in the primate temporal visual cortex. *Experimental Brain Research* **114**: 149–162.

Rolls, E.T., Treves, A., Foster, D. and Perez-Vicente, C. (1997). Simulation studies of the CA3 hippocampal subfield modelled as an attractor neural network. *Neural Networks*, in press.

Rolls, E.T., Treves, A., Tovee, M. and Panzeri, S. (1997) Information in the neuronal representation of individual stimuli in the primate temporal visual cortex. *Journal of Computational Neuroscience*, in press.

Rolls, E.T., Robertson, R.G. and Georges-François, P. (1997) Spatial view cells in the primate hippocampus. *European Journal of Neuroscience*, in press.

Rolls, E.T., Francis, S., Bowtell, R., Browning, D., Clare, S., Smith, E. and McGlone, F. (1997) Pleasant touch activates the orbitofrontal cortex. *Neuroimage* **5**: S17.

Rolls, E.T., Treves, A, Robertson, R., Georges-François, P. and Panzeri, S. (1998) Information about spatial view in an ensemble of primate hippocampal cells.

Rolls, E.T., Treves, A. and Critchley, H.D. (1998) The representation of olfactory information by populations of neurons in the primate orbitofrontal cortex.

Rolls, E.T. and Perez-Vicente, C. (1998), in preparation.

Rosch, E. (1975) Cognitive representations of semantic categories. *Journal of Experimental Psychology: General* **104**: 192–233.

Rose, D. and Dobson, V.G. (1985) *Models of the Visual Cortex.* Wiley: Chichester.

Rosenblatt, F. (1961) *Principles of Neurodynamics: Perceptrons and the Theory of Brain Mechanisms.* Spartan: Washington, DC.

Rudy, J.W. and Sutherland, R.J. (1995) Configural association theory and the hippocampal formation: an appraisal and reconfiguration. *Hippocampus* **5**: 375–389.

Rumelhart, D.E. and Zipser, D. (1985) Feature discovery by competitive learning. *Cognitive Science* **9**: 75–112.

Rumelhart, D.E., Hinton, G.E. and Williams, R.J. (1986a) Learning representations by back-propagating errors. *Nature* **323**: 533–536. (Reprinted in Anderson and Rosenfeld, 1988).

Rumelhart, D.E., Hinton, G.E. and Williams, R.J. (1986b) Learning internal representations by error propagation. Ch. 8 in *Parallel Distributed Processing: Explorations in the*

Microstructure of Cognition, Vol. 1, eds. D.E.Rumelhart, J.L.McClelland and the PDP Research Group. MIT Press: Cambridge, Mass.

Rumelhart, D.E. and McClelland, J.L. (eds.) (1986) *Parallel Distributed Processing*, (Vol. 1, Foundations; Vol. 2. Psychological and Biological Models). MIT Press: Cambridge, Mass.

Rumelhart, D.E. and Zipser, D. (1986) Feature discovery by competitive learning. Ch. 5 pp. 151–193 in *Parallel Distributed Processing*, (Vol. 1, Foundations) ed. D.E.Rumelhart and J.L.McClelland. MIT Press: Cambridge, Mass.

Rupniak, N.M.J. and Gaffan, D. (1987) Monkey hippocampus and learning about spatially directed movements. *Journal of Neuroscience* 7: 2331–2337.

Saint-Cyr, J.A., Ungerleider, L.G. and Desimone, R. (1990) Organization of visual cortical inputs to the striatum and subsequent outputs to the pallido-nigral complex in the monkey. *Journal of Comparative Neurology* 298: 129–156.

Samsonovich, A. and McNaughton, B.L. (1996) Attractor-map-based path integrator model of the hippocampus reproduces the phase precession phenomenon. *Society for Neuroscience Abstracts* 22: 734.11.

Sanghera, M.K., Rolls, E.T. and Roper-Hall, A. (1979) Visual responses of neurons in the dorsolateral amygdala of the alert monkey. *Experimental Neurology* 63: 610–626.

Schacter, G.B., Yang, C.R., Innis, N.K. and Mogenson, G.J. (1989) The role of the hippocampus-nucleus accumbens pathway in radial-arm maze performance. *Brain Research* 494: 339–349.

Schultz, W., Apicella, P., Romo, R. and Scarnati, E. (1995a) Context-dependent activity in primate striatum reflecting past and future behavioral events. Ch. 2, pp. 11–27 in *Models of Information Processing in the Basal Ganglia*, eds. J.C.Houk, J.L.Davis and D.G.Beiser. MIT Press: Cambridge, Mass.

Schultz, W., Romo, R., Ljunberg, T. Mirenowicz, J., Hollerman, J.R. and Dickinson, A. (1995b) Reward-related signals carried by dopamine neurons. Ch. 12, pp. 233–248 in *Models of Information Processing in the Basal Ganglia*, eds. J.C.Houk, J.L.Davis and D.G.Beiser. MIT Press: Cambridge, Mass.

Scott, T.R., Yaxley, S., Sienkiewicz, Z.J. and Rolls, E.T. (1986a) Gustatory responses in the frontal opercular cortex of the alert cynomolgus monkey. *Journal of Neurophysiology* 56: 876–890.

Scott, T.R., Yaxley, S., Sienkiewicz, Z.J. and Rolls, E.T. (1986b) Taste responses in the nucleus tractus solitarius of the behaving monkey. *Journal of Neurophysiology* 55: 182–200.

Scoville, W.B. and Milner, B. (1957) Loss of recent memory after bilateral hippocampal lesions. *Journal of Neurology, Neurosurgery and Psychiatry* 20: 11–21.

Segev, I., Rinzel, J. and Shepherd, G.M. (ed.) (1994) *The Theoretical Foundation of Dendritic Function. Selected Papers of Wilfrid Rall with Commentaries.* MIT Press: Cambridge, MA.

Seleman, L.D. and Goldman-Rakic, P.S. (1985) Longitudinal topography and interdigitation of corticostriatal projections in the rhesus monkey. *Journal of Neuroscience* **5**: 776–794.

Seltzer, B. and Pandya, D.N. (1978) Afferent cortical connections and architectonics of the superior temporal sulcus and surrounding cortex in the rhesus monkey. *Brain Research* **149**: 1–24.

Seltzer, B. and Pandya, D.N. (1989) Frontal lobe connections of the superior temporal sulcus in the rhesus monkey. *Journal of Comparative Neurology* **281**: 97–113.

Seress, L. (1988) Interspecies comparison of the hippocampal formation shows increased emphasis on the regio superior in the Ammon's horn of the human brain. *J Hirnforschung* **29**: 335–340.

Shannon, C.E. (1948) A mathematical theory of communication. *AT&T Bell Laboratories Technical Journal* **27**: 379–423.

Shepherd, G.M. (1988) *Neurobiology*. Oxford University Press: New York.

Shepherd, G.M. (ed.) (1990) *The Synaptic Organization of the Brain*. 3rd Edition. Oxford University Press: New York.

Siegel, M. and Auerbach, J.M. (1996) Neuromodulators of synaptic strength. Ch. 7, pp. 137–148 in *Cortical Plasticity: LTP and LTD*, eds. M.S.Fazeli and G.L.Collingridge. Bios: Oxford.

Sillito, A.M. (1984) Functional considerations of the operation of GABAergic inhibitory processes in the visual cortex. Ch. 4, pp. 91–117 in *Cerebral Cortex, Vol. 2, Functional Properties of Cortical Cells*, eds. E.G.Jones and A.Peters. Plenum: New York.

Simmen, M.W., Rolls, E.T. and Treves, A. (1996a) On the dynamics of a network of spiking neurons. In *Computations and Neuronal Systems: Proceedings of CNS95*, eds. F.H.Eekman and J.M.Bower. Kluwer: Boston.

Simmen, M.A., Treves, A., and Rolls, E.T. (1996b) Pattern retrieval in threshold-linear associative nets. *Network* **7**: 109–122.

Singer, W. (1987) Activity-dependent self-organization of synaptic connections as a substrate for learning. Pp. 301–335 in *The Neural and Molecular Bases of Learning*, eds. J.-P.Changeux and M.Konishi. Wiley: Chichester.

Singer, W. (1995) Development and plasticity of cortical processing architectures. *Science* **270**: 758–764.

Skaggs, W.E. and McNaughton, B.L. (1992) Quantification of what it is that hippocampal cell firing encodes. *Society for Neuroscience Abstracts,* **18**: 1216.

Skaggs, W.E., McNaughton, B.L., Gothard, K. and Markus, E. (1993) An information theoretic approach to deciphering the hippocampal code. Pp. 1030–1037 in *Advances in Neural Information Processing Systems*, Vol. 5, eds. S.J.Hanson, J.D.Cowan and C.L.Giles. Morgan Kaufmann: San Mateo, CA.

Skaggs, W.E., McNaughton, B.L., Wilson, M.A. and Barnes, C.A. (1996) Theta phase precession in hippocampal neuronal populations and the compression of temporal sequences. *Hippocampus* **6**: 149–172.

Sloper, J.J. and Powell, T.P.S. (1979a) A study of the axon initial segment and proximal axon of neurons in the primate motor and somatic sensory cortices. *Philosophical Transactions of the Royal Society of London, Series B* **285**: 173–197.

Sloper, J.J. and Powell, T.P.S. (1979b) An experimental electron microscopic study of afferent connections to the primate motor and somatic sensory cortices. *Philosophical Transactions of the Royal Society of London, Series B* **285**: 199–226.

Smith, M.L. and Milner, B. (1981) The role of the right hippocampus in the recall of spatial location. *Neuropsychologia* **19**: 781–793.

Smith, D.V. and Travers, J.B. (1979). A metric for the breadth of tuning of gustatory neurons. *Chemical Senses* **4**: 215–219.

Somogyi, P., Kisvarday, Z.F., Martin, K.A.C. and Whitteridge, D. (1983) Synaptic connections of morphologically identified and physiologically characterized large basket cells in the striate cortex of the cat. *Neuroscience* **10**: 261–294.

Somogyi, P. and Cowey, A.C. (1984) Double bouquet cells. Ch. 9, pp. 337–360 in *Cerebral Cortex, Vol.* **1**, *Cellular Components of the Cerebral Cortex*, eds. A.Peters and E.G.Jones. Plenum: New York.

Sompolinsky, H. (1987) The theory of neural networks: the Hebb rule and beyond. Pp. 485–527 in *Heidelberg Colloquium on Glassy Dynamics*, Vol. 275, eds. L.van Hemmen and I.Morgenstern. Springer: New York.

Sompolinsky, H. and Kanter, I. (1986) Temporal association in asymmetric neural networks. *Physical Review Letters* **57**: 2861–2864.

Squire, L.R. (1992) Memory and the hippocampus: a synthesis from findings with rats, monkeys and humans. *Psychological Review* **99**: 195–231.

Squire, L.R., Shimamura, A.P. and Amaral, D.G. (1989) Memory and the hippocampus. Ch. 12, pp. 208–239 in *Neural Models of Plasticity: Theoretical and Empirical Approaches*, eds. J.Byrne and W.O.Berry. Academic Press: New York.

Squire, L.R. and Knowlton, B.J. (1995). Memory, hippocampus, and brain systems. Ch. 53, pp. 825–837 in *The Cognitive Neurosciences*, ed. M.S.Gazzaniga. MIT Press: Cambridge, Mass.

Stent, G.S. (1973) A psychological mechanism for Hebb's postulate of learning. *Proceedings of the National Academy of Sciences of the USA* **70**: 997–1001.

Storm-Mathiesen, J., Zimmer, J. and Ottersen, O.P. (eds.) (1990) Understanding the brain through the hippocampus. *Progress in Brain Research* **83**.

Strick, P.L., Dum, R.P. and Picard, N. (1995) Macro-organization of the circuits connecting the basal ganglia with the cortical motor areas. Ch. 6, pp. 117–130 in *Models of Information Processing in the Basal Ganglia*, eds. J.C.Houk, J.L.Davis and D.G.Beiser. MIT Press: Cambridge, Mass.

Strong, S.P., Koberle, R., de Ruyter van Steveninck, R.R. and Bialek, W. (1996) Entropy and information in neural spike trains. *Physical Review Letters*, in press.

Sutherland, R.J. and Rudy, J.W. (1989) Configural association theory: the contribution of the hippocampus to learning, memory, and amnesia. *Psychobiology* **17**: 129–144.

Sutherland, R.J. and Rudy, J.W. (1991) Exceptions to the rule of space. *Hippocampus* **1**: 250–252.

Sutton, R.S. and Barto, A.G. (1981) Towards a modern theory of adaptive networks: expectation and prediction. *Psychological Review* **88**: 135–170.

Sutton, R.S. and Barto, A.G. (1990) Time-derivative models of Pavlovian reinforcement. Pp. 497–537 in *Learning and Computational Neuroscience*, eds. M.Gabriel and J.Moore. MIT Press: Cambridge, MA.

Suzuki, W.A. and Amaral, D.G. (1994a) Perirhinal and parahippocampal cortices of the macaque monkey: cortical afferents. *Journal of Comparative Neurology* **350**: 497–533.

Suzuki, W.A. and Amaral, D.G. (1994b) Topographic organization of the reciprocal connections between the monkey entorhinal cortex and the perirhinal and parahippocampal cortices. *Journal of Neuroscience* **14**: 1856–1877.

Szentagothai, J. (1968) Structuro-functional considerations of the cerebellar network. *Proceedings IEEE* **56**: 960–968.

Szentagothai, J. (1978) The neuron network model of the cerebral cortex: a functional interpretation. *Proceedings of the Royal Society of London, Series B* **201**: 219–248.

Taira, K. and Rolls, E.T. (1996) Receiving grooming as a reinforcer for the monkey. *Physiology and Behavior* **59**: 1189–1192.

Tanaka, K., Saito, C., Fukada, Y. and Moriya, M. (1990), Integration of form, texture, and color information in the inferotemporal cortex of the macaque. Ch. 10, pp. 101–109 in *Vision, Memory and the Temporal Lobe*, eds. E.Iwai and M.Mishkin. Elsevier: New York.

Tarkowski, W. and Loewenstein, M. (1993) Learning from correlated examples in a perceptron. *Journal of Physics A* **26**: 3669–3679.

Thomson, A.M. and Deuchars, J. (1994) Temporal and spatial properties of local circuits in neocortex. *Trends in Neurosciences* **17**: 119–126.

Thompson, R.F. (1986) The neurobiology of learning and memory. *Science* **233**: 941–947.

Thompson, R.F. and Krupa, D.J. (1994) Organization of memory traces in the mammalian brain. *Annual Review of Neuroscience* **17**: 519–549.

Thorpe, S.J., Rolls, E.T. and Maddison, S. (1983) Neuronal activity in the orbitofrontal cortex of the behaving monkey. *Experimental Brain Research* **49**: 93–115.

Thorpe, S.J. and Imbert, M. (1989) Biological constraints on connectionist models. Pp. 63–92 in *Connectionism in Perspective*, eds. R.Pfeifer, Z.Schreter and F.Fogelman-Soulie. Elsevier: Amsterdam.

Thorpe, S.J., O'Regan, J.K. and Pouget, A. (1989) Humans fail on XOR pattern classification problems. Pp. 12–25 in *Neural Networks: From Models to Applications*, ed. L.Personnaz and G.Dreyfus. I.D.S.E.T.: Paris.

Tippett, L.J., Glosser, G. and Farah, M.J. (1996) A category-specific naming impairment after temporal lobectomy. *Neuropsychologia* **34**: 139–146.

Toulouse, G. (1977) Theory of frustration effect in spin glasses. *Comm. Phys.,* **2**: 115–125.

Tovee, M.J. and Rolls, E.T. (1992) Oscillatory activity is not evident in the primate temporal visual cortex with static stimuli. *Neuroreport* **3**: 369–372.

Tovee, M.J., Rolls, E.T., Treves, A. and Bellis, R.P. (1993) Information encoding and the responses of single neurons in the primate temporal visual cortex. *Journal of Neurophysiology* **70**: 640–654.

Tovee, M.J., Rolls, E.T. and Azzopardi, P. (1994) Translation invariance and the responses of neurons in the temporal visual cortical areas of primates. *Journal of Neurophysiology* **72**: 1049–1060.

Tovee, M.J. and Rolls, E.T. (1995) Information encoding in short firing rate epochs by single neurons in the primate temporal visual cortex. *Visual Cognition* **2**: 35–58.

Tovee, M.J., Rolls, E.T. and Ramachandran, V.S. (1996) Visual learning in neurons of the primate temporal visual cortex. *Neuroreport* **7**: 2757–2760.

Treves, A. (1990) Graded-response neurons and information encodings in autoassociative memories. *Physical Review A* **42**: 2418–2430.

Treves, A. (1991a) Dilution and sparse encoding in threshold-linear nets. *Journal of Physics A* **23**: 1–9.

Treves, A. (1991b) Are spin-glass effects relevant to understanding realistic autoassociative networks? *Journal of Physics A* **24**: 2645–2654.

Treves, A. (1993) Mean-field analysis of neuronal spike dynamics. *Network* **4**: 259–284.

Treves, A. (1995) Quantitative estimate of the information relayed by the Schaffer collaterals. *Journal of Computational Neuroscience* **2**: 259–272.

Treves, A. (1997) On the perceptual structure of face space. *Biosystems* **40**: 189–196.

Treves, A. and Rolls, E.T. (1991) What determines the capacity of autoassociative memories in the brain? *Network* **2**: 371–397.

Treves, A. and Rolls, E.T. (1992) Computational constraints suggest the need for two distinct input systems to the hippocampal CA3 network. *Hippocampus* **2**: 189–199.

Treves, A. and Rolls, E.T. (1994) A computational analysis of the role of the hippocampus in memory. *Hippocampus* **4**: 374–391.

Treves, A. and Panzeri, S. (1995) The upward bias in measures of information derived from limited data samples. *Neural Computation* **7**: 399–407.

Treves, A., Barnes, C.A. and Rolls, E.T. (1996) Quantitative analysis of network models and of hippocampal data. Ch. 37, pp. 567–579 in: *Perception, Memory and Emotion: Frontiers in Neuroscience*, eds. T.Ono, B.L. McNaughton, S. Molotchnikoff, E.T. Rolls and H. Nishijo. Elsevier: Amsterdam.

Treves, A., Panzeri, S., Robertson, R.G., Georges-François, P. and Rolls, E.T. (1996) The emergence of structure in neuronal representations. *Society for Neuroscience Abstracts* **22**: 116.8.

Treves, A., Skaggs, W.E. and Barnes, C.A. (1996) How much of the hippocampus can be explained by functional constraints? *Hippocampus* **6**: 666–674.

Treves, A., Rolls, E.T. and Simmen, M.W. (1997) Time for retrieval in recurrent associative memories. *Physica D*, in press.

Treves, A., Panzeri, S., Rolls, E.T., Booth, M., and Wakeman, E.A. (1998) Firing rate distributions of inferior temporal cortex neurons to natural visual stimuli: an explanation. In preparation.

Tsai, K.Y., Carnevale, N.T., Claiborne, B.J. and Brown, T.H. (1994) Efficient mapping from neuroanatomical to electrotonic space. *Network* **5**: 21–46.

Tsodyks, M.V. and Feigel'man, M.V. (1988) The enhanced storage capacity in neural networks with low activity level. *Europhysics Letters* **6**: 101–105.

Tsodyks, M.V. and Sejnowski, T. (1995) Rapid state switching in balanced cortical network models. *Network* **6**: 111–124.

Turner, B.H. (1981) The cortical sequence and terminal distribution of sensory related afferents to the amygdaloid complex of the rat and monkey. Pp. 51–62 in *The Amygdaloid Complex*, ed. Y.Ben-Ari. Elsevier: Amsterdam.

Ullman, S. (1996) *High-Level Vision*. Object Recognition and Visual Cognition. Bradford/MIT Press: Cambridge, Mass.

Ungerleider, L.G. (1995) Functional brain imaging studies of cortical mechanisms for memory. *Science* **270**: 769–775.

Ungerleider, L.G. and Mishkin, M. (1982) Two cortical visual systems. In *Analysis of Visual Behaviour*, ed. D.Ingle, M.A.Goodale and R.J.W.Mansfield. MIT Press: Cambridge, Mass.

Ungerleider, L.G. and Haxby, J.V. (1994) 'What' and 'Where' in the human brain. *Current Opinion in Neurobiology* **4**: 157–165.

Van Hoesen, G.W. (1981) The differential distribution, diversity and sprouting of cortical projections to the amygdala in the rhesus monkey. Pp.77–90 in *The Amygdaloid Complex*, ed. Y.Ben-Ari. Elsevier: Amsterdam.

Van Hoesen, G.W. (1982) The parahippocampal gyrus. New observations regarding its cortical connections in the monkey. *Trends in Neurosciences* **5**: 345–350.

Van Hoesen, G.W., Pandya, N.D. and Butters, N. (1975) Some connections of the entorhinal (area 28) and perirhinal (area 35) cortices of the rhesus monkey. II. Frontal lobe afferents. *Brain Research* **95**: 25–38.

Van Hoesen, G.W., Yeterian, E.H. and Lavizzo-Mourey, R. (1981) Widespread corticostriate projections from temporal cortex of the rhesus monkey. *Journal of Comparative Neurology* **199**: 205–219.

Van Vreeswijk, C.A. and Hasselmo, M.E. (1997) Self-sustained memory states in a simple model with excitatory and inhibitory neurons. *Biological Cybernetics,* in press.

Von Bonin, G. and Bailey, P. (1947) *The Neocortex of Macaca mulatta*. University of Illinois Press: Urbana.

Wallis, G., Rolls, E.T. and Foldiák, P. (1993) Learning invariant responses to the natural transformations of objects. *International Joint Conference on Neural Networks* **2**: 1087–1090.

Wallis, G. and Rolls, E.T. (1997) Invariant face and object recognition in the visual system. *Progress in Neurobiology* **51**: 167–194.

Warrington, E.K. and Weiskrantz, L. (1973) An analysis of short-term and long-term memory defects in man. Ch. 10, pp. 365–395 in *The Physiological Basis of Memory* ed. J.A.Deutsch. Academic Press: New York.

Warrington, E.K. and Shallice, T. (1984) Category specific semantic impairments. *Brain* **107**: 829–854.

Watanabe, E. (1984) Neuronal events correlated with long-term adaptation of the horizontal vestibulo-ocular reflex in the primate flocculus. *Brain Research* **297**: 169–174.

Weisbuch, G. and Fogelman-Souliè, F. (1985) Scaling laws for the attractors of Hopfield networks. *Journal de Physique, Lettres* **2**: 337–341.

Weiskrantz, L. (1956) Behavioral changes associated with ablation of the amygdaloid complex in monkeys. *Journal of Comparative and Physiological Psychology* **49**: 381–91.

Weiskrantz, L. (1968) Emotion. Pp. 50–90 in *Analysis of Behavioral Change*, ed. L.Weiskrantz. Harper and Row: New York.

Weisskopf, M.G., Zalutsky, R.A. and Nicoll, R.A. (1993) The opioid peptide dynorphin mediates heterosynaptic depression of hippocampal mossy fiber synapses and modulates long-term depression. *Nature* **362**: 423–427.

West, M.J. and Gundersen, H.G.J. (1990) Unbiased stereological estimation of the numbers of neurons in the human hippocampus. *Journal of Comparative Neurology* **296**: 1–22.

Whittlesea, B.W.A. (1983) *Representation and Generalization of Concepts: the Abstractive and Episodic Perspectives Evaluated*. Unpublished doctoral dissertation, MacMaster University.

Wickens, J. and Kotter, R. (1995) Cellular models of reinforcement. Ch. 10, pp. 187–214 in *Models of Information Processing in the Basal Ganglia*, eds. J.C.Houk, J.L.Davis and D.G.Beiser. MIT Press: Cambridge, Mass.

Wickens, J.R., Begg, A.J. and Arbuthnott, G.W. (1996) Dopamine reverses the depression of rat corticostriatal synapses which normally follows high-frequency stimulation of cortex *in vitro*. *Neuroscience* **70**: 1–5.

Widrow, B. and Hoff, M.E. (1960) Adaptive switching circuits. In 1960 IRE WESCON Convention Record, Part 4, pp. 96–104. IRE: New York. (Reprinted in Anderson and Rosenfeld, 1988)

Widrow, B. and Lehr, M.A. (1990) 30 years of adaptive neural networks: perceptron, madeline, and backpropagation. *Proceedings of the IEEE* **78**: 1415–1442.

Williams, S. and Johnston, D. (1989) Long-term potentiation of hippocampal mossy fiber synapses is blocked by postsynaptic injection of calcium chelators. *Neuron* **3**: 583–588.

Williams, G.V., Rolls, E.T., Leonard, C.M. and Stern, C. (1993) Neuronal responses in the ventral striatum of the behaving macaque. *Behavioural Brain Research* **55**: 243–252.

Willshaw, D.J. (1981) Holography, associative memory, and inductive generalization. Ch. 3, pp. 83–104 in *Parallel Models of Associative Memory*, eds. G.E.Hinton and J.A.Anderson. Erlbaum: Hillsdale, NJ.

Willshaw, D.J. and Longuet-Higgins, H.C. (1969) The holophone—recent developments. In *Machine Intelligence* Vol 4, ed. D.Michie. Edinburgh University Press: Edinburgh.

Willshaw, D.J., Buneman, O.P. and Longuet-Higgins, H.C. (1969) Non-holographic associative memory. *Nature* **222**: 960–962.

Willshaw, D.J. and von der Malsburg, C. (1976) How patterned neural connections can be set up by self-organization. *Proceedings of The Royal Society of London, Series B* **194**: 431–445.

Willshaw, D.J. and Buckingham, J.T. (1990) An assessment of Marr's theory of the hippocampus as a temporary memory store. *Philosophical Transactions of The Royal Society of London, Series B* **329**: 205–215.

Wilson, C.J. (1995) The contribution of cortical neurons to the firing pattern of striatal spiny neurons. Ch. 3, pp. 29–50 in *Models of Information Processing in the Basal Ganglia*, ed. J.C.Houk, J.L.Davis and D.G.Beiser. MIT Press: Cambridge, Mass.

Wilson, F.A.W. and Rolls, E.T. (1990a) Neuronal responses related to the novelty and familiarity of visual stimuli in the substantia innominata, diagonal band of Broca and periventricular region of the primate. *Experimental Brain Research* **80**: 104–120.

Wilson, F.A.W. and Rolls, E.T. (1990b) Neuronal responses related to reinforcement in the primate basal forebrain. *Brain Research* **509**: 213–231.

Wilson, F.A.W. and Rolls, E.T. (1990c) Learning and memory are reflected in the responses of reinforcement-related neurons in the primate basal forebrain. *Journal of Neuroscience* **10**: 1254–1267.

Wilson, F.A.W. and Rolls, E.T. (1993) The effects of stimulus novelty and familiarity on neuronal activity in the amygdala of monkeys performing recognition memory tasks. *Experimental Brain Research* **93**: 367–382.

Wilson, F.A.W. and Rolls, E.T. (1998) The primate amygdala and reinforcement: a dissociation between rule-based and associatively-mediated memory revealed in amygdala neuronal activity. In preparation.

Wilson, F.A.W., O'Sclaidhe, S.P. and Goldman-Rakic, P.S. (1993) Dissociation of object and spatial processing domains in primate prefrontal cortex. *Science* **260**: 1955–1958.

Wilson, M.A. and McNaughton, B.L. (1994) Reactivation of hippocampal ensemble memories during sleep. *Science* **265**: 603–604.

Witter, M.P., Groenewegen, H.J., Lopes da Silva, F.H. and Lohman, A.H.M. (1989) Functional organization of the extrinsic and intrinsic circuitry of the parahippocampal region. *Progress in Neurobiology* **33**: 161–254.

Wu, X., Baxter, R.A. and Levy, W.B. (1996) Context codes and the effect of noisy learning on a simplified hippocampal CA3 model. *Biological Cybernetics* **74**: 159–165.

Yaxley, S., Rolls, E.T., Sienkiewicz, Z.J. and Scott, T.R. (1985) Satiety does not affect gustatory activity in the nucleus of the solitary tract of the alert monkey. *Brain Research* **347**: 85–93.

Yaxley, S., Rolls, E.T. and Sienkiewicz, Z.J. (1988) The responsiveness of neurones in the insular gustatory cortex of the macaque monkey is independent of hunger. *Physiology and Behavior* **42**: 223–229.

Yaxley, S., Rolls, E.T. and Sienkiewicz, Z.J. (1990) Gustatory responses of single neurons in the insula of the macaque monkey. *Journal of Neurophysiology* **63**: 689–700.

Yeo, C.H., Hardiman, M.J. and Glickstein, M. (1985) Classical conditioning of the nictitating membrane response of the rabbit. II. Lesions of the cerebellar cortex. *Experimental Brain Research* **60**: 99–113.

Yeo, C.H. and Hardiman, M.J. (1992) Cerebellar cortex and eyeblink conditioning: a reexamination. *Experimental Brain Research* **88**: 623–638.

Yuille, A.L., Kammen, D.M. and Cohen, D.S. (1989) Quadrature and the development of orientation selective cortical cells by Hebb Rules. *Biological Cybernetics* **61**: 183–194.

Zohary, E., Shadlen, M.N. and Newsome, W.T. (1994) Correlated neuronal discharge rate and its implications for psychophysical performance. *Nature* **370**: 140–143.

Zola-Morgan, S., Squire, L.R., Amaral, D.G. and Suzuki, W.A. (1989) Lesions of perirhinal and parahippocampal cortex that spare the amygdala and hippocampal formation produce severe memory impairment. *Journal of Neuroscience* **9**: 4355–4370

Zola-Morgan, S., Squire, L.R. and Ramus, S.J. (1994). Severity of memory impairment in monkeys as a function of locus and extent of damage within the medial temporal lobe memory system. *Hippocampus* **4**: 483–494.

Index